Woodhouse, Sir O.: "Social security: Policy, priorities and planning", in ILO: *Social security: Principles and practice*, ILO/FNPF South Pacific Sub-Regional Seminar, Suva, Fiji, 1983, Bangkok, Regional Office for Asia and the Pacific. 218 pp.

WHO: *Strengthening ministries of health for primary health care.* Technical Report Series No. 766, Geneva, 1988. 110 pp.

Biotechnology in Agriculture and Forestry

Volume 66

Series Editors

Prof. Dr. Jack M. Widholm (Managing Editor)
285A E.R. Madigan Laboratory, Department of Crop Sciences,
University of Illinois, 1201 W. Gregory, Urbana, IL 61801, USA

Prof. Dr. Toshiyuki Nagata
Professor and Dean, Faculty of Biological Sciences and Applied Chemistry,
Hosei University, 3-7-2 Kajino-cho, Koganei-shi, Tokyo 184-8584, Japan;
Emeritus Professor of the University of Tokyo, 7-3-1 Hongo, Bunkyo-ku,
Tokyo 113-0033, Japan

Biotechnology in Agriculture and Forestry

Further volumes can be found at springer.com.

Volume 33: Medicinal and Aromatic Plants VIII (1995)
Volume 34: Plant Protoplasts and Genetic Engineering VI (1995)
Volume 35: Trees IV (1996)
Volume 36: Somaclonal Variation in Crop Improvement II (1996)
Volume 37: Medicinal and Aromatic Plants IX (1996)
Volume 38: Plant Protoplasts and Genetic Engineering VII (1996)
Volume 39: High-Tech and Microprogation V (1997)
Volume 40: High-Tech and Microprogation VI (1997)
Volume 41: Medicinal and Aromatic Plants X (1998)
Volume 42: Cotton (1998)
Volume 43: Medicinal and Aromatic Plants XI (1999)
Volume 44: Transgenic Trees (1999)
Volume 45: Transgenic Medicinal Plants (1999)
Volume 46: Transgenic Crops II (1999)
Volume 47: Transgenic Crops II (2001)
Volume 48: Transgenic Crops III (2001)
Volumes 1–48 were edited by Y.P.S. Bajaj†
Volume 49: Somatic Hybridization in Crop Improvement II (2001)
    T. Nagata and Y.P.S. Bajaj (Eds.)
Volume 50: Cryopreservation of Plant Germplasm II (2002)
    L.E. Towill and Y.P.S. Bajaj (Eds.)
Volume 51: Medicinal and Aromatic Plants XII (2002)
    T. Nagata and Y. Ebizuka (Eds.)
Volume 52: Brassicas and Legumes: From Genome Structure to Breeding (2003)
    T. Nagata and S. Tabata (Eds.)
Volume 53: Tobacco BY-2 Cells (2004)
    T. Nagata, S. Hasezawa, and D. Inzé (Eds.)
Volume 54: Brassica (2004)
    E.C. Pua and C.J. Douglas (Eds.)
Volume 55: Molecular Marker Systems in Plant Breeding and Crop Improvement (2005)
    H. Lörz and G. Wenzel (Eds.)
Volume 56: Haploids in Crop Improvement II (2005)
    C.E. Palmer, W.A. Keller, and K.J. Kasha (Eds.)
Volume 57: Plant Metabolomics (2006)
    K. Saito, R.A. Dixon, and L. Willmitzer (Eds.)
Volume 58: Tobacco BY-2 Cells: From Cellular Dynamics to Omics (2006)
    T. Nagata, K. Matsuoka, and D. Inzé (Eds.)
Volume 59: Transgenic Crops IV (2007)
    E.C. Pua and M.R. Davey (Eds.)
Volume 60: Transgenic Crops V (2007)
    E.C. Pua and M.R. Davey (Eds.)
Volume 61: Transgenic Crops VI (2007)
    E.C. Pua and M.R. Davey (Eds.)
Volume 62: Rice Biology in the Genomics Era (2008)
    H.-Y. Hirano, A. Hirai, Y. Sano and T. Sasaki (Eds.)
Volume 63: Molecular Genetic Approaches to Maize Improvement (2009)
    A.L. Kriz and B.A. Larkins (Eds.)
Volume 64: Genetic Modification of Plants: Agriculture, Horticulture and Forestry (2010)
    F. Kempken and C. Jung (Eds.)
Volume 65: Cotton: Biotechnological Advances (2010)
    U.B. Zehr (Ed.)
Volume 66: Plant Biotechnology for Sustainable Production of Energy and Co-products (2010)
    P.N. Mascia, J. Scheffran and J.M. Widholm (Eds.)

Peter N. Mascia • Jürgen Scheffran •
Jack M. Widholm
Editors

Plant Biotechnology for Sustainable Production of Energy and Co-products

Springer

Editors
Dr. Peter N. Mascia (deceased)
1535 Rancho
Conejo Blvd.
Thousand Oaks, CA 91320, USA

Prof. Dr. Jürgen Scheffran
Institute for Geography, Klima Campus
ZMAW, Hamburg University
Bundesstr. 53
20146 Hamburg, Germany
juergen.scheffran@zmaw.de

Prof. Dr. Jack M. Widholm
University of Illinois
Department of Crop Sciences
1201 W. Gregory, ERML
Urbana, Il 61801, USA
widholm@illinois.edu

ISSN 0934-943X
ISBN 978-3-642-13439-5   e-ISBN 978-3-642-13440-1
DOI 10.1007/978-3-642-13440-1
Springer Heidelberg Dordrecht London New York

Library of Congress Control Number: 2010935443

© Springer-Verlag Berlin Heidelberg 2010
This work is subject to copyright. All rights are reserved, whether the whole or part of the material is concerned, specifically the rights of translation, reprinting, reuse of illustrations, recitation, broadcasting, reproduction on microfilm or in any other way, and storage in data banks. Duplication of this publication or parts thereof is permitted only under the provisions of the German Copyright Law of September 9, 1965, in its current version, and permission for use must always be obtained from Springer. Violations are liable to prosecution under the German Copyright Law.
The use of general descriptive names, registered names, trademarks, etc. in this publication does not imply, even in the absence of a specific statement, that such names are exempt from the relevant protective laws and regulations and therefore free for general use.

Cover design: WMXDesign GmbH, Heidelberg, Germany

Printed on acid-free paper

Springer is part of Springer Science+Business Media (www.springer.com)

This volume is dedicated to the late Peter Mascia, who had the foresight to see the future value of plant biomass and who was intimately involved in plant biotechnology and the commercial development of biomass crops. He is an editor of this volume since he helped plan a large part of it, but was unable to complete the work due to his untimely death on 6 May 2009.

Preface

This book is a collection of chapters concerning the use of biomass for the sustainable production of energy and chemicals–an important goal that will help decrease the production of greenhouse gases to help mitigate global warming, provide energy security in the face of dwindling petroleum reserves, improve balance of payment problems and spur local economic development.

Clearly there are ways to save energy that need to be encouraged more. These include more use of energy sources such as, among others, manure in anaerobic digesters, waste wood in forests as fuel or feedstock for cellulosic ethanol, and conservation reserve program (CRP) land crops that are presently unused in the US. The use of biofuels is not new; Rudolf Diesel used peanut oil as fuel in the first engines he developed (Chap. 8), and ethanol was used in the early 1900s in the US as automobile fuel [Songstad et al. (2009) Historical perspective of biofuels: learning from the past to rediscover the future. In Vitro Cell Dev Biol Plant 45:189–192). Brazil now produces enough sugar cane ethanol to make up about 50% of its transportation fuel needs (Chap. 4).

The next big thing will be cellulosic ethanol. At present, there is also the use of *Miscanthus x giganteous* as fuel for power plants in the UK (Chap. 2), bagasse (sugar cane waste) to power sugar cane mills (Chap. 4), and waste wood and sawdust to power sawmills (Chap. 7).

We have attempted to put together a distinguished group of authors to write chapters discussing many topics including the need for energy and the present problem of global warming that might be mitigated by using biomass instead of fossil fuels (Chap. 1). While ethanol is the most familiar fuel produced from biomass there are many other energy producing possibilities (Chap. 2). Of course, biomass can also be burned directly, and when mixed with coal helps decrease emissions of SO_2, NO_x, and non-renewable CO_2.

The overall general principals, possibilities and methods for designing plants for use as biomass feedstock are discussed in Chap. 3. Specific discussion of crops that produce sugar, starch or oil and trees and grasses can be found in Chaps. 4, 5, 6, 7 and 8. The general problems of invasiveness and gene dispersal and how to mitigate these problems are covered in Chaps. 9, 10 and 11. Chapter 12 describes models for integrated biorefineries that can produce many different products including industrial chemicals, and Chapter 13 describes models for the use of maize stover to supply heat and power for ethanol plants.

Other topics covered include new agricultural systems for biomass production for biofuels (Chap. 14), the life cycle analysis of biofuels (Chap. 15) and overall discussions of the many uses of biomass and possible cautions and criteria for standards for biomass sustainability and certification (Chap. 16).

Clearly, much development is still needed to fulfill the dream of the widespread use of biomass for energy and co-products. One of the biggest key questions is when will the production of cellulosic ethanol or other fuels become economically competitive with other liquid fuels. The infrastructure is in place to utilize ethanol if costs become competitive. We feel confident that progress will be made in cellulosic processing and fermentation due to the large amount of current interest and research and development funding so that this goal should be realized. Clearly, in the end, economics will decide the winners among the many crops and processes. The next decade should be exciting to see the winners and losers in the race to produce biomass for energy and co-products and to see how effective this is for the good of the world.

We have attempted to clarify the units used in the actual chapters, but the following list may be useful for comparative purposes.

Energy value of ethanol = 67% that of gasoline
1 kilogram (kg) = 2.205 pounds
1 metric ton = 1,000 kg = 1 mega gram (Mg) = 1 million g = 2,205 pounds
1 giga ton (Gt) = 1 billion metric tons
1 short ton = 2,000 pounds = 0.907 metric ton
1 hectare (ha) = 2.47 acres
1 liter (l) = 0.265 gallons
1 barrel (bbl) = 42 gallons = 158.8 l
1 meter (m) = 1.094 yard = 3.28 ft
MJ (megajoule) = million joules
BTU (British thermal unit) = 1,054.5 joules or 252 calories
KW (kilowatt) = 1,000 joules
KWH (kilowatt hour) = 3.6 MJ
TW (terawatt) = 1 million MW (megawatts)
MW (megawatt) = 1 million watts

June 2010   Jack M. Widholm

Contents

Part A Introduction to Biofuels

1 Introduction Overview: World Energy Resources and the Need for Biomass for Energy and Lower Fossil Carbon Dioxide Emissions 3
Charles E. Wyman
1.1 Introduction 3
1.2 World Dependence on Petroleum 4
1.3 Oil and Global Climate Change 7
1.4 What are our Options to Reduce Petroleum Use? 7
1.5 Why Biomass for Transportation? 8
1.6 Overview of Conversion Approaches 10
    1.6.1 Biomass Composition 10
    1.6.2 Higher Temperature Processes 11
    1.6.3 Lower Temperature Processes 13
    1.6.4 Comparison of Conversion Options 15
1.7 What is the Goal and How Much Biomass will be Needed? 16
1.8 Challenges to Commercial Applications 18
1.9 Closing Thoughts 19
References 21

2 Designing Biomass Crops with Improved Calorific Content and Attributes for Burning: a UK Perspective 25
Gordon G. Allison, Mark P. Robbins, José Carli,
John C. Clifton-Brown, and Iain S. Donnison
2.1 The Need for Non-Food Energy Crops 25
2.2 Biomass Combustion Technologies 26
    2.2.1 The Combustion Process 26
    2.2.2 Biomass as a Feedstock for Combustion 27

2.3 Lignocellulose 28
  2.3.1 Structure and Composition of the Plant Cell Wall 28
  2.3.2 Plant Cell Wall Architecture 29
2.4 The Effect of Chemical Composition on Feedstock Properties 30
2.5 Energy Crops for Combustion Processes in the European Union ... 31
  2.5.1 Miscanthus Species 33
  2.5.2 Switchgrass 35
  2.5.3 Willow and Poplar 36
  2.5.4 Reed Canary Grass 38
2.6 Technologies for Crop Design 38
  2.6.1 Modification of Hemicellulose and Cellulose 38
  2.6.2 Modification of Lignin 39
  2.6.3 Breeding Strategies 42
  2.6.4 Chemical Phenotyping and High-Throughput Screening 42
  2.6.5 Case Study: Variation in Cell Wall Composition
        Between 249 *Miscanthus* Genotypes 44
2.7 Conclusions and Future Perspectives 46
References 47

3 Designing Plants To Meet Feedstock Needs 57
Peter N. Mascia, Michael Portereiko, Mark Sorrells,
and Richard B. Flavell
3.1 Introduction 57
3.2 Feedstock Crops 58
3.3 Trait Improvement 61
3.4 Molecular Markers for Breeding and Genetic Mapping 64
3.5 Comparative Genomics 66
3.6 Heterosis 67
3.7 Improving Traits by Molecular Plant Breeding 68
3.8 Transgenic Traits 72
  3.8.1 First Generation Transgenic Traits 72
  3.8.2 Transgenic Output Traits 73
  3.8.3 Co-products 76
  3.8.4 Genetic Confinement and Prevention of Seed Formation 77
3.9 Concluding Remarks 79
References 80

Part B Specific Biofuel Feedstocks

4 Engineering Advantages, Challenges and Status of Sugarcane and other Sugar-Based Biomass Resources 87
Ricardo A. Dante, Plinio T. Cristofoletti, and Isabel R. Gerhardt
4.1 Introduction 87
  4.1.1 Sugar-Based Industry and Ethanol Uses 87
  4.1.2 Sugarcane Production System 90

Contents

	4.2 Biotechnology and Breeding Strategies for Increasing Sugarcane Sucrose Yields	91
	4.2.1 Photosynthetic Capacity of Sugarcane and the Sink–Source Relationship: What Determines Sucrose Accumulation?	91
	4.2.2 Sugarcane Biotechnology	94
	4.2.3 Molecular Markers in Sugarcane Breeding	98
	4.3 Other Sugar Crops Suitable for Ethanol Production	99
	4.4 Perspectives	101
	References	102
5	**High Fermentable Corn Hybrids for the Dry-Grind Corn Ethanol Industry**	**111**
	Joel E. Ream, Ping Feng, Inigo Ibarra, Susan A. MacIsaac, Beena A. Neelam, and Erik D. Sall	
	5.1 Introduction	111
	5.2 Value of High Fermentable Corn Hybrids	112
	5.3 Factors Influencing the Fermentability of Corn Grain	114
	5.4 Measuring Corn Grain Fermentability	116
	5.4.1 NIT Calibration	116
	5.4.2 Reference Chemistry	117
	5.4.3 NIT Calibration	118
	5.4.4 Commercial Validation of NIT Calibration	119
	5.5 Designation of High Fermentable Corn Hybrids	121
	5.6 Opportunities to Increase Corn Grain Fermentability	122
	5.7 Summary	123
	References	123
6	**Engineering Advantages, Challenges and Status of Grass Energy Crops**	**125**
	David I. Bransby, Damian J. Allen, Neal Gutterson, Gregory Ikonen, Edward Richard Jr, William Rooney, and Edzard van Santen	
	6.1 Introduction	125
	6.2 Miscanthus	126
	6.2.1 *Miscanthus* Phylogeny and Growth	127
	6.2.2 Genetic Improvement of *Miscanthus*	127
	6.2.3 Conventional Breeding Challenges	128
	6.3 Switchgrass	131
	6.3.1 Switchgrass Phylogeny and Growth	132
	6.3.2 Genetic Improvement of Switchgrass	133
	6.3.3 Conventional Breeding Challenges	134
	6.4 Sugarcane	136
	6.4.1 Sugarcane Phylogeny and Growth	136
	6.4.2 Genetic Improvement Needs	139
	6.4.3 Genetic Improvement Strategies	141

6.5	Sorghum	142
	6.5.1 Sorghum Phylogeny and Growth	142
	6.5.2 Genetic Improvement	144
6.6	Integration of Grasses into Cellulosic Biomass Supply Systems	146
6.7	Conclusions	147
	References	147

7 Woody Biomass and Purpose-Grown Trees as Feedstocks for Renewable Energy ... 155
Maud A.W. Hinchee, Lauren N. Mullinax, and William H. Rottmann

7.1	The Forest Industry and Renewable Energy	155
7.2	Biopower	158
	7.2.1 Processes for Energy Production from Woody Biomass	159
	7.2.2 Characteristics of Wood Feedstock that Impact Bioenergy Production	164
	7.2.3 Tree Species for Biopower	167
	7.2.4 Softwood Species for Bioenergy	172
7.3	Liquid Biofuels	173
	7.3.1 Cellulosic Ethanol	173
	7.3.2 Conversion Processes	175
	7.3.3 Other Cellulosic Liquid Fuels	179
	7.3.4 Feedstock Characteristics Affecting Biofuel Production	180
7.4	Purpose-Grown Trees for Renewable Energy	181
	7.4.1 Genetic Improvement for Productivity	184
	7.4.2 Genetic Improvement for Wood Properties	191
7.5	Sustainable Production of Purpose-Grown Trees	193
7.6	Conclusion	197
	References	198

8 Engineering Status, Challenges and Advantages of Oil Crops ... 209
Richard F. Wilson and David F. Hildebrand

8.1	Global Trends in Supply and Demand for Edible Oils	209
	8.1.1 Constraints on the Use of Edible Crop Products for Biofuel	209
	8.1.2 Availability and Cost of Biodiesel Feedstocks	212
	8.1.3 Sustainability	216
8.2	Technology Trends to Further Enhance the Sustainability of Edible Oils for Biofuel	218
	8.2.1 Physical Properties of Edible Oils	218
	8.2.2 Genetic Modification of the Physical Properties of Edible Oils	221
	8.2.3 Development of Markets for Edible Oils with Modified Traits	222

8.3 Advances in Genetically Modified Oil Trait Technology
    in Major Oilseed Crops 223
    8.3.1 Biological Basis for Trait Modified Oils 223
    8.3.2 Modified Oil Traits in the Commercial Pipeline 226
8.4 Advances in Genetically Modified Oil Trait Technology
    in New or Underdeveloped Oilseed Crops 229
    8.4.1 New Crop Oils for Industrial Chemicals 229
    8.4.2 Biological Basis for Industrial Oil Traits 241
8.5 Conclusions 247
References 248

Part C Mitigating Invasiveness

9 Invasive Species Biology, Ecology, Management and Risk Assessment: Evaluating and Mitigating the Invasion Risk of Biofuel Crops 263
Jacob N. Barney, Joseph M. DiTomaso
9.1 Biofuel Crops and Invasive Species 263
9.2 Invasive Species Biology and Ecology 265
9.3 Assessing the Invasive Risk of Biofuel Crops 267
    9.3.1 Risk Assessment 268
    9.3.2 Species biology 269
    9.3.3 Niche Modeling 271
    9.3.4 Propagule Biology 272
    9.3.5 Habitat Susceptibility 273
    9.3.6 Hybridization Potential 274
    9.3.7 Competitive Interactions 274
9.4 Mitigating the Invasion Risk Along the Biofuel Chain 275
    9.4.1 Crop Development 276
    9.4.2 Crop Importation and Dissemination 277
    9.4.3 Crop Production 277
    9.4.4 Feedstock Harvesting, Processing, Transport, and Storage . 278
    9.4.5 Feedstock Conversion 279
9.5 Response to Biofuel Crop Escapes 279
    9.5.1 Eradication Techniques 279
9.6 Conclusions 280
References 281

10 Gene Flow in Genetically Engineered Perennial Grasses: Lessons for Modification of Dedicated Bioenergy Crops 285
Albert P. Kausch, Joel Hague, Melvin Oliver, Lidia S. Watrud,
Carol Mallory-Smith, Virgil Meier, and C. Neal Stewart
10.1 Introduction 285
10.2 Gene Flow in Glufosinate-Resistant Grasses 287

	10.3	Gene Flow in Glyphosate-Resistant Creeping Bentgrass	289
		10.3.1 Gene Flow via Pollen in Glyphosate-Resistant Bentgrass	290
	10.4	Gene Flow via Seed Scatter	292
		10.4.1 Gene Flow via Seed Escape in Glyphosate-Resistant Bentgrass	293
	10.5	Future Impacts of Gene Flow from Glyphosate-Resistant Creeping Bentgrass	294
	10.6	Conclusions	294
	References		296

11 Genetic Modification in Dedicated Bioenergy Crops and Strategies for Gene Confinement 299

Albert P. Kausch, Joel Hague, Melvin Oliver, Yi Li, Henry Daniell, Peter Mascia, and C. Neal Stewart Jr

	11.1	Introduction	299
	11.2	Methods for Gene Confinement in Genetically Engineered Plants	300
		11.2.1 Physical, Spatial, Mechanical and Temporal Control	300
		11.2.2 Pollen Sterility	301
		11.2.3 Cytoplasmic Male Sterility, Chloroplast Transformation and Maternal Inheritance	302
		11.2.4 Seed-Based Gene Confinement	304
		11.2.5 Perceived Risks Associated with GURTs	304
		11.2.6 Gene Deletor System	309
		11.2.7 Total Sterility	309
		11.2.8 Total Sterility and Confinement Expression Systems	310
	11.3	Regulatory Issues for Perennial Bioenergy-Dedicated Crops	311
	11.4	Conclusions	311
	References		313

Part D Models for Uses of Biomass Feedstocks

12 Integrated Biorefineries–A Bottom-Up Approach to Biomass Fractionation 319

Birgit Kamm

	12.1	Introduction	319
	12.2	Biorefinery Technologies and Biorefinery Systems	321
		12.2.1 Background	321
		12.2.2 Lignocellulosic Feedstock Biorefinery	322
		12.2.3 Whole Crop Biorefinery	324
		12.2.4 Green Biorefinery	327
		12.2.5 The Two Platforms Biorefinery Concept	329
	12.3	Platform Chemicals	330
		12.3.1 Background	330

            12.3.2 The Role of Biotechnology in Production of Platform
                    Chemicals 332
            12.3.3 Green Biomass Fractionation and Energy Aspects 334
            12.3.4 Mass and Energy Flows for Green Biorefining 334
            12.3.5 Assessment of Green Crop Fractionation Processes 337
    12.4 Green Biorefinery: Economic and Ecologic Aspects 338
    References 339

13 **Heat and Power Production from Stover for Corn Ethanol Plants** .. 345
    Shahab Sokhansanj, Sudhagar Mani, Cannayen Igathinathane,
    and Sam Tagore
    13.1 Introduction 345
    13.2 Economics of Stover Supply to the Ethanol Plant 347
            13.2.1 Stover Collection 348
            13.2.2 Preprocessing 349
            13.2.3 Stover Transport 349
            13.2.4 On-Site Storage and Fuel Preparation 350
    13.3 Costs 350
            13.3.1 Cost of Biomass Collection 350
            13.3.2 Preprocessing Costs 351
            13.3.3 Transport Costs 352
            13.3.4 On-Site Fuel Storage and Preparation 353
            13.3.5 Total Cost of Biomass Fuel Delivered to the Burner .. 353
    13.4 Heat and Power Production 353
            13.4.1 Process Heat Generation 355
            13.4.2 Combined Heat and Power Generation 357
    13.5 Concluding Remarks 360
    References 361

**Part E Agricultural Fit of Biomass Crops and Lifecycle Analysis
           and Criteria**

14 **The Problem is the Solution: the Role of Biofuels in the Transition
     to a Regenerative Agriculture** 365
    Daniel G. De La Torre Ugarte and Chad C. Hellwinckel
    14.1 Introduction 365
            14.1.1 The Recent Price Bubble 366
            14.1.2 Long-Term Factors 368
    14.2 The Socio-Economic Impacts of Industrial Agriculture 370
    14.3 The Environmental Footprint of Industrial Agriculture 371
            14.3.1 Soil Loss 371
            14.3.2 Fossil Energy Dependence 372
            14.3.3 Greenhouse Gas Emissions 373

	14.4 Future Agricultural Policy: What is Needed?	374
	14.4.1 How do Agricultural Prices Impact Food Security, and Environmental Performance?	374
	14.4.2 The Role of Biofuels	377
	14.4.3 Transformative Investments in a New Agriculture	378
	14.4.4 Structural Shift	381
	14.5 Final Remarks	383
	References	383
15	**Life-Cycle Analysis of Biofuels**	**385**
	Michael Wang	
	15.1 Introduction	385
	15.2 Potential Biofuel Production Pathways	386
	15.3 Biofuel Life Cycle Analysis Boundary	387
	15.4 Life-Cycle Analysis Models for Biofuels	389
	15.4.1 The GREET Model at Argonne National Laboratory	389
	15.4.2 The Lifecycle Emissions Model at the University of California at Davis	390
	15.4.3 The GHGenius Model in Canada	391
	15.4.4 The E3 Database from Ludwig-Bölkow-Systemtechnik	391
	15.5 Life-Cycle Energy and Greenhouse Gas Emission Results of Key Biofuel Pathways with GREET Simulations	392
	15.5.1 Corn and Cellulosic Ethanol	392
	15.5.2 Sugarcane Ethanol	395
	15.5.3 Biodiesel and Renewable Diesel from Soybeans	396
	15.5.4 Corn Butanol	400
	15.6 Key Life-Cycle Analysis Issues and Uncertainties	402
	15.6.1 Direct and Indirect Land Use Changes	402
	15.6.2 Co-Product Issues for Biofuel Life-Cycle Analyses	403
	15.6.3 Other Environmental Sustainability Issues	405
	15.7 Conclusions	406
	References	407
16	**Criteria for a Sustainable Bioenergy Infrastructure and Lifecycle**	**409**
	Jürgen Scheffran	
	16.1 Introduction	409
	16.2 Optimizing Bioenergy Lifecycle and Infrastructure	411
	16.2.1 Bioenergy Supply Chain and Lifecycle	411
	16.2.2 The Integrated Biorefinery—From Feedstocks to Bioproducts	412
	16.2.3 Biomass Transportation Infrastructure	414
	16.3 Biogenic Wastes and Residues	416
	16.4 Energy Balance and Efficiency	419

16.5	Carbon Intensity and Conservation of Carbon Stocks	420
16.6	Soil Protection, Land Use and Food Security	423
16.7	Water Needs and Water Crisis	427
16.8	Wildlife, Biodiversity and Environmental Impact	430
16.9	Health, Safety and Social Criteria	432
	16.9.1 Health and Safety	432
	16.9.2 Decent Working Conditions	433
	16.9.3 Fair Feedstock Production and Land Rights	433
16.10	Sustainability Standards and Certification Schemes for Bioenergy	433
16.11	Conclusion	443
References		443

Index 449

Part A
Introduction to Biofuels

Chapter 1
Introduction Overview: World Energy Resources and the Need for Biomass for Energy and Lower Fossil Carbon Dioxide Emissions

Charles E. Wyman

1.1 Introduction

Recently, we have witnessed major swings in petroleum prices, ranging from the highest ever to the lowest in decades within months of each other, with such factors as politics, commodities trading schemes, limits in production capacity, bad weather, rapid increases in world demand, and oil cartels credited with responsibility for these huge price shifts. Regardless of the reasons, such energy price instability damages economies and has been a major contributor to the financial crises now gripping the world. The key is to find new sources of energy to get us out of this recurring dilemma, but little progress has been made in spite of several past episodes of high oil prices leading to economic recession.

Part of the problem is that we take energy supplies for granted as an inherent right that receives attention only when supplies are threatened. Maslow ranked human needs in the following hierarchy of increasing sophistication: physiological, security, loving and belonging, self esteem, and self actualization (Maslow 1943). He further stated that as each level of need is filled, we tend to take that need for granted and focus on fulfilling the next higher level need, which is more difficult to satisfy. Thus, a person without food or water is concerned only with addressing this pressing need for survival and not likely to be overly concerned with self actualization. In a similar way, we generally take energy for granted in our day-to-day lives and give it attention only when supply limitations cause inconveniences or high prices result in economic pain. Unfortunately, a long time is needed to change our energy infrastructure, and failure to devote sufficient attention in advance to assuring a long term and sustainable energy supply will lead to substantial economic, environmental, and societal disruptions that cannot be rapidly fixed.

C.E. Wyman
Ford Motor Company Chair in Environmental Engineering, Center for Environmental Research and Technology and Professor of Chemical and Environmental Engineering, Bourns College of Engineering, University of California, Riverside, California 92507, USA
e-mail: charles.wyman@ucr.edu

In this chapter, an overview of the current energy picture provides a perspective on why new, and in particular sustainable, fuels are needed. The role that cellulosic biomass can play in meeting this important need is then outlined, followed by a summary of options available to convert biomass into fuels that can substitute for petroleum-based products. Based on this background, a picture is provided of how much cellulosic biomass would be needed to make an impact on petroleum use. The chapter concludes with some aspects of the key attributes for cellulosic biomass that could enhance the impact to provide some thoughts on research and development opportunities to support the emergence of a meaningful biomass fuels industry.

1.2 World Dependence on Petroleum

Petroleum prices have a tremendous impact on our economy because of the dominant role oil plays in providing our energy. Overall, petroleum is the source of about 170 quadrillion (10^{15}) British thermal units (BTUs), or quads of energy, of the total of more than 460 quads the world uses, with coal, natural gas, hydroelectric power, nuclear energy, and geothermal and other sources providing the remaining roughly 122, 105, 29, 27, and 7 quads, respectively (US Department of Energy 2008). Figure 1.1 outlines the relative contributions of major energy sources to world uses. Over half of petroleum in this world total is now used for transportation, and demand by this sector is projected to grow rapidly as vehicle traffic increases throughout the world and even accelerates in Asia.

Similarly, the United States obtains more energy from petroleum than from any other resource, with about 40 quads of the 100 total being from this one source. However, as illustrated in Fig. 1.2, the US uses a higher portion for transportation, amounting to about 70% of the petroleum used, and the transportation system is almost totally dependent on this one resource for energy. On top of that, we must add in more than 25% more petroleum to account for that consumed in processing petroleum to fuels, with the result that production and consumption of transportation fuels accounts for about 88% of the petroleum used in the US (US Department of Energy 2008).

Petroleum is favored for transportation because of the convenience of using liquid fuels and their particular suitability for transportation. However, despite having only about 4.6% of the world's population, the United States currently consumes about 7.5 billion barrels of the close to 31 billion barrels of oil used each year around the world, i.e., nearly one-quarter (US Department of Energy 2008). In contrast to our abundant domestic supplies of coal or reserves of natural gas, which are adequate in the short term, US production of petroleum has declined steadily since 1970 and now amounts to only about one-quarter of the total we consume. Despite recent cries to drill our way out of this dilemma, proved US petroleum reserves only amount to less than 25 billion barrels in total, and the relatively large reserves in Alaska, Texas, the Gulf of Mexico, and California would only last the country about half a year each if we tried to satisfy all of our large oil

1 Introduction Overview: World Energy Resources and the Need for Biomass for Energy

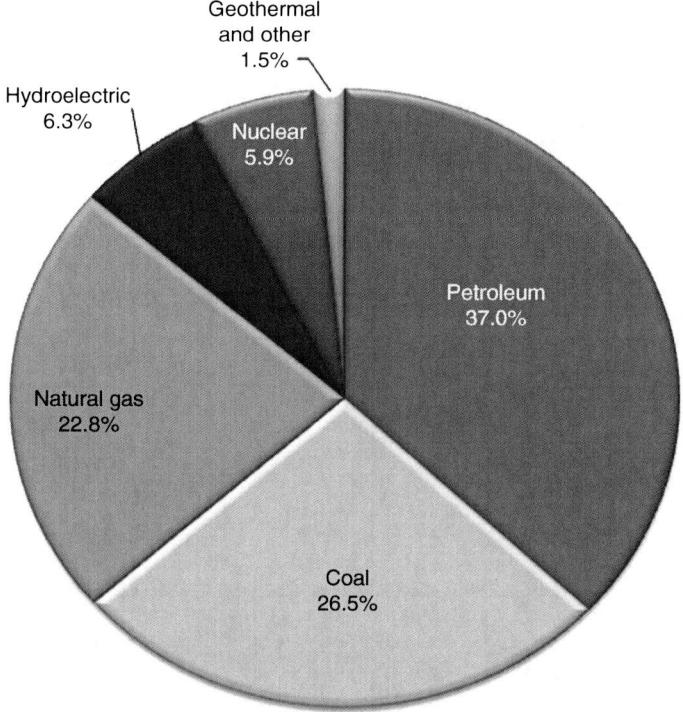

Fig. 1.1 The World is far more dependent on petroleum than any other source for total primary energy production, and non fossil energy sources provide less than 15% of the total energy consumed. Over half of the petroleum consumed is for transportation (US Department of Energy 2008)

Fig. 1.2 The United States uses a larger share of petroleum directly for transportation than most other countries, with another 18% of petroleum supplying energy for converting crude oil to fuels (US Department of Energy 2008)

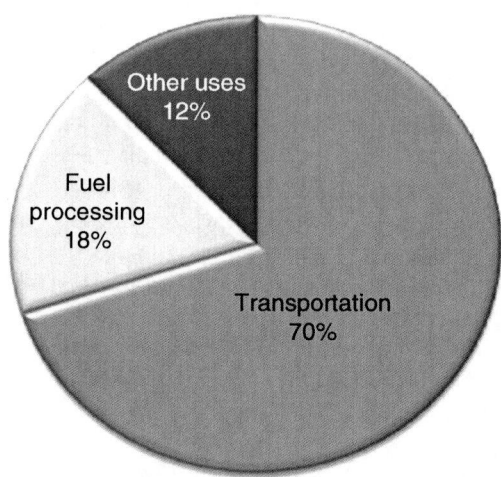

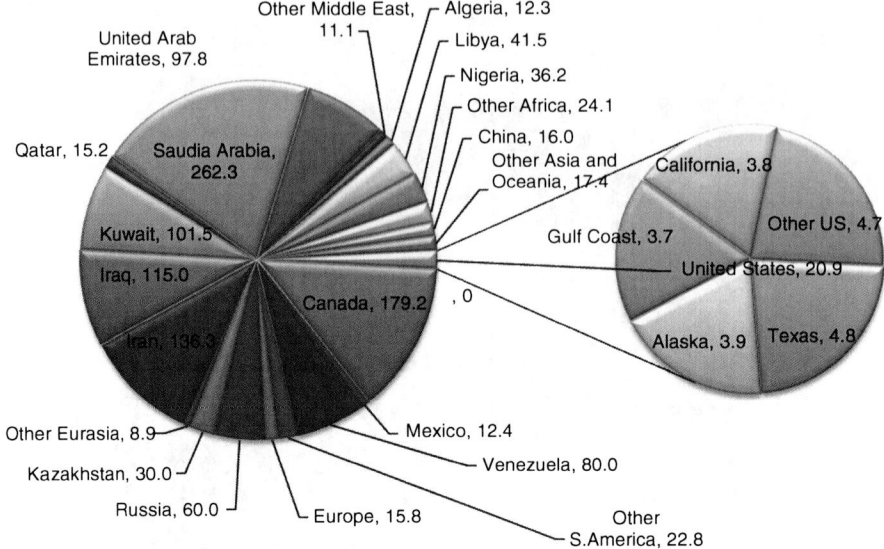

Fig. 1.3 Petroleum reserves. Known petroleum reserves in the United States are only about 2% of total world reserves and much smaller than for many other countries (amounts given in units of billions of barrels of oil). Clearly, domestic reserves are not nearly adequate to satisfy the current US demand of about 7.5 billion barrels/year for more than a few years (US Department of Energy 2008)

appetite from domestic sources. Although claims by critics that new discoveries of oil will extend supplies beyond the proved reserves are no doubt true, the rate of discovery is clearly dropping, with the result that total US production has declined from a high of 3.5 billion barrels in 1970 to about half that level now, and continues to drop. In contrast, as shown in Fig. 1.3, out of total world reserves of somewhat more than 1 trillion barrels of oil, more than 700 billion barrels are said to be located in Saudi Arabia, Iran, Iraq, Kuwait, the United Arab Emirates, and other countries of the Middle East; another 114 billion in Africa; about 100 billion in Eurasia; about 80 billion in Venezuela; and about 35 billion barrels in Asia and Oceania plus about another 180 billion barrels as tar sands in Canada. Thus, continued petroleum use will depend heavily on sources in unstable regions of the world and, although new discoveries will no doubt extend the supply, all of current world petroleum reserves would be depleted in about 30 years at current international consumption rates. Furthermore, we have now consumed about as much oil as are known to be in reserves, and a point of maximum petroleum production could be near due to declining supplies — a point known as Hubbert's peak in oil production — and political impediments to accessibility in many countries. And we should not lose sight of the fact that imported petroleum contributes more to the annual US trade deficit than any other source, with an annual cost of between US $200 and $800 billion dollars, depending on the price of oil.

1.3 Oil and Global Climate Change

In addition to concerns such a high dependence on oil should raise about balance of trade, energy security, and energy supply, burning petroleum is a major source of the carbon dioxide accumulation that leads to global climate change. Carbon dioxide concentrations have increased from about 315 ppm to about 390 ppm (US Department of Energy 2008) over the last 50 years, and this build up is predicted to cause many changes in our climate, with consequences including melting of the polar ice caps, flooding, drought, extinction of species, and disruptions in food supply. The United States and China, with almost five times as many people, each now release about 6 billion metric tons of carbon dioxide annually, more than any other country. Although clearing forests and other human activities also contribute greenhouse gases, about 85% of the US contribution comes from energy consumption, with petroleum amounting to about 43% of the amount from energy. Because of its heavy dependence on oil, transportation has become the leading emitter of carbon dioxide from energy use.

1.4 What are our Options to Reduce Petroleum Use?

The so-called "energy crises" of the 1970s and 1980s were really due to embargoes of petroleum by Organization of the Petroleum Exporting Countries (OPEC) that resulted in reduced supplies and dramatic price run ups. Electric utilities learned from this experience and virtually phased out petroleum use. Similarly, the industrial, commercial, and residential sectors have taken advantage of more abundant reserves of coal and natural gas to limit petroleum consumption. Each of these sectors can also choose among many non-fossil energy sources in the future, including geothermal, hydropower, nuclear power, photovoltaic electricity, solar thermal heat and power, and wind energy. But such sources tend to be more suitable for stationary applications, and alternatives to petroleum for mobility are not so simple to implement. As a result, the near total dependence of transportation on oil has grown substantially following the energy crises, with the result we use a much larger amount of petroleum than ever and import an even larger fraction now than at any time in the past (US Department of Energy 2008).

Serious analysis of the situation shows that the options to reduce both petroleum consumption and greenhouse gas (GHG) emissions by transportation are limited. First, we could rely much more on public transportation and drive less miles — important opportunities but ones that are counter to historic trends. In addition, many fear that reduced travel would hurt the economy. The second choice is to drive more efficient vehicles such as hybrid and lighter, more fuel efficient, cars. This path presents low hanging fruit that has not been taken advantage of sufficiently, and is synergistic with introducing new fuels as well as reducing petroleum use. The third option is to use a source other than petroleum to fuel our vehicles, but we must keep in mind that only a sustainable option will avoid GHG emissions.

Furthermore, sustainable technologies will circumvent the need to change transportation infrastructure again later. Thus, while we may wish we had other choices, we must select from this limited spectrum and should keep in mind that intelligent integration of all three is almost certainly essential to accommodate a growing population that will continue to stress the environment and exhaust finite resources.

In simplistic terms, a transition to sustainable energy appears imminently doable. The world consumes energy at a rate of about 13 terrawatts (TW = 10^{12} watts), of which about 10 TW is now from fossil fuels (Lewis and Nocera 2006). On the other hand, the sun provides about 173,000 TW or 17,000 times our fossil energy consumption. In addition, plants capture about 140 TW of energy through photosynthesis. Thus, sustainable energy sources are more than sufficient to meet human needs, but they are also diffuse and require high capital costs to capture while our society prefers low capital costs to avoid long-term commitments. An additional complication is that, because the most abundant sustainable energy produced by the wind and sun is sporadic and unavailable 24 hours per day, 7 days per week, energy storage is needed to meet current habits. Sustainable energy from biomass, hydroelectric, and geothermal is available 24/7, but accessible amounts are far more limited. Thus, considerable challenges with respect to costs, coincidence to demand, environmental issues, and siting stand in the way of all sustainable technologies.

1.5 Why Biomass for Transportation?

As illustrated in Fig. 1.4, mobile applications present the greatest challenges for sustainable energy. Mechanical energy generated by wind and hydropower cannot be used to power a vehicle directly. Similarly, the heat generated from solar, geothermal, and nuclear sources is not readily used for transportation. Batteries can store the electricity produced by such sustainable mechanical and heat sources as well as by direct solar conversion by photovoltaic devices to power vehicles. However, batteries have historically been very expensive, required a long time to charge, been very heavy, and suffered from a limited range. Electricity can also electrochemically decompose water into hydrogen for use as a fuel, but the electricity employed must be derived from sustainable or nuclear sources if we wish to avoid contributing to GHG emissions. Unless unforeseen breakthroughs in thermochemical or biological systems occur, the best routes to hydrogen will be based on electrochemical generation, but, in that case, the resulting hydrogen will be more expensive than the electricity from which it is made. Beyond that, hydrogen presents major challenges in generation, storage, transport, overall efficiency, and use (Bossel 2006).

Plants convert solar energy into chemical bond energy by photosynthesis, thereby serving as a very aesthetic device for both solar collection and energy storage. Burning biomass releases the energy captured from the sun as heat, which in turn can be employed to generate electricity, and we can also process plants to release hydrogen. However, biomass is unique among sustainable options as a resource for

1 Introduction Overview: World Energy Resources and the Need for Biomass for Energy

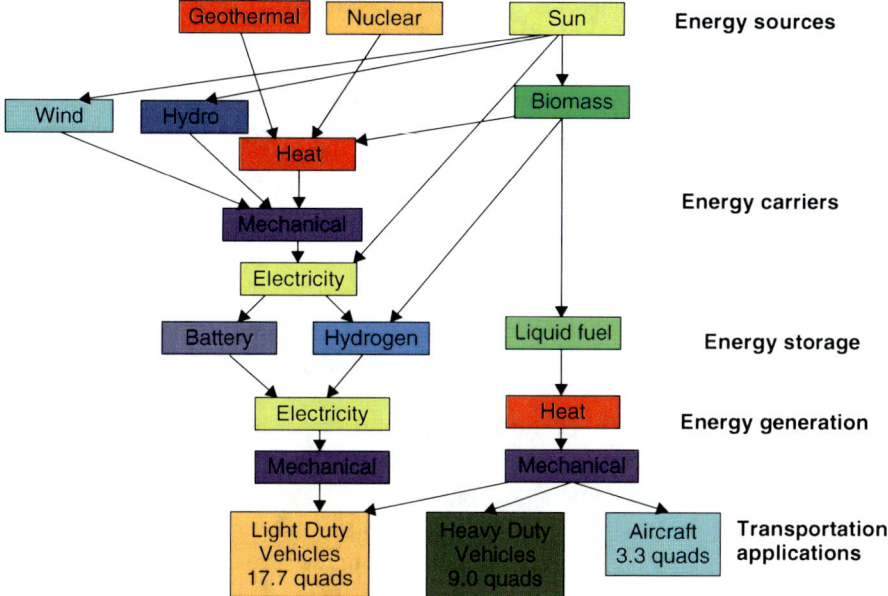

Fig. 1.4 Routes to sustainable energy. The electrical, mechanical, or thermal energy produced by most sustainable resources must be stored in batteries or converted to hydrogen to provide mobility, with the result that these options may be matched to light duty applications with short range travel. Only biomass can be converted into the liquid fuels that are vital to air travel and heavy duty applications such as trucking as well as being valuable for long distance driving. Energy consumption values shown are for the United States

making the liquid organic fuels on which our transportation sector almost totally depends, and no other option is known for making such fuels once we deplete fossil sources (Lynd 1996). Failure to recognize this fundamental truth, and wishing for magical solutions that no one has found in over 30 years of searching (although not funded at serious levels), will only magnify the stress future generations will face as they strive to find energy sources that can provide the conveniences we now take for granted. Calls to burn biomass to generate electricity are also misguided in light of the multiple options we have to produce power and the uniqueness of biomass for sustainably making liquid transportation fuels that are essential to such applications as air travel, heavy duty trucking, and long-distance travel.

Conversion of plentiful and low-cost cellulosic biomass to ethanol, diesel substitutes, and other biofuels is our only option for significantly reducing our dependence on petroleum for liquid fuels, and approaches such as growing algae to make oils could play an important role as well. However, this picture has been clouded of late by continued attacks and negative press about such questions as how effective the current commercial practice of converting corn to ethanol is in terms of saving energy, impacts on food prices, and GHG emissions. Unfortunately, a number of these reports are based on faulty analyses and sensationalism, and although corn ethanol has important limitations, it can still offer some short-term advantages compared to gasoline or other fossil fuels. On the other hand, cellulosic plants such as poplar wood and switchgrass, together with corn stover and many other agricultural residues, are far more plentiful than food crops such as the corn and cane sugar now used to make biofuels, and have a much more limited impact on the environment if sound agricultural practice is followed (Perlack et al. 2005). Furthermore, GHG emissions are far lower, and more limited fossil energy input is needed in the overall plant production, harvesting, conversion, and utilization cycle (Farrell et al. 2006; Tyson 1993; Wyman 1994). Although there are many technical, policy, and infrastructure considerations that must be addressed to use cellulosic resources wisely, we have no apparent choice but to take these on if future needs for liquid fuels are to be addressed in anything resembling a perpetual manner.

1.6 Overview of Conversion Approaches

Both thermochemical and biological routes can convert non-edible cellulosic biomass into diesel like fuels or fuels for automobile spark ignition engines. In this section, some primary options will be outlined, but the reader is advised to refer to more comprehensive sources of information for details as space is too limited here to provide the details needed for full consideration of the complex technologies and features for biomass conversion into fuels (Huber et al. 2006; Wyman et al. 1992).

1.6.1 Biomass Composition

It is useful to outline the structure of cellulosic biomass to help understand the conversion options. Cellulosic materials have evolved a complex structure to serve the structural, nutrient, and water needs of plants. Typically about 40–50% of cellulosic biomass is comprised of cellulose, which is a long chain of glucose molecules joined together by covalent bonds. These long chains align with one another through hydrogen bonding to form fibers that support the plant, and the well ordered regions of these parallel chains are highly crystalline. Hemicellulose, which makes up about 20–30% of most plants, is also a long chain of sugar molecules covalently bonded to one another but can be comprised of up to five different sugars: arabinose, galactose, glucose, mannose, and xylose. Other compounds, such as acetyl groups, are incorporated into the hemicellulose chains and, while cellulose is a straight chain, hemicellulose is branched and not crystalline. Much of the rest of the plant, about 15–30%, is primarily lignin, a phenyl propene compound that "glues" the cellulose and hemicellulose together. In addition, plants can contain various oils, free sugars, starch, and minerals, with the total amounts often small but highly dependent on the species. The dominant hemicellulose and cellulose portions are built from the chemical backbone CH_2O, "C-water," and lignin has an atomic composition of roughly $C_{10}H_{12}O_n$, with n ranging from about 3 to 4 for many types of biomass (Holtzapple 1993; Wiselogel et al. 1996).

1.6.2 Higher Temperature Processes

As shown in Fig. 1.5, biomass can be gasified by partial reaction with air, oxygen, and/or steam to produce a mixture of carbon dioxide, carbon monoxide, hydrogen, methane, and typically much lower amounts of other compounds in various proportions (Huber et al. 2006). If air is used in gasification, significant amounts of nitrogen dilute the other components and complicate conversion to fuels, and the product, often called producer gas, is usually burned to generate heat and/or electricity. Gasification of biomass in this fashion was used as recently as during World War II to power vehicles in Europe at a time when liquid fuels were in short supply. Gas mixtures rich in mostly carbon monoxide and hydrogen are called syngas, and the hydrogen content of syngas can be enriched by reaction of steam with carbon monoxide. Syngas can be converted catalytically into diesel fuel or gasoline substitutes by Fischer-Tropsch chemistry, i.e., methanol, ethanol, hydrogen, or other fuels, as well as provide a building block for many chemicals. The Germans applied coal gasification in conjunction with Fischer Tropsch catalysis to make liquid fuels and other products from syngas during World War II to compensate for limited access to petroleum. South Africa introduced coal gasification to produce liquid fuels when petroleum imports were sanctioned because of apartheid policies and continues to employ this technology today. Although biomass and coal gasification are similar in many ways, biomass is more reactive and can be gasified

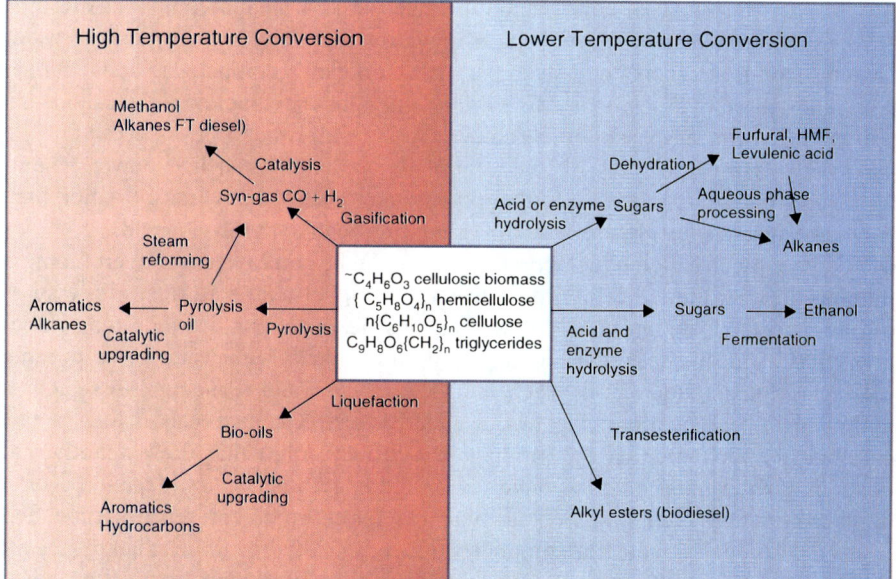

Fig. 1.5 High- and low-temperature conversion technologies. A number of conversion technologies and products can be made from biomass, with many of the leading processing routes and products illustrated here

at lower temperatures. On the other hand, the higher mineral content of biomass forms slag that can foul conventional gasification equipment. In addition, tars and other problematic components must be removed from the gas prior to introduction to catalytic conversion to reduce fouling of expensive catalysts. Most commercial biomass gasification units produce heat and electricity, but the technology could be adapted to make other products if prices are high enough to provide adequate returns on the high capital investments needed. Several companies are working on using organisms to convert syngas into ethanol.

Figure 1.5 also shows that liquids can be produced from biomass by rapid heating in the absence of air – an approach known as pyrolysis (Huber et al. 2006). However, control of the time and temperature of reaction is vital as long times at low temperatures favor formation of solids such as charcoal while heating to high temperatures for longer times produces gases. As one example, "destructive distillation" approaches were employed a century ago to generate charcoal, methanol, and other products by holding wood at low temperatures for times of about a day. On the other hand, fast heat up of biomass to moderate temperatures between about 375°C and 525°C and pressures of about 1–5 atmospheres for short reaction times forms liquids that are often termed bio-oils. Overall, pyrolysis is a relatively simple technology in that reactor requirements are moderate in terms of both volume and reaction time, and a large fraction of the energy in biomass can end up in the liquid product. Like most thermochemical methods, pyrolysis can process a wide range of feedstocks, with the primary demand simply being that they contain carbon and hydrogen and not many minerals. The composition of the liquid produced in pyrolysis is complex and contains a wide range of acids, alcohols, aldehydes, aromatic compounds, esters, and ketones, making it somewhat challenging to utilize. The high oxygen and water contents reduce the energy density of the fuel significantly compared to conventional petroleum-based hydrocarbons. More importantly, the fuel is acidic and unstable and will react to form very viscous oils and even solids if allowed to sit. As a result, while pyrolysis oils are employed for some stationary applications such as power generation, they must be upgraded by further processing to be used as a transportation fuel. Biomass must also be ground to a small size for effective pyrolysis yields.

In an approach called liquefaction, catalysts can be employed to convert biomass into bio-oil with limited solubility in water, but higher temperatures and pressures of about 250–325°C and 50–200 atmospheres, respectively, are required (Huber et al. 2006). In this case, the biomass is usually mixed with water or an organic solvent to form a slurry prior to entering the reactor, and reducing gases such as hydrogen and carbon monoxide may be added as well. Catalysts can include metals or alkali compounds, and the product is a brown liquid that flows readily and contains a wide range of compounds, with the exact distribution depending on a number of factors including biomass type, reactor environment, reaction time, and catalyst. Although liquefaction produces higher grade hydrocarbons than pyrolysis, the equipment required is very expensive due to the high temperatures and pressures applied, and these extreme operating conditions present significant challenges for solids handling. As with pyrolysis, liquefaction bio-oil requires upgrading by catalytic or other routes to produce fuels acceptable for transportation.

1.6.3 Lower Temperature Processes

Lower temperatures of 210°C and below can also be applied to breakdown biomass, but at these conditions, the long chains of hemicellulose and cellulose first form the sugars from which they are made and form little if any of the shorter molecules of carbon monoxide, hydrogen, or the oils typically formed in higher temperature reactions. Because of its amorphous nature and differences in chemical bonding, hemicellulose is broken down more easily than cellulose to release the sugars arabinose, galactose, glucose, mannose, and xylose, along with acetic acid and other compounds that make up hemicellulose through hydrolysis with water (Mosier et al. 2005). As outlined in Fig. 1.5, such reactions can generally be carried out with or without added acid or other chemicals, although sugar yields tend to be higher when about 1.0–4.0% dilute sulfuric acid or similar concentrations of other acids are used. However, temperatures that give rapid release of sugars are typically between 140°C and 200°C, and the sugars released degrade to form furfural in the case of xylose and other compounds when held at these temperatures. Thus, if we seek to maximize recovery of sugars, we are faced with a classical series reaction that demands a balance between reacting long enough to breakdown a high percentage of the hemicellulose but not so long as to degrade the sugars released. Nonetheless, yields on the order of 85–90% of theoretical are possible through optimization of time and temperature histories when using acids, with sulfuric acid often favored because of its low cost (Wyman et al. 2005, 2008). Hemicellulose sugar yields are lower when hydrolysis is conducted with just water, with one report showing yields limited to about 65% of theoretical with sugar cane bagasse as the feedstock (Heitz et al. 1991).

Although cellulose can be deconstructed to release its building block of glucose through hydrolysis as well, the crystalline structure of cellulose makes it more difficult to hydrolyze and, consequently, glucose yields are low for dilute acid hydrolysis of cellulose (Brennan et al. 1986; Grethlein and Converse 1991). While temperature has limited impact on yields from acid hydrolysis of hemicellulose, glucose yields from cellulose do increase with temperature, but unfortunately, the reaction times for the maximum yields also drop significantly to a matter of seconds at temperatures near 260°C, which favor yields of about 65% or more (Wright and D'Agincourt 1984). Such short reaction times are impractical to apply in large-scale commercial systems, and, at these near pyrolysis temperatures, performance is further compromised by the formation of tars and others problematic compounds that tend to precipitate and plug equipment. As a result, yields are limited to about 50% of the maximum possible for dilute acid hydrolysis of cellulose to glucose, hurting returns on capital. Concentrated acids will decrystallize cellulose and achieve high sugar yields, but the acid recovery required for economic and environmental viability is expensive, limiting applications to those with high co-product credits or feedstock tipping fees (Wright and Power 1987).

Cellulase enzymes can achieve high yields from cellulose, which are essential to economic viability, with over 85% of the cellulose converted to glucose. In addition, cellulase and hemicellulase enzymes will enhance yields, and therefore economics, through release of much of the residual sugar left in the hemicellulose. However, high doses of enzymes are needed because of the crystallinity of cellulose and low activity of the enzymes, with on the order of 15 filter paper units (FPU) of enzyme activity per gram of glucan in the biomass typically being applied to realize glucose yields of about 85% of theoretical (Wooley et al. 1999). Because this enzyme loading translates into about a quarter pound of enzyme for every gallon of ethanol produced, cellulase would have to be produced for less than US $1.00/pound to be economically viable. Recent statements by enzyme producers suggest current costs are about US $1.00/gallon of ethanol made, and four enzyme producing companies have been contracted by the US Department of Energy to lower the costs further, with goals on the order of US $0.50/gallon announced (Sheridan 2008).

An alternative approach finally receiving significant attention is to apply anaerobic organisms to both produce enzymes and ferment the sugars released into ethanol (Lynd et al. 2008). This consolidated bioprocessing (CBP) approach has a number of advantages, including better enzyme–organism synergy, lower cost production through avoiding aerobic enzyme production, and higher temperature operation. However, challenges include improving the selectivity of the organisms to ethanol and/or introducing enzyme production into organisms such as yeast that already produce ethanol well. In addition, the resulting systems must be robust for large-scale industrial use, and be ethanol tolerant.

Regardless of the enzyme-based technology employed, cellulosic biomass must be pretreated to overcome its natural resistance to biological attack. Hemicellulose removal with dilute acid as described above is an effective method often favored for pretreatment, but alternative technologies of ammonia fiber expansion (AFEX), controlled pH, lime, sulfur dioxide, and soaking in aqueous ammonia (SAA) have proven effective for various feedstocks (Hsu 1996; Mosier et al. 2005). In general, pretreatment has received relatively meager attention even though it impacts virtually all other processing operations, and a key need is to develop pretreatment technologies integrated into the entire process. Conventional yeast can ferment glucose and other six-carbon sugars to ethanol, and genetically modified microorganisms have been developed for fermentation of the previously difficult to ferment arabinose and xylose. After release of sugars and fermentation to ethanol, conventional distillation and dehydration technologies very similar to those applied for corn ethanol production can be employed to recover ethanol, while the unconverted fraction can be burned in a boiler to generate all of the heat and power to run the process with excess energy sold (Aden et al. 2002). Alternatively, some or all of the lignin and/or other components left after ethanol production could be converted into aromatic fuels resembling gasoline and other valuable products using gasification, liquefaction, pyrolysis, and catalytic upgrading approaches such as outlined above, although GHG emissions could become substantial if much fossil energy has to be imported to replace driving the process by lignin, thereby negating a major benefit of cellulosic ethanol production. The sugars could be fermented to other fuels such as butanol.

Recently, technology known as aqueous phase processing (APP) has been devised to catalytically convert the sugars released by acid and/or enzymes into alkanes that can be used as diesel or jet fuel substitutes (Huber et al. 2006). Zeolites and other catalysts could be applied to make aromatics from sugars, and catalytic processing could be used to make hydrogen as well. One advantage of the APP approach is that furfural, hydroxyl methyl furfural, and other soluble carbon compounds resulting from sugar degradation can also be converted into alkanes. Developers also claim easier purification of the water-insoluble products following the process, but integrated operation must be demonstrated. An important challenge with these new aqueous phase systems is to handle the impurities in the liquid without excessive catalyst fouling, with research underway to make the technology more robust for application.

Plants oils, such as those from soybean and rape seed, can be used directly as a diesel-like fuel but have high melting points that make them problematic for use as a mainstream transportation fuel. However, these triglycerides can be reacted with methanol or ethanol via transesterification to produce esters that are more suitable diesel fuels. The primary obstacles to large-scale impact are the high costs of many of the oils, such as those from soybeans or rape seeds, and the limited production potential. Recently, interest in taking advantage of the high growth rates of microalgae to produce triglycerides has reemerged, but significant engineering, site selection, and technology issues must be overcome for this immature technology to be viable for large-scale impact with low costs (Sheehan et al. 1998). For example, key biological requirements are for the algae to produce a high percentage of their body mass as oils and to be able to avoid invasion by unwanted organisms, and engineering challenges are associated with siting in suitable climates for close to year-round operation with adequate access to flat land, concentrated carbon dioxide, water, and sunlight.

1.6.4 Comparison of Conversion Options

Biological routes tend to be very specific in terms of compounds they will process and products they will form, while thermochemical technologies can transform a wide range of biomass materials into a wide range of products. Thus, biological routes can achieve high yields of targeted fuels, while thermochemical options often form co-products that must be upgraded or properly disposed of. Some of the latter are degradation products such as tars and minerals that foul catalysts and plug transfer lines and vessels and are costly to remove. Biological conversion can take advantage of the power of modern genetic engineering to significantly improve performance and lower costs but, despite many similarities to commercial sugar ethanol, corn ethanol, and forestry processing, suffers from perceived concerns about the risk of applying these new technologies for the first time. On the other hand, although thermochemical options have more limited opportunities for dramatic cost reductions, a substantial experience base exists for many technologies such as gasification that can lend confidence to scale-up to commercial scale. The lower temperatures and pressures for biological routes can reduce containment costs, although the longer reaction times can increase volumes. The higher temperatures and pressures of thermochemical options provide materials of construction and solids handling issues, but the fast reaction times reduce vessel sizes. Unit costs of capital equipment drop with increasing size for both, favoring larger scale operations, and larger scales appear particularly vital for thermochemical routes, driving up investment costs and stretching feedstock availability.

All of this said, the decision as to which of these and other conceivable options to employ will ultimately come down to cost and consumer preferences unless policies are implemented to encourage some routes over others to meet environmental or other national goals. Furthermore, we now use a number of different liquid fuels to meet the diverse needs for air, car, truck, rail, and ship transport, with about 63% of total transportation energy coming from gasoline, 21% from diesel fuel, and 11.4% as jet fuel, with much lower amounts being from aviation gasoline, lubricants, liquefied petroleum gas (LPG), and residual fuel oil (US Department of Energy 2008). Thus, several types of biofuels will no doubt be desirable, requiring a mix of technologies to make them. In addition, no one solution is likely to be adequate to meet the energy demands of a growing population faced with diminishing resources, and new combinations with other technologies, such as plug-in electric hybrids that use biofuels to support longer range driving, could well prove best. Thermochemical and biological conversion routes can also be synergistic, resulting in better economics, efficiency, and impact if integrated in a so-called biorefinery than if employed separately. In any event, liquid biofuels provide the only viable resource for sustainably powering aircraft and, most likely, heavy duty trucks, providing rapid refueling and allowing long distance travel, and it would be premature to try to eliminate options as long as no fundamental barriers are identified to meaningful commercial application.

1.7 What is the Goal and How Much Biomass will be Needed?

Biomass at US $60/dry short ton is to petroleum at about US $20/barrel on an equivalent energy basis (Lynd et al. 1999). Thus, the cost can be quite competitive and will only become more so as petroleum prices increase as supplies are diminished. However, the amount of petroleum used is huge, making its replacement challenging if no measures are taken to reduce energy use, and the problem only mounts as demand grows with increasing populations and the desire for greater mobility and other conveniences. Thus, we must find ways to cut our energy needs to more modest levels if we hope to meet the need for petroleum sustainably.

To provide an idea of the magnitude of the challenge, as transportation fuels, the United States directly consumes about 14.3 million barrels of oil per day or 5.2 billion barrels per year (US Department of Energy 2008). Assuming efficient biomass conversion technologies, we should be able to achieve yields of fuels on the order of 1.7 barrels of petroleum equivalent per short ton of biomass (Lynd et al. 2008). Thus, we would need something like 3 billion dry short tons of biomass per year. Even for a somewhat high biomass productivity of 10 short tons acre^{-1} year^{-1}, this demand would translate into a land area of 300 million acres, compared to the roughly 450 million acres now used in the United States for agriculture. Thus, it is quite unlikely that we could hope to produce enough biomass to meet all of our current petroleum needs from biofuels if the technology and use remain static.

A more thoughtful approach to energy use and biomass access can make this challenge less daunting. First, we need to move toward more efficient use of fuels, perhaps dropping our total consumption to something like 1.7 billion barrels of oil equivalent and thereby reducing biomass demand to about 1 billion short tons. Possibilities to achieve this goal include plug-in hybrid vehicles that use biofuels for longer distances, greater use of public transportation, and lighter vehicles with smaller engines. Next, we could take advantage of agricultural, forestry, and municipal wastes to generate around 600 million dry short tons or more of biomass to meet a good portion of the need, leaving about 400 million dry short tons to produce by energy crops (Perlack et al. 2005). Even for a low biomass productivity of 5 dry short tons acre^{-1} year^{-1}, the land area would be 80 million acres — a more attainable level than given above. However, with success in developing faster-growing plants, we could drop the land needed to something like 40 or perhaps even 20 million acres. More detailed scenarios and rationales for them have been presented in the "Billion Ton Biomass" study (Perlack et al. 2005).

As a side note, it is important to point out that achieving a productivity of 10 short tons acre^{-1} year^{-1} would make it possible to produce about 50 million short tons of biomass within a 50 mile radius, enough to make the equivalent of about 3.5 billion gallons of gasoline, with this distance being selected based on typical practice for hauling corn and pulp wood. This quantity of biomass would obviously drop due to demands for land for other uses such as housing, roads, golf courses, etc, but access to even 10% of the area would still allow production of about 350 million gallons of gasoline equivalent within the 50 mile radius. Higher productivity plants could raise the total to as high as 7 billion gallons of gasoline equivalent, about 5% of current gasoline consumption in the US.

An important point to make in this context is that, in addition to displacing petroleum use for transportation, substituting biofuels can reduce what one might term "parasitic petroleum" requirements for conversion of crude oil into finished products. As pointed out before, over 25% of the energy of the finished product is needed to power a petroleum processing facility, amounting to about another 1.3 billion barrels of oil per year. Because biomass conversion takes into account the generation of heat and electricity to run the process from biomass, for example, using the lignin left after carbohydrate conversion in the case of cellulosic ethanol, the need for this additional petroleum energy is avoided. Consequently, the potential grows to displacing the need for about 6.5 billion barrels per year of petroleum used in the United States, a large fraction of the total of 7.5 billion barrels we now use annually, through a combination of more efficient transportation, use of agricultural, forestry, and municipal residues, and more productive biomass.

Competing land use has recently grown in importance in the consideration of biofuels development. Attention has focused primarily on production of energy crops displacing use of land previously used for producing food and forcing farmers to clear new land to grow food, resulting in the release of carbon from the soil and from the cleared biomass (Searchinger et al. 2008). In one study, the emphasis was to show the importance of using marginal land to reduce such impacts (Fargione et al. 2008). This topic is quite complex and beyond the scope possible to handle appropriately in this overview, but one which must be taken seriously to address as fully as possible the need to improve energy security, reduce trade deficits, and reduce GHG emissions. However, it is also important to recognize the assumptions inherent in such scenarios to be sure they are consistent with expectations, the meaning of such analysis for other sustainable technologies, and the impact of possible technical evolutions in both biomass production and conversion and in changes in the fossil resource base as finite supplies of petroleum are consumed.

In summary, the following outcomes are important to consider for developing cellulosic crops that can reduce the demand for petroleum and also reduce GHG emissions:

- High productivity with growth rates of 10 or more dry short tons $acre^{-1}$ $year^{-1}$
- Drought resistance to minimize or avoid the need for irrigation
- Ability to grow on marginal land that will not compete with food production
- Low fertilizer demands to avoid runoff and nitrous oxide emissions and keep costs low
- Perennial growth to minimize disruption of the soil and also cut costs
- High lignin content when targeting thermochemical conversion
- Low lignin content and more easily processed lignin when targeting biological conversion
- Coproduction of protein to reduce competition with production of animal feed

These are challenging goals that are likely mutually exclusive in some cases but important to consider in the quest to produce biofuels on a crowded planet with increasing population, increasing per capita energy demand, limited land and water resources, and the looming consequences of global climate change.

1.8 Challenges to Commercial Applications

To this point, the energy challenges, conversion technologies, and biomass crop needs have been outlined. Although many of these options are already technically viable, concerns about taking the risk of first time implementation have stalled commercial applications. Rapid fluctuations in petroleum prices underscore the basis for this fear in that no one will invest hundreds of millions of dollars in a cellulosic biofuels facility only to see a sudden drop in oil prices wipe out their returns and likely the business. This concern is exacerbated by the fact that petroleum refining is well established technology that has benefited from a huge learning curve of over 100 years of operation and that has paid for most of its capital — definite advantages compared to fledgling sustainable technology industries.

1 Introduction Overview: World Energy Resources and the Need for Biomass for Energy   19

This situation translates into a huge advantage in a price war. Unless measures are taken to counter this reality, we are likely to see continued boom-to-bust oil and gasoline prices of likely increasing magnitudes, with limited, if any, real gains in commercializing sustainable alternatives. Furthermore, no action might be taken until high prices are sustained for long enough timeframes, likely measured in years, to convince investors of stable prices, by which time the economic consequences would be even more severe and a meaningful response would require decades to materialize due to limitations in capital and time demands for implementation.

Several strategies can help fill in the "this resulting 'valley of death'" such as use of municipal financing or other low cost debt, manufacture of valuable co-products, retrofitting of existing facilities, and employment of experienced engineers and contractors with directly relevant experience in agricultural and forestry processing (Wyman and Goodman 1993). However, even these measures are still not likely to be sufficient in light of the long time frames and great cost of conventional process scale-up regimens and investor timidity, and the only likely catalyst to overcome the "this valley of death" that now strands commercialization of sustainable transportation fuel technologies in a timeframe that can avert economic chaos is government action. But, the government must support sound and not political projects to be sure they provide real societal benefits and protect the taxpayer by not interfering with a rigorous due diligence process that can assure success for large investments. The government could provide incentives, but most investors tend not to trust government subsidies because they fear they will be withdrawn, putting huge capital investments at risk. An option favored by this author is for the government to provide a portion of the investment, say half, in sustainable energy projects to compensate for first-of-a-kind risk and then either sell off these positions or enjoy the returns once the project is successful. Alternatively, research could be aggressively funded to lower costs to be more economically attractive for risk involved, although this strategy postpones any tangible benefits for much longer and uncertain timeframes.

1.9 Closing Thoughts

Hopefully, this chapter makes it clear that the world, and the United States in particular are highly dependent on fossil fuels and particularly petroleum, deriving more energy from this one source than any other. In addition, we have used almost half of the total that is reasonably accessible, with most of the remaining reserves being located in politically volatile regions of the world. Furthermore, petroleum is the largest contributor to the GHG emissions that could lead to major environmental catastrophes. We fail to recognize the limited choices we have, and rather than tackle this dire situation with determination, we continue to wait for a miracle that is extremely unlikely to come. It is irresponsible, even unethical, to declare the problem as too difficult for us to solve and leave it to future generations to deal with our legacy saddled with even larger populations, fewer energy choices, and irreversible environmental problems.

The above concern is supported by tracing the course we have followed over the last 30 and more years toward addressing our energy issues. Soon after the United States oil production began declining after 1970, OPEC realized the power of controlling world oil production to support their political agenda and quickly cut supplies and raised prices. The result was economic and political chaos, with service stations often running out of gasoline, drivers waiting in long lines to fuel their cars, and limitations being placed on how much gasoline one could buy at a time. Against this background, President Carter declared on 18 April 1977 that: "Our decision about energy will test the character of the American people and the ability of the President and the Congress to govern. This difficult effort will be the 'moral equivalent of war'—except that we will be uniting our efforts to build and not destroy. I know that some of you may doubt that we face real energy shortages. The 1973 gasoline lines are gone, and our homes are warm again. But our energy problem is worse tonight than it was in 1973 or a few weeks ago in the dead of winter. It is worse because more waste has occurred, and more time has passed by without our planning for the future. And it will get worse every day until we act."

Under Carter's leadership, a serious effort was initiated to change this course by developing new sources of energy that would reduce our dependence on imported oil but was abandoned less than 3 years after the quoted speech was delivered. Indeed, as summarized in Table 1.1, dependence on petroleum has only increased, with atmospheric carbon dioxide levels increasing as well. Thus, despite periodic short-term calls to action and some temporary reductions in oil use in the mid-1970s resulting from high prices and scarcity, since the end of 1973, the world has consumed over another 800 billion barrels of oil of the more than 1 trillion barrels used since 1960. Furthermore, in that timeframe, the rate of consumption increased from 57.24 million barrels/day in 1973 to 84.62 million barrels/day in 2006. The result is that we now have reserves of only about 1.1–1.3 trillion barrels of oil if we include oil sands in Canada, a total that would last only about 40 years at current consumption rates. New discoveries will no doubt add to the supply, but increasing

Table 1.1 Selected facts on World and US petroleum use (US Department of Energy 2008)

	1970: Peak oil in US	1973: first OPEC oil embargo	Current time
Total world oil consumed since 1960[a]	129	188	1,018
Total US oil consumed since 1960[a]	48	65	288
US oil production rate[b]	9.64	9.21	5.10
US oil consumption rate[b]	14.70	17.31	20.69
World oil consumption rate[b]	46.81	57.24	84.62
Mauna Loa carbon dioxide concentrations (ppm)	325	330	386

OPEC Organization of the Petroleum Exporting Countries, *bbls* Barrels
[a]Billion bbls
[b]Million bbls/day

consumption will depete it even faster unless we change course. Even more importantly, world carbon dioxide emissions from energy consumption increased to 28.2 billion metric tons annually, with the United States responsible for 5.9 billion metric tons of this total, and atmospheric carbon dioxide levels measured at Mauna Loa rose from 330 ppm in 1974 to 386 ppm in 2008, a 17% increase (US Department of Energy 2008). Early transitions by consumers to buy fuel efficient vehicles gave way to cravings for sport utility vehicles (SUVs), with losses in gasoline economy and automobile companies held responsible to counter the preference of the American public for large, powerful, gas-guzzling cars. And miles driven increased. One of the few measures taken to replace oil was the implementation of cane sugar ethanol in Brazil and the now much maligned corn ethanol technology in the United States, even though the latter was very successful in meeting the goals of the 1970s and 1980s of reducing petroleum use. There have been recent forays made toward implementation of sustainable fuels technologies that promise to take us in the right direction, but it remains to be seen if these will make it to market. All this brings to mind a quote by Einstein that "Insanity is doing the same thing over and over and expecting different results." Until we change, nothing will change.

Acknowledgments The author is grateful to the Ford Motor Company for sponsoring the Chair in Environmental Engineering at the Bourns College of Engineering at the University of California, Riverside and for providing funds for our research on cellulosic ethanol. We also acknowledge support by the Bourns College of Engineering at the University of California at Riverside through the Center for Environmental Research and Technology (CE-CERT) and the Chemical and Environmental Engineering Department. Financial, moral, and technical support by the BioEnergy Science Center, a US Department of Energy Bioenergy Research Center supported by the of Biological and Environmental Research Office in the DOE Office of Science; DARPA through two different partnerships with the University of Massachusetts and Logos Technologies; Mascoma Corporation; Mendel Biotechnology; the National Institute of Standards and Technology through our partnership with Dartmouth College; the USDA National Research Initiative Competitive Grants Program; the US Department of Energy Office of the Biomass Program; and numerous past and present students, coworkers, and partners are critical to making our research on cellulosic biomass conversion possible.

References

Aden A, Ruth M, Ibsen K, Jechura J, Neeves K, Sheehan J, Wallace K, Montague L, Slayton A, Lukas J (2002) Lignocellulosic biomass to ethanol process design and economics utilizing co-current dilute acid prehydrolysis and enzymatic hydrolysis for corn stover. National Renewable Energy Laboratory, Golden, CO, NREL/TP-510-32438

Bossel U (2006) Does a hydrogen economy make sense? Proc IEEE 94:1826–1837

Brennan AH, Hoagland W, Schell DJ (1986) High temperature acid hydrolysis of biomass using an engineering scale plug flow reactor: results of low solids testing. Biotechnol Bioeng Symp 17:53–70

Fargione J, Hill J, Tilman D, Polasky S, Hawthorne P (2008) Land clearing and the biofuel carbon debt. Science 319:1235–1238

Farrell AE, Plevin RJ, Turner BT, Jones AD, O'Hare M, Kammen DM (2006) Ethanol can contribute to energy and environmental goals. Science 311:506–508

Grethlein HE, Converse AO (1991) Continuous acid hydrolysis of lignocelluloses for production of xylose, glucose, and furfural. In: Chahal DS (ed) Food, feed, and fuel from biomass. Oxford & IBH, New Delhi, pp 267–279

Heitz M, Capek-Menard E, Koeberle PG, Gagne J, Chornet E, Overend RP, Taylor JD, Yu E (1991) Fractionation of *Populus tremuloides* at the pilot plant scale: optimization of steam pretreatment using Stake II technology. Bioresour Technol 35:23–32

Holtzapple MT (1993) Chapters "Cellullose," "Hemicelluloses," and "Lignin". In: Macrae R, Robinson RK, Sadler MJ (eds) Encyclopedia of food science, food technology, and nutrition. Academic, London, pp 758–767, 2324–2334, 2731–2738

Hsu T-A (1996) Pretreatment of biomass. In: Wyman CE (ed) Handbook on bioethanol, production and utilization. Taylor & Francis, Washington, DC, pp 179–212

Huber GW, Iborra S, Corma A (2006) Synthesis of transportation fuels from biomass: chemistry, catalysts, and engineering. Chem Rev 106:4044–4098

Lewis NS, Nocera DG (2006) Powering the planet: chemical challenges in solar energy utilization. Proc Natl Acad Sci USA 103:15729–15735

Lynd LR (1996) Overview and evaluation of fuel ethanol from cellulosic biomass: technology, economics, the environment, and policy. Annu Rev Energ Environ 21:403–465

Lynd LR, Wyman CE, Gerngross TU (1999) Biocommodity engineering. Biotechnol Prog 15:777–793

Lynd LR, Laser MS, Bransby D, Dale BE, Davison BH, Hamilton R, Himmel M, Keller M, McMillan JD, Sheehan J, Wyman CE (2008) How biotech can transform biofuels. Nature Biotechnol 26:169–172

Maslow AH (1943) A theory of human motivation. Psychol Rev 50:370–396

Mosier N, Wyman CE, Dale B, Elander R, Lee YY, Holtzapple M, Ladisch M (2005) Features of promising technologies for pretreatment of lignocellulosic biomass. Bioresour Technol 96:673–686

Perlack R, Wright L, Turhollow A, Graham R, Stokes B, Erbach D (2005) Biomass as feedstock for a bioenergy and bioproducts industry: the technical feasibility of a billion-ton annual supply. Oak Ridge National Laboratory, Oak Ridge, TN

Searchinger T, Heimlich R, Houghton RA, Dong F, Elobeid AJF, Tokgoz S, Hayes D, Yu T-H (2008) Use of US croplands for biofuels increases greenhouse gases through emissions from land-use change. Science 319:1238–1240

Sheehan J, Dunahay T, Benemann J, Roessler P (1998) A look back at the US Department of energy's aquatic species program – biodiesel from algae. National Renewable Energy Laboratory, Golden, CO, NREL/TP-580-24190

Sheridan C (2008) Europe lags, US leads 2nd generation biofuels. Nat Biotechnol 26:1319–1321

Tyson KS (1993) Fuel cycle evaluations of biomass-ethanol and reformulated gasoline, vol I. National Renewable Energy Laboratory, Golden, CO, NREL/TP-463-4950 DE94000227

US Department of Energy (2008) Annual Energy Review 2007. Report DOE/EIA-0384(2007) June: Energy Information Administration, Washington, DC

Wiselogel A, Tyson S, Johnson D (1996) Biomass feedstock resources and composition. In: Wyman CE (ed) Handbook on bioethanol: production and utilization. Taylor & Francis, Washington, DC, pp 105–118

Wooley R, Ruth M, Glassner D, Sheehan J (1999) Process design and costing of bioethanol technology: a tool for determining the status and direction of research and development. Biotechnol Prog 15:794–803

Wright JD, D'Agincourt CG (1984) Evaluation of sulfuric acid hydrolysis processes for alcohol fuel production. Biotechnol Bioeng Symp 14:105–123

Wright JD, Power AJ (1987) Comparative technical evaluation of acid hydrolysis processes for conversion of cellulose to alcohol. Energy, Biomass and Wastes 10:949–971

Wyman C, Hinman N, Bain R, Stevens D (1992) Ethanol and methanol from cellulosic biomass. In: Williams R, Johansson T, Kelly H, Reddy A (eds) Fuels and electricity from renewable resources. Island, Washington, DC, pp 865–924

Wyman C, Dale B, Elander R, Holtzapple M, Ladisch M, Lee Y, Mitchinson C, Saddler J (2008) Comparative sugar recovery and fermentation data following pretreatment of poplar woody leading technologies. Biotechnol Prog 25:333–339

Wyman CE (1994) Alternative fuels from biomass and their impact on carbon dioxide accumulation. Appl Biochem Biotechnol 45–46:897–915

Wyman CE, Goodman BJ (1993) Near term application of biotechnology to fuel ethanol production from lignocellulosic biomass. In: Busche R (ed) Opportunities for innovation in biotechnology. National Institutes of Standards and Technology, Gaithersburg, MD, pp 151–190

Wyman CE, Dale BE, Elander RT, Holtzapple M, Ladisch MR, Lee YY (2005) Comparative sugar recovery data from laboratory scale application of leading pretreatment technologies to corn stover. Bioresour Technol 96:2026–2032

Chapter 2
Designing Biomass Crops with Improved Calorific Content and Attributes for Burning: a UK Perspective

Gordon G. Allison, Mark P. Robbins, José Carli, John C. Clifton-Brown, and Iain S. Donnison

2.1 The Need for Non-Food Energy Crops

Rapidly increasing energy costs, a foreseeable depletion of fossil fuel reserves and the pressing need to reduce greenhouse gas (GHG) emissions and mitigate global warming have made finding new sources of energy more urgent than ever before. Biomass, i.e. organic matter originating from plants (including algae, trees, crops and plant derived waste), has been used throughout human history as a source of heat and power. Indeed, biomass is estimated to be the fourth largest source of energy in the world, supplying 10–14% of primary energy, i.e. 46 EJ/year (Sims et al. 2006; Parikka 2004). Detailed reviews published by Sims et. al. (2006) and McKendry (2002a, b) provide thorough overviews of the different biomass feed stocks that are available currently and the diverse range of conversion technologies.

It is generally recognised that unless this potential increase in biomass production is carefully regulated there is likely to be very significant impacts on food production. At present, biofuel feedstock production occupies just 1% of cropland but the rising world population, changing diets and demand for biofuels are estimated to increase demand for cropland by between 17% and 44% by 2020 and, although sufficient suitable land is probably available, current policies do not ensure that additional production occurs in these areas (Renewable Fuels Agency 2008). Indeed, this and other reports (e.g. Davis et al. 2009; Robertson et al. 2008) highlight that, left unrestricted, production of biomass or fuels from traditional food crops will displace existing agricultural production, reduce biodiversity, and promote changes in land use that may even lead to extensive GHG emissions rather than savings (Gibbs et al. 2008; Environment Agency 2009; Upham et al. 2009; Buddenhagen

---

G.G. Allison, M.P. Robbins, J. Carli, J.C. Clifton-Brown, and I.S. Donnison
Biorenewables and Environmental Change (BEC) Division, Institute of Biological, Environmental and Rural Sciences, Aberystwyth University Gogerddan, Aberystwyth Ceredigion, Wales, SY23 3EB, UK
e-mail: Gordon.Allison@aber.ac.uk

et al. 2009). Conversely, other life cycle analysis studies have reported that, with careful consideration of the impacts of changing crops and land use, and also of the social, economic and environmental impact, the cultivation of energy crops is likely to be sustainable and have a mainly beneficial impact on GHG emissions (Hillier et al. 2009; Hastings et al. 2009; Monti et al. 2009; Haughton et al. 2009). Consequently, there is considerable pressure to identify non-food biomass crops that can meet the rising demand for biomass in a manner that is sustainable (Yuan et al. 2008; Gibbs et al. 2008; Smith 2008; Pauly and Keegstra 2008).

This chapter will focus specifically on the use and optimisation of non-food biomass crops that are suitable for cultivation in the United Kingdom and Northern Europe for energy production by combustion processes. It will briefly examine the range of combustion technologies, describe biomass composition and review how chemical composition influences the efficiency of energy conversion by combustion. It will then discuss the energy crops that are suitable for cultivation in the UK, highlight why they are fit for purpose and explore how these crops can be improved by approaches that include selective breeding and genetic manipulation (GM).

2.2 Biomass Combustion Technologies

2.2.1 The Combustion Process

There are three main thermal-conversion processes by which biomass can be converted to energy: combustion, gasification and pyrolysis; these three conversion processes are compared in Fig. 2.1. Combustion consists of burning biomass in air to convert the chemical energy stored in the biomass to heat, mechanical power or electricity (McKendry 2002b; Bridgwater 2003). Complete combustion requires sufficiently high temperature, strong turbulence of the air–gas mixture and a long residence time of the mixture in the fire chamber. Molecules of fuel are generally not reactive until they have undergone dissociation into reactive molecular fragments brought about by the high speed molecular collisions that are characteristic of high temperature. The latter two parameters increase the chance that molecules of pyrolysis gas have the opportunity to react with oxygen (Küçük and Demirbas 1997). The combustion of wood or woody biomass results in the production of hot gases, mainly carbon dioxide (CO_2) and water vapour (H_2O) through a complex series of reaction steps. An overall equation for the combustion of wood is presented below:

$$C_{42}H_{60}O_{28} + 43O_2 \rightarrow 42CO_2 + 30H_2O \qquad (2.1)$$

In contrast to gasification and pyrolysis, which produce intermediates that can be stored and used for subsequent energy or chemical production (syngas and bio-oil), the heat produced by combustion cannot be stored and must be used immediately for heat or the generation of power (Bridgwater 2003). Effective combustion of

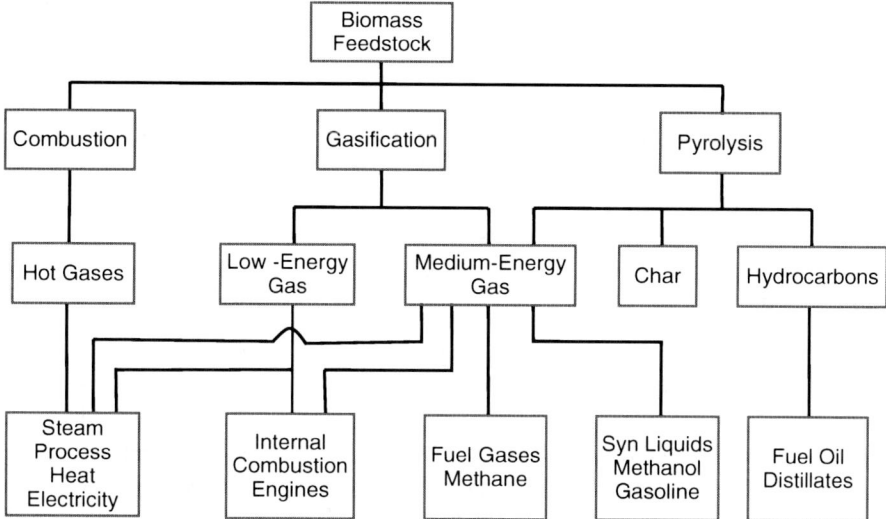

Fig. 2.1 Main thermal conversion processes for biofuels showing intermediates and final energy products (taken from McKendry 2002b)

biomass requires temperatures of approximately 800–1,000°C and, whilst it is possible to burn any type of biomass, in practice it is only feasible to burn biomass with moisture content lower than 50% due to the additional energy expenditure of drying woody biomass to below 40% moisture content, although large-scale facilities are better able to cope with higher moisture contents. An additional loss of energy yield from biomass combustion results from the need to mill feed-stocks to dimensions compatible with commercial combustion technologies (McKendry 2002b; Royal Society 2008). In this context, the scale of combustion plants range from the small domestic scale to large-scale industrial plants capable of producing 100–3,000 MW (McKendry 2002b).

2.2.2 Biomass as a Feedstock for Combustion

Biomass has much higher ratios of hydrogen:carbon and oxygen:carbon compared to fossil fuels and therefore less energy content for thermal conversion; coal, for example, contains between 75% and 90% carbon (Jenkins et al. 1998) while biomass typically has a carbon content of the order of 50% (Ptasinski et al. 2007; Obernberger et al. 2006). However, biomass fuels contain a greater proportion of volatile components than coal and therefore are more reactive at high temperature. At temperatures of around 500°C, approximately 85% of wood biomass (by weight) is converted into gaseous compounds (Ptasinski et al. 2007; McKendry 2002a) and efficiencies of 15% for small power stations and up to 30% for larger and newer plants are typical (Bridgwater 2003). However, the availability of agricultural, pulp and paper waste materials has ensured that combustion has remained commercially viable despite problems due to high levels of emissions and ash handling (Bridgwater 2003). There are two possible options for biomass and waste utilisation by combustion processes: biofuels can be burnt as a single fuel in specially designed power stations of limited capability, e.g. Slough Heat and Power (http://www.slough-heatandpower.co.uk/) and the soon to be constructed power station at Stevens Croft, Lockerbie (http://www.eon-uk.com/generation/stevenscroft.aspx), or co-combusted with coal in existing power stations, e.g. Drax power station (http://www.draxpower.com/corporate_responsibility/climatechange/cofiring/) (Department of Trade and Industry 1998). The former option requires significant financial investment, as the low-energy density of biomass would dictate the construction of new biomass-specific power stations. In addition, the associated infrastructure cost for these stations would be high as a result of the requirement for them to be decentralised in order to minimise fuel transportation costs (Carroll and Somerville 2009). Such stations would require extensive storage areas because of the seasonal availability of most biofuels (Hein and Bemtgen 1998). In contrast, many large multi-partner studies carried out in the UK for the department of Trade and Industry (Department of Trade and Industry 1998; Woods et al. 2006), the European Union (EU) funded APAS project (Activite de promotion, D'Accompagnement et de Suivi) (Hein and Bemtgen 1998), in the United States for the Department of Energy (Segrest et al. 1998) and the Alliance for Global Sustainability (Massachusetts Institute of Technology, The University of Tokyo, Chalmers University of Technology and the Swiss Federal Institute of Technology) (Leckner 2007) have found that co-combustion of coal with up to 10% of biomass is possible using existing power stations and infrastructure (Carroll and Somerville 2009). Furthermore, this is unlikely to lead to increased emissions of sulphur dioxide (SO_2), oxides of nitrogen (NO_x) or hydrochloric acid (HCl) (Department of Trade and Industry 1998). Co-combustion using a greater proportion of biomass may be possible but this is likely to require the development of new combustion systems (Huang et al. 2006). Whilst the range of biomass feed stock for co-combustion is highly diverse, e.g. wood, olive and palm residues, tall oil (a viscous yellow-black odorous liquid by-product of the Kraft process of wood pulp manufacture), sunflower and cereal pellets, sewerage sludge, waste derived fuels, tallow and biomass from dedicated energy crops (Department of Trade and Industry 1998), this chapter will focus only on the latter.

2.3 Lignocellulose

2.3.1 Structure and Composition of the Plant Cell Wall

By far the largest component of biomass from dedicated crops is lignocellulose, which forms the cell walls of plants. The composition of lignocellulose directly affects biomass quality for combustion and many efforts to improve biomass crops as feed stocks for combustion or other processes will focus on making specific modifications to cell wall composition. Cell walls are strong flexible composites of biological polymers that serve to maintain the structural integrity of the cell. The main components of cell walls, and the most abundant biopolymers on the planet, are cellulose (approximately 40–50% of most biomass by weight); hemicellulose (10–40%) and lignin (5–30% of biomass by weight; McKendry 2002a) with cellulose and lignin being the two most abundant biopolymers on Earth. Indeed lignin may account for 30% of all carbon fixed annually (Boerjan et al. 2003). Cellulose is made up of microfibrils (semi-crystalline bundles of 500–14,000 monomers of D-glucose joined linearly by β_{1-4} linkages); hemicelluloses (a hydrated matrix of cross-linked linear and branched polysaccharides composed of pentose and hexose sugars including glucose, mannose, xylose and arabinose) and pectin (structurally complex and often highly substituted linear and branched polymers rich in galacturonic acid; Somerville 2006; Mohnen 2008; Carpita and Gibeaut 1993).

Cells present in the vascular tissues of all higher plants contain high levels of lignin—a complex aromatic heteropolymer covalently bound to hemicellulose and which gives the strength and rigidity that allow plants to grow upright. Lignin also provides the vascular system with the hydrophobicity necessary for the transport of water and solutes (Vanholme et al. 2008). Lignin is formed from three hydroxycinnamyl alcohol monolignol monomers (hydroxyphenyl/ guaiacyl/ syringyl; H/G/S) differing in their degree of methoxylation (Boerjan et al. 2003; Boudet 1998). Lignin has a highly complex and somewhat random structure in which the three types of monolignol are linked by a variety of ether and carbon–carbon bonds. Current opinion holds that biosynthesis of lignin occurs in the extracellular milieu, where monolignols are oxidised by peroxide or laccase enzymes and coupled in a combinatorial fashion (Barsberg et al. 2006; Méchin et al. 2007; Morreel et al. 2004; Weng et al. 2008; Grabber 2005).

2.3.2 Plant Cell Wall Architecture

The primary cell wall is formed during cell elongation. In all dicotyledonous species (dicots) and many monocotyledonous species (type I monocots), the primary cell wall is composed primarily of cellulose microfibrils embedded in a hydrated matrix of xyloglucan hemicelluloses, pectins and structural proteins. The primary cell walls of type II monocots, i.e. grasses and related monocots (Poales), have a different composition. In this case, the major cross linking hemicellulose polymers are glucuronoarabinoxylans (GAX; Carpita and Gibeaut 1993). In addition, type II primary cell walls contain a higher proportion of cellulose and only negligible amounts of pectin (Carpita 1996).

Secondary walls are deposited during the differentiation of xylem, phloem and transfer cells once elongation is complete. Woody species and forest crops in particular, are rich in secondary cell walls. The molecular architecture of secondary walls is much less well characterised than that of primary walls (McCann and Carpita 2008; Boudet 1998). Secondary walls are generally thicker than primary walls, are enriched in xylans and cellulose, and contain only minor amounts of protein and pectin (Mellerowicz et al. 2001). Most importantly, in secondary walls, lignin replaces much of the water, making them impenetrable to solutes and enzymes (Pauly and Keegstra 2008).

Primary and secondary cell walls from grass species are also distinct from those of dicots in that they contain large amounts of cell-wall-bound hydroxycinnamic acids, namely *p*- coumaric acid (up to 3%) and ferulic acid (up to 4%; Allison et al. 2009a; Grabber et al. 1995; Waldron et al. 1996; Vogel 2008), which are bound to the arabinoxylan moieties of GAX and lignin by ether and ester bonds. Furthermore, ferulic acid forms a variety of dimers (Hatfield et al. 1999a) and, to a lesser extent, trimers through ether and ester linkages (Bunzel et al. 2003, 2004). These play an important structural role in the grass cell wall as they covalently cross-link adjacent GAX molecules by ester linkages and bind GAX to lignin by a combination of ester and ether bonds (Hatfield et al. 1999a).

2.4 The Effect of Chemical Composition on Feedstock Properties

Whilst biomass feedstocks are comprised primarily of carbon, hydrogen, oxygen and nitrogen, they contain additional components or 'impurities' that disrupt the combustion process. The presence and concentration of such substances is dependent on the plant material, agronomic and agricultural practices and geographic location and, in some cases, levels may also increase as a result of increasing crop yield (Royal Society 2008). Of particular concern for combustion efficiency are: calorific value; moisture content; the proportion of fixed carbon; and the content of ash, residues and alkaline metals (McKendry 2002a; Obernberger et al. 2006). Moisture content and carbon density have dramatic effects on calorific value, which is usually expressed as higher heating value (HHV) or lower heating value (LHV). HHV is the energy content when the material is burnt in air under standard conditions of temperature and pressure and the value includes the condensation enthalpy of water in contrast to LHV (Friedl et al. 2005). Biofuel quality could therefore be improved by breeding, agricultural or other interventions that decrease moisture content or increase carbon density — the most obvious route being to increase the proportion of lignocellulose, although, as will be discussed later, this is far from trivial. Typically, biomass fuels have moisture contents ranging from 16% to 30% and have LHVs of around 16–19 MJ/kg, in contrast to coal, which typically has a moisture content of approximately 11% and a LHV of 43 MJ/kg (McKendry 2002a).

Alkaline metals (Na, K, Mg, and Ca) occur naturally in plants and their concentration in biofuels has major effects on combustion efficiency as they are involved in ash formation and decrease ash melting point, which can in turn cause blockage, erosion and/or corrosion of equipment through processes that are now well understood (Misra et al. 1993; Lewandowski and Kicherer 1997; Jenkins et al. 1998). In addition, high levels of nitrogen, chlorine and sulphur can lead to unacceptable emissions of NO_x, HCl and SO_2 and also boiler corrosion (Obernberger et al. 2006;

Lewandowski and Kicherer 1997). Furthermore, the concentration of chlorides in the biomass has as much of an influence on the amount of alkaline metals vaporised during combustion as the concentration of the alkaline metals themselves, and is thought to act as a shuttle, transporting alkaline metals from the fuel to surfaces where they form stable sulphates (Jenkins et al. 1998). The continual removal of these minerals by harvesting can lead to soil degradation and non-sustainable production practices (El-Nashaar et al. 2009). At least one study has shown that the effect of alkaline metals in reducing conversion efficiency in several biomass fuels was much greater than the effect of differing lignin content. This latter study also showed that washing the biomass before conversion improved efficiency, presumably by leaching out chloride and alkaline metals (Fahmi et al. 2008).

Silica is another biomass component and whilst not posing a problem by itself it is involved in ash formation and is known to react with alkaline metals (Jenkins et al. 1998). It is abundant in the walls of grasses, where it is present mostly as inclusion bodies in the epidermis, periderm and other specialised root cells, rhizome and aerial shoots (Carpita 1996) and may be introduced as soil contamination during harvesting. The range of ash content in biofuels can vary between 1% and 20%, with wood typically having a low ash content of 1–2% (Misra et al. 1993); biomass from dedicated energy grass species lies within an acceptable range of 3–5% (Fahmi et al. 2008; Lewandowski and Kicherer 1997; McKendry 2002a). These levels are achieved by three key harvest management practices; namely, harvesting after senescence has occurred, harvesting after over-wintering in the case of *Miscanthus* species; or, for other grass species, allowing the mown crop to leach in the field for 1–4 weeks before baling (Cornell University 2006). These processes allow leaching of chlorine and alkaline metals so reducing ash content, and also reduce water content and the concentrations of protein and nitrogen in the foliar tissues. In addition, senescence, the natural process of winter die-back, allows nutrients and minerals to be mobilised to below-ground tissues for storage over the winter months (Jørgensen 1997). This decreases the requirement for fertiliser input and improves crop sustainability. In addition, harvesting after over-winter weathering dramatically reduces the proportion of leaf material in the biomass. Whilst leaves do not make a significant contribution to the composition of wood biomass, leaf material can make a significant contribution to grass biomass and, even after senescence, levels of ash, nitrogen, phosphorus, silica and alkaline metals are much greater in leaf material than in stem, resulting in a significant deterioration of biofuel quality (Monti et al. 2008).

2.5 Energy Crops for Combustion Processes in the European Union

Several studies have identified the importance of low external inputs as a key factor for energy crops and whilst this may result in poorer energy yields, emission balances are much more favourable (Kaltschmitt et al. 1997). In addition, suitable energy crops must be capable of growing on land that is marginally fertile in order not to displace current food production. These criteria have considerably narrowed the number of potential energy crop species suitable for production in the UK. Wood products and forest waste have obvious roles as biomass for combustion, and indeed wood chips from poplar have been found to be the most favourable when compared with other forms of bioenergy with the exception of rapeseed oil and wood chips from willow (Kaltschmitt et al. 1997). However, a study by Berndes et al. (2003), which reviewed 17 published studies on the contribution of biomass to the future global energy supply, reported that perhaps as much as half of the timber available in Europe may not be available for energy production (Berndes et al. 2003), and predicted that over the next 100 years it would be energy crops that would contribute the largest proportion to bioenergy supply.

All of the crops identified as having potential as biomass crops for Northern Europe and the UK in particular have high levels of lignocellulose (Table 2.1). Fast-growing woody C_3 crops are attractive as sources of biomass in Europe as they meet with agronomic, environmental and societal requirements for successful deployment as energy sources, and much attention has been given to short rotation willow (*Salix* spp.; Smart and Cameron 2008) and poplar (McKendry 2002a). In the UK, *Miscanthus x giganteus*, a naturally occurring sterile hybrid of the South East Asian species *Miscanthus sinensis* and *Miscanthus sacchariflorus*, and, to a lesser extent, switchgrass (*Panicum virgatum*), a native of North America, are the two herbaceous species that have received most attention as commercially viable and environmentally sustainable biomass crops for combustion (Bullard 1999; Carroll and Somerville 2009; Bouton 2008; Price et al. 2004). Both of these perennial grass species provide easily harvestable annual crops with low moisture content and high dry matter yield. Furthermore, both *M. x giganteus* and switchgrass have C_4 photosynthetic apparatus, which is common in species originating from tropical or dry locations. Both therefore have potential photosynthetic advantages over native C_3 perennial grasses, e.g. temperate forage grasses such as *Lolium*, when CO_2 is limiting, temperatures are high and water is scarce, and are able to convert a higher proportion of incident light into biomass (Ehleringer et al. 1997). In addition,

Table 2.1 Comparison of the compositions of biomass feed-stocks (from Pauly and Keegstra 2008, and IENICA 2009 crop database). Values adjusted to percentage dry weight (%DW)

Feedstock	Cellulose	Hemicellulose	Lignin	Ash	Solubles
Corn stover	39.4	33.1	14.9	ND	8.9
Wheat straw	34.9	22.5	21.3	9.4	11.9
Rice straw	41.6	31.5	12.5	14.4	ND[a]
Miscanthus	41.9	16.6	13.3	3.2	15.0
Sorghum	15.0	12.3	5.8	0.4	66.5
Switch grass	46.1	32.2	12.3	4.7	ND
Reed canary grass	28.0	22.0	14.0	8.0	28.0
Sugar cane	48.6	31.1	19.1	1.2	ND
Hardwood spp	43.3	31.8	24.4	0.5	ND
Softwood spp	40.4	31.1	28.0	0.5	ND

[a]Not determined

perennial grasses generally have lower nitrogen content and have a lower requirement for nitrogen inputs when compared with annual species. They do not require annual tilling and this allows considerable amounts of carbon to remain sequestered in the soil, reduces soil erosion and decreases the energy inputs required for the operation of heavy farming machinery (Heaton et al. 2004a).

2.5.1 Miscanthus Species

M. x giganteus is propagated from rhizomes and, in the UK, typically grows to a height of approximately 3–4 m (see Fig. 2.2a). The crop takes two to three years to establish before yields are maximised. Weeds are controlled by soil-acting herbicide application after planting and again prior to emergence in the 2nd and possibly the 3rd year (Clifton-Brown et al. 2008a). *M. x giganteus* has a large root structure that extends approximately 1.8 m below the surface, where nutrients are stored in the rhizomes over the winter months (Carroll and Somerville 2009); recent studies suggest that there is considerable potential for these roots to sequester carbon and thereby decrease GHG emissions (Hillier et al. 2009; Clifton-Brown et al. 2007). It has also been grown successfully in the US and in European locations including Turkey, Ireland, Denmark, Germany, the UK, Switzerland, Spain and Italy (Lewandowski et al. 2000; Heaton et al. 2004b; Clifton-Brown et al. 2007; Acaroglu and Semi Aksoy 2005). In England and Wales, dry matter harvestable yields have been reported to range between 6.9 and 24.1 t ha^{-1} year^{-1} when the crop is grown on arable land (Price et al. 2004). One study in Ireland on marginal land reported average autumn and spring dry matter yields of 13.4 and 9.0 t ha^{-1} year^{-1} over a period of 15 years (Clifton-Brown et al. 2007) and modelling has predicted a peak output yield across Ireland of between 16 and 26 t ha^{-1} year^{-1} (Clifton-Brown et al. 2000). Yields reported in Europe range from 4 t ha^{-1} year^{-1} in Central Germany to 44 t ha^{-1} year^{-1} in Northern Greece and Italy (Angelini et al. 2009; Lewandowski et al. 2000). This considerable range of yield is most likely due to variations between sites in temperature and rainfall as well as differences in harvesting date, phenotypic type and possibly fertiliser treatment. In the UK, yields with current varieties of *M. x giganteus* are likely to be greater in the wetter west than in the drier east of the country (McKendry 2002a). One recent estimate has put the amount of *Miscanthus* (presumably mainly *M. x giganteus*) under cultivation in the UK in 2007 at 10,000 ha (Nix 2007), although the estimated area given over to *Miscanthus* made by the UK National Non-Food Crops Centre (2009) is somewhat lower, with only 4,032 ha of *Miscanthus* being grown in the UK in 2007 with the total area of new *Miscanthus* being planted each year rising from 302 ha in 2005 to more than 2,300 ha in 2006 and 2007. Most of this *Miscanthus* is used for co-firing with coal at large power-stations although an increasingly greater proportion is being used directly for the generation of combined heat and power at biomass dedicated stations, e.g. The Bluestone Holiday Village Project in

Fig. 2.2 a Photograph of a mature stand of *Miscanthus giganteus* being harvested (Pembrokeshire, Wales, February 2008). The stand height is between 2.5 m and 3.0 m. **b** Photographs of representatives of the two best represented *Miscanthus* species in the Aberystwyth collection: *left M. sacchariflorus* (canopy height 1.83 m), *right M. sinensis* (canopy height 1.24 m)

Pembrokeshire, Wales (http://www.energycropswales.co.uk/opening_markets.php.en?subid=0). Dramatic increases in *Miscanthus* cultivation have been predicted for the next 20 years as the requirement for biofuel feedstocks increases. The area of land that is suitable for the cultivation of *M. x giganteus* in the UK alone amounts to more than 1.5 million ha (approximately 10% of agricultural land, J.C.C.-B., unpublished data), capable of yielding approximately 18.7 million t/year, and the area available in the 15 member states of the EU amounts to over 11.6 million ha, with a potential yield of more than 158 million t/year (Clifton-Brown et al. 2004). It is likely that these are indeed conservative estimates given that a primary goal of *Miscanthus* breeders is to radically increase yields by the development of "hi-tech" hybrid varieties (Hastings et al. 2009) as, despite the promising features of *M. x giganteus*, there is vast scope for genetically improving *Miscanthus* as a biomass feedstock by integrating desirable traits by exploitation of the huge genetic variation present in wild *Miscanthus* accessions, particularly those of *M. sinensis* (Stewart et al. 2009). This approach will therefore seek to increase the tolerance of *Miscanthus* to environmental stress, thereby opening up opportunity for cultivation on as yet unsuitable land.

Compared to many other lignocellulosic plants *M. x giganteus* has excellent combustion properties with low water (16–33%) and mineral content (Cl= 0.3–2.1 g kg^{-1}; N= 0.9–3.4 g kg^{-1} and K= 3.7–11.2 g kg^{-1}; Lewandowski and Kicherer 1997). Similar values were detected in a recent study in which 15 *Miscanthus* accessions were grown in five locations in Europe (Lewandowski et al. 2003). A major goal of breeding will be to increase yields at low levels of input (Moller et al. 2007). A substantial impediment preventing widespread cultivation of *M. x giganteus* in the UK is poor frost tolerance (Clifton-Brown et al. 2000; Farrell et al. 2006). An extensive breeding programme is underway at Aberystwyth University aimed at incorporating traits from wild genotypes of *M. sacchariflorus* and *M. sinensis* into new high yielding *Miscanthus* varieties (both novel hybrids as well as new varieties of *M. sinensis* and *M. sacchariflorus*) tailored for the commercial sector to improve drought and cold tolerance and stay-green characteristics (Clifton-Brown et al. 2008b). Figure 2.2b shows clearly the significant morphological differences that typically exist between these two species. *M. sacchariflorus* tends to have fewer but taller and thicker stems whilst *M. sinensis* has many smaller and thinner stems. In summary, improving the ability of new varieties to grow to high yields in the UK, and the development of varieties that could be propagated by seed are primary objectives in establishing wide-spread commercial cultivation but currently these are still some way off realisation.

2.5.2 Switchgrass

Switchgrass has received relatively little attention in Europe compared with *M. x giganteus* despite the former having been identified by the US Department of Energy as its main herbaceous dedicated energy crop because of its potential for high yields, low environmental impact and low input requirement (Bouton 2008; Carroll and Somerville 2009). It is a major component of the American prairies and many varieties grow in small dense clumps. In the US, like *M. x giganteus*, switchgrass may reach up to 3 m in height and its chemical composition and low moisture content make it ideally suited for a variety of bioenergy uses, including lignocellulosic conversion to bioethanol and combustion. Published annual switchgrass yields are sometimes lower than those of *M. x giganteus*, depending on the climate (Heaton et al. 2004a), and switchgrass is harvested annually or semi-annually (Bouton 2008). There are two distinct ecotypes available: lowland and upland, with the former, which has thicker stems, growing more sparsely as densely bunched plants. Furthermore, later maturity tends to result in higher mineral concentrations at harvest (DTI 2006). Switchgrass takes 3 years or more to reach maturity and is optimally grown as a highly managed single crop, generally sown using grassland drill; weeds are controlled using pre- and post-emergent herbicides. Commercial varieties can tolerate a wide range of soil and pH conditions and, with only limited fertiliser input, can produce a greater yield than other warm season grass species. In addition, the excellent seasonal yield distribution of switchgrass, especially for high spring yields, means that crops are of value to the live-stock industry in addition to its use as biomass (Vogel 2004). One study in Ardmore, Oklahoma, reported yields of between approximately 8 and 17 t ha^{-1} depending on the time of harvest (Bouton 2008) whilst a comprehensive comparison of switchgrass with *M. x giganteus* conducted in Illinois estimated an average yield of 10 t ha^{-1} from 77 separate observations (Heaton et al. 2004a). This latter review concluded that, in certain climates, *M. x giganteus* holds greater promise for biomass energy cropping than switchgrass. In the UK, yields of 9.63 t ha^{-1} year^{-1} have been reported for a lowland ecotype across three sites and two growth years in comparison with approximately 7 t ha^{-1} year^{-1} for typical upland varieties (DTI 2006). However, despite these lower yields and the difficulty of establishment, switchgrass is likely to have a role as a bioenergy crop in the UK as it can be sown from seed (rather than rhizomes, which require specialised equipment for planting) and harvested and bailed using equipment that is commonly available on farms familiar with growing perennial forage grasses (Vogel 2004). At present however, there is little evidence of commercial switchgrass cultivation in the UK and northern Europe, and most existing plantations are for research purposes.

2.5.3 Willow and Poplar

Willow and poplar are promising candidates for woody energy crops and have received much attention in the US (Smart and Cameron 2008; Davis 2008). In the UK, willow (Fig. 2.3) has received comparatively much greater attention due to programmes such as the European Union funded "Willow for Wales" project (http://www.willow4wales.co.uk) and the National Willows Collection at Rothamsted Research. These collections each comprise approximately 1,300 genetically characterised clones. The availability of a complete genome sequence for poplar and its role as a model organism for plant biology will no doubt facilitate the development of improved varieties for bioenergy use (Carroll and Somerville 2009). Yields of 12.4 t ha^{-1} year^{-1} and 22.5 t ha^{-1} year^{-1} have been reported for poplar grown on non-irrigated and irrigated soils (Deckmyn et al. 2004) whilst a

Fig. 2.3 Photograph of a mature stand of coppice willow planted in 2004 and cut back in 2006. The stand (approximately 3.5–4.0 m in height) represents 2 years of growth. (Courtesy of Chris Duller, Field trial co-ordinator of Willow for Wales, Pembrokeshire, Wales, February 2008)

study in Quebec found average yields of 17.3 t ha^{-1} year^{-1} for poplar and 16.9 t ha^{-1} year^{-1} for willow without fertiliser or irrigation (Labrecque and Teodorescu 2005). These yields seem comparable with those which might be expected from *M. x giganteus* but there is obvious need for additional trials that will allow a direct comparison of these species with *M. x giganteus* and switchgrass at a variety of geographical locations in northern Europe to inform which crops would be most suitable in given locations. It is highly likely that some regions will be more suitable for the growth of trees rather than energy grasses. Large-scale commercialisation of short rotation coppice willow is already practiced in northern Europe (Moller et al. 2007) but in the UK, growth of tree bioenergy crops has been slow to take off, with only very limited amounts of willow under cultivation, and poplar not being cultivated on a commercial scale at all. One problem has been that cultivation requires considerable expenditure for establishment and subsequent harvesting ties up land for cultivation for considerable periods of time; indeed removing trees from land requires considerable expenditure. Furthermore, willow and poplar demand large amounts of water, which excludes them from growth in certain areas and, in addition, both are extremely susceptible to rust (Moller et al. 2007). Lastly, the availability of cheap forest chipped waste in the UK has undermined willow as a commercial crop at present but with increasing demand for clean chipped wood of high quality for domestic heating it is likely that willow cultivation will increase.

2.5.4 Reed Canary Grass

Lastly, reed canary grass (*Phalaris arundinace*) is a rhizomatous C_3 perennial grass that warrants mention as a potential bioenergy crop. This species is distributed widely across the temperate regions of Europe, Asia and North America and can grow as high as 2 m. Reed canary grass possesses two highly desirable traits: the ability to withstand drought and also to tolerate excessive precipitation. Like switchgrass it is propagated from seed and has the flexibility to be used for animal feed as well as for biomass, but unlike switchgrass it is relatively easy to establish, with full yields being reached in fewer years. One UK study has found that the crop requires nitrogen fertilisers for optimal growth (DTI 2006). One comprehensive study of 72 accessions at five locations in the US reported yields that varied with environment (mean of 9.2 t ha^{-1} year^{-1}) and which remained high in wet locations and marginal land (Casler 2009). Yields of 10 t ha^{-1} year^{-1} have been reported in Sweden, where it is being evaluated as a bioenergy crop; however, in the UK much lower yields of approximately 4 t ha^{-1} year^{-1} (Chisholm 1994) and 5.3–5.5 t ha^{-1} year^{-1} (DTI 2006) are more typical. The ease and low cost of establishing and cultivating this crop suggest that in time there may be a role for reed canary grass as a secondary energy crop in the UK but currently there is little or no commercial cultivation of reed canary grass as a bioenergy crop (UK National Non-Food Crops Centre 2009).

2.6 Technologies for Crop Design

2.6.1 Modification of Hemicellulose and Cellulose

The modification of biomass crops for improved combustion can be divided into several key areas: (1) manipulation of the amount and structure of lignocellulose in the crop biomass; (2) altering the chemical composition of the biomass; and (3) altering quality parameters such as moisture content and particle size. Increasing cell wall polysaccharide concentrations would most likely also increase calorific value; however, efforts to modify hemicellulose or cellulose have been hampered by the extreme complexity of structural polysaccharide biosynthetic systems in plants. Hemicellulose and cellulose are synthesised in different cellular compartments by very different complex processes that are still not thoroughly understood (Somerville 2006). Cellulose is synthesised at the plasma membrane by rosette complexes that are thought to consist of 36 individual cellulose synthase proteins belonging to three or more different classes (Mutwil et al. 2008). In contrast, hemicellulose is synthesised in the Golgi, packaged into secretory vesicles and transported to the cell surface for incorporation into the cell wall matrix (Pauly and Keegstra 2008). However, the natural variability observed in the wall composition of several biomass feed stocks shown in Table 2.1 suggests that there is a great potential for altering wall composition without compromising the life cycle of the plant (Pauly and Keegstra 2008) but this goal may be difficult to achieve without a better understanding of the exact processes involved in biosynthesis.

Several investigations have shown the need for caution when altering wall composition in planta since this may cause changes that are detrimental for plant growth and lead to phenotypes including dwarfism (Desprez et al. 2007), lethality (Goubet et al. 2003) or compromised defence against pathogens (Sticklen 2006). There are several reports of increased polysaccharide concentration being effected by manipulation of growth regulators or insertion of genes to delay flowering (reviewed by Sticklen 2006). These studies, however, are still some way from being effective strategies for biomass improvement, and realistic options for biomass improvement by increasing wall polysaccharide content will depend on a more comprehensive understanding of the genes involved in cell wall biosynthesis. Over the last decade, considerable progress has been made in this area and genes have been identified that are involved in the biosynthesis of cellulose, hemicellulose and pectin, as well as genes responsible for the biosynthesis of the sugar nucleotide donors involved in polysaccharide biosynthesis (Zhong and Ye 2007; York and O'Neill 2008; Ye et al. 2006). This process has been greatly assisted by the availability of new model systems, e.g. maize (*Zea mays*), sorghum (*Sorghum bicolor*; Carpita and McCann 2008) and *Brachypodium* distachyon (Opanowicz et al. 2008), new data base resources, e.g. Maizewall (Guillaumie et al. 2007), better understanding of cell wall architecture (McCann and Carpita 2008) and by new techniques for identifying cell wall biosynthetic genes (McCann et al. 2007; Mitchell et al. 2007).

2.6.2 Modification of Lignin

In contrast to cellulose and hemicellulose, the biosynthesis of lignin is better understood (Boerjan et al. 2003) and lignin has proved to be highly amenable to manipulation by genetic engineering (Boudet 1998; Li et al. 2008, Vanholme et al. 2008; Weng et al. 2008). Several reviews have been published that describe in detail efforts to alter lignin quantitatively and qualitatively to improve the efficiency of lignocellulosic fermentation to liquid transport fuels and biorefinery intermediates (Hatfield et al. 1999b; Grabber 2005, Weng et al. 2008; Chang 2007; Sticklen 2006). Generally, this involved modifying lignocellulose to improve degradation and facilitate enzymic deconstruction, and only rarely has the focus been on increasing calorific value. However, although lignin content is positively correlated with calorific value (Demirbas 2001), there is growing evidence linking lignin concentration to soot formation (Fitzpatrick et al. 2008). Furthermore, changes in cell wall composition may affect particle size in the processed biomass, which has been shown to have implications for combustion efficiency (Bridgeman et al. 2007). Therefore, in some cases it may be desirable to breed varieties of Miscanthus with reduced lignin content.

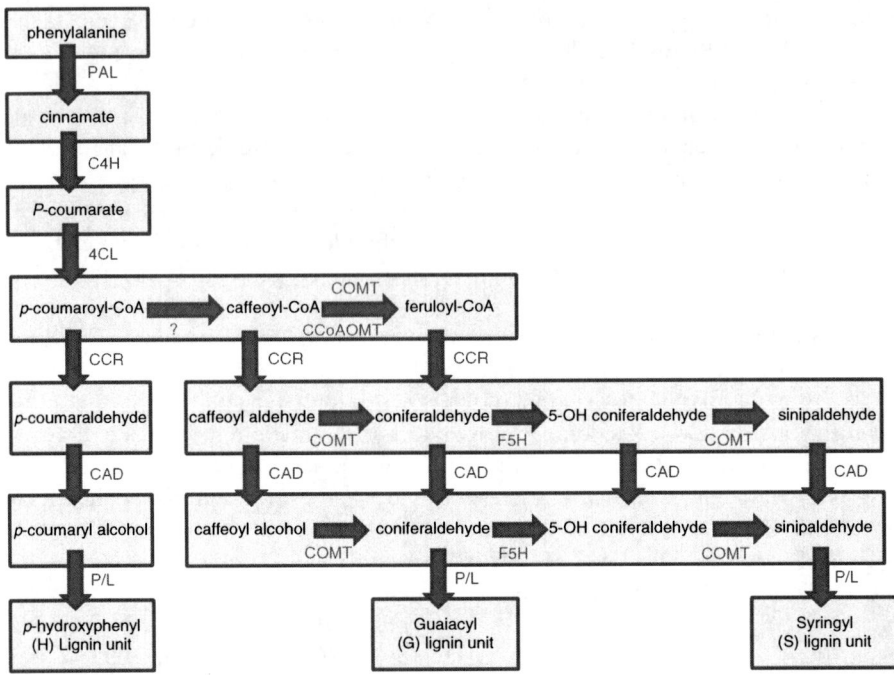

Fig. 2.4 Schematic showing the main biosynthetic pathway of monolignol biosynthesis (based on Boerjan et al. 2003). In order of appearance: *PAL* phenylalanine ammonia-lyase, *C4H* cinnamate 4-hydroxylase, *4CL* 4-coumarate:CoA ligase, *CCR* cinamoyl-CoA reductase, *COMT* caffeic acid O-methyltransferase, *CCoAOMT* caffeoyl-CoA O-methyltransferase, *CAD* cinamyl alcohol dehydrogenase, *F5H* ferulate 5-hydroxylase, *P/L* peroxidise and laccase

Our understanding of lignin biosynthesis has been furthered by the study of the brown-midrib mutants of maize (Li et al. 2008), sorghum and millet (Vogel 2008), all of which are characterised by a reddish-brown pigmentation of the leaf midrib and are associated with altered and often lowered lignin content (Barriere et al. 2004; Marita et al. 2003). Most of the work to characterise these mutations was carried out on maize, the first species in which these mutations were identified. A pathway showing the biosynthetic pathway of lignin is shown in Fig. 2.4. Two of these phenotypes are due to lowered activity of specific lignin biosynthetic genes. The *bm1* phenotype is due to a mutation affecting cinnamyl alcohol dehydrogenase (CAD), but it is not yet clear whether the mutation actually lies within this gene (Halpin et al. 1998); b*m3* mutants are defective in caffeic acid O-methyltransferase (COMT; Vignols et al. 1995). The molecular basis of the other two phenotypes *bm2* and *bm4* is not yet understood (Marita et al. 2003); *bm2* mutants contain fewer guaiacyl and syringyl residues and have altered patterns of lignin deposition (Vermerris and Boon 2001), whilst the lignin composition of *bm4* mutants resembles that of *bm2* (Barriere et al. 2004). There is now considerable evidence to suggest that lignin biosynthesis is highly plastic; normal maize lines and hybrids display substantial genetic variability for lignin and degradability traits, at times rivalling the extremes associated with *bm* mutants (Argillier et al. 1996; Deinum and Struik 1989; Dhillon et al. 1990; Jung et al. 1998; Jung and Buxtono 1994; Lundvall et al. 1994; Méchin et al. 2000; Roth et al. 1970) and given the genetic similarity of *Miscanthus* with model C_4 grass species it may be possible to breed successfully for brown-midrib traits in this energy crop.

Another, more adaptable, approach to modifying lignin content in energy crops is GM. Evidence from studies in model species, e.g. tobacco, alfalfa, maize and poplar (Vanholme et al. 2008; Li et al. 2008) suggest that altering lignin content in *Miscanthus* species and coppice tree species using molecular approaches is quite feasible, although in practice it may be easier to decrease or modify lignin rather than to increase total lignin content. Changing the expression of many of the lignin biosynthetic genes often results in altered lignin monomer composition and or reduced total lignin content (Li et al. 2008; Vanholme et al. 2008); however, all too frequently, transgenic phenotypes are less dramatic or possibly contrary to expectation. Many lignin biosynthetic genes are members of multigene families and other homologues may be involved in other important cellular processes (Campbell and Sederoff 1996); furthermore, this leads to enormous plasticity of plant metabolism, often resulting in surprising and unexpected phenotypes. For example, decreasing the activity of CAD might be expected to result in lower levels of monolignols available for incorporation into lignin and reduced lignin content. In practice, this is often not the case as other intermediates, e.g. cinnamaldehydes, may be incorporated into lignin in their place (Boerjan et al. 2003). This aside, genetic engineering can result in dramatic changes in lignin content; in one study, down-regulation of 4-coumarate: CoA ligase (4CL) in aspen was shown to reduce lignin content by up to 45% (Hu et al. 1999). Caffeoyl-CoA O-methyl transferase (CCoAOMT) seems to be a major hub in controlling lignification (Ye et al. 1994), and probably also cross-linking in grasses, making this enzyme an ideal target for digestibility improvement by lignin reduction and altered composition. Indeed, in alfalfa, reducing CCoAOMT activity to a residual 5% increased cell-wall digestibility by 34% (Guo et al. 2001). Strongly reduced lignin content with radically altered composition has also been reported in an *Arabidopsis ref8* mutant defective in *p*-coumarate 3-hydroxylase (C3H). This intervention resulted in drastically altered phenyl propanoid metabolism, the formation of lignin composed almost entirely of H units and significant developmental defects (Franke et al. 2002). Down-regulation of COMT leads to decreased synthesis of synapil alcohol and compensatory deposition of 5-hydroxyconiferyl alcohol (M15H). This alteration of lignin composition also results in the incorporation of large amounts of novel benzodioxane structures (Atanassova et al. 1995; Ralph et al. 2001, 2000). Simultaneous downregulation of 4CL and over-expression of ferulate 5-hydroxylase (F5H) has been reported to result in lower lignin content, higher S/G level, and increased cellulose in aspen plants (Li et al. 2003).

New power plants designed specifically for biomass combustion may negate the emission problems associated with elevated lignin content and make high-lignin energy crop varieties with increased calorific value practical. Although this has so far been difficult to achieve in a predictable manner by simply over-expressing biosynthetic genes, further studies using alternative gene promoter sequences and different construct architectures may have more success. Perhaps a better approach for increasing lignin content in biofuels may be to over-express appropriate regulatory genes (Weisshaar and Jenkins 1998; Zhong and Ye 2007; Patzlaff et al. 2003). This approach has been utilised with great success in other phenylpropanoid pathways, e.g. condensed tannins and anthocyanins (Robbins et al. 2003; Nesi et al. 2000; Kubo et al. 1999; Quattrocchio et al. 1999; Spelt et al. 2000) and is likely to deliver biomass with higher lignin levels, increased structural strength, and improved pest and disease resistance. In turn, these modifications would also increase the value of timber crops as construction materials, reduce the need for toxic wood preservatives and also increase the unit calorific value of the biofuel.

2.6.3 Breeding Strategies

In addition to modification approaches based upon GM, other plant breeding strategies may have value for *Miscanthus*. Building upon the range of Miscanthus biomass currently available, traditional methodologies are immediately available as *Miscanthus sinensis* has previously been bred primarily for ornamental applications, e.g. *Miscanthus sinensis* "Zebrina". Therefore, as this crop has not been subjected to selection for combustion characteristics, initial selections from germplasm (combined with accurate chemical phenotyping) may well produce improved cultivars in short- to medium-term time periods. An example of this type of approach has been outlined by Clifton-Brown et al. (2008b), who reported phenotypic variation in a replicated spaced trial containing 249 genotypes grown in Aberystwyth. Many genotypes are hard to cross for various reasons including sexual incompatibility. Clearly, future approaches for lines derived from high yielding accessions may well rely upon the identification of quantitative trait loci (QTL) for combustibility traits derived from *Miscanthus* mapping families (Atienza et al. 2002).

2.6.4 Chemical Phenotyping and High-Throughput Screening

The development of improved varieties of energy crops requires extensive phenotype analysis and therefore analytical methods that are robust, cost effective and capable of coping with large numbers of samples. However, many traditional methods that are commonly used for measuring chemical composition are time-consuming and costly (Giger-Reverdin 1995; Bridgeman et al. 2007; Friedl et al. 2005) and unsuitable for large-scale analysis at high rates of through-put. By contrast, methods based on spectroscopic analysis e.g. near infrared (NIRS) and Fourier transform infrared (FTIR) spectroscopy, offer practical solutions for the inexpensive, robust and accurate analysis of parameters including cell wall structure (Chen et al. 1998), the concentration and composition of aromatic cell wall components including lignin and hydroxycinnammic acids (Allison et al. 2009a; Stewart et al. 1997; Alves et al. 2006), cell wall carbohydrates (Fairbrother and Brink 1990), digestibility (Decruyenaere et al. 2009), nitrogen content (Gislum et al. 2004), and fixed carbon, nitrogen, alkali index and ash content (Allison et al. 2009b; Huang et al. 2007). Spectra can usually be acquired within approximately a minute, and often sample preparation is considerably simplified to drying and grinding. Early use of this approach sought to correlate the absorbance at specific wave lengths to the concentrations of specific cellular components determined using gravimetric, analytical or chromatographic techniques, e.g. lignin strongly absorbs at $1{,}510$ cm^{-1} (Monties 1989); however, for analysis of non-purified samples this simplistic approach is prone to inaccuracies caused by the presence of additional compounds with overlapping absorbencies. This problem can be overcome by using multivariate regression methods such as partial least squares or multivariate regression (Labbé et al. 2008; Gislum et al. 2004; Allison et al. 2009a, b).

Pyrolysis gas chromatography/mass spectrometry has great potential for the high throughput chemical analysis of lignocellulose composition. This approach has been used extensively by engineers to profile biomass, but much less so by biologists (del Rio et al. 2007; Fahmi et al. 2007; Galletti and Bocchini 1995). The method requires rapid gasification of biomass, usually in an oxygen-free atmosphere, separation of the pyrolysis volatiles on the gas chromatography column and detection by mass spectrometry. One particularly flexible pyrolysis unit is the CDS Pyroprobe 5200 pyrolyser; this unit is temperature programmable and pyrolysis products generated between specified temperatures are first trapped and then introduced onto the gas chromatograph. The instrument can then be heated to a higher temperature with similar sample trapping allowing the sequential analysis of cell wall components in order of thermal decomposition and therefore discrimination of products originating from hemicellulose from products originating from cellulose or lignin. Both quadrupole and ion trap gas chromatograph/ mass spectrometers have application, the former offering more quantitative data with the ability to clearly distinguish between known products and the latter offering more qualitative discrimination and identification of unknown products.

Several methods are available for determining the elemental composition of biomass. Two of these rely on analysis of hot gaseous plasmas; namely inductively coupled plasma mass spectrometry (ICP MS) or optical emission spectrometry (ICP OES), and over recent years instrument stability has improved significantly, reducing the number of calibration standards required in addition to the samples being analysed, with parallel decreases in instrument cost. Both methods require samples to be ground and digested with concentrated acids overnight but subsequent analysis is largely automated. Analysis by mass spectrometry allows quantitative data to be collected on approximately 100 elements but instrumentation is generally more expensive. In contrast, analysis of emission spectra is less costly but usually only groups of 4–5 elements can be analysed at any one time and considerable method development is often required (Conte et al. 1999). Silica and chloride present special difficulties, the former requiring samples to be dissolved in concentrated hydrofluoric acid and the latter being intractable by this approach and requiring analysis by ion chromatography. X-ray fluorescence (XRF) is an alternative approach and although there are few reports of the technique being used to analyse biomass feed-stocks (Robinson et al. 2009) it has been used to determine elemental composition in biological samples, e.g. in mycobacteria (Gresits and Könczöl 2003) and mussel shells (Kurunczi et al. 2001) and it would appear that XRF has great potential for the analysis of biomass. XRF is used widely in the cement industry to measure elemental composition — an application where the concentration of chlorine is of particular interest — and the method has the advantage of stability, thus negating the requirement for frequent recalibration. However, the precisely defined particle size required for XRF analysis requires lengthy milling to the required size. XRF is therefore a time consuming process and this obstacle would need to be addressed before it could be applied at the high rates of through-put necessary for application to chemical phenotyping on a bioenergy crop breeding programme.

2.6.5 Case Study: Variation in Cell Wall Composition Between 249 Miscanthus *Genotypes*

As part of a large growth trial at Aberystwyth we have measured the amount of cellulose, hemicellulose and lignin in several triploid *M. sacchariflorus X sinensis* hybrids, including the hybrid recognised as *M. x giganteus* (64 observations), *M. sacchariflorus* (272 observations) and *M. sinensis* (1,629 observations) over two consecutive growth years. Planted in 2005 at 1.5 m intervals in four replicate plots, the plants were harvested after over-wintering in the field in the February following the 2006 and 2007 growth years. Sampling entailed removing the entire above-ground foliage and shredding of the material through a modified forage harvester. This material was then weighed, and approximately 200 g removed for cell wall analysis. This material was oven-dried at 60°C and then ground using a rotary mill to pass through a 1 mm mesh. The results of these measurements are presented in Table 2.2. These values were predicted using partial least squares models from the NIR spectra of the samples. Spectra were collected and manipulated using standard procedures (Barnes et al. 1989) and multivariate regression models to predict neutral detergent fibre (NDF; a measure of total cell wall), acid detergent fibre (ADF; a measure of total cellulose and lignin), and acid detergent lignin (ADL) were trained and validated on compositional data obtained using standard gravimetric methods (Van Soest 1963, 1967).

Whilst *M. sacchariflorus* and *M. sinensis* have very similar mean levels of cellulose and hemicellulose, there is considerable difference between the mean ADL content of these two species. Furthermore, these data suggest that the several new *Miscanthus* hybrids are similar to *M. x giganteus* in that they have significantly

Table 2.2 Average of hemicellulose, cellulose and acid detergent lignin expressed as %DW in 254 independent accessions of mature individual genotypes of *Miscanthus sinensis x sacchariflorus* hybrids (values for *Miscanthus giganteus* are shown for the purpose of comparison), *M. sacchariflorus* and *M. sinensis* grown in 4 replicated plots at Aberystwyth University, UK and harvested after over-wintering in the field for two successive growth years. Each genotype is replicated four times within the experiment and abundance of cellulose and hemicellulose have been calculated using near infrared spectroscopy (NIRS)-predicted values of acid detergent fibre (ADF), neutral detergent fibre (NDF) and acid detergent lignin (ADL). Significant differences between the *Miscanthus* species are denoted by lower case letters

Species	Hemicellulose	Cellulose	Lignin
M. giganteus			
Mean	29.68 a	46.93 a	11.97 a
SD	2.02	2.12	0.89
Range	27.01–32.26	43.99–49.41	10.70–13.24
New *M. giganteus* hybrids			
Mean	30.54 a	46.94 a	11.47 a
SD	2.04	2.06	1.00
Range	25.99–34.36	39.48–50.57	8.84–13.50
M. sacchariflorus			
Mean	32.87 b	42.69 b	10.60 b
SD	1.66	3.10	1.62
Range	28.26–36.82	31.22–49.36	6.83–13.71
M. sinensis			
Mean	33.70 c	42.60 b	9.21 c
SD	1.47	3.14	1.05
Range	24.82–38.55	30.80–50.50	6.05–13.50

higher levels of cellulose and ADL, but lower levels of hemicellulose than *M. sacchariflorus* or *M. sinensis*. The mean concentration of ADL in the *Miscanthus* hybrids (including *M. x giganteus*) is much greater than in either *M. sacchariflorus* or *M. sinensis*. Whether this similarity between the triploid is due to a dominant ploidy effect or the result of only very similar *M. sinensis* and *M. sacchariflorus* parents being compatible is currently under investigation. Statistical analysis of these data by unbalanced analysis of variance (Table 2.2) detected significant differences between the hybrids, *M. sacchariflorus* and *M. sinensis* for cellulose ($P < 0.001$; s.e.d. 0.862), hemicellulose ($P < 0.001$; s.e.d. 0.425) and ADL ($P < 0.001$; s.e.d. 0.297). Significant differences were also detected between the two growth years for cellulose [means of 42.11 and 43.39 % dry weight (DW) for 2006 and 2007 respectively; $P < 0.001$; s.e.d. 0.280] and hemicellulose (means of 33.84 and 33.18% DW for 2006 and 2007 respectively; $P < 0.001$; s.e.d. 0.142) but not for lignin (means of 9.32 and 9.30% DW for 2006 and 2007, respectively). Analysis of these data by Pearson correlation shows that ADL shows weak positive correlation with cellulose ($R = 0.492$) and weak negative correlation with hemicellulose ($R = -0.523$). There seems to be no correlation between cellulose and hemicellulose content ($R = -0.171$). The composition in cell wall measured in *M. x giganteus* in this experiment are somewhat different from those presented in Table 2.1; the specimens analysed in this trial contained greater amounts of cellulose and hemicellulose but less lignin. This is possibly due to differences in environmental conditions and genetic differences between genotypes classified as *M. x giganteus*. Indeed, in another study (de Vrije et al. 2002), *M. x giganteus* was reported to have 38% DW cellulose, 24% DW hemicellulose and 25 % DW Klason lignin (which in grass species is typically twice the amount of ADL; Hatfield and Fukushima 2005). This study has enabled genetic mapping families to be devised in order to map QTL relating to cell wall composition and assist in the breeding of *Miscanthus* varieties with optimised cell wall composition.

2.7 Conclusions and Future Perspectives

The need for developing a sustainable non-fossil fuel economy is self evident and has been highlighted in many recent reviews, including two publications by Nicholas Stern (Stern 2007, 2009). In the EU, the undertaking to reduce GHG emissions, decrease our carbon foot-print, increase security of energy supply and to steer our economies towards sustainable economic growth is stimulating the expansion of renewable energy resources, of which biofuels are part of the overall portfolio. The development of new, and improvement of existing, energy conversion technologies will enable effective and efficient conversion of biomass feedstocks to energy, and facilitate growth of dependency on biomass as a renewable source of energy. Given the changes to climate that are forecast over the next century, a key obstacle to optimal reliance on bioenergy crops will be the inevitable competition for land use between food and fuel, and it is imperative that bioenergy crops are capable of delivering sufficient yield under local climatic and soil conditions. Furthermore, the chemical composition of the biomass must be matched to the final energy conversion process. The precise goals for each species will, to some extent, be case-specific although in all cases there will be a drive to increase sustainable yield at low levels of inputs. In cases where the bioenergy crop originated in warmer wetter climates, e.g. *Miscanthus*, there will be a need to increase cold-, frost- and drought-tolerance (Oliver 2009) and develop varieties capable of germinating from seed. Due to the sterility of *M. x giganteus*, inclusion of these traits will most likely require extensive rebreeding. For switchgrass an important goal to facilitating widespread deployment will be to improve establishment time, which with current varieties may be 3 years or greater depending on the climate. It may also be possible to develop varieties capable of being grown on more marginal land for use as a secondary bioenergy crop, or for use as silage, hay or pasture. Changing biomass chemical composition may be achievable using conventional breeding technologies such as integration of traits from other *Miscanthus* accessions, or from related species, or may involve more indirect approaches such as mutation breeding or GM, perhaps by using either regulatory transgenes (Demura and Fukuda 2007) or by the stacking of multiple transgene interventions (Halpin and Boerjan 2003) to develop stable viable phenotypes with improved lignocellulose quality traits. After many years of reluctance, the EU seems to be gradually losing its objections to the cultivation of GM crops, and it might be envisaged that, given the non-food nature of these crops, this last avenue may soon be open to the development of commercial varieties once adequate proof has been amassed of trait and crop safety to humans, live-stock and the environment. An additional hazard will be that the very traits that increase yield, improve tolerance to environmental stress and allow these crops to be grown on ever more marginal land also have the effect of making energy crops potentially invasive weeds. In a recent study in Hawaii, Buddenhagen et al. (2009), using a widely accepted weed risk assessment system, reported 70% of regionally sustainable biofuel crops had a high risk of becoming potentially invasive weeds compared to only 25% of non-biofuel species. Assuming that these proportions are not the result of geographical location, these results should alert European plant breeders to be vigilant. These concerns aside, plant breeding objectives may include the initial development of low lignin–high lignocellulose varieties for combustion in existing power stations and later, as biomass specific power stations become more wide spread, high lignin–high lignocellulose varieties. Other improvements reached by breeding or GM routes might involve changing plant architecture, not only to improve light interception, but also to alter composition by increasing or decreasing grass internode length (to modulate lignin and particle size) or reduce stand height whilst increasing stand density in the case of coppice forestry to ease extraction of the mature crop from what is all too often water-logged sites. In addition, it is possible that society will once more regard itself as being dependent on the land and that work in the agro-energy sector will provide new employment opportunities. The future for energy crops in the UK and northern Europe is therefore at once challenging, dynamic and exciting. What is important is that we currently have options that are fit-for-purpose and have no excuse to prevent us moving towards meeting our GHG reduction targets in 2020 and beyond.

Acknowledgements The authors wish to acknowledge the help of Dr. Paul Robson, and their colleagues at IBERS who are associated with or belong to the *Miscanthus* breeding programme, who were instrumental in growing, phenotyping and harvesting the *Miscanthus* trait trial. They wish to acknowledge their funders: G.A., M.R. and I. D. are funded as part of the Bioenergy Strategic Programme Grant by the Biotechnology and Biological Science Research Council; J.C.-B. is funded jointly by the Biotechnology and Biological Science Research Council and The Department of Environment, Food and Rural Affairs, and J.C. is funded by the Engineering and Physical Sciences Research Council as part of Supergen Bioenergy (http://www.supergen-bioenergy.net/). Special thanks also to Mrs. Catherine Morris and the staff of the Analytical Chemistry Unit who carried out the chemical analysis of plant samples and to Mrs. Pauline Rees Stevens for proof-reading this manuscript.

References

Acaroglu M, Semi Aksoy A (2005) The cultivation and energy balance of *Miscanthus x giganteus* production in Turkey. Biomass Bioenergy 29:42–48

Allison GG, Thain SC, Morris P, Morris C, Hawkins S, Hauck B, Barraclough T, Yates N, Shield I, Bridgwater AV, Donnison IS (2009a) Quantification of hydroxycinnamic acids and lignin in perennial forage and energy grasses by Fourier-transform infrared spectroscopy and partial least squares regression. Bioresour Technol 100:1252–1261

Allison GG, Morris C, Hodgson E, Jones J, Kubacki M, Barraclough T, Yates N, Shield I, Bridgwater AV, Donnison IS (2009b) Measurement of key compositional parameters in two species of energy grass by Fourier transform infrared spectroscopy. Bioresour Technol 100:6428–6433

Alves A, Schwanninger M, Pereira H, Rodrigues J (2006) Calibration of NIR to assess lignin composition (H/G ratio) in maritime pine wood using analytical pyrolysis as the reference method. Holzforschung 60:29–31

Angelini LG, Ceccarini L, Nassi o Di Nasso N, Bonari E (2009) Comparison of *Arundo donax* L, *Miscanthus x giganteus* in a long-term field experiment in Central Italy: analysis of productive characteristics and energy balance. Biomass Bioenergy 33:635–643

Argillier O, Barrière Y, Lila M, Jeanneteau F, Gélinet K, Ménanteau V (1996) Genotypic variation in phenolic components of cell-walls in relation to the digestibility of maize stalks. Agronomie 16:123–130

Atanassova R, Favet N, Martz F, Chabbert B, Tollier MT, Monties B, Fritig B, Legrand M (1995) Altered lignin composition in transgenic tobacco expressing O-methyltransferase sequences in sense and antisense orientation. Plant J 8:465–477

Atienza SG, Satovic Z, Petersen KK, Dolstra O, Martin A (2002) Preliminary genetic linkage map of *Miscanthus sinensis* with RAPD markers. Theor Appl Genet 105:946–952

Barnes RJ, Dhanoa MS, Lister SJ (1989) Standard normal variate transformation and de-trending of near-infrared diffuse reflectance spectra. Appl Spectrosc 43:772–777

Barriere Y, Ralph J, Mechin V, Guillaumie S, Grabber JH, Argillier O, Chabbert B, Lapierre C (2004) Genetic and molecular basis of grass cell wall biosynthesis and degradability. II. Lessons from brown-midrib mutants. C R Biol 327:847–860

Barsberg S, Matousek P, Towrie M, Jørgensen H, Felby C (2006) Lignin radicals in the plant cell wall probed by Kerr-gated resonance Raman spectroscopy. Biophys J 90:2978–2986

Berndes G, Hoogwijk M, van den Broek R (2003) The contribution of biomass in the future global energy supply: a review of 17 studies. Biomass Bioenergy 25:1–28

Boerjan W, Ralph J, Baucher M (2003) Lignin biosynthesis. Annu Rev Plant Biol 54:519–546

Boudet A-M (1998) A new view of lignification. Trends Plant Sci 3:67–71

Bouton J (2008) Improvement of switchgrass as a bioenergy crop. In: Vermerris W (ed) Genetic improvement of bioenergy crops. Springer, New York, pp 295–308

Bridgeman TG, Darvell LI, Jones JM, Williams PT, Fahmi R, Bridgwater AV, Barraclough T, Shield I, Yates N, Thain SC, Donnison IS (2007) Influence of particle size on the analytical and chemical properties of two energy crops. Fuel 86:60–72

Bridgwater AV (2003) Renewable fuels and chemicals by thermal processing of biomass. Chem Eng J 91:87–102

Buddenhagen CE, Chimera C, Clifford P (2009) Assessing biofuel crop invasiveness: a case study. Plos ONE 4:e5261 http://www.plosone.org

Bullard M (1999) MAFF final report for project NF0403:*Miscanthus* agronomy (for fuel and industrial uses) www.ienica.net/usefulreports/miscanreport.pdf

Bunzel M, Ralph J, Funk C, Steinhart H (2003) Isolation and identification of a ferulic acid dehydrotrimer from saponified maize bran insoluble fiber. Eur Food Res Technol 217:128–133

Bunzel M, Funk C, Steinhart H (2004) Semipreparative isolation of dehydrodiferulic and dehydrotriferulic acids as standard substances from maize bran. J Sep Sci 27:1080–1086

Campbell MM, Sederoff RR (1996) Variation in lignin content and composition (mechanisms of control and implications for the genetic improvement of plants). Plant Physiol 110:3–13

Carpita NC (1996) Structure and biogenesis of the cell walls of grasses. Annu Rev Plant Physiol Plant Mol Biol 47:445–476

Carpita NC, Gibeaut DM (1993) Structural models of primary cell walls in flowering plants: consistency of molecular structure with the physical properties of the walls during growth. Plant J 3:1–30

Carpita NC, McCann MC (2008) Maize and sorghum: genetic resources for bioenergy grasses. Trends Plant Sci 13:415–420

Carroll A, Somerville C (2009) Cellulosic biofuels. Annu Rev Plant Biol 60:165–182

Casler MD, Cherney J, Brummer E (2009) Biomass yield of naturalized populations and cultivars of reed canary grass. Bioenergy Res 2:165–173

Chang MCY (2007) Harnessing energy from plant biomass. Curr Opin Chem Biol 11:677–684

Chen L, Carpita NC, Reiter W-D, Wilson RH, Jeffries C, McCann MC (1998) A rapid method to screen for cell-wall mutants using discriminant analysis of Fourier transform infrared spectra. Plant J 16:385–392

Chisholm CJ (1994) Reed canary grass. In: Chisholm CJ (ed) Towards a UK research strategy for alternative crops. Silsoe Research Institute, Ministry of Agriculture Fisheries and Food, Agricultural Development and Advisory Service, UK

Clifton-Brown JC, Neilson B, Lewandowski I, Jones MB (2000) The modelled productivity of *Miscanthus* x *giganteus* (*Greef et Deu*) in Ireland. Ind Crops Prod 12:97–109

Clifton-Brown JC, Stampfl PF, Jones MB (2004) *Miscanthus* biomass production for energy in Europe and its potential contribution to decreasing fossil fuel carbon emissions. Glob Change Biol 10:509–518

Clifton-Brown JC, Breuer J, Jones MB (2007) Carbon mitigation by the energy crop *Miscanthus*. Glob Change Biol 13:2296–2307

Clifton-Brown JC, Chiang Y-C, Hodkinson TR (2008a) *Miscanthus*: genetic resources and breeding potential to enhance bioenergy production. In: Vermis W (ed) Genetic improvement of bioenergy crops. Springer, New York, pp 273–294

Clifton-Brown JC, Robson P, Allison G, Lister S, Sanderson R, Hodgson E, Farrar K, Hawkins S, Jensen E, Jones S, Huang L, Roberts P, Youell S, Jones B, Wright A, Valantine J, Donnison I (2008b) Miscanthus: breeding our way to a better future. In: Booth E, Green M, Karp A, Shield I, Stock D, Turley D (eds) Biomass and energy crops III. Association of Applied Biologists, Warwick, pp 199–206

Conte RA, van Veen EH, de Loos-Vollebregt MTC (1999) Fast survey analysis of biomass by-product samples based on ICP optical emission spectra. Fresenius J Anal Chem 364:666–672

Cornell University (2006) Ash content of grasses for biofuels Bioenergy Information Sheet. http://grassbioenergy.org/downloads/Bioenergy_Info_Sheet_5.pdf

Davis JM (2008) Genetic improvement of poplar (*Populus* spp.) as a bioenergy crop. In: Vermerris W (ed) Genetic improvement of bioenergy crops. Springer, New York, pp 377–396

Davis SC, Anderson-Teixeira KJ, DeLucia EH (2009) Life-cycle analysis and the ecology of biofuels. Trends Plant Sci 14:140–146

De Vrije T, de Haas GG, Tan GB, Keijsers ERP, Claassen PAM (2002) Pretreatment of *Miscanthus* for hydrogen production by *Thermotoga elfii*. Int J Hydrogen Energy 27:1381–1390

Deckmyn G, Laureysens I, Garcia J, Muys B, Ceulemans R (2004) Poplar growth and yield in short rotation coppice: model simulations using the process model SECRETS. Biomass Bioenergy 26:221–227

Decruyenaere V, Lecomte P, Demarquilly C, Aufrere J, Dardenne P, Stilmant D, Buldgen A (2009) Evaluation of green forage intake and digestibility in ruminants using near infrared reflectance spectroscopy (NIRS): developing a global calibration. Anim Feed Sci Technol 148:138–156

Deinum B, Struik PC (1989) Genetic variation in digestibility of forage maize (*Zea mays* L.) and its estimation by near infrared reflectance spectroscopy (NIRS). An analysis. Euphytica 42:89–98

Del Rio JC, Gutierrez A, Rodriguez IM, Ibarra D, Martinez AT (2007) Composition of non-woody plant lignins and cinnamic acids by Py-GC/MS, Py/TMAH and FTIR. J Anal Appl Pyrolysis 79:39–46

Demirbas A (2001) Relationships between lignin contents and heating values of biomass. Energy Convers Manage 42:183–188

Demura T, Fukuda H (2007) Transcriptional regulation in wood formation. Trends Plant Sci 12:65–70
Department of Trade and Industry (1998) Volume 3: Converting wood fuel to energy. http://www.berr.gov.uk/files/file14937.pdf
Desprez T, Juraniec M, Crowell EF, Jouy H, Pochylova Z, Parcy F, Höfte H, Gonneau M, Vernhettes S (2007) Organization of cellulose synthase complexes involved in primary cell wall synthesis in *Arabidopsis thaliana*. Proc Natl Acad Sci USA 104:15572–15577
Dhillon BS, Paul C, Zimmer E, Gurrath PA, Klein D, Pollmer WG (1990) Variation and covariation in stover digestibility traits in diallele crosses of maize. Crop Sci 30:931–936
DTI (2006) A trial of the suitability of switchgrass and reed canary grass as biofuel crops under UK conditions. www.berr.gov.uk/files/file34815.pdf
Ehleringer JR, Cerling TE, Helliker BR (1997) C_4 photosynthesis atmospheric CO_2 and climate. Oecologia 112:285–299
El-Nashaar HM, Griffith SM, Steiner JJ, Banowetz GM (2009) Mineral concentration in selected native temperate grasses with potential use as biofuel feedstock. Bioresour Technol 100:3526–3531
Environment Agency (2009) Biomass — carbon sink or carbon sinner? www.environment-agency.gov.uk/static/documents/Biomass__carbon_sink_or_carbon_sinner_summary_report.pdf
Fahmi R, Bridgwater AV, Thain SC, Donnison IS, Morris PM, Yates N (2007) Prediction of Klason lignin and lignin thermal degradation products by Py-GC/MS in a collection of *Lolium* and *Festuca* grasses. J Anal Appl Pyrolysis 80:16–23
Fahmi R, Bridgwater AV, Donnison I, Yates N, Jones JM (2008) The effect of lignin and inorganic species in biomass on pyrolysis oil yields quality and stability. Fuel 87:1230–1240
Fairbrother TE, Brink GE (1990) Determination of cell wall carbohydrates in forages by near infrared reflectance spectroscopy. Anim Feed Sci Technol 28:293–302
Farrell AD, Clifton-Brown JC, Lewandowski I, Jones MB (2006) Genotypic variation in cold tolerance influences the yield of *Miscanthus*. Ann Appl Biol 149:337–345
Fitzpatrick EM, Jones JM, Pourkashanian M, Ross AB, Williams A, Bartle KD (2008) Mechanistic aspects of soot formation from the combustion of pine wood. Energy Fuels 22:3771–3778
Franke R, Hemm MR, Denault JW, Ruegger MO, Humphreys JM, Chapple C (2002) Changes in secondary metabolism and deposition of an unusual lignin in the *ref8* mutant of *Arabidopsis*. Plant J 30:47–59
Friedl A, Padouvas E, Rotter H, Varmuza K (2005) Prediction of heating values of biomass fuel from elemental composition. Anal Chim Acta 544:191–198
Galletti GC, Bocchini P (1995) Pyrolysis/gas chromatography/mass spectrometry of lignocellulose. Rapid Commun Mass Spectrom 9:815–826
Gibbs HK, Johnston M, Foley JA, Holloway T, Monfreda C, Ramankutty N, Zaks D (2008) Carbon payback times for crop-based biofuel expansion in the tropics: the effects of changing yield and technology. Environ Res Lett 3:034001
Giger-Reverdin S (1995) Review of the main methods of cell wall estimation: interest and limits for ruminants. Anim Feed Sci Technol 55:295–334
Gislum R, Micklander E, Nielsen JP (2004) Quantification of nitrogen concentration in perennial ryegrass and red fescue using near-infrared reflectance spectroscopy (NIRS) and chemometrics. Field Crops Res 88:269–277
Goubet F, Misrahi A, Park SK, Zhang Z, Twell D, Dupree P (2003) AtCSLA7 a cellulose synthase-like putative glycosyltransferase is important for pollen tube growth and embryogenesis in *Arabidopsis*. Plant Physiol 131:547–557
Grabber JH (2005) How do lignin composition structure and cross-linking affect degradability? A review of cell wall model studies. Crop Sci 45:820–831
Grabber JH, Hatfield RD, Ralph J, Zon J, Amrhein N (1995) Ferulate cross-linking in cell-walls isolated from maize cell-suspensions. Phytochemistry 40:1077–1082
Gresits I, Könczöl K (2003) Determination of trace elements in *Mycobacterium fortuitum* by x-ray fluorescence spectrometry. X-Ray Spectrom 32:413–417

Guillaumie S, San-Clemente H, Deswarte C, Martinez Y, Lapierre C, Murigneux A, Barriere Y, Pichon M, Goffner D (2007) MAIZEWALL. Database and developmental gene expression profiling of cell wall biosynthesis and assembly in maize. Plant Physiol 143:339–363

Guo D, Chen F, Inoue K, Blount JW, Dixon RA (2001) Downregulation of caffeic acid 3-O-methyltransferase and caffeoyl CoA 3-O-methyltransferase in transgenic alfalfa: impacts on lignin structure and implications for the biosynthesis of G and S lignin. Plant Cell 13:73–88

Halpin C, Boerjan W (2003) Stacking transgenes in forest trees. Trends Plant Sci 8:363–365

Halpin C, Holt K, Chojecki J, Oliver D, Chabbert B, Monties B, Edwards K, Barakate A, Foxon GA (1998) Brown-midrib maize (*bm1*) — a mutation affecting the cinnamyl alcohol dehydrogenase gene. Plant J 14:545–553

Hastings A, Clifton-Brown J, Wattenbach M, Mitchell CP, Stampfl P, Smith P (2009) Future energy potential of *Miscanthus* in Europe. Glob Change Biol Bioenergy 1:180–196

Hatfield RD, Fukushima RS (2005) Can lignin be accurately measured? Crop Sci 45:832–839

Hatfield RD, Ralph J, Grabber JH (1999a) Cell wall cross-linking by ferulates and diferulates in grasses. J Sci Food Agric 79:403–407

Hatfield RD, Ralph J, Grabber JH (1999b) Cell wall structural foundations: molecular basis for improving forage digestibilities. Crop Sci 39:27–37

Haughton AJ, Bond AJ, Lovett AA, Dockerty T, Sünnenberg G, Clark SJ, Bohan DA, Sage RB, Mallot MD, Mallot VE, Cunningham MD, Riche AB, Shield IF, Finch JW, Turner MM, Karp A (2009) A novel integrated approach to assessing social economic and environmental implications of changing rural land-use: a case study of perennial biomass crops. J Appl Ecol 46:315–322

Heaton E, Voigt T, Long SP (2004a) A quantitative review comparing the yields of two candidate C-4 perennial biomass crops in relation to nitrogen, temperature and water. Biomass Bioenergy 27:21–30

Heaton EA, Long SP, Voigt TB, Jones MB, Clifton-Brown J (2004b) *Miscanthus* for renewable energy generation: European Union experience and projections for Illinois. Mitig Adapt Strateg Glob Change 9:433–451

Hein KRG, Bemtgen JM (1998) EU clean coal technology—co-combustion of coal and biomass. Fuel Process Technol 54:159–169

Hillier J, Whittaker C, Dailey G, Aylott M, Casella E, Richter GM, Riche A, Murphy R, Taylor G, Smith P (2009) Greenhouse gas emissions from four bio-energy crops in England and Wales: integrating spatial estimates of yield and soil C balance in life cycle analyses. Glob Change Biol Bioenergy 1:267–281

Hu WJ, Harding SA, Lung J, Popko JL, Ralph J, Stokke DD, Tsai CJ, Chiang VL (1999) Repression of lignin biosynthesis promotes cellulose accumulation and growth in transgenic trees. Nat Biotechnol 17:808–812

Huang C, Han L, Liu X, Yang Z (2007) Proximate analysis and calorific value estimation of rice straw by near infrared reflectance spectroscopy. Waste Manage 29:1793–1797

Huang Y, McIlveen-Wright D, Rezvani S, Wang YD, Hewitt N, Williams BC (2006) Biomass co-firing in a pressurized fluidized bed combustion (PFBC) combined cycle power plant: a techno-environmental assessment based on computational simulations. Fuel Process Technol 87:927–934

IENICA (2009) Crops database. http://www.ienica.net/cropsdatabase.htm

Jenkins BM, Baxter LL, Miles TR Jr, Miles TR (1998) Combustion properties of biomass. Fuel Process Technol 54:17–76

Jørgensen U (1997) genotypic variation in dry matter accumulation and content of N, K and Cl in *Miscanthus* in Denmark. Biomass Bioenergy 12:155–169

Jung H-JG, Buxtono DR (1994) Forage quality variation among maize inbreds: relationships of cell-wall composition and in-vitro degradability for stem internodes. J Sci Food Agric 66:313–322

Jung HG, Mertens DR, Buxton DR (1998) Forage quality variation among maize inbreds: in vitro fiber digestion kinetics and prediction with NIRS. Crop Sci 38:205–210

Kaltschmitt M, Reinhardt GA, Stelzer T (1997) Life cycle analysis of biofuels under different environmental aspects. Biomass Bioenergy 12:121–134

Küçük MM, Demirbas A (1997) Biomass conversion processes. Energy Convers Manage 38:151–165

Kubo H, Peeters AJM, Aarts MGM, Pereira A, Koornneef M (1999) ANTHOCYANINLESS2, a homeobox gene affecting anthocyanin distribution and root development in *Arabidopsis*. Plant Cell 11:1217–1226

Kurunczi S, Török S, Chevallier P (2001) A micro-XRF study of the element distribution on the growth front of mussel shell (species of *Unio Crassus* Retzius). Microchim Acta 137:41–48

Labbé N, Lee S-H, Cho H-W, Jeong MK, André N (2008) Enhanced discrimination and calibration of biomass NIR spectral data using non-linear kernel methods. Bioresour Technol 99:8445–8452

Labrecque M, Teodorescu TI (2005) Field performance and biomass production of 12 willow and poplar clones in short-rotation coppice in southern Quebec (Canada). Biomass Bioenergy 29:1–9

Leckner B (2007) Co-combustion a summary of technology. http://www.energy-pathways.org/pdf/R5_co-combustion.pdf

Lewandowski I, Kicherer A (1997) Combustion quality of biomass: practical relevance and experiments to modify the biomass quality of *Miscanthus x giganteus*. Eur J Agron 6:163–177

Lewandowski I, Clifton-Brown JC, Scurlock JMO, Huisman W (2000) *Miscanthus*: European experience with a novel energy crop. Biomass Bioenergy 19:209–227

Lewandowski I, Clifton-Brown JC, Anderson B, Basch B, Christian DG, Jørgensen U, Jones MB, Riche AB, Schwartz KU, Tayebi K, Teixeira F (2003) Environment and harvest time affects the combustion qualities of *Miscanthus* genotypes. Agron J 95:1274–1280

Li L, Zhou Y, Cheng X, Sun J, Marita JM, Ralph J, Chiang VL (2003) Combinatorial modification of multiple lignin traits in trees through multigene cotransformation. Proc Natl Acad Sci USA 100:4939–4944

Li X, Weng J-K, Chapple C (2008) Improvement of biomass through lignin modification. Plant J 54:569–581

Lundvall JP, Buxton DR, Hallauer AR, George JR (1994) Forage quality variation among maize inbreds: in vitro digestibility and cell-wall components. Crop Science 34:1672–1678

Marita JM, Vermerris W, Ralph J, Hatfield RD (2003) Variations in the cell wall composition of maize brown midrib mutants. J Agric Food Chem 51:1313–1321

McCann MC, Carpita NC (2008) Designing the deconstruction of plant cell walls. Curr Opin Plant Biol 11:314–320

McCann MC, Defernez M, Urbanowicz BR, Tewari JC, Langewisch T, Olek A, Wells B, Wilson RH, Carpita NC (2007) Neural network analyses of infrared spectra for classifying cell wall architectures. Plant Physiol 143:1314–1326

McKendry P (2002a) Energy production from biomass (part 1): overview of biomass. Bioresour Technol 83:37–46

McKendry P (2002b) Energy production from biomass (part 2): conversion technologies. Bioresour Technol 83:47–54

Méchin V, Argillier O, Menanteau V, Barrière Y, Mila I, Pollet B, Lapierre C (2000) Relationship of cell wall composition to in vitro cell wall digestibility of maize inbred line stems. J Sci Food Agric 80:574–580

Méchin V, Baumberger S, Pollet B, Lapierre C (2007) Peroxidase activity can dictate the in vitro lignin dehydrogenative polymer structure. Phytochemistry 68:571–579

Mellerowicz EJ, Baucher M, Sundberg B, Boerjan W (2001) Unravelling cell wall formation in the woody dicot stem. Plant Mol Biol 47:239–274

Misra MK, Ragland KW, Baker AJ (1993) Wood ash composition as a function of furnace temperature. Biomass Bioenergy 4:103–116

Mitchell RAC, Dupree P, Shewry PR (2007) A novel bioinformatics approach identifies candidate genes for the synthesis and feruloylation of arabinoxylan. Plant Physiol 144:43–53

Mohnen D (2008) Pectin structure and biosynthesis. Curr Opin Plant Biol 11:266–277
Moller R, Toonen M, van Beilen J, Salentijn E, Clayton D (2007) Crop platforms for cell wall biorefining: lignocellulose feedstocks. http://www.epobio.net/pdfs/0704LignocelluloseFeedstocksReport.pdf
Monti A, Di Virgilio N, Venturi G (2008) Mineral composition and ash content of six major energy crops. Biomass Bioenergy 32:216–223
Monti A, Fazio S, Venture G (2009) Cradle-to-farm gate life cycle assessment in perennial crops. Eur J Agron 31:77–84
Monties B (1989) Lignins. In: Harborne JB (ed) Plant phenolics. Academic, London, pp 113–157
Morreel K, Ralph J, Kim H, Lu F, Goeminne G, Ralph S, Messens E, Boerjan W (2004) Profiling of oligolignols reveals monolignol coupling conditions in lignifying poplar xylem. Plant Physiol 136:3537–3549
Mutwil M, Debolt S, Persson S (2008) Cellulose synthesis: a complex complex. Curr Opin Plant Biol 11:252–257
National Non-Food Crops Centre (2009) Area statistics for non-food crops. http://www.nnfcc.co.uk/metadot/index.pl?id=2179;isa=Category;op=show
Nesi N, Debeaujon I, Jond C, Pelletier G, Caboche M, Lepiniec L (2000) The TT8 gene encodes a basic helix-loop-helix domain protein required for expression of DFR and BAN genes in *Arabidopsis siliques*. Plant Cell 12:1863–1878
Nix J (2007) Farm management pocketbook. Imperial College at Wye, University of London, pp 71–72
Obernberger I, Brunner T, Bärnthaler G (2006) Chemical properties of solid biofuels — significance and impact. Biomass Bioenergy 30:973–982
Oliver RJ, Finch JW, Taylor G (2009) Second generation bioenergy crops and climatic change: a review of the effects of elevated atmospheric CO_2 and drought on water use and implications for yield. Glob Change Biol Bioenerg 1:97–114
Opanowicz M, Vain P, Draper J, Parker D, Doonan JH (2008) *Brachypodium distachyon*: making hay with a wild grass. Trends Plant Sci 13:172–177
Parikka M (2004) Global biomass fuel resources. Biomass Bioenergy 27:613–620
Patzlaff A, McInnis S, Courtenay A, Surman C, Newman LJ, Smith C, Bevan MW, Mansfield S, Whetten RW, Sederoff RR, Campbell MM (2003) Characterisation of a pine MYB that regulates lignification. Plant J 36:743–754
Pauly M, Keegstra K (2008) Cell-wall carbohydrates and their modification as a resource for biofuels. Plant J 54:559–568
Price L, Bullard M, Lyons H, Anthony S, Nixon P (2004) Identifying the yield potential of *Miscanthus x giganteus*: an assessment of the spatial and temporal variability of *M. x giganteus* biomass productivity across England and Wales. Biomass Bioenergy 26:3–13
Ptasinski KJ, Prins MJ, Pierik A (2007) Exergetic evaluation of biomass gasification. Energy 32:568–574
Quattrocchio F, Wing J, van der Woude K, Souer E, de Vetten N, Mol J, Koes R (1999) Molecular analysis of the anthocyanin2 gene of petunia and its role in the evolution of flower color. Plant Cell 11:1433–1444
Ralph J, Lapierre C, Marita JM, Kim H, Lu FC, Hatfield RD, Ralph S, Chapple C, Franke R, Hemm MR, Van Doorsselaere J, Sederoff RR, O'Malley DM, Scott JT, MacKay JJ, Yahiaoui N, Boudet AM, Pean M, Pilate G, Jouanin L, Boerjan W (2000) Elucidation of new structures in lignins of CAD- and COMT-deficient plants by NMR. Phytochemistry 57:993–1003
Ralph J, Lapierre C, Lu FC, Marita JM, Pilate G, Van Doorsselaere J, Boerjan W, Jouanin L (2001) NMR evidence for benzodioxane structures resulting from incorporation of 5-hydroxyconiferyl alcohol into lignins of O-methyltransferase-deficient poplars. J Agric Food Chem 49:86–91
Renewable Fuels Agency (2008) The Gallagher review of the indirect effects of biofuels production http://www.renewablefuelsagency.org/_db/_documents/Report_of_the_Gallagher_review.pdf

Robbins MP, Paolocci F, Hughes J-W, Turchetti V, Allison G, Arcioni S, Morris P, Damiani F (2003) *Sn* a maize bHLH gene modulates anthocyanin and condensed tannin pathways in *Lotus corniculatus*. J Exp Bot 54:239–248

Robertson GP, Dale VH, Doering OC, Hamburg SP, Melillo JM, Wander MM, Parton WJ, Adler PR, Barney JN, Cruse RM, Duke CS, Fearnside PM, Follett RF, Gibbs HK, Goldemberg J, Mladenoff DJ, Ojima D, Palmer MW, Sharpley A, Wallace L, Weathers KC, Wiens JA, Wilhelm WW (2008) Sustainable biofuels redux. Science 322:49–50

Robinson JM, Barrett SR, Nhoy K, Pandey RK, Phillips J, Ramirez OM, Rodriguez RI (2009) Energy dispersive X-ray fluorescence analysis of sulfur in biomass. Energy Fuels 23:2235–2241

Roth LS, Marten GC, Compton WA, Stuthman DD (1970) Genetic variation of quality traits in maize (*Zea mays* L.) forage. Crop Sci 10:365–367

Royal Society (2008) Sustainable biofuels: prospects and challenges. http://royalsociety.org/displaypagedoc.asp?id=28914

Segrest SA, Rockwood DL, Stricker JA, Green AES (1998) Biomass co-firing with coal at lakeland utilities. http://www.treepower.org/papers/energycrops.pdf

Sims REH, Hastings A, Schlamadinger B, Taylor G, Smith P (2006) Energy crops: current status and future prospects. Glob Change Biol 12:2054–2076

Smart LB, Cameron KD (2008) Genetic improvements of willow (*Salix* spp.) as a dedicated bioenergy crop. In: Vermerris W (ed) Genetic improvement of bioenergy crops. Springer, New York, pp 347–376

Smith AM (2008) Prospects for increasing starch and sucrose yields for bioethanol production. Plant J 54:546–558

Somerville C (2006) Cellulose synthesis in higher plants. Annu Rev Cell Dev Biol 22:53–78

Spelt C, Quattrocchio F, Mol JNM, Koes R (2000) Anthocyanin 1 of petunia encodes a basic helix-loop-helix protein that directly activates transcription of structural anthocyanin genes. Plant Cell 12:1619–1631

Stern N (2007) The economics of climate change: the Stern review. Cambridge University Press, Cambridge, UK

Stern N (2009) Blueprint for a safer planet: how to manage climate change and create a new era of progress and prosperity. Random House, London

Stewart D, Yahiaoui N, McDougall GJ, Myton K, Marque C, Boudet AM, Haigh J (1997) Fourier-transform infrared and Raman spectroscopic evidence for the incorporation of cinnamalde-hydes into the lignin of transgenic tobacco (*Nicotiana tabacum* L.) plants with reduced expression of cinnamyl alcohol dehydrogenase. Planta 201:311–318

Stewart JR, Toma Y, Fernández FG, Nishiwaki A, Yamada T, Bollero G (2009) The ecology and agronomy of *Miscanthus sinensis* a species important to bioenergy crop development in its native range in Japan: a review. Glob Change Biol 1:126–153

Sticklen M (2006) Plant genetic engineering to improve biomass characteristics for biofuels. Curr Opin Biotechnol 17:315–319

Upham P, Thornley P, Tomei J, Boucher P (2009) Substitutable biodiesel feedstocks for the UK: a review of sustainability issues with reference to the RTFO. J Cleaner Product 17:S37–S45

Van Soest PJ (1963) The use of detergents in the analysis of fibrous feeds: II. A rapid method for the determination of fibre and lignin. J Assoc Off Agric Chem 46:829–835

Van Soest PJ (1967) Development of a comprehensive system of feed analyses and its application to forages. J Anim Sci 26:119–128

Vanholme R, Morreel K, Ralph J, Boerjan W (2008) Lignin engineering. Curr Opin Plant Biol 11:278–285

Vermerris W, Boon JJ (2001) Tissue-specific patterns of lignification are disturbed in the brown midrib2 mutant of maize (*Zea mays* L.). J Agric Food Chem 49:721–728

Vignols F, Rigau J, Torres MA, Capellades M, Puigdomenech P (1995) The brown midrib 3 (*Bm3*) mutation in maize occurs in the gene encoding caffeic acid O-methyl transferase. Plant Cell 7:407–416

Vogel J (2008) Unique aspects of the grass cell wall. Curr Opin Plant Biol 11:301–307

Vogel KP (2004) Switchgrass. In: Moser LE, Burson BL, Sollenberger LE (eds) Warm-season (C_4) grasses. American Society of Agronomy, Madison, WI, pp 561–588

Waldron KW, Parr AJ, Ng A, Ralph J (1996) Cell wall esterified phenolic dimers: identification and quantification by reverse phase high performance liquid chromatography and diode array detection. Phytochem Anal 7:305–312

Weisshaar B, Jenkins GI (1998) Phenylpropanoid biosynthesis and its regulation. Curr Opin Plant Biol 1:251–257

Weng J-K, Li X, Bonawitz ND, Chapple C (2008) Emerging strategies of lignin engineering and degradation for cellulosic biofuel production. Curr Opin Biotechnol 19:166–172

Woods J, Tipper R, Brown G, Diaz-Chavez R, Lovell J, de Groot P (2006) Evaluating the sustainability of co-firing in the UK. http://www.dti.gov.uk/files/file34448.pdf

Ye ZH, Kneusel RE, Matern U, Varner JE (1994) An alternative methylation pathway in lignin biosynthesis in *Zinnia*. Plant Cell 6:1427–1439

Ye ZH, York WS, Darvill AG (2006) Important new players in secondary wall synthesis. Trends Plant Sci 11:162–164

York WS, O'Neill MA (2008) Biochemical control of xylan biosynthesis—which end is up? Curr Opin Plant Biol 11:258–265

Yuan JS, Tiller KH, Al-Ahmad H, Stewart NR, Stewart CN Jr (2008) Plants to power: bioenergy to fuel the future. Trends Plant Sci 13:421–429

Zhong R, Ye Z (2007) Regulation of cell wall biosynthesis. Curr Opin Plant Biol 10:564–572

Chapter 3
Designing Plants To Meet Feedstock Needs

Peter N. Mascia, Michael Portereiko, Mark Sorrells, and Richard B. Flavell

3.1 Introduction

Plants selected to meet feedstock needs have to satisfy a large number of criteria. Above all, feedstocks need to provide an economical and sustainable basis for the industries they are designed to serve. Knowledge of the value chains of an industry and all the economic factors that feedstocks can influence are often hard to discover, and so designing a feedstock to meet all the essential criteria adequately is difficult. In general, feedstocks need to have traits that enable them to be grown to high biomass on soils that may not be the best, they need to be resilient to biotic and abiotic stresses, be easy to cultivate, have readily available supplies of seed or propagule, require few inputs to produce high biomass, be adapted to grow close to where the industries are located, be easily harvested, stored and transported and have moisture contents suitable for burning in biopower boilers or be capable of being efficiently converted into sugar or other molecules (Vermerris 2009).

Feedstock species are many and varied, as are the industries and processes in the bioenergy field and the environments in which the feedstocks are to be grown (El Bassam 1998). This chapter reviews some of these feedstocks, the traits that they need to be acceptable and, especially, the use of biotechnology in aiding the improvement of feedstocks to possess these traits.

P.N. Mascia (deceased), M. Portereiko, and R.B. Flavell
Ceres, 1535 Rancho Conejo Boulevard, Thousand Oaks California, 91320, USA
e-mail: rflavell@ceres.net

M. Sorrells
Department of Plant Breeding and Genetics, Cornell University, Ithaca, NY 14853-1902, USA

3.2 Feedstock Crops

Plants have provided feedstocks for bioenergy in all societies from the beginning of agriculture and even long before that. This is still the case in many places in the developing world today where fossil fuels have not been used so extensively for rural populations. However, the new focus on developing energy from renewable plant biomass to provide energy security and to help control greenhouse gas accumulation, has created substantial interest in a new and extensive agriculture based on many crops (Heaton et al. 2008; Perlack et al. 2005; Jessup 2009; Tilman et al. 2009). In the United States, this agriculture could be as large as, or even larger than, corn agriculture is today, but the growth in all the aspects necessary for this to come about is well beyond the scope of this chapter. However, it should be noted that all the services and decision-making—local, national and global—necessary to support this agricultural revolution will need to co-evolve if the new feedstocks are to enter commerce on a large scale and be used as major sources of biofuels and other products. This revolution requires political, industrial and scientific progress to fulfill its vision.

The primary criterion for a feedstock crop is usually the yield of biomass that can be routinely and sustainably produced in the relevant location. Choice of feedstocks is influenced heavily by the environment in which they are to be grown. A second criterion is often the suitability of the feedstock for the industrial process. Tropical climates where there is extended sunshine and adequate rainfall are best for production of biomass. Many tropical species, including sugarcane (El Bassam 1998), illustrate this point. Yet, many countries outside the tropics wish to use plant feedstocks and therefore more temperate species are being used and considered. Tropical and more temperate C4 grasses in which more photosynthate is converted to biomass than in C3 species have an obvious advantage. This comes about because C4 species have maximum rates of photosynthesis in the range of 70–100 mg CO_2 dm^2 h^{-1} with light saturation at 1.0–1.4 cal cm^2 min^{-1} total radiation, while C3 species have equivalent rates in the range of 15–30 mg CO_2 dm^2 h^{-1} with light saturation at 0.2–0.6 cal cm^2 min^{-1} (El Bassam 1998). In C3 species the first product of photosynthesis is a three-carbon organic acid, whereas in C4 plants the first products are four-carbon organic acids. The C3 pathway is generally adapted to be optimal at lower temperatures than the C4 pathway; C4 species have higher rates of CO_2 exchange. C4 plants are also more drought tolerant and more efficient in their utilization of nitrogen. Such C4 species include maize, switchgrass, miscanthus, sorghum and other similar grasses—see Fig. 3.1 (Wang et al. 2009). Some are perennial; these are favored because their root structures bring enhanced resilience to stresses such as drought, they do not need replanting every year, and, in some cases, on senescence they transport remaining nitrogen and other molecules into the roots to conserve them for the following year. In addition to all these advantages, the flexibility that comes with an annual such as high biomass sorghum helps with local farming needs and crop rotations. Some of the characteristics of switchgrass and miscanthus, two current favorites for biomass production in the US, are outlined in Table 3.1.

Fig. 3.1 Some favorite species for high biomass production

Table 3.1 Characteristics of two C4 grasses: switchgrass and miscanthus

Switchgrass	Miscanthus
Plant characteristics:	Plant characteristics:
Perennial warm season grass	Perennial warm season grass, several species
Native to North America	Native to Asia
High-yield (6–12 tons in mid-to-low latitudes)	Very high-yield
Reaches full yield in 3rd growing season	Reaches full yield after 2–3 growing seasons
Low input requirements	Low input requirements
Strong net energy balance	Strong net energy balance
Optimal harvest after fall senescence to permit nutrient recycling	Optimal harvest during winter to permit nutrient recycling
Main concerns:	Main concerns:
Stand establishment	High establishment cost
Yield	Broad adaptation
	Management know-how
	Non-native

In the US, where corn grain starch has been used over the past 10 years for conversion into sugar and then distillation to ethanol, corn has become the predominant biofuel feedstock to date (see Chap. 5). Thus, all the traits that have been bred into corn over the past 100 years to provide very high yields in the Midwest are benefitting directly the feedstock needs of the ethanol biofuels industry. However, the corn-based process from ground preparation to ethanol is not energetically favorable because of the inputs that are used to produce high yielding corn grain. It is also often stated that because corn is a food crop it should not be used for biofuel production because sooner or later energy requirements will divert supplies away from food needs and increase food prices. Sugarcane is another well-established feedstock for sugar production followed by conversion into biofuel ethanol, and this is the feedstock of choice in Brazil and other tropical environments that are suited to this crop (see Chap. 4). In the US, sugarcane is confined to the extreme south because of its tropical nature and it is not an extensive crop in Europe at all. Many have considered using the stover, cobs or stalks of other grain crops such as wheat, barley, rice and oats as cheap sources of biomass. However, the low density of their availability in the field and their value in returning organic matter to the fields mitigate against them being available in quantities sufficient for the large-scale sustainable biofuel production envisaged, especially in the US.

Sweet sorghum is a particularly interesting feedstock (Vermerris et al. 2007; Rooney et al. 2007). Such plants can accumulate sugars, mostly sucrose, but also some glucose and fructose, in the stalk, and the sugars are extracted readily in a way similar to that by which sugar is extracted from sugarcane. While the sugar from sweet sorghum is not easily crystallized to table sugar quality, as a source of cheap sugar for biofuels and chemicals the crop has many advantages over sugarcane. It is adapted to a wide range of environments compared with the cold-sensitive sugarcane. Its sugar production occurs over a 70- to 140-day growth cycle compared with a year or more for sugarcane. It is produced from seed compared with vegetative cane cuttings and, being seed-produced annually, does not suffer from the viruses problematical in vegetative sugarcane. These differences and others make sweet sorghum a less costly feedstock than sugarcane. Sweet sorghum is a diploid and its complete DNA sequence is available (Paterson et al. 2009b). It is much easier to breed than sugarcane, which is a polyploid species with aneuploidy and longer generation times. Furthermore, especially useful high yielding hybrids of sweet sorghum have been developed by Ceres and others. Sweet sorghum appears to be a crop very suited to biotechnological improvement. The genetic control of its traits is being studied using molecular markers (Murray et al. 2009), and its similarity to corn suggests that its development as a feedstock could benefit from much of the progress made in corn biotechnology. The overall sugar yields, the timing of harvesting and the stability of the sugar in the stem are important factors for sweet sorghum, and substantial variation in all these attributes can be found in breeding material.

Wood is a major feedstock used around the world for many processes. Various tree species such as poplar, willow, and eucalyptus are being used to support some initial biofuel biorefineries, and thinnings and chippings from forestry are likely to be used for some time (see Chap. 7). However, overall, trees are unlikely to be able to compete with dedicated grasses when biofuel industries reach the envisaged scale. Nevertheless, the study of molecular genetics of traits in tree species is progressing rapidly (Krutovsky et al. 2009; Pavy et al. 2008). In summary, many new crops are being considered as sources of feedstocks and biotechnological opportunities are being applied to relevant species. The reader is directed to El Bassam (1998) for a comprehensive account of potential feedstock species.

To support any one industry/biorefinery, the use of several crops appears desirable in order to provide sustainable supplies year after year, allow for crop rotations to preserve land quality, minimize problems of disease, extend the growing season and thus feedstock supply and reduce storage problems. Also, many different versions (varieties) of any one feedstock need to be bred for use in different environments. While the sugarcane industry is based on one species, this leaves

3 Designing Plants To Meet Feedstock Needs 61

the factories unable to work all year round, crops have to be replaced regularly because of diseases and viruses, and there is a significant turnover of varieties. There is now discussion of the possibility that sweet sorghum will be introduced into sugarcane regions to extend the processing season and reduce the risks to the industry as it greatly increases biofuel production.

3.3 Trait Improvement

As noted above, the list of traits essential for a successful variety to sustain a biofuels or bioproducts industry is long and complex (Table 3.2). To optimize so many traits is always a huge problem in plant breeding. The challenges involved in enhancing the traits in all species being considered for feedstocks are massive. The characteristics of a variety or population of outbreeding feedstock plants are the result of genotype x environment x harvesting/plant development, which, because of all the variables in the environments and harvesting conditions, bring substantial complexity and uncertainty. The process of plant improvement is always incomplete and therefore always ongoing. In outline, the process begins by making a large collection of germplasm of the selected species and evaluating this for all or most of the desired traits. The breeder does this in nurseries located in several places but centered around the environments in which the crop is likely to be commercialized. For feedstocks, biomass yield is usually the most important trait and so this and the major components of yield are the breeder's first focus. There are usually very large variations in yield and so selection in the nursery is readily possible but often complex given the diversity involved (Fig. 3.2). For biomass, crop height, number of tillers, canopy volume, thickness of tillers, stem structure, leaf size and number are primary components of yield, and since these traits are under separate genetic control there is usually a huge variation in plant architecture available. Overall

Table 3.2 Valuable traits for improvement

Value	Trait for enhancement
Increase biomass; increase yield potential; lower production and transport costs; increase carbon sequestration	Architecture; canopy structure; photosynthesis; flowering time
Protect yield in stresses and on marginal land	Drought tolerance; heat tolerance; cold tolerance; salt tolerance; disease resistance; heavy metal tolerance; pH tolerance; root structure
Reduce cost of inputs	Nitrogen use efficiency; water use efficiency; reduced greenhouse gas emissions; seed propagation
Increase yield in industrial processes; reduce capital and operational costs of refineries	Composition; conversion to sugars; higher heating values; reduced Cl, K and other metals
Enhance overall economics	Addition of co-products

Fig. 3.2 Heterogenity amongst switchgrass accessions. This immense diversity poses both opportunities and difficulties for the breeder

architecture influences the amount of photosynthesis occurring in the plant and hence the overall yield. The timing of flowering is usually also a critical trait. Plants often flower in response to daylength and/or temperature and once the flowering process begins then vegetative growth usually slows or stops, thus ending the phase of increasing biomass. If grain is produced, then much of the biomass is broken down and sugars and nitrogen compounds are transported to build the grain and may be lost for feedstock purposes. Thus plants that do not flower in the place where they are being grown for feedstocks are often especially useful because they sustain biomass increases over longer periods of the growing season.

Breeders often seek to make crosses to introduce new traits, to create and find new combinations of traits, and to create and select heterotic hybrids. In outbreeding species such as switchgrass and miscanthus, populations are created, evaluated, selected and propagated as populations (Bouton 2007). For examples of breeding strategies the reader is referred to specialist and crop-specific reviews, e.g., for warm season grasses, see Vogel (2000) and Vogel and Burson (2004), and for maize, Hallauer and Carena (2009). Discussions of the protocols of breeding programs for crop enhancement are beyond the scope of this chapter.

The composition of the biomass may be crucial for increasing the economics of the value chain. Where feedstocks are being designed for conversion of lignocellulose to sugars and then to ethanol or other molecules by fermentation, then having cell walls that are readily broken down to sugars by enzymes or simple treatments is a major advantage. This, then, is a trait on which much research is being focused (Shi et al. 2007; Penning et al. 2009; Gomez et al. 2008). For thermochemical processing or for burning in biopower boilers, other properties are of higher priority, namely a higher heating value and low concentrations of deleterious elements such as chlorides, potassium and other heavy metals. These traits have not been a regular part of many breeding programs and so are new targets. While the biomass produced per acre is usually the first metric of the breeder, subsequent metrics will often be the amount of product per acre; e.g., if the feedstock is being used for sugar production then the relevant metric is tons of sugar per acre after extraction/processing. It could also be the amount of ethanol per acre if this is the product. To aid the conversion of lignocellulose in cell walls to sugar using industrial cellulase and hemicellulase cocktails, a goal of the breeder and agronomist must be to make the plant cell walls more amenable to easy degradation by mild acid, alkali or heat treatments combined with cellulases/hemicellulases, under conditions that mimic the likely industrial processes (Wyman et al. 2005). There is variation between plants in the amounts of glucan in switchgrass cell walls, for example, and the ease with which it is converted to sugars. Some lines have more complex walls such that no matter how much enzyme cocktail is added the sugars remain complexed in cellulose. On the other hand, some lines are much more easily converted into sugars. Some of this variation is connected to the environments in which the plants are grown and the stage of harvesting (Sarath et al. 2008). The goals of the breeders are to combine high accessibility of the sugars with high biomass yields to make a significant difference to the overall economics of value chains. If less enzyme could be used or if the pretreatments were less expensive then costs would be decreased significantly.

There is some urgency for these improvements of most biomass feedstock crops. What is the basis for believing that such improvements are possible? This comes from the successes of making improvements in other crops over the past 50 years. The major increases in corn and wheat in the EU and US and rice in China are shown in Fig. 3.3. The results imply that substantial improvements can be made without developing new technologies. Perhaps 50% of the advances in these crop yields have come from better agronomy, use of machinery and other factors in agricultural production. The increases from breeding (Hallauer and Carena 2009) have involved improvements in vegetative biomass production, plant architecture,

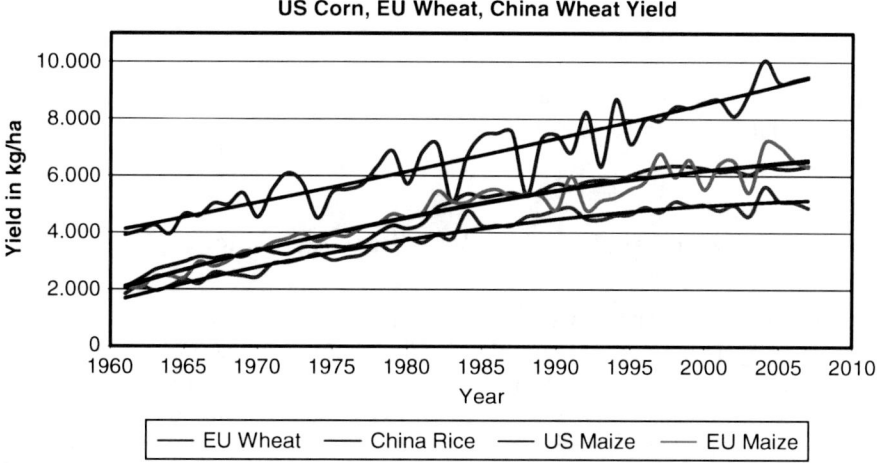

Fig. 3.3 Average US corn, EU corn and wheat and China rice yields since 1960
Source: USDA

better canopy structure and energy capture, enabling higher density planting and increasing the harvest index. Better stress tolerances have been selected, including drought tolerance while needs for nitrogen fertilizers per unit of grain yield have been reduced. What have we learnt from these breeding programs that can be applied to new feedstock species to bring these into economic reality sooner rather than later? Corn breeding in the US has been designed around the production of hybrids. Breeders selected particular inbreds that were especially able to combine to give heterotic hybrids, and that strategy led to inbreds being bred and selected into two groups displaying the complementary combining ability. The breeding progress has certainly also come from better understanding of the genetic basis of some traits. It has also embodied large increases in scale, much larger testing in the field and all this has been made possible by the advent of statistical procedures and powerful data handling systems. In the past 10 years the leading breeding companies have deployed molecular markers to help in backcrossing some traits into elite germplasm, and, since 1996, the use of powerful transgenes (Eathington et al. 2007). In wheat and rice breeding, equivalent knowledge has played a major part in the yield improvements but only in Chinese rice breeding have hybrids played a major role. Similar stories exist for many other crops but the depth of information and degrees of success and progress are less than for corn. Corn in the US and wheat in the UK set the standards for improvements in national yields to very high levels. While past plant breeding has taught us much, current biotechnology is providing new methods that will make a huge difference to plant breeding when applied on the right scale. Such methods provide exquisite insight into genetic variation and breeding and can reduce some of the major problems to manageable proportions. These new methods include the use of DNA sequencing to define the whole genome, genome-wide assessments of the expression of every gene, micro RNA and even of specific transposable elements (Varshney et al. 2009; Schnable et al. 2009). When combined with knowledge of the phenotypes, i.e., measurement of the traits in the field and laboratory, they are pinpointing the genetic basis of the trait variation that the breeder is exploiting as well as teaching us how to select new traits. These methods will be featured in the remainder of this chapter together with the potential role of transgenes in trait improvement.

3.4 Molecular Markers for Breeding and Genetic Mapping

The field of finding genetic markers is accelerating rapidly due to "next generation sequencing technologies" (Lister et al. 2009; Varshney et al. 2009). These enable millions of short sequences to be revealed in a single analysis of genomic DNA or cDNAs. When these short DNA sequences from different individuals are mapped back on to a sequenced genome then it is possible to find sequence polymorphisms in or around specific genes. These polymorphisms can be base changes or insertions/deletions. The prior existence of a sequenced reference genome helps this approach enormously. However, polymorphisms can also be found, especially

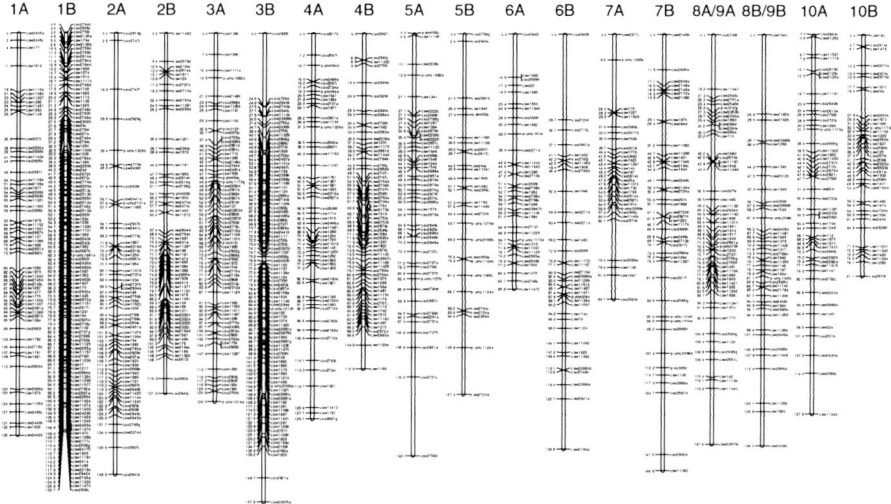

Fig. 3.4 Molecular marker map for switchgrass. The tetraploid has two sets of chromosomes that can be distinguished by molecular polymorphisms (Ceres, unpublished)

using longer sequencing reads, by using a closely related sequenced genome as a reference to identify the origin of the sequences being compared. With today's sequencing technologies, species' genomes can be sequenced readily with decreasing cost and so there is every reason to expect complete genome sequences to exist for all species in which serious breeding programs are being undertaken. Mapping fragments to genes is straightforward when the genome is diploid and contains few duplicated genes. However, plant genomes usually contain many duplications and families of genes (Schnable et al. 2009; Paterson et al. 2009b). These create difficulties in defining the exact origin of DNA fragments and so many potential polymorphisms have to be abandoned when there is ambiguity about their exact origins and which pairs of sequences are true alleles. Yet, with this approach, it is readily possible to uncover thousands of such polymorphisms rapidly. A molecular marker map of switchgrass is show in Fig. 3.4 (Ceres, unpublished). Having such a large number of polymorphic markers to distinguish parents and progeny of crosses is revolutionizing the mapping of traits in genomes. To score polymorphisms in plants, additional sequencing rounds can be performed or the sequences can be converted to single polymorphism assays such as *Taqman* by Applied Biosystems (Foster City, CA). Alternatively, the polymorphisms can be converted into features on hybridization chips (e.g., NimbleGen, Roche, Madison, WI) and the latter used as templates for hybridization of DNA from other individuals. When hybridization conditions are selected such that only precisely identical sequences give full hybridization, then it is possible to score allelic similarities and differences on thousands of genes and multiple DNA samples simultaneously. The adoption of these technologies is removing the problem of not having enough markers for plant genomes. However the traits themselves have to be measured in

3.5 Comparative Genomics

The extensive molecular mapping that has taken place in corn, rice, wheat and other monocot members of the grass family and the renewed emphasis on mapping trait genes is resulting in extensive genetic maps in these and other species. The maps define where allelic variation for traits is located on the chromosomes. During the divergent evolution of the grass species there has been extensive conservation of the positions of genes with respect to one another, even though there have been many other changes in families of repeated sequences in the genomes giving rise to major chromosomal size differences. Therefore, using the positions of the genes, the chromosomes can be aligned physically and genetically–see Fig. 3.5. Early comparative maps (e.g., Hulbert et al. 1990; Ahn et al. 1993; Kurata et al. 1994; Foote et al. 1997; Van Deynze et al. 1995; Devos and Gale 2000; Gale and Devos 1998) greatly underestimated the complexity of genome relationships. Those low resolution comparative maps are biased by the use of predominately single copy probes

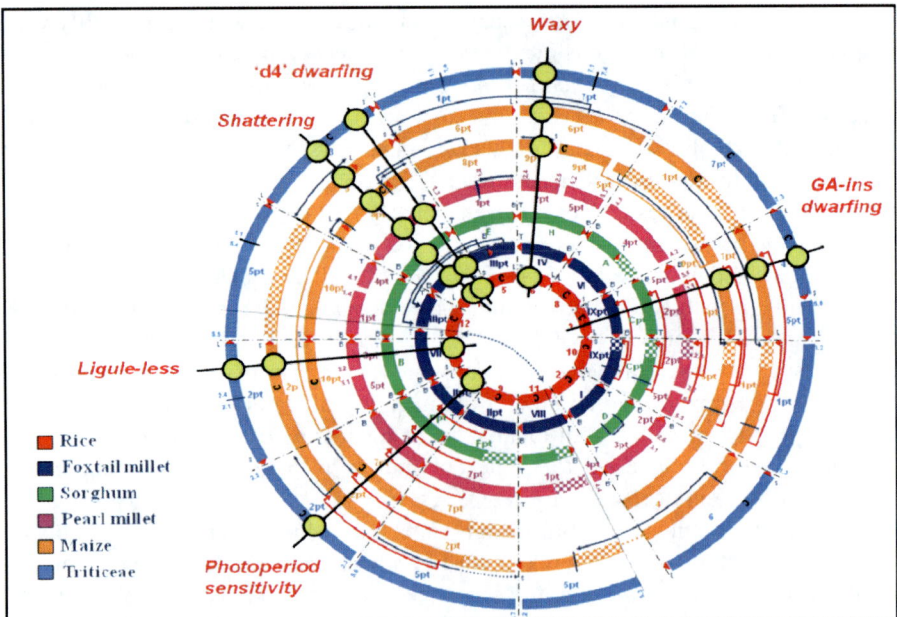

Fig. 3.5 Comparative maps of grass genomes. The genomes of the species have been aligned based on gene synteny. The alignments coupled with trait mapping reveal approximately common positions for genetic variation affecting the traits. Courtesy of J. Snape, John Innes Centre, UK

that do not sample multicopy regions, simplifying assumptions about colinearity, and overemphasizing gene-rich regions (Bennetzen 2000; Gaut 2001, 2002). While comparing the maps of different grasses, the recognition of conserved linkage blocks and their relationships with rice linkage groups has led to hypotheses about the basic organization of the ancestral grass genome (Devos and Gale 1997; Gale and Devos 1998). From this, the genetic basis of phenotypic variation mapped in one species can be predicted to be located similarly in syntenic positions in another species. The mapping of maize and sorghum chromosomes is already detailed and is progressing rapidly (Schnable et al. 2009; Paterson et al. 2009b). These maps may provide considerable help for bringing predictions of trait maps of grasses such as switchgrass and miscanthus into being very rapidly, as soon as the primary sequences are aligned (Paterson et al 2009a). Sequencing of switchgrass and miscanthus genomes is likely to be completed in the next few years. As genomes are sequenced then the divergence in gene content that has occurred during evolution can be analyzed. Recently, Wang et al. (2009) have described a comparative analysis of the genes associated with C4 photosynthesis in grasses but it is too soon to be able to define the origin of this trait in genetic terms.

3.6 Heterosis

The combining of different genomes into hybrids is a common source of yield gains. It is likely to be useful in breeding of energy crops. The term 'heterosis" was coined by G.H. Shull (Shull 1908–1914) to describe the increased vigor observed from heterozygosity (Shull 1952). Heterosis, or hybrid vigor, is defined as the positive difference in a measured phenotype of the offspring relative to the mean of the two parents (Lamkey and Edwards 1999).

The classic explanations for heterosis involve "dominance" and "overdominance" (Crow 1948)—two phenomena that center on the complementation of deleterious alleles and allelic interactions, respectively. Two observations suggest that these early explanations for heterosis are incomplete. First, the magnitude of heterosis has not diminished over time, but, rather, has actually increased slightly (East 1936; Duvick 1999). Second, both heterosis, and its reverse, inbreeding depression, have different magnitudes depending on ploidy levels (Levings et al. 1967; Mok and Peloquin 1975; Groose et al. 1988; Bingham et al. 1994). In today's world of molecular biology and genomics, new ideas have sprung forth to try to explain the phenomenon of heterosis. Recent expression analysis data, for example, suggests that hybridization that results in heterotic crosses can induce changes in the regulation of gene expression throughout the genome (Song and Messing 2003; Osborn et al. 2003). In the analysis by Swanson-Wagner et al. (2009), many of the gene expression changes in the hybrids are programmed by eQTLs (loci that contain a gene that regulates the expression of other genes localized elsewhere in the genome) originating from the male parent, implying that some form of imprinting may be underlying heterosis. In the last few years several studies have attempted to dissect the molecular mechanisms of heterosis and inconsistencies in the results and conclusions drawn from these studies have been reviewed by Hochholdinger and Hoecker (2007). The molecular data available thus far do not indicate a simple correlation between any one of the genetic hypotheses and the molecular events leading to heterosis. Future progress in the integration of genomic tools and the mapping and cloning of complex heterosis-associated quantitative trait loci (QTLs) might allow simultaneous identification of multiple key, genetically unrelated genes that contribute to heterosis. These results could provide a major step forward in creating a molecular-based hypothesis for this phenomenon.

Regardless of the molecular characterization of this phenomenon, heterosis has been identified in corn, sorghum, wheat, and even turf grasses. Heterosis has been utilized to enormous benefit to increase yield in agriculture. It is likely that heterosis will also prove advantageous for increasing biomass production, and studies using the model species Arabidopsis should prove valuable for dissecting the molecular pathways that lead to the enhancement (Barth et al. 2003; Meyer et al. 2004). Biomass enhancement as a result of hybrid crosses has long been known in many grass species, including wheat (Freeman 1919). Recent findings suggest that heterosis can be achieved in switchgrass (Martinez-Reyna and Vogel 2008); however, heterotic groups, the stocks from which parents produce offspring with hybrid vigor, have largely been unidentified. The relatively few generations of selection for existing varieties of switchgrass and *Miscanthus*, the out-crossing nature of these species, and the relatively little interest in the subject until recently, are likely key contributors to the paucity of heterotic groups. For crops like corn and sorghum, however, for which many inbred lines have been developed, heterotic groups for traits like biomass are likely to be developed rapidly, using reciprocal recurrent selection methods.

3.7 Improving Traits by Molecular Plant Breeding

Some species, such as switchgrass and miscanthus, proposed for biofuel production have received little attention from plant breeders and thus present opportunities for rapid advances using molecular approaches compared to well developed crops such as corn. Yet, leading corn breeding companies are predicting that they will double grain yields by 2030 with less nitrogen and water per unit of yield. This is an exciting proposition and sets even higher goals for other energy crops (Bouton 2007; Jakob et al. 2009). It is worth noting that with this goal the amount of corn grown today for feed and food will be achieved on half the current acreage, leaving the possibility that the other half will be available for corn-derived ethanol or other dedicated energy crops.

Crop improvement methods can be affected by the reproductive biology of a species, whether inbreeding or outbreeding. However, all breeding programs share the common goals of the discovery and characterization of alleles and the recombination and selection of genotypes with superior alleles at multiple loci. Variation in

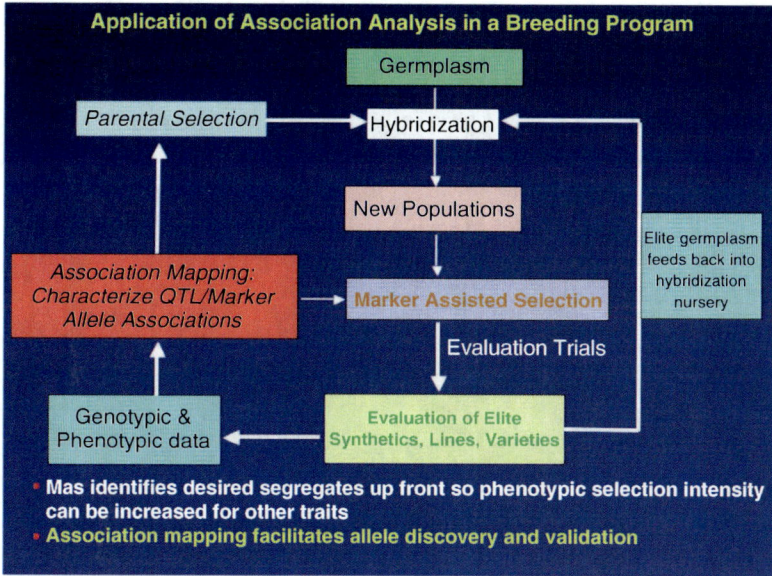

Fig. 3.6 Association breeding. Diagram shows the flow of germplasm in a breeding program and the use of association mapping for allele discovery and characterization

some traits is due to one or two alternate alleles and so the preferred one is easily selected in breeding programs. However, many traits have a complex genetic basis and so the selection of better forms is very difficult and very inefficient. There is seemingly endless variation in breeding methods but this review will consider three categories of molecular approaches; marker-assisted selection, association breeding (Breseghello and Sorrells 2006), and genomic (genome-wide) selection (Meuwissen et al. 2001). Marker-assisted selection has been used successfully in many species for backcrossing one or a few major genes into an elite line or variety (Holland 2004; reviewed by Xu and Crouch 2008). Backcrossing is the most conservative of breeding methods and is limited to a few target genes with large effects. However, genes with small effects underlie many of the most important traits and they often determine the success of a variety. Association breeding is a forward breeding strategy that superimposes association mapping methods on the structure of a conventional breeding program (Fig. 3.6). In this method, advanced trial data is combined with marker genotypes and subjected to association mapping methods for discovering new QTL and for validating previously identified QTL. That information is then used for parental selection and marker-assisted selection in segregating populations. Marker assisted selection in early generations identifies desired segregants up front so that phenotypic selection intensity can be increased for other traits.

Breeding programs are dynamic, complex entities with new germplasm being introduced each season and less desirable materials being discarded. Consequently,

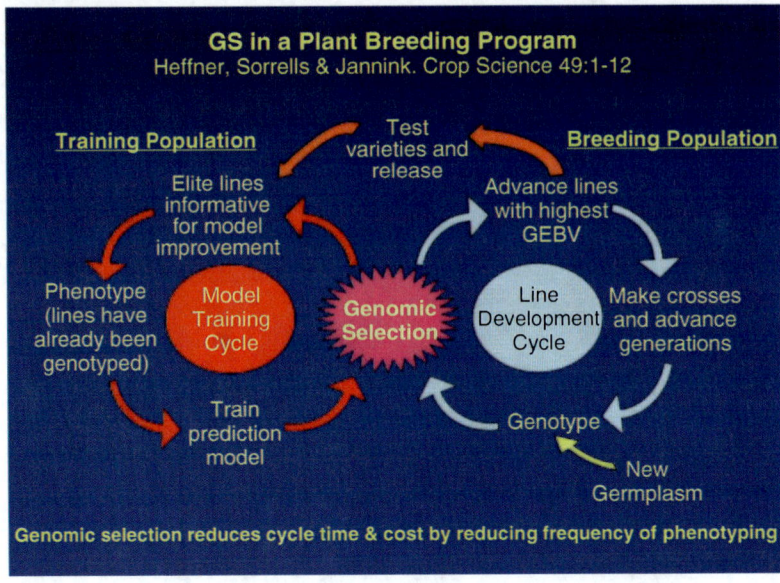

Fig. 3.7 Genomic selection (GS) showing the interaction between the training and breeding populations. *GEBV* Genomic estimated breeding value

association breeding requires frequent evaluation of marker/phenotype relationships. This is in part because marker/QTL allele relationships can be different for newly introduced materials but may also result from recombination. Accurate detection and estimation of QTL effects are required to prioritize or weight QTL (Lande and Thompson 1990) and ensure progress from selection. Lastly, population structure is more pronounced in a breeding program because multiple selections from the same crosses are often being evaluated at the same time. Statistical methods are constantly being developed that provide for improved type I error control (Yu et al. 2006), and family-based or stratified sampling can be used to counter the effects of imperfect population structures. Genomic selection (GS; Meuwissen et al. 2001) takes a substantially different approach that relies on a large number of molecular markers covering the entire genome. The assumption is that there will be at least one marker in linkage disequilibrium with all the QTL affecting the trait of interest. A training population is phenotyped and genotyped for the purpose of building a prediction model that captures the total additive genetic variance. The prediction model is then used to predict the best individuals in a population that has been genotyped but not phenotyped (Fig. 3.7; Heffner et al. 2009). Using the prediction model, a genomic estimated breeding value (GEBV) for each individual is calculated by summing the marker effects for that genotype. The prediction model is used to impose multiple generations of selection with rapid generation cycling. This approach obviates the need for determining if a QTL is significant and captures QTL with small effects. Of course, one of the critical components is the prediction model chosen to predict breeding values. Model

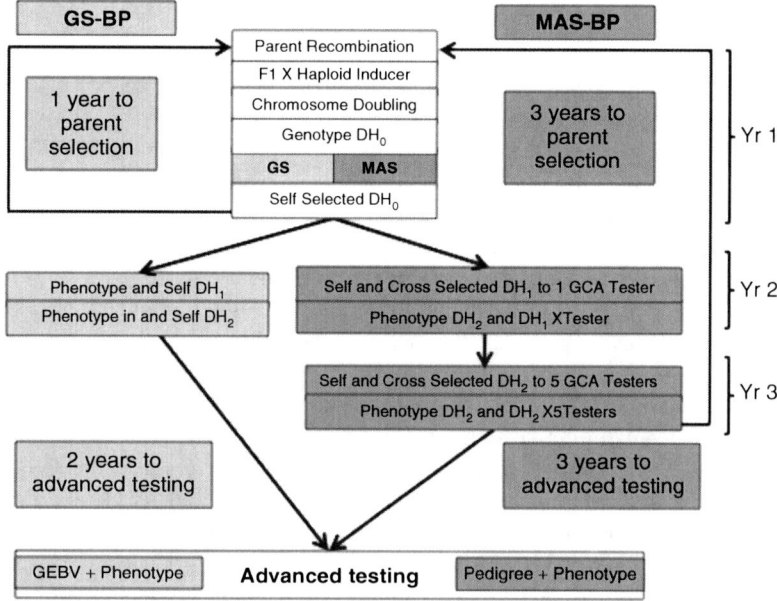

Fig. 3.8 Example of a comparison of maize marker-assisted selection (MAS) and genomic selection (GS) breeding program (BP) schemes. The GS selection cycle length is 1 year, whereas the MAS selection cycle length is 3 years. *Light shading* GS stages, *dark shading* MAS-BP stages, *not shaded* stages common to both programs

performance is based on the correlation between the estimated breeding value and the true breeding value. In practice, there may not be one prediction model that works well in all situations or for all traits. A second critical component is the phenotyping of the training population, which must be representative of the breeding population. The target production region must be adequately sampled by phenotyping in representative environments. It may be desirable to evaluate some traits in artificial environments to increase their heritabilities. In any event, all important traits should be adequately evaluated. Because GS treats markers as random effects, GEBVs can be combined in a selection index for multiple traits. The prediction model must be periodically evaluated and updated as the germplasm changes. Optimum strategies for incorporating genomic selection in a breeding program will differ for different species and also the resources available to the program (Heffner et al. 2010; Fig. 3.8). However, computer simulations suggest that genomic selection can result in gains from selection that are two- to three-fold greater than for conventional phenotypic selection (Heffner et al. 2010; Zhong et al. 2009). In summary, genomic selection captures small-effect QTL, can increase gain from selection, and can reduce advanced testing. However GS requires a large number of markers and accurate prediction models. The most important advantages are reductions in the length of the selection cycle and the associated phenotyping cost.

3.8 Transgenic Traits

Some traits are simply not in the species, or are too complex to bring into elite cultivars. Such scenarios provide the drivers for the discovery and deployment of transgenes. To find genes that are capable of improving defined traits is not easy but useful candidates are being published regularly from laboratories all over the world. Discovery has to be followed by the long and complex steps of transformation, event selection, field trialing, event selection, deregulation, introgression and commercialization. The expense of completing these steps are major deterrents to solving trait problems using transgenes but, on the other hand, such genes can be very valuable and enable a feedstock to be used where without them this would not be possible. It is interesting to speculate to what extent the deployment of transgenes has boosted average yields across the world. Consider the data in Fig. 3.3. While not suggesting causality, it is interesting to note that the increases in rice yield in China and wheat yield in the EU over time have slowed down, while corn yield in the US, where transgenic traits were widely and rapidly adopted, has continued to rise dramatically. It is noteworthy that in the EU, where transgenics have not been adopted, the corn graph shows a downward trend, with the trendline overlapping almost perfectly with the China rice graph. Unlike the others, the US corn yield improvement graph has an upward inflection. Hence, the use of molecular marker technologies does not account for the US yield graph since the same breeding technologies are used for European corn. The inflection point occurs at the same time as the introduction of transgenic corn in the US in the mid 1990s. Troyer and Mascia (1998) and others asserted that the rate of corn improvement would increase due to the use of genomics and biotech traits. Ten years later, this prediction seems to have been borne out. Monsanto estimates that the rate of corn improvement has doubled due to the application of modern breeding methods. While we cannot see the impact of this work on the yield graph, this is probably due to the adoption of ultra high throughput marker technologies not having impacted the varieties to date. Given the historical lack of improvement in genetics and agronomy for several important biomass species such as switchgrass and miscanthus, application of these new tools to biomass crops will result in relatively rapid biomass yield and quality improvement.

3.8.1 First Generation Transgenic Traits

When looking at the first generation transgenic traits in row crops, there are two classes of traits; herbicide tolerant, primarily Roundup tolerant; and insect tolerant, all Bt insect tolerant. The insects currently controlled are borers and root worm. When, and by how much, will these impact energy crops?

Transgenic herbicide tolerance might provide some value to aid in establishment of energy crops. Certainly weeds are a significant problem in getting some grasses like switchgrass and sorghum established and the problem is likely to be worse with miscanthus. There are, however, several concerns associated with the use of herbicide tolerance. In addition to being a wild species, switchgrass is commonly planted on conservation reserve program (CRP) land and along roadways in the US. Hence there is a likelihood of trait transfer from cultivated to wild switchgrass. Miscanthus is not a native species. Herbicide tolerance may make it more invasive, which could have negative consequences. Sorghum can cross with Johnson grass at some significant frequency. Thus there would be the risk of herbicide resistance being transferred from cultivated sorghum to a weedy species. Further, other crops such as corn and soy are largely Roundup Ready, and are becoming more Liberty tolerant. Addition of these traits to other species could create management problems for corn and soy agriculture. Hence, transgenic herbicide tolerance in energy crop grasses could be beneficial but may be outlawed for environmental and other reasons. In any case it would need substantial stewardship (see below). Insect resistant corn containing a gene from the soil bacterium Bt (*Bacillus thuringiensis*) has been available since 1996. Bt bacteria produce proteins that are toxic to specific insects. When these genes are broadly expressed by plants, they protect the plants from insect damage. According to the US Department of Agriculture (USDA), "plantings of Bt corn grew from about 8 percent of US corn acreage in 1997 to 26 percent in 1999, then fell to 19 percent in 2000 and 2001, before climbing to 29 percent in 2003 and 57 percent in 2008". The expansion in acreage is due largely to the introduction of rootworm resistant corn in 2003. In much of the Corn Belt, root worm damage is greater than European corn borer damage. Bt corn adoption is likely to fluctuate, but has probably reached an equilibrium based on where insect protection is needed in the corn belt. Now there is a nine gene stack coming in corn where most of the genes are different forms of Bt genes. Insect tolerance may prove to be important in energy crops. Energy crops are likely to be susceptible to borers and root worms. However, it will be some time before large-scale monocultures of energy crops are established and we determine the economic impact of borer and root worm damage.

In summary, the first generation traits that have had such a significant impact on maize improvement may not be the first ones commercialized in energy crops grasses. As these crops expand in acreage, opportunities may emerge for first generation traits.

3.8.2 Transgenic Output Traits

While the patent literature has many examples of output traits in maize (Troyer and Mascia 1998), little progress has been made in the commercialization of these, so they do not provide a useful benchmark for energy crops. It seems likely that output traits will have an impact on biofuel production. We discuss a few examples here. There are two general categories of output traits: (1) modulation of traits that are endogenous, and (2) traits that are not endogenous to the crop, such as industrial enzymes and plastics.

3.8.2.1 Endogenous Traits

Currently, the most widely studied endogenous traits are components of biomass and tolerances to stresses. Many laboratories have published genes that are capable of enhancing such critical traits. Most of these have come from the study of Arabidopsis or rice. The sources of transgene have usually been the same species but with their level of expression greatly increased in most if not all tissues. This strategy has been based on the hypothesis that variation in traits is readily created by changing the expression of certain genes. Trait changes have also been created by "knock-out" insertions. These have involved insertions by simple T-DNAs in Arabidopsis and rice as well as by inserted transposable elements. Arabidopsis is a dicot and rice is a monocot and so these have served as models for both kingdoms. Brachypodium (http://brachypodium.pw.usda.gov) is now being developed as another C3 grass model species given the many advantages it has for laboratory-based studies. Some companies, including Ceres, Mendel Biotechnology and Monsanto have undertaken large-scale screening of traits resulting from the misexpression of thousands of genes in Arabidopsis. Many transgenes have also been misexpressed in rice and their effects studied in the field (Fig. 3.9). The added genes have come from several species. From all these studies genes that affect many traits including flowering time, salt-, heat-, high- and low-light-tolerances, water use efficiency, nitrogen use efficiency, many architectural traits and overall biomass production have been defined. Some of these are being transferred to switchgrass and the new traits recognized. Rice therefore appears to be a useful genetic model for testing effects of transgenes for grasses, even C4 grasses, at least for some traits. Numerous transgenes have been successfully tested in tree species, especially poplar. Thus it is beginning to appear that valuable traits can be enhanced in energy crops, over time, where the economics is appropriate. It remains to be seen what the

Fig. 3.9 Comparison of field-grown rice plants illustrating effect of adding an Arabidopsis gene under the control of a broadly active promoter that stimulates height and biomass accumulation without significantly affecting flowering time. *WT* Non transgenic parent (Ceres, unpublished)

deregulation of such traits may entail, given the concerns over invasiveness, and transfer of traits to wild plants and non genetically modified (gmo) crops. For this reason a valuable trait may be forms of sterility that contain the transgenes within the crop and prevent seeds being formed. Prevention of seed formation would eliminate the risk that seeds would give rise to volunteer plants (see below).

Considerable effort is being focused on manipulating the cell wall structure to improve processing economics. A trait most likely to have impact in this regard is modulation of the lignin in the stalk. The transgenic approach to stalk modulation for animal feeding has not yet progressed commercially; however, low lignin brown midrib mutant corn and sorghum are available commercially. The composition of feedstocks needs to be considered in the light of the industrial processes for which the feedstock will be used. Feedstock biomass that is to be used for conversion to sugars for biofuels or to support the growth of microorgansisms for bioproduct production need to be able to be converted efficiently and cheaply. This is difficult because the cellulose is complexed with lignin and hemicelluloses in cell walls, especially secondary cell walls (Himmel et al. 2007; Grabber 2005; Anderson and Akin 2008). For use in thermochemical conversions to biofuels, higher heating values and other parameters are more important. Higher heating value is related to lignin content so for these uses higher lignin values are advantageous. There is now a major focus on discovering the molecular genetic structure of plant cell walls and the factors that affect their use as feedstock (McCann and Carpita 2008; Shi et al. 2007; Penning et al. 2009; Gomez et al. 2008). From this will come many ways to change the structure of cell walls transgenically, as well as by selecting natural variants, and hopefully enhance the value of cell walls for industrial sources of sugars and lignin.

The structures of cell walls are based on cellulose fibrils linked with hemicelluloses. The hemicelluloses are decorated with various sugar complexes and these stimulate linkages with lignin (Himmel et al. 2007; Hisano et al. 2009; Grabber 2005). The resulting complexes interfere with the release of polysaccharides, absorb the enzymes during saccharification or reduce the efficiency of saccharification in other ways (Grabber 2005). Thus feedstock is usually pretreated with acid, alkali, steam or ammonia to release the polysaccharides before enzyme treatment (Wyman et al. 2005; Lau et al. 2009). Various studies have led to reductions in lignin to investigate if this enhances sugar release during saccharification. (Chen and Dixon 2007; Chapple et al. 2007; Ralph et al. 2006). Alfalfa plants transgenically downregulated for enzymes in the lignin biosynthesis pathway illustrated that lines with least lignin had the highest amount of carbohydrates, reflecting the compensating use of carbon. However, in switchgrass accessions the total amount of glucan is correlated with the amount of lignin (Ceres, unpublished). In the alfalfa transgenics there was a strong negative correlation between the lignin concentration and the amount of sugar released by enzymatic hydrolysis (Chen and Dixon 2007), supporting the view that lignin is a source of major difficulty in releasing sugars from cell walls in plants using enzymes. There is variation in feedstocks in the extent to which pretreatments stimulate release of sugars by enzymes. This variation is in part genetic and in part due to environmental and developmental factors

(Sarath et al. 2008). The effects of variation in lignin on growth and development in feedstock crops has yet to be investigated in detail.

During cell wall formation, xylans are cross-linked with ferulate monomers into a complex array of dimers and trimers and by extensive copolymerization of these ferulated xylans into lignin. This has led to the hypothesis that breeding for or making transgenics that possess fewer ferulate linkages should make the lignocellulose more suitable for easy saccharification. Consistent with this idea, Buanafina et al. (2007) have shown that expression of a fungal ferulic acid esterase in tall fescue increases cell wall digestibility. Overall, therefore, it appears that engineering the plant cell walls by altering the interactions between hemicellulose, lignin and cellulose microfibrils should result in reducing the need for costly enzymes in the saccharification process.

3.8.2.2 Novel, Non-Endogenous Traits

The concept of making proteins and enzymes in plants has a relatively long history (Mooney 2009). A recent example that relates to the biofuels arena is the addition by Syngenta of genes into corn encoding two enzymes for modification of seeds. The first is a chimeric thermostable alpha-amylase (AMY797E) derived from three alpha-amylase genes originating from three hyperthermophilic microorganisms of the archael order *Thermococcal* accompanied by the second gene, which encodes a phosphomannose isomerase enzyme used as a selectable marker. The product concept is that grain from the deregulated event will be the source of an alpha-amylase enzyme in the dry-grind ethanol process. This amylase will break down starch in the dry milling process and will replace the use of microbially produced alpha-amylase. Alpha-amylase catalyzes the hydrolysis of starch into starch fragments that range in size from 5 to 50 glucose units. These are then broken down to glucose using a conventional glucoamylase enzyme. The idea is to save on enzyme production costs by using high enzyme whole grain as an enzyme additive. This effectively removes all the cost of enzyme purification. Gene fragments encoding the alpha-amylases were created from organisms found deep in the Pacific Ocean at 90°C, pH 6.5. The resulting enzymes were combined to create a library of recombinant alpha-amylase enzymes that were then screened to identify a high level of activity. The resultant enzyme contains four fragments from BD5031, two fragments from BD 5064 and three fragments from BD 5063. The coding sequences in corn are driven by the gamma-zein corn seed storage promoter to ensure high level expression in endosperm (Richardson et al. 2002). The enzyme cocktail was developed by Diversa (San Diego, CA).

3.8.3 Co-products

There are many high volume co-products from crops that are in commerce today or are being developed. Starch and cellulose as well as proteins (soy protein, wheat gluten) and natural rubber are among principal ones that have been exploited extensively (Mooney 2009). A different example comes from sugarcane processing. After the sugar is extracted the remaining bagasse, consisting mainly of lignocellulose, is burnt to make electricity to return to the grid and to run the whole industrial plant. In this example, electricity is the co-product.

Production of biodegradable plastics, such as polyhydroxyalkanoates (PHA and PHB), in plants is a major goal. PHAs are made in certain bacteria (*Alcaligenes* and *Ralsonia*) and accumulate in inclusion bodies. Production of PHBs was first achieved in plants by transforming the genes encoding acetoacetyl CoA reductase and PHB synthase from *Ralsonia* into Arabidopsis. PHB accumulated in the nucleus, cytosol and vacuoles but only to low levels. Highest levels were found in Arabidopsis chloroplasts. Levels were not appreciably increased in tobacco, cotton, potato or maize above those seen in Arabidopsis leaf chloroplasts. Metabolix has transferred the genes to switchgrass (Somleva et al. 2008). The genes carried chloroplast targeting sequences to ensure the bacterial biosynthetic enzymes enter the chloroplasts. Constitutive and light-inducible promoters were used. Up to 4% dry weight PHB accumulated in the leaves and 1.2% in whole tillers. Importantly, the growth of the plant was not compromised. There is thus promise that this crop selected as a feedstock for bioenergy production can be dual purpose, and that industrial plants can be a source of PHB and the lignocellulose a source of biofuels. Readers are referred to reviews on this broad topic of co-products, e.g., Mooney (2009).

There are many other possible compounds that could theoretically be made in plants. The idea of adding value by adding co-products that piggyback on existing production systems may be best adapted to energy crops, where the products are not used for food or feed. For their use in extensive agriculture many issues, such as their impact on wildlife, must be addressed.

3.8.4 Genetic Confinement and Prevention of Seed Formation

Total sterility may prove to be the only acceptable route to the release of certain transgenically improved varieties that meet with the requirements of political agencies and those concerned with environmental/ecological impacts. Invasive plant species can cause significant environmental impacts, including the loss of native flora and fauna. *Miscanthus giganteus*, a promising energy crop candidate due to its high biomass yield, perennial nature, and cold hardiness, is thought to be minimally invasive as it is a sterile triploid. However, for genetic improvement and practical agronomic purposes, seed-based varieties are desirable. Plants of the *Miscanthus* genus are not native to the United States and the invasiveness potential for a drought-tolerant and cold-hardy variety must be considered high until proven otherwise. Switchgrass (*Panicum virgatum*) is native to the US east of the Rocky Mountains from Texas to Canada. While the invasiveness risk of switchgrass in its native habitat, per se, is considered to be low, transgenic switchgrass would have ample chance to cross-pollinate with wild relatives, which raises the potential for environmental impact. In addition, it has been suggested that switchgrass may prove invasive in regions where it has not previously been introduced (Barney and Ditomaso 2008), even without the presence of transgenes. Given the risks associated with invasiveness, careful attention should be given to minimizing the potential gene flow of species, whether transgenic or not, into novel environments (Wolt 2009).

The most prevalent methods of gene flow from field-based transgenic species into the environment are pollen flow from transgenic plants to native species and the dispersal of transgenic seed away from the field plots (see Chaps. 9–11). Methods of limiting such dispersal include pollen containment (e.g., physical barriers, large distances between crops, and/or the planting of crops where no native species exist), and actions to limit seed dispersal (e.g., field maintenance to watch for volunteer crops, sealed seed transport containers, and barriers to limit animal access to the crops). Implementation of a sterility system that eliminates both male and female gamete production would minimize the risks of invasiveness in both these instances.

Sterility can be achieved through several means, including both traditional and transgenic approaches (Mascia and Flavell 2004; Daniell 2002). One method to repress reproductive development is the use of daylength-sensitive plant material. Plants have evolved to switch from vegetative to reproductive development under optimal environmental conditions. Vegetative growth occurs when the apical meristem at the internodes produce new stem material. This switch requires the plant to respond to inputs that include daylength, time from germination, temperature, and environmental stressors, including drought. When the reproductive switch is induced, the meristem becomes determinant and produces reproductive material, i.e., reiteration of vegetative nodes and internodes ceases. By moving a plant to an environment in which it has not been adapted, it is possible to create a scenario in which the plant never experiences the appropriate flowering time cues, and thus remains in vegetative growth throughout the entire growing season. One example of this trait is a variety of tropical maize. Maize is a short-day grass, i.e., it flowers under conditions where daylength is short. Grown at higher latitudes where the daylength can be quite long, tropical maize fails to flower until the end of the growing season when frost eliminates the potential of pollen or seed dispersal (http://www.aces.uiuc.edu/news/stories/news4169.html). Daylength-sensitive varieties produce considerably more biomass and available resources, such as sugars, for the biofuels stream since they do not convert energy into seed product. Two caveats to this approach to sterility are that (1) it is environmentally regulated, and thus careful analysis of each putative growing area must be performed; and (2) the plants do not naturally senesce in the field, creating downstream processing issues.

Cytoplasmic male sterility (CMS) is commonly used in breeding programs where the source of the pollen needs to be managed. CMS is a condition under non-nuclear genetic control that results in the inability of the male gametophyte (pollen) to contribute to sexual reproduction. Unlike the flowering time approach mentioned above, plants with CMS mutations still undergo the vegetative-to-reproductive transition, but without an ample supply of pollen, no seed set occurs.

From a biomass perspective, this has potential value in that carbon resources are not transported to the reproductive structures for grain filling; rather, it can be used to increase usable sugars in the stalk of the plant. For genetic confinement purposes, pollen escape is minimized, and for non-native species (e.g., *Miscanthus* in the US), seed set would be negligible. CMS systems are rarely 100% effective due to genetic modifiers in different germplasm, environmental factors, and the fact that sperm can occasionally contain cytoplasmic DNA (Kuroiwa et al. 1993; Weider et al. 2009; Yu and Russell 1994). In addition, for a native plant like switchgrass, this methodology may prove of limited value because of the alternative sources of viable pollen that could fertilize the crop. Pollen from wild relatives enables seed production that can escape into the environment and produce transgenic volunteers.

Transgenic technologies can also be used to create sterility. Genetic pathways leading to the development of sporophyte reproductive structures, the resultant gametophytes, and seed are becoming increasingly well-known. By taking advantage of this information, gain-of-function and loss-of-function systems can be utilized to selectively disrupt plant sexual reproduction while minimizing the impact on vegetative growth. Despite the obvious advantages of using this type of technology to limit transgene spread into the environment, the opportunities for the use of this type of technology may be restricted. Genetic use restriction technology, or GURT, is the name given to proposed methods for restricting the use of genetically modified plants by causing second generation seeds to be sterile (Hills et al. 2007). One type of GURT initially developed by the USDA and multinational seed companies has not been commercialized anywhere in the world due to opposition from farmers, indigenous peoples, civil society and some governments. The opposition to GURT technology may be limited to food crops, and the utility for biomass crops that do not enter the food chain may be invaluable. The whole topic of risk assessment of transgenic crops is one that will inevitably generate controversy; regulators will hopefully retain a science-based risk assessment process with the goals of harmonizing standards internationally (Wolt 2009).

3.9 Concluding Remarks

There are great expectations that more crops are going to serve as feedstocks for many more products, especially biofuels, to help with fuel security, decrease greenhouse gas production and stimulate agriculture. Many crops and genetic variants of these crops will be required to enable these industries to grow in many locations. The economics of the value chains will be crucial. To increase feedstock outputs and decrease costs requires many improvements in many feedstock traits. Thus, much plant breeding is required and this breeding needs to be much more efficient than has been the case historically. This is a large challenge, especially in crops that have little breeding history. The application of state-of-the-art genetics, marker-assisted breeding, association breeding and genomic selection, exploiting gene-trait associations, can bring about increases in the rate of feedstock improvement, especially if comparative genomics can be exploited to enable gene-trait knowledge in one crop, such as corn, can be used to aid other related grasses. Certain traits can also be introduced using transgenes and, providing they can be deregulated, can make a significant contribution to productivity. Where co-products can be exploited, the economics of the value chain may be improved considerably. Where feedstocks serve as sources of lignocellulose for biofuels, improvements in process-specific compositions are required to optimize their cost-effectiveness, especially if sugars are to be produced. Crops such as sugarcane and sweet sorghum already offer an efficient supply of cheap sugar because of the accumulation of soluble sugar in the green material. As with all crops, it is not only breeding that has to be improved but also agronomy. Above all, the new biotechnological advances need to be integrated with established crop improvement systems and industrial needs.

References

Ahn S, Anderson JA, Sorrells ME, Tanksley SD (1993) Homoeologous relationships of rice, wheat and maize chromosomes. Mol Gen Genet 241(5–6):483–490
Anderson WF, Akin DE (2008) Structural and chemical properties of grass lignocellulose related to conversion for biofuels. J Ind Microbiol Biotechnol 35:355–366
Barney JN, Ditomaso JM (2008) Nonnative species and bioenergy: are we cultivating the next invader? BioSci 58:64–70. doi: 10.1641/b580111
Barth S, Busimi AK, Utz HF, Melchinger AE (2003) Heterosis for biomass yield and related traits in five hybrids of *Arabidopsis thaliana* L. Heynh. Heredity 91:36–42
Bennetzen JL (2000) Comparative sequence analysis of plant nuclear genomes microcollinearity and its many exceptions. Plant Cell 12(7):1021–1030
Bingham ET, Groose RW, Woodfield DR, Kidwell KK (1994) Complementary gene interactions in alfalfa are greater in autotetraploids than diploids. Crop Sci 34:823–829
Bouton JH (2007) Molecular breeding of switchgrass as a bioenergy crop. Curr Opin Genet Dev 17:553–558
Breseghello F, Sorrells ME (2006) Association analysis as a strategy for improvement of quantitative traits in plants. Crop Sci 46:1323–1330
Buanafina MM, Langdon T, Hauck B, Dalton S, Morris P (2007) Expression of a fungal ferulic esterase increases cell wall digestibility of tall fescue (*Festuca arundinacea*). Plant Biotechnol J 6(3):264–280. doi: 10.1111/j1467–7652
Chapple C, Ladisch M, Melian R (2007) Loosening lignin's grip on biofuel production. Nat Biotechnol 25:746–747
Chen F, Dixon RA (2007) Lignin modification improves fermentable yields for biofuel production. Nat Biotechnol 25:759–761
Crow JF (1948) Alternative hypotheses of hybrid vigor. Genetics 33:477–487
Daniell H (2002) Molecular strategies for gene containment in transgenic crops. Nat Biotechnol 20:581–586
Devos KM, Gale MD (2000) Genome relationships. Plant Cell 12(5):637–646
Duvick DN (1999) Heterosis: feeding people and protecting natural resources. In: Coors JG, Pandey S (eds) Genetics and exploitation of heterosis in crops. American Society of Agronomy, Madison, WI, pp 19–30
East EM (1936) Heterosis. Genetics 21:375–397
Eathington SR, Crosbie TM, Edwards MD, Reiter RS, Bull JK (2007) Molecular markers in a commercial breeding program. Crop Sci 47:S154–S163

3 Designing Plants To Meet Feedstock Needs 81

El Bassam N (1998) Energy plant species. James & James, London
Foote T, Roberts M, Kurata N, Sasaki T, Moore G (1997) Detailed comparative mapping of cereal chromosome regions corresponding to the Ph1 locus in wheat. Genetics 147:801–807
Freeman GF (1919) Heredity of quantitative characters in bread wheat. Genetics 4:1–93
Gale MD, Devos KM (1998) Comparative genetics in the grasses. Proc Natl Acad Sci USA 95:1971–1974
Gaut BS (2001) Patterns of chromosomal duplication in maize and their implications for comparative maps of the grasses. Genome Res 11:55–66
Gaut BS (2002) Evolutionary dynamics of grass genomes. New Phytol 154(1):15–28
Gomez LD, Steele-King CG, McQueen-Mason SJ (2008) Sustainable liquid biofuels from biomass; the writing is on the walls. New Phytol 178(3):473–485. doi: 10.1111/j1469-8137
Grabber JH (2005) How do lignin composition structure and cross linking affect degradability? A review of cell wall model studies. Crop Sci 45:820–831
Groose RW, Kojis WP, Bingham ET (1988) Combining ability differences between isogenic and tetraploid alfalfa. Crop Sci 28:7–10
Hallauer AR, Carena MJ (2009) Maize breeding. In: Carena MJ (ed) Handbook of plant breeding, vol 3. Cereals. Springer, New York, pp 1–96
Heaton EA, Mascia PN, Flavell R, Thomas S, Long PS, Dohleman FG (2008) Energy crop development: current progress and future prospects. Curr Opin Biotechnol 19:202–209
Heffner EL, Sorrells ME, Jannink JL (2009) Genomic selection for crop improvement. Crop Sci 49:1–12
Heffner EL, Lorenz AJ, Jannink JL, Sorrells ME (2010) Plant breeding with genomic selection: potential gain per unit time and cost. Crop Sci (in press)
Hills M, Hall L, Arnison P, Good A (2007) Genetic use restriction technologies (GURTs). Strategies to impede transgene movement. Trends Plant Sci 12:177–183
Himmel ME, Ding SY, Johnson DK, Adney WS, Nimlos MR, Brady JW, Foust TD ((2007) Biomass recalcitrance: engineering plants and enzymes for biofuels production. Science 315 (5813):804–807
Hisano H, Nandakumar R, Wang ZY (2009) Genetic modification of lignin biosynthesis for improved biofuel production. In Vitro Cell Dev Biol Plant 45:306–313
Hochholdinger F, Hoecher N (2007) Towards the molecular basis of heterosis. Trends Plant Sci 12:4427–432
Holland JB (2004) Implementation of molecular markers for quantitative traits in breeding programs—challenges and opportunities. In: New directions for a diverse planet: Proceedings of the 4th International Crop Sci Congress, Brisbane Australia, p 26
Hulbert SH, Richter TE, Axtell JD, Bennetzen JL (1990) Genetic mapping and characterization of sorghum and related crops by means of maize DNA probes. Proc Natl Acad Sci USA 87 (11):4251–4255
Jakob K, Zhou F, Paterson AH (2009) Genetic improvement of C4 grasses as cellulosic biofuel feedstocks. In Vitro Cell Dev Biol Plant 45:291–305
Jessup RW (2009) Development and status of dedicated energy crops in the United States. In Vitro Cell Dev Biol Plant 45:282–290
Krutovsky KV, St. Clair JB, Saich R, Hipkins VD, Neale DB (2009) Estimation of population structure in coastal Douglas-fir [*Pseudotsuga menziesii* (Mirb.) Franco var. *menziesii*] using allozyme and microsatellite markers. Tree Genet Genomes 5:641–658
Kurata N, Nagamura Y, Yamamoto K, Harushima Y, Sue N, Wu J, Antonio BA, Shomura A, Shimizu T, Lin SY, Inoue T, Fukuda A, Shimano T, Kuboki Y, Toyama T, Miyamoto Y, Kirihara T, Hayasaka K, Miyao A, Monna L, Zhong HS, Tamura Y, Wang ZX, Momma T, Umehara Y, Yano M, Sasaki T, Minobe Y (1994) A 300 kilobase interval genetic map of rice including 883 expressed sequences. Nat Genet 8:365–372
Kuroiwa T, Kawazu T, Uchida H, Ohta T, Kuroiwa H (1993) Direct evidence of plastid DNA and mitochondrial DNA in sperm cells in relation to biparental inheritance of organelle DNA in *Pelargonium zonale* by fluorescence/electron microscopy. Eur J Cell Biol 62:307–313

Lamkey KR, Edwards JW (1999) Quantitative genetics of heterosis. In: Coors JG, Pandey S (eds) Genetics and exploitation of heterosis in crops. American Society of Agronomy, Madison, WI, pp 31–48

Lande R, Thompson R (1990) Efficiency of marker-assisted selection in the improvement of quantitative traits. Genetics 124:743–756

Lau MW, Gunawan C, Dale B (2009) The impacts of pretreatment on the germentability of pretreated lignocellulose biomass: a comparative evaluation between ammonia fiber expansion and dilute acid pretreatment. Biotech Biofuels 2:30

Levings CS, Dudley JW, Alexander DE (1967) Inbreeding and crossing in autotetraploid maize. Crop Sci 7:72–73

Lister R, Gregory BD, Ecker JR (2009) Next is now: new technologies for sequencing of genomes; transcriptomes; and beyond. Curr Opin Plant Biol 12:1–12

Martinez-Reyna JM, Vogel KP (2008) Heterosis in switchgrass: spaced plants. Crop Sci 48:1312–1320

Mascia PN, Flavell RB (2004) Safe and acceptable strategies for producing foreign molecules in plants. Curr Opin Plant Biol 7:189–195

McCann MC, Carpita NC (2008) Designing the deconstruction of plant cell walls. Curr Opin Plant Biol 11:314–320

Meuwissen THE, Hayes BJ, Goddard ME (2001) Prediction of total genetic value using genome-wide dense marker maps. Genetics 157:1819–1829

Meyer RC, Törjék O, Becher M, Altmann T (2004) Heterosis of biomass production in Arabidopsis. Establishment during early development. Plant Physiol 134:1813–1823

Mok DWS, Peloquin SJ (1975) Breeding value of 2n pollen (diplandroids) in tetraploid x diploid crosses in potatoes. Theor Appl Genet 46:307–314

Mooney BP (2009) The second green revolution? Production of plant-based biodegradable plastics. Biochem J 418(2):219–232

Murray SC, Rooney WL, Hamblin MT, Mitchell SE, Kresovich S (2009) Sweet sorghum genetic diversity and association mapping for Brix and height. Plant Genome 1:48–62

Osborn TC, Pires JC, Birchler JA, Auger DL, Chen ZJ, Lee HS, Comai L, Madlung A, Doerge RW, Colot V, Martienssen RA (2003) Understanding mechanisms of novel gene expression in polyploids. Trends Genet 19:141–147

Paterson AH, Bowers JE, Feltus FA, Tang H, Lin L, Wang X (2009a) Comparative genomics of grasses promises a bountiful harvest. Plant Physiol 149:125–131

Paterson AH, Bowers JE, Bruggmann R, Dubchak I, Grimwood J, Gundlach H, Haberer G, Hellsten U, Mitros T, Poliakov A, Schmutz J, Spannagl M, Tang H, Wang X, Wicker T, Bharti AK, Chapman J, Feltus FA, Gowik U, Grigoriev IV, Lyons E, Maher CA, Martis M, Narechania A, Otillar RP, Penning BW, Salamov AA, Wang Y, Zhang L, Carpita NC, Freeling M, Gingle AR, Hash CT, Keller B, Klein P, Kresovich S, McCann MC, Ming R, Peterson DG, Rahman MU, Ware D, Westhoff P, Mayer KFX, Messing J, Rokhsar DS (2009b) The *sorghum bicolor* genome and the diversification of grasses. Nature 457:551–556

Pavy N, Pelgas B, Beauseigle S, Blais S, Gagnon F, Gosselin I, Lamothe M, Isabel N, Bousquet J (2008) Enhancing genetic mapping of complex genomes through the design of highly-multiplexed SNP arrays: application to the large and unsequenced genomes of white spruce and black spruce. BMC Genomics 9:21

Penning BW, Hunter III CT, Tayengwa R, Eveland AL, Dugard CK, Olek AT, Vermerris W, Koch KE, McCarty DR, Davis MF, Thomas SR, McCann MC, Carpita NC (2009) Genetic resources for maize cell wall biology. Plant Physiol 151:1703–1728

Perlack RD, Wright LL, Turhollow AF, Graham RL, Stokes BJ, Erbach DC (2005) Biomass as feedstock for a bioenergy and bioproducts industry: the technical feasibility of a billion-ton annual supply. USDA report, Oak Ridge National Laboratory, TN, pp 1–78

Ralph J, Akiyama T, Kim H, Lu F, Schatz PF, Marita JM, Ralph SA, Reddy MSS, Chen F, Dixon R (2006) Effects of coumarate 3-hydroxylase down regulation on on lignin structure. J Biol Chem 281:8843–8853

Richardson TH, Tan X, Frey G, Callen W, Cabell M, Lam D, Macomber J, Short JM, Robertson DE, Miller C (2002) A novel, high performance enzyme for starch liquefaction: discovery and optimization of a low pH, thermostable alpha-amylase. J Biol Chem 277:26501–26507

Rooney WL, Blumenthal J, Bean B, Mullet JE (2007) Designing sorghum as a dedicated bioenergy feedstock. Biofuels Bioprod Biorefining 1:147–157

Sarath G, Akin DE, Mitchell RB, Vogel KP (2008) Cell wall composition and accessibility to hydrolytic enzymes is differentially altered in divergently bred switchgrass (*Panicum virgatum* L.) genotypes. Appl Biochem Biotechnol 150:1–14

Schnable PS, Ware D, Fulton RS, Stein JC, Wei F, Pasternak S, Liang C, Zhang J, Fulton L, Graves TA, Minx P, Reily AD, Courtney L, Kruchowski SS, Tomlinson C, Strong C, Delehaunty K, Fronick C, Courtney B, Rock SM, Belter E, Du F, Kim K, Abbott RM, Cotton M, Levy A, Marchetto P, Ochoa K, Jackson SM, Gillam B, Chen W, Yan L, Higginbotham J, Cardenas M, Waligorski J, Applebaum E, Phelps L, Falcone J, Kanchi K, Thane T, Scimone A, Thane N, Henke J, Wang T, Ruppert J, Shah N, Rotter K, Hodges J, Ingenthron E, Cordes M, Kohlberg S, Sgro J, Delgado B, Mead K, Chinwalla A, Leonard S, Crouse K, Collura K, Kudrna D, Currie J, He R, Angelova A, Rajasekar S, Mueller T, Lomeli R, Scara G, Ko A, Delaney K, Wissotski M, Lopez G, Campos D, Braidotti M, Ashley E, Golser W, Kim H, Lee S, Lin J, Dujmic Z, Kim W, Talag J, Zuccolo A, Fan C, Sebastian A, Kramer M, Spiegel L, Nascimento L, Zutavern T, Miller B, Ambroise C, Muller S, Spooner W, Narechania A, Ren L, Wei S, Kumari S, Faga B. Levy M, McMahan L, Van Buren P, Vaughn MW, Ying K, Yeh CT, Emrich SJ, Jia Y, Kalyanaraman A, Hsia AP, Barbazuk WB, Baucom RS, Brutnell TP, Carpita NC, Chaparro C, Chia JM, Deragon JM, Estill JC, Fu Y, Jeddeloh JA, Han Y, Lee H, Li P, Lisch DR, Liu S, Liu Z, Nagel DH, McCann MC, SanMiguel P, Myers AM, Nettleton D, Nguyen J, Penning BW, Ponnala L, Schneider KL, Schwartz DC, Sharma A, Soderlund C, Springer NM, Sun Q, Wang H, Waterman M, Westerman R, Wolfgruber TK, Yang L, Yu Y, Zhang L, Zhou S, Zhu Q, Bennetzen JL, Dawe RK, Jiang J, Jiang N, Presting GG, Wessler SR, Aluru S, Martienssen RA, Clifton SW, McCombie WR, Wing RA, Wilson RK (2009) The B73 maize genome: complexity, diversity and dynamics. Science 326(5956):1112–1115

Shi C, Uzarowska A, Ouzunova M, Landbeck M, Wenzel G, Lübberstedt T (2007) Identification of candidate genes associated with cell wall digestibility and eQTL (expression quantitative trait loci) analysis in a Flint x Flint maize recombinant inbred line population. BMC Genomics 8:22

Shull GH (1952) Beginnings of the heterosis concept. In: Gowen JW (ed) Heterosis: a record of researches directed toward explaining and utilizing the vigor of hybrids. Iowa State College Press, Ames, pp 14–48

Somleva MN, Snell KD, Beaulieu JJ, Peoples OP, Garrison BR, Patterson NA (2008) Production of polyhydroxybutyrate in switchgrass, a value-added co-product in an important lignocellulosic biomass crop Plant Biotechnol J 6(7):633–678

Song R, Messing J (2003) Gene expression of a gene family in maize based on noncollinear haplotypes. Proc Natl Acad Sci USA 100:9055–9066

Swanson-Wagner RA, DeCook R, Jia Y, Bancroft T, Ji T, Zhao X, Nettleton D, Schnable PS (2009) Paternal cominance of trans-eQTL influences gene expression patterns in maize hybrids. Science 326(5956):1118–1120

Tilman D, Socolow R, Foley JA, Hill J, Larson E, Lynd L, Pacala S, Reilly J, Searchinger T, Somerville C, Williams R (2009) Beneficial biofuels—the food, energy and environment trilemma. Science 325:270–271

Troyer AF, Mascia PN (1998) Key technologies impacting corn genetic improvement—past, present and future. Maydica 44:55–68

Van Deynze AE, Nelson JC, Yglesias ES, Harrington SE, Braga DP, McCouch SR, Sorrells ME (1995) Comparative mapping in grasses. Wheat relationships. Mol Gen Genet 248(6):744–754

Varshney RK, Nayak SN, May GD, Jackson SA (2009) Next generation sequencing technologies and their implications for crop genetics and breeding. Trends Biotechnol 27:522–530

Vermerris W (2009) Genetic improvement of bioenergy crops. Springer, New York

Vermerris W, Saballos A, Ejeta G, Mosier N, Ladisch M, Carpita N (2007) Molecular breeding to enhance ethanol production from corn and sorghum stover. Crop Sci 47 S3:S142–S153

Vogel KP (2000) Improving warm-season grasses using selection, breeding, and biotechnology. In: Moore KJB Anderson B (eds) Native warm season grasses: research trends and issues. CSSA Spec Publ 30 CSSA and ASA, Madison, WI

Vogel KP, Burson B (2004) Breeding and genetics. In: Moser LE, Sollenberger L, Burson B (eds) Warm-season (C4) grasses. ASA Monogr 45 ASA, CSSA and SSSA, Madison, WI

Wang X, Gowik U, Tang H, Bowers J, Westhoff P, Paterson A (2009) Comparative genomic analysis of C4 photosynthetic pathway evolution in grasses Genome Biol 10: R68

Weider C, Stamp P, Christov N, Husken A, Foueillassar X, Camp K-H, Munsch M (2009) Stability of cytoplasmic male sterility in maize under different environmental conditions. Crop Sci 49:77–84. doi: 10.2135/cropsci2007120694

Wolt JD (2009) Advancing environmental risk assessment for transgenic biofeedstock crops. Biotechnol Biofuels 2:27

Wyman CE, Dale BE, Elander RT, Holtzapple M, Ladisch MR, Lee YY (2005) Coordinated development of leading biomass pretreatment technologies. Bioresour Technol 96:1050–1966

Xu YB, Crouch JH (2008) Marker-assisted selection in plant breeding: from publications to practice. Crop Sci 48:391–407

Yu HS, Russell SD (1994) Occurrence of mitochondria in the nuclei of tobacco sperm cells. Plant Cell 6:1477–1484. doi: 10.1105/tpc6101477

Yu J, Pressoir G, Briggs WH, Bi IV, Yamasaki M, Doebley JF, McMullen MD, Gaut BS, Nielsen DM, Holland JB (2006) A unified mixed-model method for association mapping that accounts for multiple levels of relatedness. Nat Genet 38:203–208

Zhong SQ, Dekkers JCM, Fernando RL, Jannink JL (2009) Factors affecting accuracy from genomic selection in populations derived from multiple inbred lines: a barley case study. Genetics 182:355–364

Part B
Specific Biofuel Feedstocks

Chapter 4
Engineering Advantages, Challenges and Status of Sugarcane and other Sugar-Based Biomass Resources

Ricardo A. Dante, Plinio T. Cristofoletti, and Isabel R. Gerhardt

4.1 Introduction

4.1.1 Sugar-Based Industry and Ethanol Uses

Global warming mitigation requires new renewable sources of energy to replace oil consumption worldwide. One of the most promising and available renewable sources is ethanol, a high octane fuel that can be used in today's automobiles. Ethanol is an alternative fuel that dates back to the 1800s (Songstad et al. 2009) and today is obtained mainly through fermentation processes from starch- and sucrose-accumulating crops. Besides these sources, efforts are being made globally to develop technologies for ethanol production from lignocellulosic materials to enhance the feedstock supply available for ethanol production. Today, ethanol production represents only 0.7% of the energy equivalent in oil production, or 3% of the global transportation fuel supply (Coyle 2007). However, ethanol has a high capacity to penetrate the market as a gasoline blend (E3–E85) or even as E100. The most successful example of fuel ethanol utilization is in Brazil, where the only automotive fuels available are blends of up to E25 or hydrous ethanol (approximately 95% ethanol and 5% water). Although many uncertainties surround ethanol supply and demand in the next decade, it is clear that, under current mandates, ethanol production will increase, with output in 2015 being forecast to be over 115 billion L (30.4 billion gallons), rising about 135% over 2006 levels (Licht 2006). To satisfy this growth in the worldwide demand for renewable sources of fuel, such as

---

R.A. Dante, P.T. Cristofoletti, and I.R. Gerhardt*
Canavialis e Alellyx, Via Anhanguera, km 104, Condomínio TechnoPark, Rua James Clark Maxwell, 320/360, 13069-380 Campinas, SP, Brazil
e-mail: ricardo.a.dante@alellyx.com.br; plinio.t.cristofoletti@alellyx.com.br
*Present Address:
Embrapa Florestas, Estrada da Ribeira, km 111, Caixa Postal 319, 83411-000 Colombo, PR, Brazil
e-mail: isabel.gerhardt@cnpf.embrapa.br

those based on sugars, three non-mutually exclusive scenarios are possible: (1) increasing the production area, (2) increasing productivity on existing farmland, and (3) improving and adopting new industrial processes. Increasing production area would not require direct technology investments. However, it would require increased logistical effort, along with a demand for the latest agricultural practices and improved cultivars if it is to meet worldwide demand (Goldemberg and Guardabassi 2009). Improvements in agricultural productivity can be achieved by extensive adoption of current technologies in agriculture and by the introduction of more productive cultivars obtained via breeding and genetic modification. In addition, new industrial processes can be incorporated in mills for production of ethanol from lignocellulosic materials.

Productivity enhancement of energy crops will probably be similar to that achieved for corn in the last few decades in the US. In this crop, continuous gains have been achieved, and average US yields are still expected to double in the next two decades through improvements in agronomics, breeding, and biotechnology (Edgerton 2009). Sugarcane and other biomass-accumulating grasses can be expected to have a larger yield enhancement, since they are in the early stages of genetic improvement. Breeding and biotechnology, and their combination, will contribute to increasing productivity by the introduction of new genes affecting yield, drought- and cold-tolerance, nitrogen use, disease- and insect-resistance and other traits positively.

Among all crops currently used worldwide for biofuel purposes, corn (maize; *Zea mays*) and sugarcane account for most of the ethanol produced. Both species are C4 grasses of the Andropogoneae tribe. These species very efficiently synthesize and store non-structural carbohydrates that are economical for ethanol production through fermentation. The major source of sugars in corn is the starch accumulated in the grain, while in sugarcane it is the sucrose accumulated in stalks. The use of corn and sugarcane for ethanol today reflects the differences between the most successful programs for renewable fuel production: the US and Brazil, respectively (Table 4.1). Corn is at the base of the food and feed industry worldwide, especially in the US. Sugarcane is currently cultivated in almost 23 million ha in tropical and subtropical regions, one-third of this area being in Brazil. Sugarcane is used mostly for raw sugar production, corresponding to around two-thirds of world sugar production (FAOSTAT 2007). Comparisons between corn and sugarcane productivity showed the largest production of ethanol per hectare in sugarcane (Table 4.1). Ethanol produced from corn yields $4.0 \text{ m}^3 \text{ ha}^{-1} \text{ year}^{-1}$, while sugarcane produces $6.6 \text{ m}^3 \text{ ha}^{-1} \text{ year}^{-1}$.

Table 4.1 also illustrates that the land used today for ethanol production is relatively small considering the global arable land available (14,111 million ha). The total agricultural land used in Brazil and the US for all purposes is 263.5 million ha and 411 million ha, respectively. In Brazil, only 4 million ha arable land (or 6%) is used for ethanol production with around 197 million ha used for pastures (FAOSTAT 2007), with some pasture land available immediately to be converted in sugarcane cropland. In addition, sugarcane ethanol production fulfills approximately 50% of Brazilian transportation fuel needs, and corresponds to

Table 4.1 Estimated ethanol yields in the 2008/2009 crop season using different sources of feedstock for the two most important global producers

	Corn, United States[a]	Sugarcane, Brazil[b]
Total harvest area (million hectares)	31.8	7.4
Total production (metric tons)	307.3	569.0
Total harvest area for ethanol (%)	29	56
Yield of feedstocks (metric ton/hectare)	9.7	75.0
Feedstock ethanol production (liter/kilogram feedstock)	0.41	0.10
Total ethanol production (billion liters)	37.2	27.5
Ethanol production (cubic meter/hectare)	4.0	6.6
Arable land (million hectares)[c]	173.1	66.5

[a]FAPRI—Food and Agricultural Policy Research Institute—(2009)
[b]UNICA—União das Indústrias de Cana-de-Açúcar—(2009)
[c]FAOSTAT (2007)

16% of all energy used in the country. As a comparison, approximately 9 million ha of the arable land (or 5%) in the US is used for ethanol production from corn (Table 4.1).

A comparison of energy return on investment (EROI) of these two crops reveals the efficiency of sugarcane for ethanol production. This calculation takes into account the ratio of the energy recovered after the industrial process by the input energy for ethanol production. For sugarcane ethanol, this ratio was calculated as 9.2–10.2 (Macedo and Cortez 2000; Goldemberg 2007). The value obtained with current production methods of corn ethanol is 1.05–1.35, but this value is expected to rise as industrial processes and the use of lignocellulosic materials are optimized (Farrell et al. 2006). Corn ethanol produced via current technologies is a first step toward building the infrastructure for next-generation ethanol. Taking into account current technologies and economy, including the cost of production and subsidies (Goldemberg and Guardabassi 2009), it is clear that sugarcane has an enormous advantage over corn for ethanol production. This fact may change considerably in the future with the addition of corn cellulosic feedstock for fuel ethanol production.

The industry for sugarcane processing in Brazil is well established (Nass et al. 2007), operating with production costs that are competitive with oil products such as gasoline, and using part of the bagasse biomass to generate a surplus of electricity that is an additional source of income for the mills. The environment in Brazil for the development of this industry is unique and results from the governmental program ProÁlcool during the 1970s and 1980s that supported the development of fuel ethanol (Nass et al. 2007; Matsuoka et al. 2009). Current ethanol production receives no subsidies. Recently, a new wave of growth, driven by the need for oil independence from imports, renewable sources of energy and flex fuel cars, which represent the majority of new cars sold in Brazil [UNICA (União das Indústrias de Cana-de-Açúcar) 2009], is changing the industry and increasing its competitiveness. Although this is a very particular scenario compatible with the middle-sized Brazilian economy (Nass et al. 2007; Matsuoka et al. 2009), ethanol as fuel can be deployed and adapted to meet the needs of different countries: this includes adapting to the local availability of sugar resources and even lignocellulosic materials in the near future. Besides Brazil and the US, this path is now being taken by other countries that are adopting ethanol for transportation or discussing the use of ethanol in gasoline blends.

Similar to Brazil, other tropical and subtropical countries could benefit from sugarcane production for ethanol. The generation of more productive cultivars via breeding as well as biotechnology suited for new areas and geographies, adapted to poor soil and drought conditions, could contribute to this scenario. To meet such a challenging goal, several efforts are being taken to develop the basis of sugarcane biotechnology, such as high throughput transformation, genome sequencing, as well as studies aimed at gaining a deeper understanding of sugarcane physiology and responses to different environments. New cultivars are being shaped to new agricultural frontiers by several breeding programs that are incorporating molecular markers for assisted selection. In addition, the upcoming technology of cellulosic ethanol will further drive breeding programs for developing high-fiber content cultivars (Matsuoka et al. 2009). All these technologies combined will confer on sugarcane the position of a global crop helping to mitigate the impacts of global warming.

4.1.2 Sugarcane Production System

Sugarcane is usually propagated vegetatively through stalk cuttings, carrying axillary buds, from healthy plants. These buds can be propagated in nurseries or by tissue culture in biofactories that increase the throughput of clonal propagation. In general, one stalk can produce between 10–20 new plants by cuttings. In biofactories, during the same period, hundreds of plants can be produced by using optimized growth conditions. Upon plantlet establishment in the field, axillary buds in basal nodes grow into a large number of tillers that develop into new stalks. Following harvesting, axillary buds located at the base of the harvested stalks grow into new stalks, making sugarcane a semi-perennial crop. Crops thus generated are usually referred to as ratoon crops. Usually four to six crops are harvested from the same plant. Under optimal growth conditions, stalks are harvested after 12–18 months, producing an average of 85 metric tons (1,000 kg = 1 Mg) of stalks/ha, from which around 12.5 t sugar can be extracted (average values in Central and Southern Brazil). Mild drought and cold stresses that coincide with the late stage of sugarcane crop development restrict vegetative growth and maximize sucrose accumulation. During plant development, sucrose concentration increases basipetally along the sugarcane stalk, while the profile of glucose and fructose concentrations is reversed. The crop harvesting season lasts several months; during this period different cultivars are harvested sequentially as they reach maximal sucrose content. At this mature (or ripened) stage, sugarcane reaches its economical point of harvesting, at which sugar content is high enough to permit harvesting and transportation of stalks to the mills for processing into raw sugar and ethanol. During each harvest cycle a yield decline is observed. When yields become economically unsustainable, the field is renewed by planting or, ideally, by rotating the cropland with leguminous plants such as peanuts or soybeans.

4.2 Biotechnology and Breeding Strategies for Increasing Sugarcane Sucrose Yields

Sugarcane (*Saccharum* spp.) originated in South and Southeast Asia and has been cultivated for its high sugar content for hundreds of years. Modern cultivars of sugarcane are highly polyploid and aneuploid, with about 120 chromosomes, produced by interspecific hybridizations among members of the genus *Saccharum*, mostly *S. officinarum* and *S. spontaneum*, and to a lesser extent, *S. robustum*, *S. barberi*, and *S. sinense* (reviewed by D'Hont et al. 2008). The monoploid genome size of sugarcane is approximately 920 Mbp (D'Hont et al. 2008), which is more than twice the size of the rice (*Oryza sativa*) genome (389 Mbp) and slightly larger than that of sorghum (*Sorghum bicolor*; 760 Mbp). Initiatives to understand the complex sugarcane genome have been made in recent years, based largely on the production of expressed sequence tag (EST) databases. The Brazilian SUCEST, the largest of such EST projects, produced a database with 238,000 ESTs from 26 sugarcane cDNA libraries representing several tissues and different developmental stages (Vettore et al. 2001; Vettore et al. 2003). Transcripts obtained in this project cover possibly around 80% of the sugarcane transcriptome. In the near future, next-generation sequencing technologies are expected to allow the complete sequence of the sugarcane genome to become available through organized efforts involving groups in South Africa, Australia, Brazil and the US (Sugarcane Genome Sequencing Initiative 2009). Ever-increasing knowledge of the sugarcane genome and transcriptome will continue to foster breeding and biotechnology approaches aiming at increasing sucrose and biomass yields.

4.2.1 Photosynthetic Capacity of Sugarcane and the Source–Sink Relationship: What Determines Sucrose Accumulation?

Sugarcane is a very photosynthetically efficient plant that has the remarkable ability to accumulate sucrose in mature stalk internodes to levels of approximately 0.7 M (Moore 1995). As sugarcane becomes not only one of the world's main sources of sugar and ethanol but also a potential feedstock for cellulosic biofuel, many efforts are underway to increase biomass production and sugar accumulation. One way to achieve that is by increasing photosynthetic rates, translocation and partition of photosynthates. Sugarcane, like other important crops such as corn and sorghum, has developed anatomical and biochemical mechanisms to achieve high photosynthetic efficiency. By fixing CO_2 through the C4 photosynthetic pathway, these plants are able to effectively utilize atmospheric CO_2 concentration to increase the rate of photosynthesis. Despite the fact that C4 photosynthesis is close to saturation under current atmospheric CO_2 concentration, systematic positive responses of C4 species to elevated CO_2 concentration have been observed (Wand et al. 1999).

Altered biomass accumulation and carbohydrate metabolism were observed in sugarcane in plants growing in an enriched CO_2 atmosphere. Aiming to characterize the leaf photosynthetic mechanism of sugarcane grown at double-ambient (~720 ppm) CO_2, Vu et al. (2006) have shown that leaf CO_2 exchange rates (CER) at 7, 14 and 32 days after leaf emergence were 20, 7 and 10% greater than that of ambient CO_2-grown plants. They also observed an up-regulation of key photosynthetic enzymes, such as ribulose-1,5-biphosphate carboxylase-oxygenase (Rubisco), NADP-malate dehydrogenase (NADP-MDH), pyruvate orthophosphate dikinase (PPDK), and sucrose-phosphate synthase (SPS) in young developing leaves under elevated growth CO_2. At final harvest, these plants presented increased plant biomass and sucrose production, indicating an acclimation of sugarcane to elevated CO_2 concentration, especially at the early stages of leaf development. Greater CER and Rubisco activity in young expanding leaves of sorghum grown at elevated CO_2 also resulted in plants with increased dry matter (Prasad et al. 2009), suggesting that such alterations, even when they happen in early stages of plant development, can contribute to biomass accumulation.

Other photosynthesis proteins, such as those belonging to the electron transport system of the leaves, may also be involved in photosynthetic efficiency in sugarcane grown under elevated CO_2. According to De Souza et al. (2008), gene expression of subunit N of the photosystem I reaction center, protein K of photosystem II, and ferredoxin I was elevated in sugarcane grown in higher CO_2 atmosphere (~720 ppm). The observed increase of 30% in photosynthesis, the accumulation of 40% more biomass, and the increase of 29% in stalk sucrose content in comparison with plants grown at ambient CO_2 were attributed to a direct relationship between molecular and physiological responses.

The response of sugarcane to elevated CO_2 does not seem to be affected by other environmental stress conditions. Two different sugarcane cultivars grown at elevated CO_2 and high temperature exhibited increases of 50%, 26%, 84% and 124% in leaf area, leaf dry weight, stalk dry weight and stalk juice volume, respectively, compared with plants grown at ambient CO_2 and ambient temperature, as well as a two- to three-fold increase in stalk soluble solids (Vu and Allen Jr 2009a). When 4-month-old sugarcane suffered a 13-day period of drought stress, those grown at elevated CO_2 showed lower leaf conductance and transpiration rate and greater water-use efficiency, leading to no effect on photosynthesis for at least an extra day (Vu and Allen Jr 2009b).

Photosynthetic efficiency is not just a matter of light and CO_2 availability, but is controlled by a whole plant source-sink balance (Paul and Foyer 2001). The observation that *S. spontaneum*, which exhibits low sucrose content in the stalk, has a 30% higher photosynthetic rate than higher sucrose-accumulating *Saccharum* spp. hybrids suggests that source-sink relationships are crucial in shaping sucrose accumulation in stalks (Irvine 1975).

The increased photosynthetic capacity in sugarcane as a consequence of the increase in sink strength was demonstrated by McCormick et al. (2006). By shading all but the third fully expanded leaf in 9- to 12-month-old sugarcane, these authors promoted an overall increase in plant sink size, transforming what were originally source leaves into sinks. In unshaded leaves, there was a significant increase in photosynthetic rate, carboxylation efficiency and electron transport rate. In further studies, the same group observed that decreased photosynthetic rates are correlated with higher concentrations of leaf hexose and sucrose (McCormick et al. 2008a, b), which suggests that attempts to enhance sucrose accumulation in stalks should consider that sugarcane leaves are able to respond to increased demand, independent of leaf age (McCormick et al. 2008c).

Corroborating the idea that sink demand can play a major role not only in photosynthesis but also in leaf senescence control in sugarcane, Wu and Birch (2007) demonstrated that transgenic plants expressing a sucrose isomerase targeted to the vacuole of parenchyma cells exhibited accumulation of isomaltulose without a reduction in sucrose content. As a consequence of the apparent increased demand, transgenic sugarcane lines showed twice as much total sugar concentration in juice than control plants, and the enhanced total sugar accumulation was accompanied by elevated leaf photosynthetic rate, increased sugar transport to the stalk, and a delay in leaf senescence by 15–20 days.

Even if sucrose accumulation in sugarcane stalks is driven by a mechanism that associates photosynthesis and sink demand, there is other evidence indicating that differences in sucrose content can be the result of photoassimilate partitioning between the stalk and other organs in the plant (Inman-Bamber et al. 2008). A comparison made among two low-sucrose clones and two high-sucrose clones grown in a greenhouse showed that variation in sucrose accumulation is related more to the architecture of the plant than to the stalk sucrose content controlling the major sink. By limiting expansive growth and maintaining photosynthesis (no significant difference in net photosynthesis was observed between low and high clones), both low and high sucrose clones were able to increase sucrose accumulation, with low sucrose clones presenting some internodes with the same sucrose amount as high sucrose clones (Inman-Bamber et al. 2009). The authors observed that low sucrose clones produced more stalks and allocated more dry matter to leaves than high sucrose clones. The additional demand for structural carbohydrates in low sucrose clones could be one of the reasons for the lower sucrose accumulation in the stalk. A proposed model to determine the contribution of each trait revealed that stalk number, plant extension rate and photosynthesis were most responsible for the variation in sucrose accumulation rate and content between low and high sucrose clones (Inman-Bamber et al. 2009). It is expected that genetic modification of genes affecting aspects identified by above-mentioned studies will possibly increase sugar yield. Gene functional studies will allow not only a better understanding of sugar accumulation in sugarcane, but also make possible more productive cultivars.

4.2.2 Sugarcane Biotechnology

Transgenic sugarcane has been produced routinely in many laboratories since the early 1990s. Currently, many cultivars can be transformed, and several biolistic and *Agrobacterium*-mediated protocols of relatively high efficiency are available (Bower and Birch 1992; Manickavasagam et al. 2004; Arencibia and Carmona 2006). Engineered traits have included herbicide-, virus-, and insect-resistance, enhanced sugar content and drought tolerance, flowering control, and molecular farming of biopolymers and pharmacological proteins (reviewed by Lakshmanan et al. 2005; D'Hont et al. 2008). Collectively, to date, the most commonly and efficaciously targeted traits are herbicide resistance (Gallo-Meagher and Irvine 1996; Falco et al. 2000; Leibbrandt and Snyman 2003) and insect herbivory resistance via expression of *Bacillus thuringiensis* Cry proteins (Arencibia et al. 1997; Braga et al. 2001, 2003; Weng et al. 2006), proteinase inhibitors (Falco and Silva-Filho 2003; Christy et al. 2009) and lectins (Nutt et al. 1999). Herbicide and insect resistance most likely will be the first traits to be commercialized and bring to sugarcane the benefits observed in several other major crops.

Frequent occurrence of transgene silencing and lack of a suite of robust promoters for controlling transgene expression have hindered sugarcane biotechnology and analysis of gene function. These constraints usually require screening of a prohibitively high number of events for the identification of those with appropriate expression levels. Transgene silencing is probably exacerbated by biolistics, the most commonly used sugarcane transformation method, since it usually generates complex transgene integration sites. Corn Ubi-1 (Christensen and Quail 1996) has been the most frequently used and dependable nearly constitutive promoter in sugarcane, yet it often fails to sustain high expression levels throughout the sugarcane growth cycle (Wei et al. 2003; Wang et al. 2005). Reports indicate that promoters derived from banana streak virus (Schenk et al. 2001), sugarcane bacilliform virus (Braithwaite et al. 2004) and a tandem CaMV-35S:corn Ubi-1 promoter (Groenewald et al. 2000; Groenewald and Botha 2008) could provide relatively high levels of expression in mature plants. Several endogenous sugarcane promoters have been isolated aiming at constitutive, and tissue- or stage-specific expression, but these promoters frequently fail to drive transgene expression beyond callus or plantlet stage in sugarcane (Wei et al. 2003; Mudge et al. 2008). However, stable stalk-preferred transgene expression driven by endogenous promoters in sugarcane was reported recently (Damaj et al. 2010). Adoption of *Agrobacterium*-mediated methods for sugarcane transformation, in addition to the manipulation of silencing mechanisms combined with the identification of regulatory elements for gene expression, will likely be necessary to overcome current limitations.

Despite the biological and commercial importance of sucrose accumulation in sugarcane, few transgenic strategies targeting genes involved directly with carbohydrate metabolism that have successfully generated genotypes with higher sucrose content or yields have been reported. Nonetheless, as discussed in Sect. 4.2.1, recent studies indicate that sugarcane plants have enough photosynthetic capacity to support additional growth and sugar accumulation. These findings highlight the great potential of transgenic manipulation of key regulators of source-sink relationships and their underlying sensing mechanisms for enhancing sugarcane biomass and sucrose yields.

4.2.2.1 Increasing Sucrose and Modified Sugar Accumulation via Transgenic Approaches

A rare successful attempt to increase sugarcane sucrose content by altering the expression of an endogenous gene central to carbohydrate metabolism was targeted at pyrophosphate: fructose 6-phosphate 1-phosphotransferase (PFP)—an important glycolytic enzyme (Groenewald and Botha 2008). In sugarcane internodes, PFP activity is inversely correlated with sucrose content and directly related to respiration activity (Whittaker and Botha 1999). Accordingly, down-regulation of PFP activity, via antisense or co-suppression strategies, enhances sucrose accumulation in immature internodes of several transgenic lines (Groenewald and Botha 2008). In addition, fiber content is increased in PFP-down-regulated lines, possibly caused by the diversion of carbon units from glycolysis and respiration toward storage and structural carbohydrates.

Expression of a highly efficient, vacuole-targeted, sucrose isomerase (SI) from the enterobacteria *Pantoea dispersa* resulted in the accumulation of isomaltulose in addition to sustained levels of sucrose, nearly doubling the total amount of sugars stored in mature stalks (Wu and Birch 2007). SI-overexpressing lines with enhanced sugar accumulation also showed increased photosynthesis, sucrose transport and sink strength. Similarly, plants overexpressing a sucrose::sucrose 1-fructosyltransferase (1-SST) from the artichoke *Cynara scolymus* accumulated 1-kestose while sucrose levels were not modified, with consequent increased total sugar accumulation in stalks (Nicholson 2007). Lines accumulating high levels of 1-kestose did so apparently at the expense of insoluble carbohydrates. These results suggest that accumulation of modified sugars overrides sensing mechanisms that limit sucrose accumulation in sugarcane stalks, and demonstrate the great potential for increasing the accumulation of sugars via heterologous overexpression of a single, gain-of-function gene.

4.2.2.2 Bagasse Uses

Sugarcane bagasse, the fibrous residue generated after stalks are crushed to extract their juice, is a major source of fuel for power co-generation at sugar mills. Fiber and sucrose contents are inversely related traits inherited from *S. officinarum* and *S. spontaneum* progenitors, respectively (Jackson 2005). Bagasse contains lignocellulosic fiber originating from vascular, parenchyma, and rind cells and insoluble materials found in stalks. The amount of lignocellulosic materials in sugarcane is cultivar-dependent, but it makes up approximately two-thirds of all energy accumulated in sugarcane plants, half of which is in the bagasse and half in the straw (leaves and tops) that are usually left in the field or sometimes burned prior to manual harvesting. The harvested stalks are transported to the sugar mills, where the bagasse is produced after juice extraction. Currently, sugar mills burn part of the bagasse to supply energy for sugar and ethanol processing, which makes the mills energy independent.

Depending on the efficiency of the industrial process, surplus bagasse can become available for other uses. One such application for bagasse is its use to generate electricity that is sold as a secondary product. As an available lignocellulosic material at the sugar mills, bagasse is interesting for other applications, such as production of cellulosic ethanol, pyrolysis, briquettes or gasification. Compared with other biomass, bagasse is available at almost no cost to the industrial unit. However, compared with other biomasses, such as corn stover, bagasse conversion to ethanol is poorly studied, with limited reports focused on steam explosion pretreatment (Pandey et al. 2000; Sendelius 2005). As a comparison, a very broad evaluation was made to select the best pretreatment of corn stover for cellulosic ethanol production (Wyman et al. 2005). Similarly, hydrolysis and fermentation of bagasse have not been investigated sufficiently (Pandey et al. 2000). Despite the paucity of studies on bagasse utilization, biotechnology-based optimization of bagasse and sugarcane straw is economically promising, and will enable their use as a source of sugars for new applications such as cellulosic ethanol.

Sugarcane bagasse has a composition similar to other monocot biomass, ranging around 25% hemicelluloses, 41% cellulose and 23% lignin on a dry weight basis (Pessoa Jr et al. 1997; Pandey et al. 2000). Because of the high conservation of cell wall composition in grasses, the bagasse is expected to be very recalcitrant to enzymatic hydrolysis (Akin 2007). Sugarcane primary cell wall belongs to the type II category, which is characterized by cellulose fibers enclosed by glucuronoarabinoxylans and by high levels of ferulic acid (Vogel 2008). Cross-linking of ferulic acid, hemicelluloses and lignin decreases the access of hydrolytic enzymes. Plants with a decreased amount of cross-linked ferulic acid have a higher level of digestibility (Casler et al. 2008). Cell wall modification by up- or down-regulation of genes associated with cell wall deposition allows the understanding of fundamental aspects of cell wall biology and its optimization for enzymatic hydrolysis and cellulosic ethanol applications. This strategy has been demonstrated successfully by altering the expression of multiple genes functioning in lignin biosynthesis in wood tissue (Li et al. 2008; Weng et al. 2008) and also in alfalfa (Chen and Dixon 2007). Alfalfa biomass thus modified can be digested efficiently even without pretreatment, improving its utilization for cellulosic ethanol and animal feed (Chen and Dixon 2007). Several strategies for biomass optimization proposed for corn stover (Akin 2007) are possibly suitable for sugarcane and other grasses. In this regard, a number of genes related to cell wall metabolism has been already identified in the sugarcane transcriptome (Ramos et al. 2001; Lima et al. 2001; Casu et al. 2007).

The expression of cell wall hydrolytic enzymes in planta is another strategy for biotechnological improvement since it has the potential to reduce the enzyme needs by combining microbial and endogenous enzyme production. The feasibility of expressing high levels of cellulolytic enzymes in several tissues has been demonstrated in several plant species (Sticklen 2008). Studies identified an increase in the levels and simultaneous production of highly efficient enzymes with different specificities as subjects for further research. For example, the thermophilic E1 enzyme from *Acidothermus cellulolyticus*, with high activation energy, has been produced in rice at levels of almost 5% of total soluble protein (Oraby et al. 2007) and in corn seed at about 16% of total soluble protein (Hood et al. 2007), which is close to the levels needed for complete hydrolysis without addition of microbially produced endoglucanase. However, complete hydrolysis would require a synergistic combination of several cellulolytic enzymes. Besides the use of cellulolytic enzymes, cell wall hydrolysis could be enhanced by the expression of lignolytic enzymes such as phenolic acid esterases (Akin 2007). Further studies are required to identify the optimal combination of enzymes for bagasse hydrolysis, as well as the most appropriate compartment in stalk cells for enzyme storage and possible modifications required in the current industrial processes (Sainz 2009). Taking into account all these considerations, it is especially important to overcome the lack of a robust transgene expression technology in sugarcane, as discussed above.

4.2.2.3 New Applications for Fermentable Sugars

The emerging area of synthetic biology will allow production of new molecules from carbohydrate sources utilizing metabolically engineered bacteria or fungi. In the future, industries based on sugar- and starch-accumulating crops will increasingly allow the production of other valuable products, such as bioplastics and hydrocarbons, in addition to fuel ethanol (Ragauskas et al. 2006). As an effective and low-cost source of sucrose, sugarcane is also set to be one the first sources of these materials. This emerging application combines the large availability of carbohydrates derived from the sugarcane industry with the utilization of microbial fermentative or non-fermentative processes. One such example is the synthesis of higher and branched alcohols, such as butanol and isobutanol, from glucose by engineering a pathway utilizing 2-keto acid intermediates in *Escherichia coli* (Atsumi et al. 2008). Higher alcohols are promising alternatives to oil-based fuels. Compared to ethanol, higher alcohols have a higher energy density and lower hygroscopicity, but their large-scale adoption is currently precluded by their costly production. An even more promising strategy for obtaining next-generation biofuels from sugarcane is through engineering of the mevalonate pathway to produce isoprenoids in microorganisms. Efficient production of isoprenoid precursors has been already achieved using engineered *E. coli* hosts (Martin et al. 2003). This metabolic modification enables the production of branched-chain and cyclic alkanes, alkenes and alcohols of different sizes with diverse structural and chemical properties. These compounds can be used to produce gasoline precursors, additives or substitutes for diesel and jet fuel. In

Brazil, a commercial deployment of this technology, utilizing sugarcane-derived products, is in progress (Amyris Biotechnologies 2009).

4.2.3 Molecular Markers in Sugarcane Breeding

Most modern sugarcane cultivars derive from *S. officinarum* ($2n=8x=80$) by *S. spontaneum* ($2n=5x$ to $16x=40$ to 128) crosses (n and x are the haploid and monoploid numbers, respectively; reviewed by D'Hont et al. 2008). Two subsequent backcrosses to *S. officinarum* and selection were made for combined high sucrose and disease resistance characteristics from *S. officinarum* and *S. spontaneum*, respectively—a process known as sugarcane 'nobilisation'. Consequently, modern cultivars obtained by this process have inherited most of their chromosomes from *S. officinarum*, and a minority from *S. spontaneum* or resulting from recombination between progenitor chromosomes. Current sugarcane breeding programs rely mostly on crossing elite parental cultivars followed by selection of superior progeny—a process that takes approximately one decade until commercial release. Although breeding programs have continuously produced high yielding cultivars in the last decades (Jackson 2005; Matsuoka et al. 2009), exceedingly long selection processes, combined with high genomic complexity and small gene pool, could preclude further yield enhancement.

The development of molecular and statistical tools for the identification of quantitative trait loci (QTL) allow the utilization of markers linked to QTL for breeding purposes, or marker-assisted selection (MAS; Bernardo 2008). The highly polyploid, aneuploid and heterozygous nature of sugarcane imposes severe complications on genetic mapping, QTL identification and, consequently, MAS. High polyploidy and heterozygosity allow small individual contribution of multiple alleles present at a single hom(e)ologous locus to affect phenotypic variation. Also, in comparison to diploid species, determining the number and frequency of allele types at a single locus requires more sophisticated methods (Cordeiro et al. 2006). In addition, most traits of interest seem to be of quantitative nature, and QTL identified so far make only small contributions to phenotypic variation (Hoarau et al. 2002; McIntyre et al. 2005; Reffay et al. 2005; Aitken et al. 2006, 2008). Loci with dominant alleles present in single copies are more easily mapped in sugarcane. Accordingly, so far only three genes have been mapped (Daugrois et al. 1996; Raboin et al. 2006). Among them is *Bru1*, a major gene controlling the important rust-resistance trait, identified as a dominant, Mendelian gene in the progeny of cultivar R570 (Asnaghi et al. 2000, 2004). Despite the difficulties in mapping genes in sugarcane, it is possible to take advantage of comparative mapping with other Poaceae species for which high density maps and sequenced genomes are available, such as rice, corn and especially sorghum, whose genome shows the most colinearity with that of sugarcane (Grivet et al. 1994; Dufour et al. 1997; Glaszmann et al. 1997; Guimarães et al. 1997; Ming et al. 1998; Jannoo et al. 2007), to facilitate mapping in sugarcane. In this regard, mapping of *Bru1* with the help of sorghum-derived markers illustrates the importance of comparative genomics and the application of synteny for genetic mapping in species such as sugarcane bearing complex genomes. Numerous genetic maps have been generated for sugarcane, based on a variety of molecular marker types, among them restriction fragment length polymorphisms (RFLPs), amplified fragment length polymorphisms (AFLPs) and, more recently, simple sequence repeats (SSRs; reviewed by Cordeiro et al. 2007).

ESTs are an important resource for the generation of molecular markers such as SSRs and single nucleotide polymorphisms (SNPs; Cordeiro et al. 2001; Lima et al. 2002; Grivet et al. 2003; Rossi et al. 2003; Pinto et al. 2004, 2006; Aitken et al. 2005; Da Silva and Bressiani 2005; Garcia et al. 2006; Oliveira et al. 2007, 2009). EST collections have also allowed the development of tools for transcript profiling of stress and phytohormone responses, signal transduction, and sugar accumulation (Casu et al. 2003, 2004; Nogueira et al. 2003, 2005; Papini-Terzi et al. 2005, 2009; Calsa and Figueira 2007; Rocha et al. 2007; Schlögl et al. 2008). Results from these analyses indicated candidate genes controlling these various biological processes for further marker generation and functional analysis via transgenic approaches.

Large breeding programs will be likely based on MAS, since it can greatly accelerate breeding rates and high-throughput genotyping is becoming less expensive than phenotyping (Bernardo 2008). MAS is used routinely in breeding programs at Monsanto (Eathington et al. 2007; Edgerton 2009). Sugarcane breeding, which takes an exceedingly long time, could benefit greatly from MAS strategies. Initial efforts have been made at marker-assisted QTL introgression in sugarcane (Aitken et al. 2002). Due to the genetic complexity of sugarcane, deployment of MAS requires the discovery of a number of SNPs that is not met by the current availability of public sugarcane ESTs, which were produced largely by cDNA cloning and dideoxy sequencing. Recently, second-generation sequencing efforts for the identification of SNPs in sugarcane were reported (Bundock et al. 2009). Second generation sequencing, in conjunction with high-throughput technologies for SNP genotyping and array-based platforms (Wenzl et al. 2004; Syvänen 2005; Gupta et al. 2008), will accelerate the rate of marker discovery and validation. Large amounts of marker and phenotypic data need to be combined for the identification of marker trait associations (Bernardo 2008) that, ultimately, will allow the utilization of MAS to accelerate sugarcane breeding programs.

4.3 Other Sugar Crops Suitable for Ethanol Production

Other sugar-accumulating crops can expand ethanol production to other geographies where cultivation of sugarcane, currently the sugar crop used most widely for this purpose, is not viable. Sorghum and sugar beet are among the most promising of such crops. In addition to yielding large amounts of biomass and sugar, sorghum and sugar beet have the advantage of being more adapted than sugarcane to water-limited and temperate environments, respectively.

Sorghum (*Sorghum bicolor*) is a C4 grass of the Andropogoneae tribe, and a close relative of sugarcane. Sorghum originated from East Africa and, despite being a tropical species, can be cultivated in temperate climates (Gnansounou et al. 2005). Different types of sorghum are cultivated for grain, fiber or sugar accumulated in stalks (Woods 2001). Sweet sorghum has high photosynthetic efficiency, being able to assimilate large amounts of carbon (50 g m^{-2} day^{-1}) and to accumulate sucrose in stalk juice to around 17% on a weight basis (Prasad et al. 2007). It has a crop cycle shorter than that of sugarcane (3–5 months vs 12–18 months, respectively). Some varieties can yield as much as 100 t/ha above-ground biomass at the end of the crop cycle (Freelman et al. 1986; Woods 2001). However, the greatest differential of sweet sorghum in comparison to other energy crops is its water use efficiency. For the production of one unit of above-ground biomass, sweet sorghum requires one-third of the amount of water when compared to sugarcane, and half of the water required by corn (Woods 2001; Saballos 2008). Collectively, these characteristics indicate that sweet sorghum has a great potential to be used as a biofuel crop, especially in water-limited regions.

The production of transgenic sorghum is not easily attained, due to difficulties in regeneration and transformation processes (Howe et al. 2006). Transgenic plants are generated mainly by *Agrobacterium*-mediated or biolistic transformation using immature embryos as explant sources. The successful use of explants such as immature inflorescence and shoot meristem has also been demonstrated (reviewed by Godwin 2004; Jogeswar et al. 2007). There are reports of transgenic grain sorghum transformed with reporter genes, and herbicide-, insect- and disease-resistance genes (Zhu et al. 1998; Gao et al. 2005; Girijashankar et al. 2005), but it was only recently that the generation of the first transgenic sweet sorghum was announced by an Australian group (University of Queensland 2009).

The recent release of the complete sorghum genome sequence (Paterson et al. 2009), associated with already existing high-density genetic maps (reviewed by Saballos 2008), will also contribute to conventional and molecular breeding programs aimed at improving sweet sorghum as a bioenergy crop. In addition, the use of natural and induced mutant populations is also an important tool in chasing useful traits in sorghum, as demonstrated by the *brown midrib* (*bmr*) family of cell wall mutants (Porter et al. 1978). The recent identification of some *bmr* genes (Bout and Vermerris 2003; Saballos et al. 2009; Sattler et al. 2009) could allow their use to generate sweet sorghum lines with increased cell wall digestibility. This strategy could enable not only the production of ethanol, utilizing the juice extracted from the stalk, but also the lignocellulosic material from sweet sorghum bagasse.

Sugar beet (*Beta vulgaris*) belongs to Amaranthaceae family and is the second largest source of raw sucrose worldwide, accounting for 227 million t or approximately 14% of the amount produced from sugarcane in 2008/2009 (FAPRI 2009). Sugar beet is a crop adapted mainly to temperate climates and is cultivated mostly in Europe, the US and Russia (FAOSTAT 2007). The sugar beet tap root is the main storage organ from which sucrose is extracted. This species is grown from seeds and has a biannual cycle but, for optimal sucrose yields, it is usually harvested around 200 days after seeds are sown. Sugar beet has been bred for accumulation of sucrose early in its growth cycle since the 1700s by Franz Carl Achard, considered the founder of the sugar beet industry (Francis 2006). High sucrose accumulating varieties have been selected, yielding, on average, 49.7 t roots/ha and 7.8 t sucrose/ha (USDA 2006). A major difference between sugar beet and sugarcane with respect to raw sugar industrial processing is related to the energetic balance at the mills. While in the sugarcane industry, a byproduct, bagasse, is available to supply energy for the process, in the sugar beet-based industry no by-product is available as fuel, being necessary an energetic input, usually from a fossil source. Sugar beet pulp and molasses generated after sugar extraction are byproducts of this industry, and are used for animal feed. Ethanol production from sugar beet has a positive energy balance of 2.1 considering the economical environment in Europe (Goldemberg 2007; Licht 2006). However, the development of a sugar beet-based ethanol industry is just initiating, affected mostly by the implementation of the Common Market Organization reforms of the sugar market in Europe (FAPRI 2009; Licht 2006).

Several methods of genetic transformation have been tested in sugar beet, e.g., *Agrobacterium*-mediated transformation of shoot-base tissues, cotyledonary-node explants, shoot explants or embryogenic callus, polyethylene glycol-mediated transformation of guard-cell protoplasts (Ivic-Haymes and Smigocki 2005) and also plastid transformation (De Marchis et al. 2009). However, sugar beet is considered a species recalcitrant to genetic transformation, because of low reproducibility and genotype dependence. Despite these difficulties, sugar beet has been genetically transformed with the aim of expressing resistance to the herbicides glufosinate and glyphosate (De Marchis et al. 2009). Also, sugar beet has been modified genetically for the production of different types of fructans via expression of onion fructosyltransferases (Weyens et al. 2004) and 1-SST from *Helianthus tuberosus* (Sévenier et al. 1998), without consequent modification of total storage carbohydrate content and basic physiological processes in the plant. Although these reports indicate that 1-SST overexpression in sugar beet does not increase total sugar accumulation as observed in sugarcane (Nicholson 2007), they highlight the potential of genetic modification strategies for altering sugar metabolism toward the utilization of sugar beet for ethanol production.

4.4 Perspectives

Similar to other major crops, such as corn, soybeans and cotton, incorporation of MAS in breeding programs, in association with genetically engineered traits, will allow the enhancement of sucrose and biomass yields in sugarcane. The first generation of genetically modified sugarcane cultivars, targeting qualitative traits such as herbicide and insect resistance, is expected to reach commercialization in the short term. Modification of the chemical composition of bagasse is anticipated to make feasible its utilization for cellulosic ethanol production. Recent studies have shown that sugarcane photosynthetic potential has not been utilized fully by conventional breeding. Successful marker-assisted breeding and genetic modification approaches aimed at further enhancement of sugarcane sucrose and biomass yields are currently challenged by few existing marker trait associations and the lack of a robust toolbox for controlling gene expression, respectively. Ongoing efforts at dissecting sugarcane physiology, molecular biology, genetics and genomics are paving the way for the successful utilization of such approaches in this complex crop. Similarly, biotechnology approaches could benefit other sucrose-accumulating crops, such as sweet sorghum and sugar beet, toward their utilization for biofuel production.

Acknowledgments We wish to thank several colleagues at Monsanto who made helpful contributions to the content or editing of this review.

References

Aitken K, Jackson P, McIntyre C (2005) A combination of AFLP and SSR markers provides extensive map coverage and identification of homo(eo)logous linkage groups in a sugarcane. Theor Appl Genet 110:789–801

Aitken KS, Jackson PA, McIntyre CL, Piperidis G (2002) Marker assisted introgressing of high sucrose genes in sugarcane. In: Proceedings of the 12th Australasian Plant Breeding Conference, Perth, Australia, 15–20 September 2002. p 120

Aitken KS, Jackson PA, McIntyre CL (2006) Quantitative trait loci identified for sugar related traits in a sugarcane (*Saccharum* spp.) cultivar × *Saccharum officinarum* population. Theor Appl Genet 112:1306–1317

Aitken KS, Hermann S, Karno K, Bonnett GD, McIntyre LC, Jackson PA (2008) Genetic control of yield related stalk traits in sugarcane. Theor Appl Genet 117:1191–1203

Akin D (2007) Grass lignocellulose. Appl Biochem Biotechnol 137–140:3–15

Amyris Biotechnologies (2009) Amyris Brasil, www.amyris.com/index.php?option=com_content&task=view&id=69&Itemid=257. Cited 28 September 2009

Arencibia AD, Carmona ER (2006) Sugarcane (*Saccharum* spp.). Methods Mol Biol 344:227–235

Arencibia A, Vázquez RI, Prieto D, Téllez P, Carmona ER, Coego A, Hernández L, Riva GADl, Selman-Housein G (1997) Transgenic sugarcane plants resistant to stem borer attack. Mol Breeding 3:247–255

Asnaghi C, Paulet F, Kaye C, Grivet L, Deu M, Glaszmann JC, D'Hont A (2000) Application of synteny across Poaceae to determine the map location of a sugarcane rust resistance gene. Theor Appl Genet 101:962–969

Asnaghi C, Roques D, Ruffel S, Kaye C, Hoarau J-Y, Télismart H, Girard JC, Raboin LM, Risterucci AM, Grivet L, D'Hont A (2004) Targeted mapping of a sugarcane rust resistance gene (*Bru1*) using bulked segregant analysis and AFLP markers. Theor Appl Genet 108:759–764

Atsumi S, Hanai T, Liao JC (2008) Non-fermentative pathways for synthesis of branched-chain higher alcohols as biofuels. Nature 451:86–89

Bernardo R (2008) Molecular markers and selection for complex traits in plants: learning from the last 20 years. Crop Sci 48:1649–1664

Bout S, Vermerris W (2003) A candidate-gene approach to clone the sorghum *Brown midrib* gene encoding caffeic acid *O*-methyltransferase. Mol Genet Genomics 269:205–214

Bower R, Birch RG (1992) Transgenic sugarcane plants via microprojectile bombardment. Plant J 2:409–416

Braga DPV, Arrigoni EDB, Burnquist WL, Silva-Filho MC, Ulian EC (2001) A new approach for control of *Diatraea saccharalis* (Lepidoptera: Crambidae) through the expression of an insecticidal CryIa(b) protein in transgenic sugarcane. Proc Int Soc Sugar Cane Technol 24:331–336

Braga DPV, Arrigoni EDB, Silva-Filho MC, Ulian EC (2003) Expression of the Cry1Ab protein in genetically modified sugarcane for the control of *Diatraea saccharalis* (Lepidoptera: Crambidae). J New Seeds 5:209–221

Braithwaite KS, Geijskes RJ, Smith GR (2004) A variable region of the sugarcane bacilliform virus (SCBV) genome can be used to generate promoters for transgene expression in sugarcane. Plant Cell Rep 23:319–326

Bundock PC, Eliott FG, Ablett G, Benson AD, Casu RE, Aitken KS, Henry RJ (2009) Targeted single nucleotide polymorphism (SNP) discovery in a highly polyploid plant species using 454 sequencing. Plant Biotechnol J 7:347–354

Calsa T, Figueira A (2007) Serial analysis of gene expression in sugarcane (*Saccharum* spp.) leaves revealed alternative C4 metabolism and putative antisense transcripts. Plant Mol Biol 63:745–762

Casler MD, Jung HG, Coblentz WK (2008) Clonal selection for lignin and etherified ferulates in three perennial grasses. Crop Sci 48:424–433

Casu RE, Grof CPL, Rae AL, McIntyre CL, Dimmock CM, Manners JM (2003) Identification of a novel sugar transporter homologue strongly expressed in maturing stem vascular tissues of sugarcane by expressed sequence tag and microarray analysis. Plant Mol Biol 52:371–386

Casu RE, Dimmock CM, Chapman SC, Grof CP, McIntyre CL, Bonnett GD, Manners JM (2004) Identification of differentially expressed transcripts from maturing stem of sugarcane by in silico analysis of stem expressed sequence tags and gene expression profiling. Plant Mol Biol 54:503–517

Casu RE, Jarmey JM, Bonnett GD, Manners JM (2007) Identification of transcripts associated with cell wall metabolism and development in the stem of sugarcane by Affymetrix GeneChip Sugarcane Genome Array expression profiling. Funct Integr Genomics 7:153–167

Chen F, Dixon RA (2007) Lignin modification improves fermentable sugar yields for biofuel production. Nat Biotechnol 25:759–761

Christensen A, Quail P (1996) Ubiquitin promoter-based vectors for high-level expression of selectable and/or screenable marker genes in monocotyledonous plants. Transgenic Res 5:213–218

Christy L, Arvinth S, Saravanakumar M, Kanchana M, Mukunthan N, Srikanth J, Thomas G, Subramonian N (2009) Engineering sugarcane cultivars with bovine pancreatic trypsin inhibitor (aprotinin) gene for protection against top borer (*Scirpophaga excerptalis* Walker). Plant Cell Rep 28:175–184

Cordeiro G, Casu R, McIntyre C, Manners J, Henry R (2001) Microsatellite markers from sugarcane (*Saccharum* spp.) ESTs cross transferable to erianthus and sorghum. Plant Sci 160:1115–1123

Cordeiro G, Eliott F, McIntyre C, Casu R, Henry R (2006) Characterisation of single nucleotide polymorphisms in sugarcane ESTs. Theor Appl Genet 113:331–343

Cordeiro G, Amouyal O, Eliott F, Henry R (2007) Sugarcane. In: Kole C (ed) Pulses, sugar and tuber crops, vol 3. Springer, Heidelberg, pp 175–203

Coyle W (2007) The future of biofuels: a global perspective. http://www.ers.usda.gov/AmberWaves/November07/Features/Biofuels.htm Cited 19 September 2009

Damaj MB, Kumpatla SP, Emani C, Beremand PD, Reddy AS, Rathore KS, Buenrostro-Nava MT, Curtis IS, Thomas TL, Mirkov TE (2010) Sugarcane *DIRIGENT* and *O-METHYLTRANSFERASE* promoters confer stem-regulated gene expression in diverse monocots. Planta, doi: 10.1007/s00425-010-1138-5

Da Silva JA, Bressiani JA (2005) Sucrose synthase molecular marker associated with sugar content in elite sugarcane progeny. Genet Mol Biol 28:294–298

Daugrois JH, Grivet L, Roques D, Hoarau JY, Lombard H, Glaszmann JC, D'Hont A (1996) A putative major gene for rust resistance linked with a RFLP marker in sugarcane cultivar 'R570'. Theor Appl Genet 92:1059–1064

De Marchis F, Wang Y, Stevanato P, Arcioni S, Bellucci M (2009) Genetic transformation of the sugar beet plastome. Transgenic Res 18:17–30

De Souza AP, Gaspar M, da Silva EA, Ulian EC, Waclawovsky AJ, Nishiyama MY Jr, dos Santos R, Teixeira MM, Souza GM, Buckeridge MS (2008) Elevated CO_2 increases photosynthesis, biomass and productivity, and modifies gene expression in sugarcane. Plant Cell Environ 31:1116–1127

D'Hont A, Souza GM, Menossi M, Vincentz M, Van-Sluys M-A, Glaszmann JC, Ulian E (2008) Sugarcane: a major source of sweetness, alcohol, and bio-energy. In: Moore PH, Ming R (eds) Genomics of tropical crop plants, vol 1. Springer, New York, pp 483–513

Dufour P, Deu M, Grivet L, D'Hont A, Paulet F, Bouet A, Lanaud C, Glaszmann JC, Hamon P (1997) Construction of a composite sorghum genome map and comparison with sugarcane, a related complex polyploid. Theor Appl Genet 94:409–418

Eathington SR, Crosbie TM, Edwards MD, Reiter RS, Bull JK (2007) Molecular markers in a commercial breeding program. Crop Sci 47:S154–S163

Edgerton MD (2009) Increasing crop productivity to meet global needs for feed, food, and fuel. Plant Physiol 149:7–13

Falco M, Silva-Filho M (2003) Expression of soybean proteinase inhibitors intransgenic sugarcane plants: effects on natural defense against *Diatrea saccharalis*. Plant Physiol Biochem 41:761–766

Falco MC, Tulmann Neto A, Ulian EC (2000) Transformation and expression of a gene for herbicide resistance in a Brazilian sugarcane. Plant Cell Rep 19:1188–1194

FAOSTAT (2007) FAO (United Nations Food and Agricultural Organization), Rome. http://faostat.fao.org/default.aspx Cited 17 September 2009

Food and Agricultural Policy Research Institute (2009) FAPRI 2009 U.S. and World Agricultural Outlook. http://www.fapri.iastate.edu/outlook/2009/ Cited 18 September 2009

Farrell AE, Plevin RJ, Turner BT, Jones AD, O'Hare M, Kammen DM (2006) Ethanol can contribute to energy and environmental goals. Science 311:506–508

Francis SA (2006) The development of sugarbeet. In: Draycott AP (ed) Sugar beet. Wiley-Blackwell, Oxford, pp 9–29

Freelman KC, Braodhead DM, Zummo N, Westbrook FE (1986) Sweet sorghum culture and syrup production. USDA Agriculture Handbook, Number 611, United States Department of Agriculture, Washington, DC

Gallo-Meagher M, Irvine JE (1996) Herbicide resistant transgenic sugarcane plants containing the *bar* gene. Crop Sci 36:1367–1374

Gao ZS, Jayaraj J, Muthukrishnan S, Claflin L, Liang GH (2005) Efficient genetic transformation of sorghum using a visual screening marker. Genome 48:321–333

Garcia AAF, Kido EA, Meza AN, Souza HMB, Pinto LR, Pastina MM, Leite CS, Silva JAG, Ulian EC, Figueira A, Souza AP (2006) Development of an integrated genetic map of a sugarcane (*Saccharum* spp.) commercial cross, based on a maximum-likelihood approach for estimation of linkage and linkage phases. Theor Appl Genet 112:298–314

Girijashankar V, Sharma HC, Sharma KK, Swathisree V, Prasad LS, Bhat BV, Royer M, San Secundo B, Narasu ML, Altosaar I, Seetharama N (2005) Development of transgenic sorghum for insect resistance against the Spotted Stem Borer (*Chilo partellus*). Plant Cell Rep 24:513–522

Glaszmann JC, Dufour P, Grivet L, D'Hont A, Deu M, Paulet F, Hamon P (1997) Comparative genome analysis between several tropical grasses. Euphytica 96:13–21

Gnansounou E, Dauriat A, Wyman CE (2005) Refining sweet sorghum to ethanol and sugar: economic trade-offs in the context of North China. Bioresour Technol 96:985–1002

Godwin ID (2004) Sorghum genetic engineering: current status and prospectus. In: Seetharama N, Godwin I (eds) Sorghum tissue culture and transformation. Oxford & IBH, New Delhi, pp 1–8

Goldemberg J (2007) Ethanol for a sustainable energy future. Science 315:808–810

Goldemberg J, Guardabassi P (2009) Are biofuels a feasible option? Energy Policy 37:10–14

Grivet L, D'Hont A, Dufour P, Hamon P, Roques D, Glaszmann JC (1994) Comparative genome mapping of sugar cane with other species within the Andropogoneae tribe. Heredity 73: 500–508

Grivet L, Glaszmann JC, Vincentz M, da Silva F, Arruda P (2003) ESTs as a source for sequence polymorphism discovery in sugarcane: example of the *Adh* genes. Theor Appl Genet 106:190–197

Groenewald J-H, Botha FC (2008) Down-regulation of pyrophosphate: fructose 6-phosphate 1-phosphotransferase (PFP) activity in sugarcane enhances sucrose accumulation in immature internodes. Transgenic Res 17:85–92

Groenewald J-H, Hiten NF, Botha FC (2000) The introduction of an inverted repeat to the 5' untranslated leader sequence of a transgene strongly inhibits gene expression. Plant Cell Rep 19:1098–1101

Guimarães CT, Sills GR, Sobral BWS (1997) Comparative mapping of Andropogoneae: *Saccharum* L. (sugarcane) and its relation to sorghum and maize. Proc Natl Acad Sci USA 94:14261–14266

Gupta P, Rustgi S, Mir R (2008) Array-based high-throughput DNA markers for crop improvement. Heredity 101:5–18

Hoarau J-Y, Grivet L, Offmann B, Raboin L-M, Diorflar J-P, Payet J, Hellmann M, D'Hont A, Glaszmann JC (2002) Genetic dissection of a modern sugarcane cultivar (*Saccharum* spp.). II. Detection of QTLs for yield components. Theor Appl Genet 105:1027–1037

Hood EE, Love R, Lane J, Bray J, Clough R, Pappu K, Drees C, Hood KR, Yoon S, Ahmad A, Howard JA (2007) Subcellular targeting is a key condition for high-level accumulation of cellulase protein in transgenic maize seed. Plant Biotechnol J 5:709–719

Howe A, Sato S, Dweikat I, Fromm M, Clemente T (2006) Rapid and reproducible *Agrobacterium*-mediated transformation of sorghum. Plant Cell Rep 25:784–791

Inman-Bamber NG, Bonnett GD, Spillman MF, Hewitt ML, Jackson J (2008) Increasing sucrose accumulation in sugarcane by manipulating leaf extension and photosynthesis with irrigation. Aust J Agric Res 59:13–26

Inman-Bamber NG, Bonnett GD, Spillman MF, Hewitt ML, Xu J (2009) Source-sink differences in genotypes and water regimes influencing sucrose accumulation in sugarcane stalks. Crop Pasture Sci 60:316–327

Irvine J (1975) Relations of photosynthetic rates and leaf canopy characters to sugarcane yield. Crop Sci 15:671–676

Ivic-Haymes SD, Smigocki AC (2005) Biolistic transformation of highly regenerative sugar beet (*Beta vulgaris* L.) leaves. Plant Cell Rep 23:699–704

Jackson PA (2005) Breeding for improved sugar content in sugarcane. Field Crops Res 92:277–290

Jannoo N, Grivet L, Chantret N, Garsmeur O, Glaszmann JC, Arruda P, D'Hont A (2007) Orthologous comparison in a gene-rich region among grasses reveals stability in the sugarcane polyploid genome. Plant J 50:574–585

Jogeswar G, Ranadheer D, Anjaiah V, Kavi Kishor PB (2007) High frequency somatic embryogenesis and regeneration in different genotypes of *Sorghum bicolor* (L.) Moench from immature inflorescence explants. In Vitro Cell Dev Biol Plant 43:159–166

Lakshmanan P, Geijskes R, Aitken K, Grof C, Bonnett G, Smith G (2005) Sugarcane biotechnology: the challenges and opportunities. In Vitro Cell Dev Biol Plant 41:345–363

Leibbrandt NB, Snyman SJ (2003) Stability of gene expression and agronomic performance of a transgenic herbicide-resistant sugarcane line in South Africa. Crop Sci 43:671–677

Li X, Weng J-K, Chapple C (2008) Improvement of biomass through lignin modification. Plant J 54:569–581

Licht FO (2006) World ethanol markets—the outlook to 2015. Agra Informa, Tunbridge Wells, UK

Lima DU, Santos HP, Tiné MA, Molle FRD, Buckeridge MS (2001) Patterns of expression of cell wall related genes in sugarcane. Genet Mol Biol 24:191–198

Lima MLA, Garcia AAF, Oliveira KM, Matsuoka S, Arizono H, de Souza CL Jr, de Souza AP (2002) Analysis of genetic similarity detected by AFLP and coefficient of parentage among genotypes of sugar cane (*Saccharum* spp.). Theor Appl Genet 104:30–38

Macedo IC, Cortez LAB (2000) Sugar-cane industrial processing in Brazil. In: Rosillo-Calle F, Bajay SV, Rothman H (eds) Industrial uses of biomass energy. Taylor-Francis, London, pp 140–154

Manickavasagam M, Ganapathi A, Anbazhagan VR, Sudhakar B, Selvaraj N, Vasudevan A, Kasthurirengan S (2004) *Agrobacterium*-mediated genetic transformation and development of herbicide-resistant sugarcane (*Saccharum* species hybrids) using axillary buds. Plant Cell Rep 23:134–143

Martin VJ, Pitera DJ, Withers ST, Newman JD, Keasling JD (2003) Engineering a mevalonate pathway in *Escherichia coli* for production of terpenoids. Nat Biotechnol 21:796–802

Matsuoka S, Ferro J, Arruda P (2009) The Brazilian experience of sugarcane ethanol industry. In Vitro Cell Dev Biol Plant 45:372–381

McCormick AJ, Cramer MD, Watt DA (2006) Sink strength regulates photosynthesis in sugarcane. New Phytol 171:759–770

McCormick AJ, Watt DA, Cramer MD (2008a) Changes in photosynthetic rates and gene expression of leaves during a source sink perturbation in sugarcane. Ann Bot 101:89–102

McCormick AJ, Cramer MD, Watt DA (2008b) Regulation of photosynthesis by sugars in sugarcane leaves. J Plant Physiol 165:1817–1829

McCormick AJ, Cramer MD, Watt DA (2008c) Culm sucrose accumulation promoter physiological decline of mature leaves in ripening sugarcane. Field Crops Res 108:250–258

McIntyre CL, Whan VA, Croft B, Magarey R, Smith GR (2005) Identification and validation of molecular markers associated with Pachymetra Root Rot and Brown Rust resistance in sugarcane using map- and association-based approaches. Mol Breeding 16:151–161

Ming R, Liu SC, Lin YR, da Silva J, Wilson W, Braga D, van Deynze A, Wenslaff TF, Wu KK, Moore PH, Burnquist W, Sorrells ME, Irvine JE, Paterson AH (1998) Detailed alignment of *Sorghum* and *Saccharum* chromosomes: comparative organization of closely related diploid and polyploid genomes. Genetics 150:1663–1882

Moore PH (1995) Temporal and spatial regulation of sucrose accumulation in the sugarcane stem. Aust J Plant Physiol 22:661–679

Mudge SR, Osabe K, Casu RE, Bonnett GD, Manners JM, Birch RG (2008) Efficient silencing of reporter transgenes coupled to known functional promoters in sugarcane, a highly polyploid crop species. Planta 229:549–558

Nass LL, Pereira PAA, Ellis D (2007) Biofuels in Brazil: an overview. Crop Sci 47:2228–2237

Nicholson TL (2007) Carbon turnover and sucrose metabolism in the culm of transgenic sugarcane producing 1-kestose. MSc Thesis. University of Stellenbosch, Matieland, South Africa

Nogueira FTS, De Rosa VE Jr, Menossi M, Ulian EC, Arruda P (2003) RNA expression profiles and data mining of sugarcane response to low temperature. Plant Physiol 132:1811–1824

Nogueira FTS, Schlögl PS, Camargo SR, Fernandez JH, Vicente E, De Rosa J, Pompermayer P, Arruda P (2005) SsNAC23, a member of the NAC domain protein family, is associated with cold, herbivory and water stress in sugarcane. Plant Sci 169:93–106

Nutt KA, Allsopp PG, McGhie TK, Shepherd KM, Joyce PA, Taylor GO, McQualter RB, Smith GR (1999) Transgenic sugarcane with increased resistance to canegrubs. Proc Aust Soc Sugar Cane Technol 21:171–176

Oliveira KM, Pinto LR, Marconi TG, Margarido GRA, Pastina MM, Teixeira LHM, Figueira AV, Ulian EC, Garcia AAF, Souza AP (2007) Functional integrated genetic linkage map based on EST-markers for a sugarcane (*Saccharum* spp.) commercial cross. Mol Breeding 20:189–208

Oliveira KM, Pinto LR, Marconi TG, Mollinari M, Ulian EC, Chabregas SM, Falco MC, Burnquist W, Garcia AAF, Souza AP (2009) Characterization of new polymorphic functional markers for sugarcane. Genome 52:191–209

Oraby H, Venkatesh B, Dale B, Ahmad R, Ransom C, Oehmke J, Sticklen M (2007) Enhanced conversion of plant biomass into glucose using transgenic rice-produced endoglucanase for cellulosic ethanol. Transgenic Res 16:739–749

Pandey A, Soccol CR, Nigam P, Soccol VT (2000) Biotechnological potential of agro-industrial residues. I: sugarcane bagasse. Bioresour Technol 74:69–80

Papini-Terzi F, Rocha F, Vencio R, Felix J, Branco D, Waclawovsky A, Del Bem L, Lembke C, Costa M, Nishiyama M, Vicentini R, Vincentz M, Ulian E, Menossi M, Souza G (2009) Sugarcane genes associated with sucrose content. BMC Genomics 10:120

Papini-Terzi FS, Rocha FR, Nicoliello Vencio RZ, Oliveira KC, de Maria Felix J, Vicentini R, de Souza Rocha C, Quirino Simões AC, Ulian EC, Marli Zingaretti di Mauro S, Maria Da Silva A, Alberto de Braganca Pereira C, Menossi M, Souza GM (2005) Transcription profiling of signal transduction-related genes in sugarcane tissues. DNA Res 12:27–38

Paterson AH, Bowers JE, Bruggmann R, Dubchak I, Grimwood J, Gundlach H, Haberer G, Hellsten U, Mitros T, Poliakov A, Schmutz J, Spannagl M, Tang H, Wang X, Wicker T, Bharti AK, Chapman J, Feltus FA, Gowik U, Grigoriev IV, Lyons E, Maher CA, Martis M, Narechania A, Otillar RP, Penning BW, Salamov AA, Wang Y, Zhang L, Carpita NC, Freeling M, Gingle AR, Hash CT, Keller B, Klein P, Kresovich S, McCann MC, Ming R, Peterson DG, Mehboob-ur-Rahman, Ware D, Westhoff P, Mayer KF, Messing J, Rokhsar DS (2009) The *Sorghum bicolor* genome and the diversification of grasses. Nature 457:551–556

Paul MJ, Foyer CH (2001) Sink regulation of photosynthesis. J Exp Bot 52:1383–1400

Pessoa A Jr, Mancilha IM, Sato S (1997) Acid hydrolysis of hemicellulose from sugarcane bagasse. Braz J Chem Eng 14:309–312

Pinto LR, Oliveira KM, Ulian EC, Garcia AAF, de Souza AP (2004) Survey in the sugarcane expressed sequence tag database (SUCEST) for simple sequence repeats. Genome 47:795–804

Pinto LR, Oliveira KM, Marconi T, Garcia AAF, Ulian EC, Souza APd (2006) Characterization of novel sugarcane expressed sequence tag microsatellites and their comparison with genomic SSRs. Plant Breed 125:378–384

Porter KS, Axtell J D, Lechtenberg VL, Colenbrander V F (1978) Phenotype, fiber composition, and in vitro dry matter disappearance of chemically induced *brown midrib* (*bmr*) mutants of sorghum. Crop Sci 18:205–208

Prasad P, Vu J, Boote K, Allen L (2009) Enhancement in leaf photosynthesis and upregulation of Rubisco in the C4 sorghum plant at elevated growth carbon dioxide and temperature occur at early stages of leaf ontogeny. Funct Plant Biol 36:761–769

Prasad S, Singh, Anoop, Jain N, Joshi H C (2007) Ethanol production from sweet sorghum syrup for utilization as automotive fuel in India. Energy Fuel 21:2415–2420

Raboin L, Oliveira K, Lecunff L, Telismart H, Roques D, Butterfield M, Hoarau J, D'Hont A (2006) Genetic mapping in sugarcane, a high polyploid, using bi-parental progeny: identification of a gene controlling stalk colour and a new rust resistance gene. Theor Appl Genet 112:1382–1391

Ragauskas AJ, Williams CK, Davison BH, Britovsek G, Cairney J, Eckert CA, Frederick WJ Jr, Hallett JP, Leak DJ, Liotta CL, Mielenz JR, Murphy R, Templer R, Tschaplinski T (2006) The path forward for biofuels and biomaterials. Science 311:484–489

Ramos RLB, Tovar FJ, Junqueira RM, Lino FB, Sachetto-Martins G (2001) Sugarcane expressed sequences tags (ESTs) encoding enzymes involved in lignin biosynthesis pathways. Genet Mol Biol 24:235–241

Reffay N, Jackson PA, Aitken KS, Hoarau J-Y, D'Hont A, Besse P, McIntyre CL (2005) Characterisation of genome regions incorporated from an important wild relative into Australian sugarcane. Mol Breeding 15:367–381

Rocha F, Papini-Terzi F, Nishiyama M, Vencio R, Vicentini R, Duarte R, de Rosa V, Vinagre F, Barsalobres C, Medeiros A, Rodrigues F, Ulian E, Zingaretti S, Galbiatti R, Almeida R, Figueira A, Hemerly A, Silva-Filho M, Menossi M, Souza G (2007) Signal transduction-related responses to phytohormones and environmental challenges in sugarcane. BMC Genomics 8:71

Rossi M, Araujo PG, Paulet F, Garsmeur O, Dias VM, Chen H, Van Sluys M-A, D'Hont A (2003) Genomic distribution and characterization of EST-derived resistance gene analogs (RGAs) in sugarcane. Mol Genet Genomics 269:406–419

Saballos A (2008) Development and utilization of sorghum as a bioenergy crop. In: Vermerris W (ed) Genetic improvement of bioenergy crops. Springer, New York, pp 211–248

Saballos A, Ejeta G, Sanchez E, Kang C, Vermerris W (2009) A genome-wide analysis of the cinnamyl alcohol dehydrogenase family in sorghum (*Sorghum bicolor* (L.) Moench) identifies SbCAD2 as the *Brown midrib6* gene. Genetics 181:783–795

Sainz M (2009) Commercial cellulosic ethanol: the role of plant-expressed enzymes. In Vitro Cell Dev Biol Plant 45:314–329

Sattler SE, Saathoff AJ, Haas EJ, Palmer NA, Funnell-Harris DL, Sarath G, Pedersen JF (2009) A nonsense mutation in a cinnamyl alcohol dehydrogenase gene is responsible for the sorghum *brown midrib6* phenotype. Plant Physiol 150:584–595

Schenk PM, Remans T, Sági L, Elliott AR, Dietzgen RG, Swennen R, Ebert PR, Grof CPL, Manners JM (2001) Promoters for pregenomic RNA of banana streak badnavirus are active for transgene expression in monocot and dicot plants. Plant Mol Biol 47:399–412

Schlögl PS, Nogueira FTS, Drummond R, Felix JM, de Rosa VE Jr, Vicentini R, Leite A, Ulian EC, Menossi M (2008) Identification of new ABA- and MEJA-activated sugarcane bZIP genes by data mining in the SUCEST database. Plant Cell Rep 27:335–345

Sendelius J (2005) Steam pretreatment optimisation for sugarcane bagasse in bioethanol production. MSc Thesis. Lund University, Lund, Sweden

Sévenier R, Hall RD, van der Meer I, Hakkert HJ, van Tunen AJ, Koops AJ (1998) High level fructan accumulation in a transgenic sugar beet. Nat Biotechnol 16:843–846

Songstad DD, Lakshmanan P, Chen J, Gibbons W, Hughes S, Nelson R (2009) Historical perspective of biofuels: learning from the past to rediscover the future. In Vitro Cell Dev Biol Plant 45:189–192

Sticklen MB (2008) Plant genetic engineering for biofuel production: towards affordable cellulosic ethanol. Nat Rev Genet 9:433–443

Sugarcane Genome Sequencing Initiative (2009) In: Proceedings of the Plant and Animal Genome XVIII Conference, San Diego, CA

Syvänen A-C (2005) Toward genome-wide SNP genotyping. Nat Genet 27:S5–S10

UNICA (2009) União das Indústrias de Cana-de-açúcar. Dados e Cotações—Estatísticas. http://www.unica.com.br/dadosCotacao/estatistica/ Cited 18 September 2009

University of Queensland (2009) UQ researchers produce world's first transgenic sweet sorghum. http://www.uq.edu.au/news/index.html?article=20025 Cited 13 November 2009

USDA (2006) The economic feasibility of ethanol production from sugar in the United States. US Department of Agriculture. http://www.usda.gov/oce/reports/energy/EthanolSugarFeasibilityReport3.pdf Cited 18 October 2009

Vettore AL, da Silva FR, Kemper EL, Arruda P (2001) The libraries that made SUCEST. Genet Mol Biol 24:1–7

Vettore AL, da Silva FR, Kemper EL, Souza GM, da Silva AM, Ferro MIT, Henrique-Silva F, Giglioti EA, Lemos MVF, Coutinho LL, Nobrega MP, Carrer H, Franca SC, Bacci M Jr, Goldman MHS, Gomes SL, Nunes LR, Camargo LEA, Siqueira WJ, Van Sluys M-A, Thiemann OH, Kuramae EE, Santelli RV, Marino CL, Targon MLPN, Ferro JA, Silveira HCS, Marini DC, Lemos EGM, Monteiro-Vitorello CB, Tambor JHM, Carraro DM, Roberto PG, Martins VG, Goldman GH, de Oliveira RC, Truffi D, Colombo CA, Rossi M, de Araujo PG, Sculaccio SA, Angella A, Lima MMA, de Rosa Jr. VE, Siviero F, Coscrato VE, Machado MA, Grivet L, Di Mauro SMZ, Nobrega FG, Menck CFM, Braga MDV, Telles GP, Cara FAA, Pedrosa G, Meidanis J, Arruda P (2003) Analysis and functional annotation of an expressed sequence tag collection for tropical crop sugarcane. Genome Res 13:2725–2735

Vogel J (2008) Unique aspects of the grass cell wall. Curr Opin Plant Biol 3:301–307

Vu JCV, Allen LH Jr, Gesch RW (2006) Up-regulation of photosynthesis and sucrose metabolism enzymes in young expanding leaves of sugarcane under elevated growth CO_2. Plant Sci 171:123–131

Vu JCV, Allen LH Jr (2009a) Growth at elevated CO_2 delays the adverse effects of drought stress on leaf photosynthesis of the C4 sugarcane. J Plant Physiol 166:107–116

Vu JCV, Allen LH Jr (2009b) Stem juice production of the C4 sugarcane (*Saccharum officinarum*) is enhanced by growth at double-ambient CO_2 and high temperature. J Plant Physiol 166:1141–1151

Wand S, Midgley G, Jones M, Curtis PS (1999) Responses of wild C4 and C3 grass (Poaceae) species to elevated atmospheric CO_2 concentration: a meta-analytic test of current theories and perceptions. Glob Change Biol 5:723–741

Wang ML, Goldstein C, Su W, Moore PH, Albert HH (2005) Production of biologically active GM-CSF in sugarcane: a secure biofactory. Transgenic Res 14:167–178

Wei H, Wang ML, Moore PH, Albert HH (2003) Comparative expression analysis of two sugarcane polyubiquitin promoters and flanking sequences in transgenic plants. J Plant Physiol 160:1241–1251

Weng J-K, Li X, Bonawitz ND, Chapple C (2008) Emerging strategies of lignin engineering and degradation for cellulosic biofuel production. Curr Opin Biotechnol 19:166–172

Weng L-X, Deng H, Xu J-L, Li Q, Wang L-H, Jiang Z, Zhang HB, Li Q, Zhang L-H (2006) Regeneration of sugarcane elite breeding lines and engineering of stem borer resistance. Pest Manag Sci 62:178–187

Wenzl P, Carling J, Kudrna D, Jaccoud D, Huttner E, Kleinhofs A, Kilian A (2004) Diversity arrays technology (DarT) for whole genome profiling of barley. Proc Natl Acad Sci USA 101:9915–9920

Weyens G, Ritsema T, Van Dun K, Meyer D, Lommel M, Lathouwers J, Rosquin I, Denys P, Tossens A, Nijs M, Turk S, Gerrits N, Bink S, Walraven B, Lefèbvre M, Smeekens S (2004) Production of tailor-made fructans in sugar beet by expression of onion fructosyltransferase genes. Plant Biotechnol J 2:321–327

Whittaker A, Botha FC (1999) Pyrophosphate: D-fructose-6-phosphate 1-phosphotransferase activity patterns in relation to sucrose storage across sugarcane varieties. Physiol Plant 107:379–386

Woods J (2001) The potential for energy production using sweet sorghum in southern Africa. Energ Sustain Develop 5:31–38

Wu L, Birch RG (2007) Doubled sugar content in sugarcane plants modified to produce a sucrose isomer. Plant Biotechnol J 5:109–117

Wyman CE, Dale BE, Elander RT, Holtzapple M, Ladisch MR, Lee YY (2005) Coordinated development of leading biomass pretreatment technologies. Bioresource Technol 96:1959–1966

Zhu H, Muthukrishnan S, Krishnaveni S, Wilde G, Jeoung JM, Liang GH (1998) Biolistic transformation of sorghum using a rice chitinase gene. J Genet Breed 52:243–252

Chapter 5
High Fermentable Corn Hybrids for the Dry-Grind Corn Ethanol Industry

Joel E. Ream, Ping Feng, Iñigo Ibarra, Susan A. MacIsaac, Beena A. Neelam, and Erik D. Sall

5.1 Introduction

The world is looking for renewable sources of fuel to prepare for the day when fossil fuels are no longer feasible. The most accessible US sources of carbon-based energy—oil, coal and natural gas—have already been tapped. The cost of imported oil has increased substantially, has been volatile and has been used as a political lever. These factors have driven interest in quickly identifying domestic, sustainable energy sources. Significant advances are being made toward generating power from wind and solar energy. Until these sources are more widespread and economical, converting the stored solar energy from biomass into energy is an area of active pursuit. Two current fuel choices from plants are biodiesel—typically from oilseed crops like soybeans, canola/rapeseed or other crops like palm and jatropha—and ethanol. Humans have known for a very long time how to produce ethanol from plant products, especially grains and grapes, by fermentation with yeast. The stored sugars are used as an energy source for the yeast, with ethanol and carbon dioxide as products. Besides its well-known characteristics upon ingestion, ethanol is combustible. The energy released upon combustion is harnessed to do work such as propel automobiles. The simplest and most efficient way to produce significant amounts of ethanol from plant products is to start with plants with a high content of simple or complex carbohydrates. To be most economical, the plant source of ethanol is grown typically near where the ethanol will be produced from it. Sugarcane has a high

---

J.E. Ream, B.A. Neelam, and E.D. Sall
Monsanto Company, St. Louis, MO 63167, USA
e-mail: joel.e.ream@monsanto.com; beena.a.neelam@monsanto.com; erik.d.sall@monsanto.com

P. Feng and I. Ibarra
Monsanto Company, Ankeny, IA 50021, USA
e-mail: ping.feng@monsanto.com; INIGO.IBARRA@Monsanto.com

S.A. MacIsaac
Monsanto Company, North Carolina Research Center, Kannapolis, NC 28081, USA
e-mail: susan.a.macisaac@monsanto.com

content of sucrose and grows in warm climates (i.e., tropical/subtropical climates); therefore, it is used in a vigorous ethanol industry in Brazil—a country that can grow a large crop of sugarcane. In the US, grains (corn, wheat, barley, rye, etc.) have been the traditional source of ethanol for the making of alcoholic beverages. In these grains, the sugars are typically stored as complex carbohydrates (e.g., starch) that need to be broken down enzymatically to simple sugars to serve as food for the ethanol-producing yeast. The largest-acreage grain crop in the US is corn, so starch-filled corn grain is a logical source for the US commercial ethanol effort. There is a high interest in converting non-food crops into ethanol. The focus here is converting the complex polymer cellulose—a common constituent of the support infrastructure of plants (stems, leaves, roots, wood, etc.)—to ethanol. It is more technically challenging to convert cellulose to sugars than starch.

Ethanol production in the US has been around for many years, with the first big push starting in the late 1970s to early 1980s. Gradual increases in production continued during the late 1980s and into the 1990s. Government policies (i.e., the Renewable Fuel Standard that included an ethanol mandate) and financial incentives drove the rapid development of the US capacity to produce ethanol for automobile fuel from corn starting in the late 1990s through the 2000s. The process used to extract ethanol from corn is called dry-grind to distinguish it from the long-standing corn process used to produce starch and other products, called wet milling. In dry-grind, the corn grain is ground, mixed with water and enzymes (α-amylase and glucoamylase) to convert the starch to simple sugars, which are then fermented to ethanol by adding yeast. The resultant ethanol is purified and blended with gasoline, thereby reducing some of the need for petroleum oil. As the dry-grind ethanol industry has expanded rapidly to meet the increased demand for biofuels, interest in identifying the best sources for corn-grain-based ethanol has increased. Getting the most ethanol from a unit of corn grain helps make the best use of corn grain, increases the efficiency of ethanol production and lowers the cost of ethanol production. To achieve these benefits, the ethanol industry needs to be able to source the higher fermenting corn hybrids. This chapter will describe the technology for measuring corn grain fermentability and its use in selecting corn hybrids that are the best candidates for commercial ethanol production.

5.2 Value of High Fermentable Corn Hybrids

Our society looks to biofuels as a sustainable alternative energy source to complement other alternatives (wind, solar, etc.) to decrease our dependence on finite, and increasingly difficult to extract, fossil-based fuels. Plant-based ethanol production is a technically accessible source of biofuel, as is biodiesel. Until the production of significant quantities of ethanol from cellulose is realized, ethanol production from corn grain and sugarcane are the most prominent sources. In the US, producing ethanol from corn grain is driven by the US position as the major producer of corn. Producing ethanol from corn grain decreases US dependence on foreign oil while providing benefits to consumers, farmers and the agricultural industry, especially in rural areas where opportunities for economic growth are limited.

Our society also depends on corn grain for food, principally as a source of animal feed. One of the products of dry-grind ethanol production is distillers grains with solubles, which can be fed either as wet distillers grains with solubles (WDGS) or, more commonly, dried distillers grain with solubles (DDGS). This is the protein- and oil-enriched solid material left over from the distillation of the ethanol from the fermentation tank. DDGS is a good source of cattle feed and, in part, compensates for corn grain diverted from animal feed to ethanol production. Continual yield increases in corn production, coupled with more efficient production of ethanol from corn grain, will enable the production of corn grain-based ethanol without having a substantial impact on the supply available for animal feed. The more long-term focus on biofuels development from non-food sources—cellulosics, for example—will further mitigate the potential for a competition between food and fuel from corn. For consumers, a supplemental supply of energy can lower gasoline prices at the gas pump and potentially reduce rapid gasoline supply and price fluctuations due to geopolitical influences on oil supply and price. Consumers also receive an indirect benefit from a robust agricultural community in the form of a reliable food supply and lower prices.

United States growers rely on a strong demand for the corn grain they produce. This demand keeps the value of their grain sufficiently high to maintain their business and invest in new technologies for producing more grain. A reliable income allows farmer investment in more efficient farm machinery and advanced corn hybrids that enable more abundant and efficient production of corn grain for both food and fuel. The development of the corn ethanol industry has helped the corn grower by providing a new customer for their harvested grain.

This value to society, consumers and corn growers—including the policy and financial encouragement of the US government—drove a rapid rise in the number of dry-grind ethanol companies. To be successful, any company needs to make enough profit to repay the costs of their capital investment in their production plant and to reward their investors. Ethanol production costs include energy, water, yeast, enzymes, and the cost of the feedstock, i.e., corn grain. Besides securing a continuous supply of corn for the best price, the most successful corn ethanol producer needs to glean the maximum amount of ethanol from each bushel of corn purchased. This becomes increasingly important as the cost of corn increases. As with most new industries, there will be a period of rapid expansion. As the market becomes filled, those companies best equipped to maintain sustained profitability will survive. This leads to a drive to purchase grain that has the potential to efficiently produce the most ethanol; that drive will increase in a more competitive ethanol market and higher corn prices.

Commercial ethanol plants rely on (1) society/government support for development of alternative fuels; (2) production plants that can efficiently convert corn grain to ethanol that are close to large volumes of reasonably priced corn grain; and (3) the technology to produce the most ethanol possible from the grain they source. As corn prices increase, it becomes increasingly important to get the most gallons of ethanol out of each bushel purchased and processed at a commercial ethanol plant. For these reasons, seed companies have surveyed their corn hybrid products to identify those that will produce the most ethanol per bushel in addition to the highest yield of grain per acre. By helping farmers and ethanol companies to know which corn hybrids produce the most ethanol per bushel, seed companies can develop consistent customers for their corn seed. Growers will typically prioritize their seed purchase on higher value traits like yield; with that equal, the inherent fermentability of a particular grain has the potential to increase their likelihood of marketing that grain upon harvest. Society in general, consumers, agricultural communities, farmers and ethanol companies gain value from a robust corn ethanol industry. The increased gallons of ethanol that can be efficiently produced from a bushel of corn grain helps realize that value.

5.3 Factors Influencing the Fermentability of Corn Grain

Not all corn grain ferments the same. Figure 5.1 shows the distribution of ethanol yields for 231 corn hybrids over a single season. From this data, the majority of hybrids yield between 14.2 and 15.2% (volume to volume) ethanol, but there are hybrids yielding as low as 13.0% and higher than 15.6%. An understanding of what factors influence the fermentability and, more importantly to an ethanol company, how to source hybrids with the highest ethanol producing potential, increase our opportunities to produce more ethanol from corn grain. Factors affecting corn grain fermentation potential include genetics, environment, agricultural practices and storage conditions/grain quality.

There are genetic differences that account for different fermentabilities of corn grain. The main source of fermentable sugars in corn grain is starch. Corn hybrids vary in their starch content as well as other major constituents such as protein and oil; however, there is more to fermentability than starch content. There is no direct correlation between starch content and ethanol yield (Singh and Graeber 2005). What else is going on? Obviously, there are other influencing factors. The starch

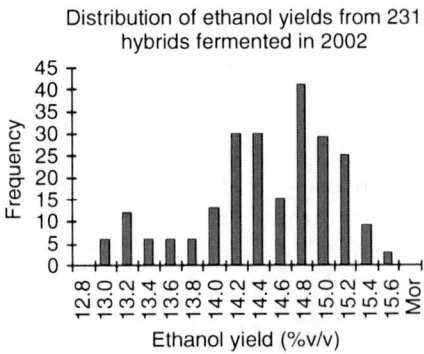

Fig. 5.1 Distribution of ethanol yields from 231 corn hybrid fermented in 2002 as measured by near infrared transmittance spectroscopy (NIT)

present in corn grain has to be accessible to the hydrolyzing enzymes before it is reduced to sugars for fermentation by yeast. The accessibility of the starch to hydrolyzing enzymes may vary between grain samples, and this accessibility is likely to be an important component of differential fermentability of different corn grain samples. Within corn grain, there are differences in the packing of starch grains and, possibly, with protein matrices around these granules. This could alter the accessibility of the starch to the enzymes and yeast and, therefore, affect the final ethanol yield under standard fermentation conditions. Other factors could be corn constituents other than starch that support the active growth of the yeast during fermentation. It is possible there are differences between corn hybrids in yeast nutrition factors such as amino acids and vitamins. These are genetic factors. Such factors mean that corn hybrids selected to be higher fermenting one year have a reasonable, but not absolute, chance of being higher fermenting in subsequent years. A big variable in predicted fermentability is the environment.

The fermentability of corn grain varies with the environmental conditions under which it was produced. As shown in Fig. 5.2, the ethanol yield from corn hybrids varies by maturity group and season. Some growing seasons, e.g., 2005 and 2007, provided environmental conditions conducive to higher ethanol yields. Besides the annual trend, environmental variability within a single growing season appears to affect the fermentability of single hybrids grown in different locations (Singh and Graeber 2005). Environment is known to affect grain components such as starch, protein and oil. Starch composition is an important factor for ethanol production but, since corn grain fermentability is not correlated directly with starch content, the environment must affect other factors as well. Stress (water, temperature, insects, disease, etc.) may influence these non-starch components of corn grain fermentability.

Not only the environment, as caused by geography and weather, but also the agronomic practices used to produce corn grain can affect its fermentability. Besides hybrid selection, one of the grower-controlled variables is the application of nitrogen. Corn grain yield increases with applied nitrogen. Ethanol yield has been shown also to vary with nitrogen application, with one study showing significant reductions

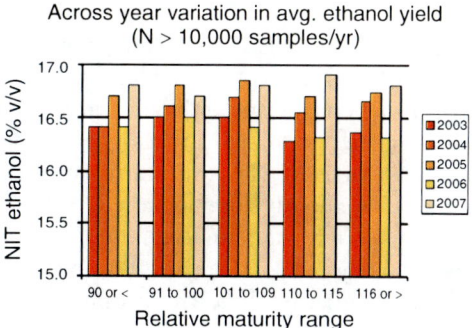

Fig. 5.2 Across-year variation in average ethanol yield by corn hybrid maturity range over five growing seasons. $N > 10,000$ samples/year

in ethanol yield with higher-than-normal and lower-than-normal nitrogen applications (Reicks et al. 2009).

Another factor influencing corn grain fermentability is storage conditions. Corn grain is supplied to ethanol plants year-round. This means freshly harvested corn grain can sit for up to a year before being converted to ethanol. Corn grain is typically stored in bins and elevators as well as in covered piles. The Monsanto research group has observed that, in general, lower quality corn grain gives lower ethanol yields. They have observed an interesting trend in the predicted fermentability of corn grain delivered to ethanol plants over the course of a year. In late spring/early summer, the fermentability of delivered corn grain as measured by near infrared transmittance spectroscopy (NIT) begins to decrease. The hypothesis for this phenomenon is that the decline is due to decreased grain quality from prolonged storage since the fall harvest. The fermentability numbers typically increase again in fall when the fresh grain begins to be delivered. Fungus and insect damage may decrease the starch content. In addition, fungal infections could potentially produce products that inhibit yeast growth, thereby decreasing ethanol yield. Finally, grain drying practices can also have a significant impact on the fermentability of corn. Drying corn grain at too high a temperature (a maximum temperature of more than 140° Fahrenheit, for example) will typically result in lower fermentability of the grain.

There is variability in the fermentability of corn grain—by hybrid, season, location, agronomic practices and storage conditions. For commercial ethanol production, it is important to understand these factors and strive to increase the overall fermentability of corn grain. In order to observe and take advantage of improvements in corn fermentability, however, the producers of corn hybrid seed and corn ethanol plants need the capability to readily and accurately measure the fermentability of corn grain.

5.4 Measuring Corn Grain Fermentability

5.4.1 NIT Calibration

The best way to measure the fermentability of corn grain is to actually ferment it to ethanol. This is possible in a laboratory using agitated flasks or small-scale fermentors. This is not convenient for broad-scale use due to the need for a laboratory environment, trained personnel, laboratory instruments and the relatively long fermentation times (about 54 h). The corn grain needs to be ground, mixed with enzymes and yeast, mixed continuously and monitored for fermentation products. The most convenient way to measure the fermentability of corn hybrids is by near infrared transmittance spectroscopy (NIT). There are two modes of infrared spectroscopy, reflectance (NIR) and transmittance (NIT). The transmittance mode is used routinely to measure grain composition in broad array of crops in non-laboratory environments—scale-houses at grain elevators, for example. This method measures the infrared light transmitted by a corn grain sample and translates this into an estimate of a component of the grain such as protein, oil, moisture or, in this application, fermentability. NIT takes advantage of the change in infrared irradiation due to chemical constituents in the corn grain. It does not need to know what those chemical constituents are. This is helpful for predicting fermentability because, as discussed above, corn grain components other than starch account for its fermentability. For whole grains, transmission rather than reflectance mode has been shown to be the near infrared method of choice and offers the convenience of rapidly and non-destructively scanning the intact grain sample. Typically, a sample of corn grain of about a half-pound (227 g) is poured into a NIT instrument. The Foss Model 1241 Grain Analyzer® (Foss in North America, Eden Prairie, MN) is an example of a NIT instrument used broadly in the commercial corn ethanol industry. The corn grain sample is scanned and the transmitted infrared spectrum is compared to those in a previously loaded calibration. The spectrometer compares the infrared spectrum transmitted from the presented sample to spectra in the calibration and displays ethanol fermentability based on the loaded model, which converts that spectrum into predicted ethanol yield. The instrument output is provided in units of choice—gallons of ethanol per bushel, for example. The advantage of NIT is that it is easy, fast and non-destructive; hence, it can be implemented readily to provide fast results in non-laboratory environments. For example, NIT measurement of grain constituents (moisture, protein, oil) is a common component of commodity grain delivery at the scale-house. In the ethanol industry, NIT instruments are used to measure the predicted fermentability of grain coming into ethanol plants, as well as the moisture, starch, protein and oil. Furthermore, the same technology can be used to estimate the major components of the DDGS, the solid product after corn grain fermentation, which helps in marketing this material to animal feed customers.

5.4.2 Reference Chemistry

Near infrared transmittance spectroscopy is easy to use and fast, but it is not a direct measurement of fermentability. Because NIT does not actually measure specific elements of grain composition, but only the change in the infrared spectrum as a result of transmitting through or reflecting off of a grain sample, it is susceptible to other factors that influence the NIT spectrum. Water also absorbs NIT energy and needs to be accounted for. Material other than corn grain in a sample—foreign matter like dust, dirt, plant debris, etc.—also provides non-ethanol-related influences on the collected NIT spectra. NIT measurements are also sensitive to temperature. All of these factors need to be accounted for in the calibration, the critical component of NIT measurement that translates the sample spectrum into the fermentation potential and minimizes the interference of moisture, temperature and debris.

Building a good calibration is a critical element in developing a robust NIT method. A NIT calibration is built by developing a mathematical model that translates the infrared spectral signal of a sample into a direct predicted fermentability number (gallons of ethanol per bushel, for example). To do this, a direct method to measure fermentability—one that is highly quantifiable and not subject to indirect factors—is needed to build that correlation. This method is called the reference method because it accurately measures the constituent of interest. The better the reference chemistry used to build the NIT calibration, the better NIT will be for predicting fermentability of a test sample of corn grain.

A typical reference procedure for measuring corn grain fermentability is flask fermentation. Under carefully controlled and replicated conditions, corn grain is ground, weighed, treated with starch-hydrolyzing enzymes (α-amylase and glucoamylase) and yeast, and fermented with agitation for about 54 h. Historically, two methods have been used to measure the ethanol produced in laboratory flask fermentations to quantify corn grain fermentability. One way is to measure the ethanol produced directly using high performance liquid chromatography (HPLC). The other method is to measure the weight loss of the fermentation flask as a measure of the other product of fermentation, carbon dioxide. NIT calibrations used in the corn ethanol industry have been made using both reference methods. The potential advantage of the HPLC method is that it measures ethanol, the fermentation product of interest, directly. The carbon dioxide weight loss method has the theoretical potential to be less accurate due to other factors during the fermentation contributing to weight loss other than carbon dioxide emission. A study comparing the two methods with five corn hybrids showed them to give similar results (Lemuz et al. 2009).

5.4.3 NIT Calibration

The accuracy of NIT to predict the fermentability of a particular corn grain sample depends on the diversity of samples used to build the calibration as well as the reference chemistry used to establish the correct value. Typically, a large number of diverse samples are used, over multiple seasons, to build a good calibration filled with sufficient sample diversity. Samples being analyzed by NIT that are quite different from those used to build the calibration are less likely to yield an accurate measurement than those more similar to the ones used to build the calibration. Therefore, the calibration needs to be built with a wide variety of corn grain—different hybrids grown under a wide variety of environmental conditions. The more corn samples the calibration sees, the more likely it is to recognize the spectra and the better it will predict fermentability. In addition, a NIT calibration is typically updated over time as new hybrids are introduced and more variability is encountered. At some point, after several seasons of calibration-development and refinement, there is little increase in the accuracy of calibration with the addition of new samples. At this time, it has pretty much "seen" all the variation it will encounter and can do a good job of predicting the fermentability of any corn sample measured. In a process called chemometrics the collected NIT spectra and

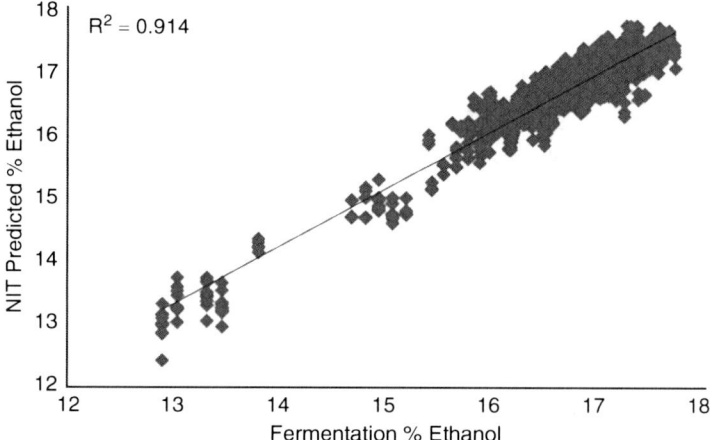

Fig. 5.3 Correlation between ethanol yield as predicted by NIT and actual ethanol yield as measured by flask fermentation

corresponding reference chemistry ethanol values are analyzed to identify the mathematical model that best correlates them. Once identified, this model becomes the calibration. Figure 5.3 shows the correlation between actual fermented ethanol values of a large series of grain samples versus the predicted ethanol values as predicted by NIT using a calibration built over five corn seasons.

5.4.4 Commercial Validation of NIT Calibration

It is one thing to develop a good NIT calibration. It is another to use it to make economically important decisions. Current NIT calibrations predicting corn grain fermentability are built using reference chemistry based on laboratory flask fermentations. These flask fermentations are developed as a surrogate for the fermentation process in a commercial dry mill corn ethanol plant (typically 50–100 million gallons of ethanol per year). The assumption is that the calibration built based on these laboratory flask fermentations will translate into industrial-scale fermentations. Commercial ethanol plants need assurance that this assumption is valid before making substantial grain sourcing decisions based on it. This assumption has been rigorously tested by the Monsanto research group by comparing the ethanol yield of grain as predicted by their NIT calibration to the actual ethanol produced at six different commercial ethanol plants. Grain representative of each fermentor load was collected, measured by NIT and compared to the actual ethanol produced from the commercial fermentation. This research showed that the ethanol yield predicted by their specific NIT calibration was positively correlated with the

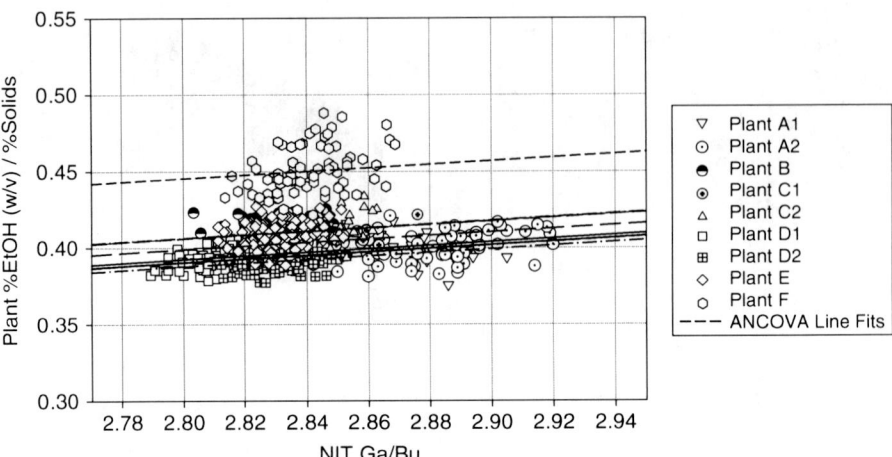

Fig. 5.4 Correlation between ethanol yield as predicted by NIT and actual yield measured at six different commercial ethanol plants. Data points are solids-normalized fermentation yields (after screening 1,150 data points for outliers and special causes) for 628 fermentation batches from six different ethanol plants obtained over a 180-day period between March 2009 and August 2009. Data within a given set are from fermentations run to the same batch time (+/– 1 h). Data sets from the same plant with different numbers (e.g., A1, A2) represent data obtained from different time periods at the same plant. NIT-predicted gallons/bushel (Gal/Bu) ethanol yield was measured by NIT using the Monsanto HFCpEtOH calibration on a representative sample going into the fermentor, obtained by automated grain sampling immediately before the mills. Regression lines were calculated by analysis of covariance (ANCOVA). This regression slope is significant ($P=0.0001$) and indicates a 0.8 (–0.3, +0.6, 95% CL) relative percentage increase in yield per 1% increase in NIT Gal/Bu

ethanol yield measured at the ethanol plants, with a high statistical significance (Fig. 5.4). In addition, the slope of the resultant curve was near one, supporting a strong correlation between the NIT calibration and the actual commercial ethanol yield. This study demonstrated that, for at least one NIT calibration (the Monsanto "HFCpEtOH" calibration)—built using flask fermentations and ethanol measurement by HPLC as the reference chemistry—a good estimate of the actual ethanol yield that is realized at a commercial ethanol plant is possible. This is important as it provides confidence to commercial ethanol plants that they can predict their plant productivity based on the measurement of sourced corn grain fermentability using a technology, NIT, that can be implemented easily at their corn ethanol plant. This provides the ethanol plant operator the opportunity to make decisions on sourcing corn grain that maximizes their ethanol yield and, therefore, profitability. For example, if corn ethanol plant grain merchants knew the predicted fermentability of corn grain from different suppliers, they could make purchase decisions based on these values to optimize the ethanol produced per bushel of purchased corn grain. Another way to leverage this type of information is to purchase corn hybrids that have been shown to provide, on average, a higher ethanol yield than other hybrids.

5.5 Designation of High Fermentable Corn Hybrids

Due to the commercial ethanol industry's interest in getting the most ethanol from each load of corn it purchases, and, since there is variability in the fermentability of different hybrids of corn, corn seed companies have measured the fermentability of many different corn hybrids and published these in their seed catalogs. This helps the grower select corn hybrids that would be attractive for a major customer of harvested corn grain, commercial ethanol plants, as well as contain the other agronomic features they desire to optimize their yield. This process of designating higher fermenting corn hybrids—i.e. those that can be promoted or offered to growers as having the potential to produce more ethanol than other hybrids—must take into account environmental conditions during production of the corn grain as well as the genetics of the hybrids. Since the environment is a significant factor, corn hybrids need to be tested for fermentation potential under a wide variety of growing conditions to be confident they will deliver higher ethanol to the ethanol plants. Typical corn hybrid testing programs for yield cover broad geographic and environmental diversity. This infrastructure can be used to test for the potential to produce ethanol.

For the designation of Monsanto Processor Preferred® HFC (high fermentable corn) hybrids, a particular hybrid needs to have demonstrated higher fermentability, as measured by NIT, in many locations and over multiple years. Compared to what? As discussed previously, the fermentability of corn changes with environment, so there is no absolute measurement threshold, like starch content, to exceed in order to be designated. This means the hybrids need to outperform other hybrids in the same geography, which requires comparing the fermentability within similar maturity groups, understanding what the average is and identifying those hybrids that significantly exceed that average. This testing program increases the probability—but does not guarantee—that a designated higher fermentable hybrid will yield higher ethanol per bushel when delivered to an ethanol plant. This point is demonstrated in Table 5.1, which shows the relative ranking of nine hybrids for predicted

Table 5.1 Relative ethanol yield ranking of corn hybrids by year

Hybrid	Relative ranking[a]			Composite
	2007	2006	2005	
A	1	2	6	3
B	2	4	2	2
C	3	6	5	5
D	4	2	1	1
E	5	1	4	4
F	6	9	9	8
G	7	5	3	6
H	8	8	7	7
I	9	7	8	8

[a] ≥ 10 samples were analyzed for each hybrid for each year. Ranking was based on ethanol yield as measured by near infrared transmittance spectroscopy (NIT)

ethanol yield as measured by NIT over 3 years. Despite year-to-year variations in relative ranking, the higher fermenting hybrids are grouped higher than the lower fermenting hybrids. A farmer or corn ethanol plant that selects a designated higher fermenting hybrid will increase their probability of producing or purchasing grain that produces more ethanol per bushel than non-designated hybrids. Furthermore, that confidence is increased when the designated hybrids were selected using a NIT calibration that has been shown to correlate positively and directly with large-scale, commercial ethanol production. Since higher ethanol per bushel yields translate to more efficient ethanol production, there is an interest in increasing the amount of ethanol that can be produced per unit of corn grain.

5.6 Opportunities to Increase Corn Grain Fermentability

There are two fundamental ways to increase the fermentability of corn grain. First, you can use the fermentability, as measured by NIT for example, as a breeding selection trait. As with yield and other agronomic traits, higher overall fermentability can be achieved, even if you do not know what the reason is for the difference. Yield will always be a much more important trait in selection criteria but, yield being equal, those hybrids with higher fermentability could be advanced. The second approach is to select for, or introduce genes for, specific attributes that have been shown to increase fermentability. This requires good information on what those attributes are. Starch is obvious and already part of the standard breeding information for selections. Again, starch content is not correlated directly with fermentability in corn grain. There is as yet no good understanding of why some hybrid grain samples yield higher ethanol than others. Starch granule packing, protein matrices, yeast nutrition factors and other components could all be involved in fermentability. If these factors were identified, then targeted approaches to modify them could be applied to increase corn grain fermentability.

Another aspect of the fermentability of corn hybrids is the efficiency of ethanol production. Corn has recently been genetically modified to express the starch hydrolyzing enzyme α-amylase, which may decrease the amount of added enzymes during processing of this corn grain to ethanol.

The benefit of directly selecting or creating corn hybrids for higher fermentation has to be weighed against the return. The underlying factors involved in fermentation are not well understood. The processes (enzymes, yeast strains, processing technology, etc.) used by ethanol plants to produce ethanol change over time. In addition, there is a concerted effort to replace ethanol production from corn grain with cellulosic sources. The complexity of the trait, its value relative to other attributes and the uncertainty around long-term utility work against a substantial effort to proactively create higher fermenting corn varieties. Furthermore, since the value of the fermentability trait is considerably lower than the value of yield to the grower, hybrid selection will always favor yield. Evaluating existing hybrids, selected for more valuable and durable traits such as yield and agronomic traits for their fermentation potential is the strategy that best balances these dynamics. These selection drivers also work for the corn ethanol industry since an abundant supply of high quality corn grain is needed to sustain current and future ethanol production aspirations.

5.7 Summary

Biofuels like corn ethanol help provide a sustainable and secure non-petroleum-based source of energy. The dry-grind ethanol industry is the customer for about one-third of US-produced corn grain. It is in this industry's best interests to source corn grain that will give the maximum gallons of ethanol per bushel. Corn seed companies have evaluated corn hybrids for their potential to produce ethanol and designated these in their product offerings as a resource to help growers identify hybrids that would be the best fit for ethanol plant customers. The heart of identifying higher fermenting hybrids is a rapid, non-destructive near infrared method. This is an indirect measurement of fermentability, which requires a robust calibration built around direct reference chemistry. In at least one case, a NIT calibration has been shown to correlate highly to commercial ethanol plant data. This provides corn ethanol companies the confidence needed to use this calibration to source higher fermenting corn grain. Besides genetics, the environment is a significant factor influencing the fermentability of corn grain. The process of designating a corn hybrid as higher fermenting requires a robust testing method that accounts for these environmental effects. Higher fermenting hybrids, carefully selected based on multiple seasons and geographic diversity with a commercially validated NIT calibration built with a high diversity of relevant corn samples, can increase the efficiency of ethanol production from corn grain. High yields of high fermenting corn hybrids will be needed to meet the growing needs for both fuel and food.

References

Lemuz CR, Dien BS, Singh V, McKinney J, Tumbleson ME, Rausch KD (2009) Development of an ethanol yield procedure for dry-grind corn processing. Cereal Chem 86:355–360
Reicks G, Woodard HJ, Bly A (2009) Improving the fermentation characteristics of corn through agronomic and processing practices. Agron J 101:201–206
Singh V, Graeber JV (2005) Effect of corn hybrid variability and planting location on dry grind ethanol production. Trans ASA 48:709–714

Chapter 6
Engineering Advantages, Challenges and Status of Grass Energy Crops

David I. Bransby, Damian J. Allen, Neal Gutterson, Gregory Ikonen, Edward Richard Jr, William Rooney, and Edzard van Santen

6.1 Introduction

Key criteria needed for crops for which the intended use is energy production include high yield potential with low input (fertilizer and water) requirements; resistance to diseases, pests and drought; adaptation to a wide range of soils and climates; and biomass composition that is optimized for the intended conversion technology. Of these criteria, yield (weight of annual biomass production per unit area) might be considered the most important because of the limited amount of land available in relation to the demand for energy, and the fact that yield strongly affects the cost of, and economic return from, an energy crop (Bransby et al. 2005). Energy crops can be divided into two broad categories: herbaceous crops and woody crops. Most of the species under development in the herbaceous crop category are warm season grasses, which have the advantage of the more efficient C4 photosynthetic pathway (Moore et al. 2004) relative to the C3 pathway common

---

D.I. Bransby and E. van Santen
Department of Agronomy and Soils, Auburn University, 202 Funchess Hall, AL 36849, USA
e-mail: bransdi@auburn.edu; evsanten@acesag.auburn.edu

D.J. Allen
Mendel BioEnergy Seeds, 4846 East 450 North, Lafayette, IN 47905, USA
e-mail: dallen@MendelBio.com

N. Gutterson and G. Ikonen
Mendel Biotechnology, Inc., 3935 Point Eden Way, Hayward, CA 94545, USA
e-mail: neal@MendelBio.com; gikonen@MendelBio.com

E. Richard Jr
USDA-ARS Sugarcane Research Unit, 5883 USDA Road, Houma, LA 70360, USA
e-mail: edward.richard@ars.usda.gov

W. Rooney
Department of Soil and Crop Sciences, 370 Olsen Blvd., College Station, TX 77843, USA
e-mail: wlr@tamu.edu

to trees. Other advantages of grasses over woody crops include a shorter period from planting to attaining full yield (4–8 months for annuals, and 2–3 years for perennials), the ability for some to be sun- or air-dried to a relatively low moisture content (10–15%), and higher conversion efficiency for biochemical conversion processes because of lower concentrations of lignin. On the other hand, woody crops can be "stored on the stump" and harvested year round, whereas grasses are usually harvested once a year over a relatively short period, and harvested biomass then needs to be stored for up to 12 months on farms and delivered as needed to biorefineries.

While some grasses are able to produce starch (corn and sorghum), and others sugar (sweet sorghum and sugarcane), all grasses produce a high proportion of ligno-cellulosic (or cellulosic) biomass. The focus of this chapter is primarily on the latter as an energy feedstock; sugar- and starch-based biomass resources are the subjects of Chaps. 4 and 5, respectively. However, where appropriate, reference is made to benefits of producing both cellulosic biomass and sugar or starch as feedstocks for liquid fuel production from the same crop. Because genetic modification is the most efficient and least expensive approach to improving the beneficial traits for energy crops listed above, a considerable proportion of this chapter is focused on modern crop breeding and genetic improvement approaches. In particular, progress in, and future prospects for, genetic improvement of *Miscanthus*, switchgrass, sugarcane and sorghum are discussed as examples, recognizing that other species could offer similar potential as biomass feedstocks. In addition, possible approaches for integrating grasses into cellulosic biomass supply systems are described. For example, while the low input requirements of perennials such as Miscanthus and switchgrass are attractive features for cellulosic biomass crops, annuals such as sorghum also offer distinct advantages, including opportunities for rotation with other annual crops and rapid initiation of dedicated feedstock supply systems. Although the principles presented in this chapter generally have international relevance, examples used to illustrate these principles are drawn largely from North America.

6.2 Miscanthus

One of the leading candidates to meeting biomass demand for power generation and biofuels production is *Miscanthus*, a genus of perennial C4 grasses native to China and Japan, varieties of which have been grown in the United States and Europe as ornamentals for many decades. *Miscanthus* is essentially undomesticated, yet one species, *M. x giganteus*, already outperforms alternative biomass crops like switchgrass across much of the US and Europe (Heaton et al. 2004, 2008). Modern breeding approaches, including identification of quantitative trait loci (QTL), marker-assisted selection, and transgenic technologies, offer the potential for *Miscanthus* domestication at an unprecedented rate (Jakob et al. 2009).

6.2.1 Miscanthus *Phylogeny and Growth*

The genus *Miscanthus* belongs to the subtribe *Saccharinae* of the tribe *Andropogoneae* in the family *Poaceae* and was first described by Andersson (1855). *Miscanthus* is closely related to other genera of the "*Saccharum* complex" (*Saccharum*, *Erianthus*, *Sclerostachya* and *Narenga*; Amalraj and Balasundaram 2006). The taxonomy of *Miscanthus* has been described by Honda (1930), Keng (1932) and Ohwi (1965). Wide hybridization between *Miscanthus* species, ecological races and between *Miscanthus* and related species from the *Saccharinae* subtribe has led to new forms and makes the taxonomy of the genus complex (Hodkinson et al. 1997; Clifton-Brown et al. 2008). Variation of ploidy levels from 2x to 6x (Hodkinson et al. 1997; Lee 1964) further complicates taxonomic studies. Molecular phylogenetics have been applied recently and a broad sense definition of the *Miscanthus* genus was used to recognize 11–12 species with a basic chromosome number of 19 (Clifton-Brown et al. 2008). The genus *Miscanthus* is characterized by high genetic diversity within and between species. Some species are distributed over a wide area from the Far East and South-east Asia to the Pacific Islands (Clifton-Brown et al. 2008). As a result, many endemic races have arisen with relevance to important traits for breeding such as cold-, drought- and salt-tolerance, and also resistance to pests and diseases. *Miscanthus* x *giganteus*, the most widely grown biomass species of *Miscanthus*, is a sterile triploid resulting from an inter-specific cross between *M. sacchariflorus* and *M. sinensis* (Hodkinson et al. 1997; Clifton-Brown et al. 2008; Greef and Deuter 1993).

Miscanthus requires 3–5 years to establish and achieve maximum yield. In a comprehensive review of published studies, Heaton et al. (2004) found 3rd year peak biomass of an *M. x giganteus* clone—here referred to as cv. Illinois (Heaton et al. 2008)—to average 22 Mg ha^{-1} [10 dry tons (2,000 pounds or 907 kg) acre^{-1}], more than double that of switchgrass at 10 Mg ha^{-1} (4 tons acre^{-1}). In side-by-side research-scale trials in Illinois, *M. x giganteus* cv. Illinois had average yields after establishment of 30 Mg ha^{-1} (13 tons acre^{-1}) at harvest, three times that observed in a locally adapted switchgrass accession, Cave-in-Rock (Heaton et al. 2008). There is much less data on yields of other *Miscanthus* species, but averaging across trials in five countries in Europe, 3rd year biomass yields of 10–20 Mg ha^{-1} (4–9 tons acre^{-1}) were reported for eight different *M. sinensis* genotypes (Clifton-Brown et al. 2001).

6.2.2 *Genetic Improvement of* Miscanthus

Genetic improvement of *Miscanthus* is focused primarily on biomass yield, including yield stability via environmental stress tolerance. While transgenesis will no doubt be applied in *Miscanthus* improvement programs, there is abundant opportunity for improvement using conventional and advanced breeding methods (Jakob et al. 2009), which is the subject of this section. Species of interest for breeding of high biomass varieties are *M. sacchariflorus* (2x and 4x), *M. sinensis* (2x), *M. condensatus* (2x) and *M. floridulus* (2x). The role of two of these species has already been evaluated, as noted above, in their contribution to the widely studied *M. x giganteus* clones.

6.2.3 Conventional Breeding Challenges

There are two major limitations for *Miscanthus* breeding. First, with yield not maturing until the 3rd year, selection cycles are considerably longer than for an annual crop program. Second, *Miscanthus* demonstrates a high degree of self-incompatibility (Chiang et al. 2003), preventing a simple route to generating genetically uniform inbred and hybrid lines, which is a useful approach to elimination of deleterious alleles to accelerate genetic improvement. The capture of non-additive genetic variation in perennial grasses is possible, nonetheless, with examples including inbred orchardgrass (Nilsson-Leissner 1942) as well as non-inbred, single cross switchgrass with biomass gains up to 66% (Taliaferro et al. 1999). Inbred generations of switchgrass have been successfully produced (Taliaferro 2002), suggesting the value of creating a hybrid *Miscanthus* system.

6.2.3.1 Trait Targets

Intrinsic Yield and Flowering Time

Culms (grass stems or canes) make up more than 90% of mature *M. x giganteus* cv. Illinois biomass yield at harvest (Christian et al. 2008; Jezowski 2008). The major factors that impact yield include properties of rhizomes, such as size, angle (clumping vs spreading form), nutrient content, and number and growth of winter-surviving rhizome shoot initials, as well as properties of the developing shoot, such as leaf elongation rate, photosynthetic capacity, and leaf angle, size, and number. As a general matter, plant height is correlated with biomass across perennial grasses, and specifically in *M. sinensis* in conjunction with basal perimeter measurements, to yield high correlation coefficients to yield for basal area (0.97) and above ground biomass (0.92; Hirata et al. 1999).

In temperate areas, *M. x giganteus* cv. Illinois flowers much later in the season than switchgrass (Heaton et al. 2009), which may contribute significantly to the greater biomass yield of cv. Illinois in comparison with switchgrass ecotypes. In grasses, flowering typically terminates culm growth. Additional growth is initiated as a new tiller at the base of the plant, which experiences very low light intensity in a mature canopy, limiting the rate of photosynthetic productivity. Further delaying, or eliminating, flowering in production areas is an attractive strategy for further increasing *Miscanthus* yield.

Drought and Water Use Efficiency

When *M. x giganteus* yields in all published trials were correlated with major environmental factors, in season water supply (precipitation and irrigation) had a larger impact than temperature or nitrogen supply (Heaton et al. 2004). In response to drought stress, *M. x giganteus* and a *M. sacchariflorus* genotype exhibited premature senescence of older leaves (Clifton-Brown and Lewandowski 2000a). This firing of leaves in the deepest shade of the canopy is an efficient method to limit water loss, with a smaller impact on photosynthesis, but it cannot be reversed when conditions improve without new leaf growth. In contrast, the *M. sinensis* variety in this experiment limited water consumption primarily through lower stomatal conductance, rather than leaf firing, which could reduce plant photosynthesis more dramatically in the short term, but be more quickly reversed when water availability returns.

Cold Temperature Adaptation

M. x giganteus has shown higher productivity in cool temperate regions, such as England (52°N), than any other C4 ("warm-season") grass (Beale and Long 1995). The minimum temperature that induced shoot emergence in 50% of plants was 7–11°C across several *Miscanthus* varieties (Farrell et al. 2006). *M. x giganteus* was able to produce fully photosynthetically competent leaves while growing at 11–14°C, something other C4 crops like maize cannot achieve (Naidu et al. 2003). While winter kill has been observed in *Miscanthus* trials in Europe (Clifton-Brown and Lewandowski 2000b; Pude et al. 1996), little loss of *M. x giganteus* has been reported in the much colder winters observed in Illinois (Heaton et al. 2008).

Pest and Disease Resistance

Large increases in the land cultivated with *Miscanthus* will ultimately raise the disease pressure in these plantations. So far, only a few pests and diseases have been reported to affect *Miscanthus* yield in energy plantations, but there are numerous pathogens of wild and ornamental *Miscanthus*.

Fungal pathogens are common in causing diseases on C4 grasses, including such related crops as sugarcane, sorghum and corn. In *M. x giganteus* trials, *Fusarium* was believed to be the source for overwintering problems in Europe, with *F. culmorum* and *F. graminearum* reducing survival as well as above- and below-ground biomass accumulation of young *Miscanthus* plants (Gossmann 2000; Thinggaard 1997). *Fusarium miscanthii* has been identified from *Miscanthus* straw in Denmark (Gams et al. 1999), and *Miscanthus* blight has been observed on ornamental *M. sinensis* (O'Neill and Farr 1996). *Miscanthus* in its native area hosts many of the pathogenic fungi that attack sugarcane and corn.

Miscanthus streak (Christian et al. 1994) and barley yellow dwarf virus have been found on *M. sacchariflorus* and, when transmitted by *Rhopalosiphum maidis* (corn leaf aphid), reduced above-ground biomass of *Miscanthus* by 23% (Huggett et al. 1999). The corn leaf aphid is also a vector for maize dwarf mosaic virus and mosaic virus of sugarcane, causing damage in both crops as well as in sorghum and other grasses (Teakle et al. 1989). *Miscanthus* is also a host of the sugarcane silk/woolly floss aphid, *Ceratovacuna lanigera Zehntner* (Florida Division of Agriculture 2002). The Asian miscanthus aphid, *Melanaphis sorini*, has spread from Japan and Taiwan to Florida and California but seems to be restricted to *Miscanthus* only (Halbert and Remaudiere 2000).

Nematodes are also able to transmit viruses, but play a more crucial role as direct pathogens, reducing plant yield. A cyst nematode, *Afenestrata orientalis*, was identified from *M. purpurecens* in Russia (Subbotin et al. 2001), and has been found on ornamental *M. sinensis* in Florida and California (Florida Division of Agriculture 2005). Rhizome propagation is one source of spreading nematodes, and this problem could be ameliorated if *Miscanthus* was propagated by seed or sterile in vitro methods. Rapid crop rotations as a tool to minimize nematode populations cannot be applied to perennial crops.

Given the pests and diseases already identified on *Miscanthus*, it should not be surprising if further pathogen infections arise in energy plantations, particularly if single clones are planted on large acreages. An integrated pest management program will be needed, including regular monitoring, plant and equipment hygiene during propagation and distribution, and new registered pesticides. At least as important, however, is identification of sources of crop genetic resistance to economically relevant pests and diseases as they appear. To date there are no studies describing pest- or disease-resistant genotypes of *Miscanthus*, let alone the underlying loci. This is a gap that needs to be addressed in the next several years to provide sources of variation for improved line development.

6.2.3.2 Advanced Breeding: Molecular Markers, Marker-Assisted Selection and QTLs

Molecular markers are useful in two ways: first, for genetic fingerprinting to identify sources of genetic variation for a breeding program; and second, for efficient introgression of useful traits based on linked markers (marker-assisted selection). In addition, genetic fingerprinting will no doubt reveal misidentified *Miscanthus* accessions for which reclassification will be necessary to ensure efficient use of these lines for breeding (Hodkinson et al. 2002). Marker-assisted selection has proven to be a powerful tool for increasingly rapid crop improvement (Babu et al. 2004; Varshney et al. 2006). This procedure will be particularly important in *Miscanthus* as many of the desired traits are expressed only after 2–3 years. Once robust associations between phenotypes and molecular markers have been established, selection of desired progeny from a cross, and dismissal of the undesired progeny will be possible in the seedling stage, considerably speeding breeding progress.

The immediate application of marker-assisted selection in *Miscanthus* is limited not only by the absence of sufficient sequence data but also, more importantly, by the lack of good phenotype data in populations segregating for desirable traits. Nonetheless, further development of sequencing and array-based technology (Gupta et al. 2008; West et al. 2006) will enable marker discovery likely before the entire *Miscanthus* genome is sequenced. Different ploidy levels in *Miscanthus*, self incompatibility, and the prospect of highly repetitive sequences, as commonly observed in other grasses, will pose challenges for the mapping process of *Miscanthus*. However, a preliminary *M. sinensis* map has already been created based on restriction fragment length polymorphisms (Atienza et al. 2002), following a crossing design suggested for outcrossing species (Song et al. 1999). Whereas single nucleotide polymorphism (SNP) and simple sequence repeat (SSR) de novo discovery will take some effort, markers from close relatives of *Miscanthus* like sorghum, sugarcane and corn could be readily exploited for the identification of trait-marker associations. The vast number of SSRs and SNPs available in these crops should easily allow the identification of suitable markers in *Miscanthus*. In fact, corn and sugarcane SSRs have been used to assess genetic diversity of *Miscanthus* with a success rate of up to 75% (Hernández et al. 2001; Cordeiro et al. 2003).

Quantitative trait loci for plant height, stem diameter (Atienza et al. 2003a), yield (Atienza et al. 2003b), and combustion quality parameters (Atienza et al. 2003c, d) have been identified in a single *M. sinensis* cross. In addition to extending this approach to additional populations, environments and traits, QTLs mapped in other grasses can inform a candidate gene/loci approach in *Miscanthus*. While no QTLs have been identified yet for flowering time in *Miscanthus*, application of knowledge about those in corn (Salvi et al. 2002), sorghum (Rooney and Aydin 1999) and sugarcane (Ming et al. 2002) should allow accelerated identification of markers associated with flowering time (Jakob et al, 2009). Another *Miscanthus* breeding goal is to establish propagation from seeds. As in other undeveloped crops, seed shattering is common in *Miscanthus*. A SNP responsible for the reduction of seed shattering during domestication has been identified in rice (Konishi et al. 2006; Li et al. 2006), and a comparative genomics approach could easily be applied to test for and select related alleles in *Miscanthus*.

6.3 Switchgrass

Switchgrass (*Panicum virgatum* L.) emerged as the leading North American native species for biofuel production, based on 25 years of research sponsored by the US Department of Energy (McLaughlin and Kszos 2005; McLaughlin et al. 2006; Sanderson et al. 1996). Currently available cultivars, although developed for forage purposes, already possess high dry matter yield capability, cut either once or twice per year. Like most forage species, with the possible exception of alfalfa (*Medicago sativa* L.), switchgrass is non-domesticated, with little history of intensive breeding. Because it is native to North America, switchgrass may be more accepted by producers, ecologists and environmentalists than non-native alternatives. Gonzalez-Hernandez et al. (2009) outlined the potential concerns when non-native species are grown on marginal lands for biofuel production. Jakob et al. (2009) suggested that reduced or no fecundity as found in current *Miscanthus* biomass germplasm would simultaneously maximize biomass yield and minimize undesirable invasiveness. However, this argument may not convince environmentalists intent on minimizing the risk of new crops becoming invasive. Native prairie species, such as switchgrass, have a different set of problems, particularly when they are highly allogamous and have anemochorous pollen dispersal. Improved cultivars with highly specialized traits have the potential to contaminate the native gene pool, thereby raising strong objections from ecologists, restoration biologists and environmental advocacy groups.

6.3.1 Switchgrass Phylogeny and Growth

Switchgrass phylogeny and growth have been described extensively by Vogel (2004) and reviewed most recently by Bouton (2007). *Panicum virgatum* L. is a member of the Paniceae tribe of the Panicoideae subfamily within the Poaceae. It is one of the dominant tall grasses of the native North American prairie and can be found from Canada to Texas east of the Rocky Mountains. It is highly allogamous due to an effective 2-locus gametophytic incompatibility system (Martinez-Reyna and Vogel 2002). It is composed of two ecotypes that are defined by their chromosome number $2n = 4x = 36$ for the Lowland type and $2n = 8x = 72$ for the Upland type. In general, the Upland type is found in more northern latitudes, whereas the Lowland type is more common in the South. These two types differ not only in chromosome number but also in the cpDNA trnL intron (Missaoui et al. 2006) such that a simple test could be developed to distinguish the two types without having to resort to chromosome counting, either by traditional cytological techniques or flow cytometry. Bouton (2007) pointed out that no natural hybrids are known to occur between these two groups. Therefore, they appear to be reproductively isolated, possibly due to a post-fertilization barrier similar to the EBN system in the genus *Solanum* (Martinez-Reyna and Vogel 2006). Lowland types are much more productive than Upland types, yielding up to 35.1 Mg dry matter ha^{-1} in a 20-year Auburn University small plot study (Table 6.1). Lowland types also had much smaller variation in rank than Upland types, except for the lowest yielding Upland cv. Blackwell. Interestingly, the mean 20-year yield of the highest yielding cultivar exceeded the highest yield of F_1 hybrids in a Nebraska study (Vogel and Mitchell 2008), indicating that location has a large effect on yield potential of this biomass crop.

Table 6.1 Yield of dry biomass and yield statistics for eight switchgrass cultivars in a 20-year Auburn University small plot study

Type	Cultivar	Yield (Mg ha^{-1})				Yield ranking within years							
		Mean	Max	Min	%^{a}	1	2	3	4	5	6	7	8
Lowland	Alamo	23.4	35.1	15.4	228	19	1	–	–	–	–	–	–
Lowland	Kanlow	18.7	27.2	12.6	215	–	16	3	1	–	–	–	–
Upland	Cave-in-Rock	14.7	22.8	6.9	328	1	–	7	5	3	2	2	–
Upland	Summer	14.5	21.3	7.8	275	–	2	4	5	8	1	–	–
Upland	Kansas Native	14.2	22.0	6.0	368	–	1	4	4	5	1	4	1
Upland	Pathfinder	13.1	19.8	7.2	277	–	–	2	5	2	8	1	2
Upland	Trailblazer	11.8	16.4	6.6	248	–	–	–	1	1	6	6	6
Upland	Blackwell	11.3	16.6	5.6	298	–	–	–	–	–	2	7	11

^{a}Maximum yield as a percent of minimum yield

6.3.2 Genetic Improvement of Switchgrass

The most appropriate breeding methodology employed obviously depends on the reproductive system of a species, and on the envisoned cultivar type. Population improvement strategies such as phenotypic or genotypic recurrent selection are a natural option for cross-pollinated, seed-propagated species such as switchgrass. While procedures involving selfing are suited to some highly domesticated, self-pollinated crops such as wheat (*Triticum aestivum* L.) or soybean [*Glycine max* (L.) Merr.], this approach is probably not feasible for switchgrass (Martinez-Reyna and Vogel 2002). Consequently, modification of the reproductive system has been mentioned as a possible breeding goal.

Aims for switchgrass breeding programs will depend somewhat on the intended end use of the biomass. If switchgrass is to be co-fired with coal in traditional coal-fired power plants, gasified to produce synthesis gas, or subject to catalytic conversion, then biomass composition is relatively unimportant. However, breeding efforts could still be directed at minimizing the content of minerals that might be abrasive or induce slagging in boilers. In contrast, efficiency of biochemical conversion processes, such as enzymatic hydrolysis and fermentation to ethanol or other compounds, are highly sensitive to feedstock composition (especially lignification). Therefore, if the end use is conversion by one of these processes, improving biomass composition should definitely be a breeding goal.

Regardless of the breeding goal, there are agronomic and plant breeding strategies to improving usable product, the former possibly being more easiliy achievable in terms of both time and resources. However, Sanderson et al. (2006) pointed out that it would take a concerted effort among scientists, extension staff, and producer-cooperators to develop profitable and sustainable management practices for each agro-ecoregion.

Jessup (2009) lists ten criteria for viable dedicated energy crops (DEC), of which switchgrass fulfills all but one: vigorous establishment. Among the criteria, he identified seed resources as being critical, and implied propagation by seed (as opposed to vegetative propagation) was desirable. These two "critical features" are mutually exclusive. It is hard to imagine switchgrass having the speed of establishment of annual ryegrass (*Lolium multiflorum* Lam.). Similarly, rapid establishment from seed and non-invasiveness are somewhat incompatible. Nevertheless, the "critical features" laid down by Jessup (2009) may serve as a useful road map for biomass crop improvement.

6.3.3 Conventional Breeding Challenges

The current major limitation to genetic improvement of switchgrass by conventional breeding is the long time it takes for a stand to reach its full productive potential, which is two or three years (Parrish and Fike 2005). Establishment failure has emerged as one of the leading problems with switchgrass, particularly in regions where there is competition from weedy C4 grasses such as large carbgrass [*Digitaria sanguinalis* (L.) Scop.]. While genetic solutions may be envisioned, effective crop management strategies to address this problem have not been fully explored and are likely to make progress much more easily and quicker than breeding approaches.

6.3.3.1 Trait Targets

Stand establishment

Once established, switchgrass is one of the most drought-tolerant grasses known. In the 20-year Auburn University study a total crop failure was not observed as the least productive cultivar still yielded 5.6 Mg ha^{-1} in the worst year. During the same time period, non-irrigated cotton and maize—the dominant row crops in the region—were subject to several total crop failures. Even in drought years, switchgrass can produce appreciable biomass. However, it is prone to stand failure when inadequate moisture is available during establishment.

Selection for shoot number at the seedling stage did not improve seedling vigor and establishment under field conditions (Smart et al. 2003a, b, 2004). This suggests a possible advantage from developing techniques to establish this species vegetatively. The argument that this would not be feasible on a large scale can be refuted based on the fact that millions of acres of pine trees (*Pinus elliottii* Engelm.) and sterile hybrid bermudagrass (*Cynodon dactylon*) are established with seedlings and sprigs, respectively. Pine trees undergo only two thinnings and one final harvest during their 30-year lifespan, whereas switchgrass stands can be harvested for at least 20 consecutive years without stand decline, if properly managed. Establishment costs are thus amortized over many years. Furthermore, the higher cost of this procedure compared to planting seed might be largely offset with government financial assistance programs for establishment of perennial energy crops.

The approach to vegetative establishment would involve simple biotechnology, either through generation of plantlets grown from embryoids, or plantlets grown from nodal cuttings. Work at the University of Tennessee demonstrated regeneration of switchgrass plants from mature caryopses and meristematic leaf explants 20 years ago (Denchev and Conger 1994, 1995; Gupta and Conger 1998, 1999) and produced 500 progeny from a single switchgrass parent plant nodal culture (Alexandrova et al. 1996a, b). However, further refinement of these procedures would be necessary to adapt them to commercial scale operations.

This approach also has other advantages, such as deployment of genetically homogeneous, but heterozygous F_1 hybrid plantlets directly for production purposes, or deployment of parental clones for large-scale production of F_1 hybrid seed. Furthermore, it would be conducive to the development of a separate industry segment to provide plantlets, which could be an attractive local economic development opportunity.

Biomass Yield

Increasing total biomass yield will, for the foreseeable future, be the major breeding objective for switchgrass. Detailed heritability studies (Das et al. 2004) confirm the existence of heritable genetic variation in yield. Work at Auburn University has shown that HS-families from a collection of wild switchgrass accessions from the southeastern US exceed the highest yielding commercially available cultivar (Alamo) by as much as 25% (E. van Santen, unpublished data). Clearly, there is plenty of genetic variation present for continued improvement in switchgrass yield over a relatively long period of time (Bouton 2007).

Natural heterotic groups of switchgrass seem to exist as F_1 hybrids between the 4x lowland cv. Kanlow and the 4x upland cv. Summer, which exhibited mid-parent heterosis both in spaced-plant as well as simulated sward conditions (Vogel and Mitchell 2008; Martinez-Reyna and Vogel 2008). These authors also outlined a scheme to produce commercial F_1 hybrids through propagation of parental clones, relying on the strong SZ gametophyte self-incompatibility to prevent the production of selfed seed. In this regard, it should be recognized that the heterotic groups in maize (*Zea mays* L.) are the result of human-mediated crop evolution through intensive breeding, and are not the result of a natural event, and this has resulted in modern maize being largely a crop created by humans. Provided that switchgrass undergoes a similarly long period of genetic improvement, there is no reason to assume that heterotic groups could not be created.

Pest and Disease Resistance

Although no major devastating pest or disease issues have been reported in switchgrass to date, there are initial indications that diseases not only reduce yield, but also change nutrient utilization efficiency (W. Wood, E. van Santen, unpublished data).

Switchgrass in plots where disease was controlled through regular fungicide applications reached a maximum yield of 300% above the zero N control at 120 lbs N/acre, whereas the maximum yield without fungicide application (285%) was achieved at the highest N rate in the study (150 lbs N/acre), thus demonstrating a N x fungicide interaction. Bouton (2007) pointed out that disease pressure is likely to increase as larger contiguous stands of switchgrass are established, and a similar argument could be made for pests.

6.3.3.2 Advanced Breeding: Molecular Markers, Marker-Assisted Selection and QTLs

Molecular breeding of switchgrass is still in its infancy. Currently, there is only a single marker 4x population available (Bouton 2007). The polyploid nature of switchgrass creates some challenges for use of advanced breeding procedures, but poses no absolute barrier to employing molecular techniques. Results from early stage experiments, such as *Agrobacterium*-mediated transformation (Somleva et al. 2002) or expressed sequence tag (EST) markers, have been reported (Tobias et al. 2005). If successful, research to modify the lignin pathway in switchgrass (Bouton 2007) will certainly have an impact on efficiency of enzyme-based conversion processes, as it addresses one of the fundamental limitations of these procedures.

6.4 Sugarcane

Sucrose and other sugars from sugarcane and starch from corn are the primary substrates used in the commercial production of ethanol today, as their conversion is simple, economical, and has been done for centuries (see Chaps. 4, 5). Further interest in using sugarcane for biofuel production has increased with the emerging promise of economically feasible cellulosic biofuel technologies. Utilization of the entire above-ground sugarcane plant and the development of high fiber/low Brix types of sugarcane as a potential bioenergy feedstock using these cellulosic conversion technologies has been reviewed (Alexander 1985, 1991; Coombs 1984; Tew and Cobill 2008). In the following analysis, the focus is on the possibility of growing sugarcane as a bioenergy crop outside of traditional cane growing areas where temperatures may be colder, where soil moisture may be lacking or excessive, or where soils may not be suited for the sustainable production of many of the annual food crops.

6.4.1 Sugarcane Phylogeny and Growth

Before one can speculate on the utility of sugar cane as a dedicated feedstock for the production of bioenergy and what the future may hold for this crop, it is important to understand a little about the crop, which, if dedicated solely to the production of energy can be called "energy cane". Sugarcane is a large-stature, jointed grass that is cultivated as a perennial row crop, primarily for its ability to store sucrose in the stem, in approximately 80 countries in tropical, semi-tropical, and sub-tropical regions of the world (Tew 2003). It is one of the most efficient C4 grasses in the world, with an estimated energy in:energy out (I/O) ratio of 1:8 when grown for 12 months under tropical conditions and processed for ethanol instead of sugar (Bourne 2007; Heichel 1974; Macedo et al. 2004; Muchow et al. 1994). Under more temperate environments, where temperature and sunlight are limited, I/O ratios of 1:3 are easily obtainable with current sugarcane cultivars if ethanol production from both sugar and cellulosic biomass is the goal (Tew and Cobill 2008).

To be sustainable, a dedicated bioenergy crop has to produce high and consistent yields economically. In addition to contributing to the efficiency of the crop, the length of the growing season also impacts the number of ratoon sugarcane crops that can be harvested from a planting. Ratoon crops of sugarcane generally produce higher yields when the preceding year's crop of a crop cycle is harvested late in the growing season. Allowing the crop to mature naturally by delaying harvest can also increase the number of ratoon crops. In reality, growers expect at least three or four annual fall harvests from a single planting. As such, sugarcane is generally grown as a monoculture, with fields being replanted every 4 or 5 years. Sugarcane is vegetatively planted by laying 1.8- to 2.4-m long stalks end-to-end in a planting furrow along the row and covering with 5–10 cm soil; 1 ha seedcane can plant 6–10 ha, depending on the length and number of stalks at the time of harvesting the "seedcane" for planting, and the number of stalks per meter of row being planted. When harvesting seedcane for planting, stalks are cut at the soil surface and at the last mature node at the top of the stem. An alternate method of planting is to plant 30- to 45-cm stalk pieces (billets) that can be harvested with the same chopper harvester used to harvest sugarcane for delivery to sugar mills. This requires more seed cane, reducing the ratio for planting to 3–4 ha planted per ha harvested for seedcane. Once the stalks are planted, new plants emerge from the axillary buds located in the nodal regions along the stalk. Growers produce most of their own seedcane for planting; hence, planting is generally done a few weeks prior to the beginning of the harvest season to insure that stalks are plentiful and tall. Vegetative planting is often considered a drawback to the planting of this crop by growers who are accustomed to planting large areas of seeded crops relatively quickly with one tractor and one planter. Vegetative planting of sugarcane is an expensive process as it requires considerable labor and equipment, is relatively slow, and requires that the grower plant sugarcane that would normally be sent to the raw sugar factory for processing. However, vegetative planting also has advantages, especially when planting must be done under conditions of less than ideal seedbed preparation. Thus, the planting of sugar cane as a dedicated bioenergy crop on less productive lands would be seen as a good alternative.

Sugar cane, once delivered to the raw sugar factory for milling, is separated into its water, Brix (soluble solids of which approximately 80% is sucrose), fiber (bagasse) and sediment (ash, soil, etc.) fractions. The bagasse fraction in commercial sugar varieties consists of 38% cellulose, 19% hemicellulose, 22% lignin, 4% protein, and 3% ash, with the remaining 14% consisting of sugar, soil from harvesting, and other types of solids (Baoder and Barrier 1990). This average fiber yield does not take into consideration leafy material removed in the field during the harvesting process, or the conditions under which the crop was harvested. When sugar cane is harvested mechanically for sugar without a preharvest burning of the standing cane, 4–6 Mg ha^{-1} leaf litter is deposited back on the soil surface (Richard 1999; Viator et al. 2009a, b). Moreover, in Louisiana, where sugar cane is grown on mineral soils, if the crop is harvested under wet and muddy conditions it is not uncommon to see 5–10% contamination with soil. Sugar and fiber levels in the harvested cane stalks are generally dependent on the variety, the length of the growing season, the amount of extraneous matter present, and the harvesting conditions.

Growers are paid on the amount of Brix (sucrose and molasses) delivered to the raw sugar factory less a processing fee. Costs associated with the delivery of the sugarcane stalks to the raw sugar factory are paid by the raw sugar factory. Since the leafy material adds to harvesting and shipping costs and affects sugar recovery, the majority of the leaves are removed in the field, either mechanically during harvest, or by burning prior to harvest. Currently, the bagasse has value to the raw sugar factory as it is burned in the factory's boilers to generate the steam and electricity needed to process the cane into its saleable sugar and molasses components. The amount of bagasse needed to power a biorefinery would be considerably lower if the need for sugar crystallization is eliminated, because more would be available for conversion to ethanol. Currently, the post-harvest leaf litter is burned because it inhibits the growth of the subsequent ratoon crop (Richard 1999; Viator et al. 2006, 2009a, b). There would be some value in collecting the leaf litter with the stalks in a cellulosic conversion process (Dawson and Boopathy 2007).

Sugar cane production is generally concentrated on farms located less than 100 km from the raw sugar factory to minimize transportation costs. The crop is susceptible to the rapid spread of a number of bacterial, fungal, and viral pathogens that can be spread easily by machinery and wind currents. These pathogens can affect the yield and ratooning ability (number of yearly harvests per planting) of the crop. Race changes of some of these pathogens are common and the industry is always susceptible to new diseases. Insects, primarily stalk borers, grubs, and aphids, also plague the industry. The compactness of the industry and the fact that the crop is grown continuously as a monoculture makes sugarcane especially vulnerable to the rapid spread of diseases and insects.

As with most crops, sugarcane yields are influenced by temperature and soil moisture during the growing season. Under tropical conditions, the new growing season for sugar cane begins after the previous crop is harvested. In more temperate regions where the crop is harvested in the fall, the crop must go through a period of winter dormancy after harvest, with the new growing season beginning essentially after the last freeze event of the new year. Temperature, especially cool/cold temperatures at the start of the growing season, influences the emergence and early establishment of the crop. In the more temperate regions where sugarcane is grown, there is also the concern regarding stalk-killing freezes of −5°C or lower during the harvest season. When a mature stalk freezes and splits open, bacteria capable of converting the sugars to starch enter. The presence of starch affects sugar crystallization. However, the impact of this starch on the process in a biorefinery where ethanol is the primary product is not known. Where freezing is a concern, raw sugar processors start the milling season before the crop physiologically matures to ensure that the milling season is completed before the historical date for these killing freezes.

The timely availability of soil moisture during the growing season has a great influence on yield. Dry soils at the beginning of the growing season encourage tillering, which ultimately results in more stalks being produced. Dry conditions at the end of the growing season, i.e., start of the milling season, slows growth and promotes natural maturation and sugar accumulation, and improves harvesting and milling efficiencies. More sugar with less harvested tons equates to enhanced efficiency during harvesting and processing where sugar is the targeted saleable commodity. Adequate soil moisture during the middle of the growing season results in increases in stalk diameter and length. In Louisiana, for instance, the time of active stalk growth, i.e., the "grand growth period" is in June, July and August when stalk lengths can increase by 18–25 cm/week.

6.4.2 Genetic Improvement Needs

The success of the sugar cane industry has, and continues to be, dependent on the development of new hybrids with superior yields and increased resistance to many of the abiotic and biotic stresses previously mentioned. This formula will not change if the crop is being grown as a dedicated feedstock for the production of liquid biofuels or electricity. Successful hybridization begins with the introgression of desirable traits from the wild relative of sugarcane, *Saccharum spontaneum*. Early generation progeny from these crosses with elite sugarcane clones exhibit high levels of hybrid vigor, which translates into increased cold tolerance, greater ratooning ability, enhanced levels of moisture, increased insect and disease tolerance, and more efficient nutrient utilization (Legendre and Burner 1995). Much of the vigor of these early generation hybrids is lost in a conventional breeding program for sugar, as progeny from these crosses must be backcrossed with elite high sugar-producing clones three to four times before a commercial sugarcane variety can be produced. These early generation hybrids would be considered ideal candidates as dedicated cellulosic biomass crops, i.e., energy canes. Most of these varieties can produce over 30 dry Mg ha^{-1} annually over four fall harvests, with about 20 Mg ha^{-1} being fiber and 10 Mg ha^{-1} being Brix (Anonymous 2007).

The theoretical maximum for above-ground sugarcane biomass (total solids) yield is estimated to be 140 Mg ha^{-1} annually (Loomis and Williams 1963). This is dependent on temperature and sunlight, and would probably occur under tropical conditions. Sugarcane breeding programs have reported sugar yield gains in the order of 1–2% per year (Edme et al. 2005). The economic sustainability of growing energy cane in non-traditional cane growing regions will require further biomass yield gains of this magnitude, or greater, with a goal of ensuring that the I/O ratio of 1:8 projected for tropical countries can be met and ultimately exceeded under non-traditional cane growing conditions. Sugarcane varietal development using conventional breeding and selection techniques takes 12–13 years from the time a cross is made, until a variety is released for planting. Since the crop is propagated vegetatively, it takes an additional 3–4 years to have a sufficient supply of material from a newly released variety to supply a biorefinery for processing. Therefore, geneticists will need a clear signal from bio-processors as to the value of the sugar, leaf litter, bagasse, and the water that this crop produces.

Yield consistency over multiple yearly harvests is one of the advantages of growing sugarcane for sugar, and this will be even more important when it is grown as a dedicated bioenergy feedstock in more temperate climates. The fact that vegetative planting lends itself to planting in seedbeds that would discourage the planting of seeded crops, especially small seeded grass crops, was discussed earlier. Another advantage is the fact that the crop is planted in the late summer where summer and fall rains help establishment. Furthermore, the new crop is harvested in the fall and early winter in the year following planting. Rainfall has a minimal effect on the harvesting operation and the quality of the harvested crop. In contrast, for seeded crops, too much or too little rainfall, or cold temperatures at planting, affect crop emergence. Likewise, wet conditions during harvest affects grain quality in grain crops. In both scenarios, yields can be significantly reduced or the crop could be completely lost.

Allowing the crop to desiccate in the field and perhaps devoid itself of some of its leaves and moisture, as is proposed for many of the perennial grasses being considered for biofuels, is not an option for this crop, as the new growing season should begin as soon after harvest as possible. Consequently, the crop will have to be harvested green and dewatered so that the fiber can be stored and processed later in the year. The value of this liquid is in question because it will add to transportation costs. However, if water is needed for the digestion of the fiber or the maintenance of the bagasse under anaerobic conditions to minimize deterioration during outside storage, it would be present at no additional charge. What is also overlooked is the fact that the water contains sugar that is easily and much more cheaply converted to ethanol. Furthermore, in some conversion processes the yeast used in fermentation needs a substrate to grow and multiply on, and sucrose is an ideal substrate. Conceivably, the biorefinery would have two processes for the production of biofuel, with one having sugar obtained from de-watering at the biorefinery as the feedstock, and the other fiber (bagasse). Sugar, in addition to being an excellent substrate to grow yeast, can also be converted easily into other liquid transportation fuels such as jet fuel and diesel. Economics would have to be considered with these options.

Typically, sugarcane contains 30–32% solids (fiber and Brix). It may be that two types of energy cane will have to be developed, i.e., Type I and Type II as proposed by sugarcane geneticists (Tew and Cobill 2008). Type I would have slightly more fiber (14–18%) than conventional sugarcane varieties and approximately the same amount of Brix (14–18%). Type II would have 20–26% fiber on a fresh weight basis, with 6–12% Brix. Type II generally corresponds to the earlier generation hybrids (F_1 and BC_1 hybrids) that have higher biomass yields and more stress tolerance, and hence are better adapted to climates outside the traditional cane-growing areas. In more temperate climates, it would be possible for growers to grow both Type I (BC_2 and B_3 hybrids) and Type II varieties, with the Type I varieties being harvested early in the fall and the Type II varieties in late-fall and early-winter. Type II varieties would have the longer growing season; hence, they would be expected to have the greatest I/O energy ratios.

Having said this, the greatest needs to make energy cane a suitable feedstock for the cellulosic industry and extend its range of geographic distribution outside of traditional sugar cane-growing areas are: cold tolerance for expansion outside of tropical areas, drought- and flood-(saturated soil) tolerance, as this crop will probably be grown on marginal soils that may be prone to flooding, or where irrigation is difficult, insect and disease resistance, and a further exploitation of some varieties of sugar cane that encourages symbiotic relationships with nitrogen-fixing bacteria.

6.4.3 Genetic Improvement Strategies

In its present from, sugarcane (*Saccharum* spp.; $2n = 100 - 130$) is a genetically complex crop with a genomic makeup that results from successful interspecific hybridization efforts, involving primarily *S. officinarum* and *S. spontaneum* (Tew and Cobill 2008). Improvement of sugarcane for increased energy efficiency and adaptability to a wide range of environments is considered by many geneticists as synonymous with "genetic base broadening", i.e., utilization of wild *Saccharum* germplasm, particularly *S. spontaneum* in sugar cane breeding programs (Ming et al. 2006). *S. spontaneum*, considered a noxious weed in the US because it produces both viable seed and rhizomes, can be found in the continents of Africa, Asia, and Australia in environments ranging from the equator to the foothills of the Himalayas. This makes it an excellent source of a number of valuable genes (Mukherjee 1950; Panje and Babu 1960; Panje 1972; Roach 1978). Through the use of photoperiod facilities, *Saccharum* hybrids can also be crossed successfully with their close relatives *Erianthus* (Cai et al. 2005) and *Miscanthus* (Lo et al. 1986). Success with these crosses has further expanded the genetic diversity of this crop, and should open the door to further improvements in production efficiency. The rich source of genetic diversity and the plasticity of autopolyploid genomes also offers a wealth of opportunities for the application of genomics and related technologies to increase biomass production of sugarcane (Lam et al. 2009).

Many of the advances in sugarcane yields over the years have come from the development of improved varieties through conventional breeding programs. By enhancing the level of stress tolerance through conventional breeding techniques, the geographic area of distribution can be expanded to more temperate regions.

In the US, for instance, it is conceivable that the area devoted to this crop could be tripled, thus making it a more attractive market for biotechnology companies with proprietary genes to further enhance the level of stress tolerance, or introducing genes for the production of saleable byproducts without the labeling restrictions encountered in food crops. Another advantage of this crop is the fact that it does not flower naturally in sub-tropical and temperate climates, because relatively long nights (>12.5 h) are needed to promote floral initiation. Generally, sugarcane is harvested prior to this time, or in temperate climates, it is exposed to a frost that kills the terminal meristem, thus preventing flowering. If it does flower in commercial fields located in temperate regions, the pollen that is produced is not viable due to the cold temperatures, thus minimizing the chance of inadvertent cross pollination.

6.5 Sorghum

Sorghum evolved and was domesticated in arid areas of Northeastern Africa; it has been found in archaeological excavations estimated to be over 6,000 years old (Kimber 2000). After domestication, sorghum spread across Africa and into the continent of Asia through traditional trade routes. As the crop moved, new races were selected with specific adaptation to the new region. The species is relatively new to the Americas and Australia, arriving in the past 200–300 years. As a consequence of domestication and distribution, sorghum is an extremely diverse species with a wide range of variation within domesticated lines. This variation has resulted in (or is the result of) many different end uses.

6.5.1 Sorghum Phylogeny and Growth

Sorghum is most widely known as a cereal grain crop. It is the fifth most widely grown and produced cereal crop in the world. However, in many regions of the world, sorghum is equally if not more important as a forage crop. While accurate statistics for forage use are not available, it is very likely that sorghum's use as forage exceeds its production as a cereal grain. In addition to forage and grain, sorghum types high in stalk sugar content, and extremely lignified types (for structural building) have been grown throughout the world.

Given the current interest in bioenergy, sorghum is now being developed as a potential bioenergy crop. This designation is not new; sorghum was mentioned prominently as a potential energy crop over 20 years ago (Burton 1986). The interest in the crop is justifiable, based on several independent factors that separately indicate good potential but, when combined, clearly designate sorghum as a superior choice for bioenergy production. These factors include yield potential and composition, water-use efficiency and drought tolerance, established production systems, and the potential for genetic improvement using both traditional and genomic approaches.

Whether measured in grain yield or total biomass yield, sorghum is a highly productive C4 photosynthetic species that is well adapted to warm growing regions. The optimum type of sorghum to be grown for biofuels production is highly dependent on the type of conversion process that will be used. Hybrids of grain sorghum will provide starch for conversion, while sweet sorghum accumulates sugar in the stalk that could be used for production of liquid fuels. Finally, cellulose is produced by all types of sorghum, and specific genotypes are being developed to maximize this attribute. Each type thus fits a different production system; no other species has the flexibility to produce large quantities of starch, sugar or cellulose.

Sweet sorghums accumulate high levels of sugar in the stalk of the plant. Initially identified and used as alternative sugar sources, they are very amenable for conversion to ethanol, using methodology similar to that used in production of ethanol from sugarcane. In the mid-1970s significant research was conducted to explore the development of sweet sorghum as a feedstock for liquid biofuels and electric power production (McBee et al. 1987). Breeding programs were initiated to develop high yielding sweet sorghum specifically for ethanol production. Hallam et al. (2001) compared perennial grasses with annual row crops and found that sweet sorghum had the highest yield potential, averaging over 35 Mg ha^{-1} (dry weight basis), and also performed well when intercropped with legume species.

Specific types of photoperiod sensitive (PS) sorghums are very efficient at producing biomass; primarily structural carbohydrates. These high biomass sorghums are inherently an attractive bioenergy crop due to their high yield potential and growth habit, which allows more flexible management of the crop. They often produce biomass yields in excess of 30 Mg ha^{-1} (dry weight), and genetic modification could extend the potential of these types of hybrids to a wide range of environments. The unique feature of these sorghums is strong photoperiod sensitivity; they have long periods of vegetative growth. Under irrigation in the Texas panhandle, McCollum et al. (2005) reported yield of commercial PS sorghum hybrids as high as 80 Mg ha^{-1} (65% moisture) from a single harvest. In subtropical and tropical conditions, single cut yields are generally lower, likely due to increased night temperatures, but cumulative yields are higher due to the ratoon potential of the crop in these environments. Total biomass yields as high as 30 Mg ha^{-1} (dry weight basis) were reported near College Station, Texas (Blumenthal at al. 2007).

Composition of sorghum is highly dependent on the type that is produced; i.e., grain sorghum, sweet sorghum, forage and cellulosic (high biomass) sorghum. Sorghum grain is high in starch, with lower levels of protein, fat and ash (Rooney 2004). Significant variation in the composition of grain is controlled by both genetic and environmental factors, making consistency in composition a function of the environment at the time of production, and these factors influence ethanol production (Wu et al. 2007). Juice extracted from sweet sorghum is predominantly sucrose, with variable levels of glucose and fructose, and in some genotypes, small amounts of starch are detectable (Clark 1981; Billa et al. 1997). In forage, PS, and high biomass sorghums, the predominant compounds that are produced are structural carbohydrates (lignin, cellulose and hemi-cellulose; McBee et al. 1987; Monk et al. 1984). Amaducci et al. (2004) reported that the environment influences sucrose, cellulose and hemicellulose concentrations, while lignin content remains relatively constant.

6.5.2 Genetic Improvement

There are significant opportunities for the further improvement of sorghum as a bioenergy crop. The established history of sorghum provides an immediate breeding and seed production infrastructure. In addition, the genetics of the crop are relatively simple when compared to other prominent bioenergy crops. To that point, sorghum genome research has advanced to where the generation of superior biomass sorghum genotypes can be addressed using genome scale analysis in conjunction with other systems-based approaches. Significant research has been completed to build integrated genetic, physical, and comparative maps of the sorghum genome (Klein et al. 2000; Menz et al. 2002; Bowers et al. 2003), to validate map-based gene cloning in sorghum (Klein et al. 2005), and to carry out an in-depth gene expression analysis (Pratt et al. 2005; Buchanan et al. 2005). Finally, the completion of an 8X whole genome shotgun sequence of sorghum is another major milestone that will facilitate genetic modification of the crop (Paterson et al. 2009). Ongoing breeding and research will utilize this emerging sorghum genome information and technology platform to advance the understanding of the genetic and biochemical basis of superior sorghum biomass generation.

There are several traits of specific importance to sorghum improvement as it relates to bioenergy production. These include, but are certainly not limited to, maturity and height, drought tolerance, pest tolerance and/or resistance, and composition and/or quality. Improvements in these areas will increase yield potential, protect existing yield potential, and enhance conversion efficiency during processing.

Whether the target is a sweet sorghum or a high biomass PS sorghum, the immediate need is the development of hybrid versions of both of these crops. Currently, commercial quantities of hybrid seed are available for forage sorghums, some of which have useful application in bioenergy production, but both sweet and PS sorghum have added benefits that are further enhanced by hybridization. Furthermore, hybrid seed production systems allow for the production of commercial quantities of seed on these types. It would be difficult, if not impossible, to produce commercial quantities of seed of traditional varieties. Production of sweet sorghum and PS hybrids uses the same technology employed for generation of standard sorghum hybrids, but the germplasm is modified.

The renewed interest in bioenergy has increased research activities in sweet sorghums, especially in the development of hybrids. Hybrid development is expected to result in modest yield increases (Clark 1981) and, more importantly, make the logistics of seed production feasible. Currently, sweet sorghum hybrid production is limited by a paucity of grain type seed parents with high stem sugar content. These types are critical to maintain high sugar yields in the hybrids. Several groups have been developing these seed parents; experimental hybrids derived from them are now in experimental testing. In addition, there is a real need for further development of pollinator parents. These first generation sweet sorghum hybrids are slightly higher in yield than traditional varieties, but seed production yields are much higher and the seed is easier to harvest. Currently, existing sweet sorghum varieties are used as pollinator parents, but complementary selection of the pollinator parent with the seed parent will result in second generation sweet sorghum hybrids that are easier to produce and are expected to have even higher yields.

Production of PS sorghum hybrids utilizes the Ma5/Ma6 photoperiod sensitivity genes (Rooney and Aydin 1999). These two gene loci interact epistatically to produce a PS hybrid. Using this system, it is possible to use the photoperiod insensitive parental lines to produce PS hybrids. Consequently, seed production of these hybrids occurs in traditional seed production regions. Marker-assisted breeding (MAB), using markers associated with photoperiod sensitivity, is crucial for the conversion of newer and higher yielding parental lines for the production of PS hybrids.

While the reason for developing bioenergy feedstocks is to produce renewable fuel, one of the critical components in their production will be water. Thus, both drought tolerance and water-use efficiency is critical as many of these feedstocks will be produced in marginal environments where rainfall is limited and irrigation is either too expensive or would deplete water reserves. Sorghum is more drought-tolerant than many other biomass crops. Depending on the type of biomass production in sorghum, both pre- and post- flowering drought tolerance mechanisms will be important. In sweet sorghum, both traits are important but there has been little research into the impact of drought stress on sweet sorghum productivity. For high biomass photoperiod sensitive sorghums, preflowering drought tolerance is critical because, in most environments, this germplasm does not transition to the reproductive phase of growth. Each type of tolerance is associated with several phenotypic and physiological traits; these relationships have been used to fine map QTL associated with both pre- and post-flowering drought tolerance. Traits that have been associated with drought resistance include heat tolerance, osmotic adjustment (Basnayake et al. 1995), transpiration efficiency (Muchow et al. 1996), rooting depth and patterns (Jordan and Miller 1980), epicuticular wax (Maiti et al. 1984) and stay-green (Rosenow et al. 1983). Combining both phenotypic and MAB approaches should enhance drought tolerance breeding in bioenergy sorghums.

Disease and insect resistance have always been important traits in traditional sorghum breeding programs. However, bioenergy sorghums will likely be grown in different environments compared to traditional grain sorghum production environments, and this will likely mean that the nature of pests and diseases will shift. For example, grain weathering resistance is critical in grain sorghum, but it will be of diminished importance in a crop for which total biomass is the primary yield component. Likewise, any disease that destroys the whole plant must be mitigated with either genetic resistance or management practices.

There has been little to no research into the composition of both sweet sorghum and energy sorghum. Murray et al. (2008) indicated that there was no correlation between biomass yield and composition in sweet sorghum, suggesting that there is an opportunity to improve both biomass yield and composition in the crop. Corn (2009) indicated that both genotype and environment influence composition of both juice and bagasse from sweet sorghum, but with adequate testing, further improvements could be made. Initial analysis of composition in a wide range of photoperiod sensitive sorghum hybrids reveals significant variation for composition of lignin, cellulose and hemicelluloses (D. Packer, personal communication). Further analysis is needed to partition this variation into genotypic and environmental variation. Even if just a portion of this variation is due to genetic effects, then there is a real opportunity to manipulate composition and optimize it to the specifications of different end users.

The molecular genetic resources available in the sorghum species are the most advanced among all of the potential energy crops. These tools can be readily applied to the improvement of sorghum for biofuel production. By combining these molecular genetic resources with traditional breeding approaches, it should be possible to rapidly develop and deploy improved dedicated bioenergy sorghums that meet the needs of both crop and biofuel producers.

6.6 Integration of Grasses into Cellulosic Biomass Supply Systems

Biomass supply systems can be considered at several levels, including national and local, biorefinery-based scales. On a national scale, Perlack et al. (2005) assessed the feasibility of the US producing a billion tons of cellulosic biomass annually by 2030. Their analysis suggested that, with reasonable adjustments to current agricultural and forestry practices, it might be possible to produce 1.36 billion t per year. This estimate included 377 million t from 55 million acres of perennial herbaceous biomass crops, implying an average annual yield of 6.85 t/acre. Based on current status and future prospects of the candidate grass energy crops discussed above, this goal appears to be entirely feasible, and possibly conservative. In addition, the study suggested that 428 million t of crop residues, (mainly corn stover and wheat straw, which is also grass biomass) could be produced, bringing the total from grasses to approximately 800 million t annually. However, in contrast to the production estimate for perennial crops, estimates of the availability of crop residues seems to be somewhat optimistic.

At the local biorefinery scale, grass energy crops are attractive from several points of view. Because they are low in lignin relative to wood, they result in higher conversion efficiency for biochemical conversion processes like enzymatic hydrolysis and fermentation to ethanol. Therefore, they may be the preferred feedstock for biorefineries of this type. In cases where the conversion process is less sensitive to lignin content, as is the case with thermochemical processes like gasification or pyrolysis, it may be possible to mix grass with wood in the feedstock supply system. Advantages of using this approach might include a reduction in the average hauling distance for feedstock compared to supplying wood alone, the lower moisture content of dried grass offsetting the relatively high moisture content of wood chips, relatively quick establishment of a feedstock supply with annuals, with integration of perennials into the system over time, and environmental attractions such as increased biodiversity.

Feedstock price has a critical impact on economic viability of cellulosic biomass supply systems. Under current economic conditions, grass biomass is more expensive than wood, which is more expensive than municipal solid waste (MSW). Therefore, use of wood and/or the cellulosic fraction of MSW in mixed cellulosic biomass supply systems can offset the higher cost of grass biomass. Government incentives are another option for offsetting the cost of establishment and production, and the opportunity cost related to the lag between planting and reaching full yield.

6.7 Conclusions

Warm season C4 grasses offer considerable potential as cellulosic biomass crops, due largely to their efficient C4 photosynthetic pathway facilitating high yields. Species such as sugarcane and sweet sorghum also provide sugar as a feedstock for production of liquid biofuels. Relative to woody crops, grasses have a shorter period from planting to attaining full yield, the ability for some to be sun- or air-dried to a relatively low moisture content, and higher conversion efficiency for biochemical conversion processes because of lower concentrations of lignin. Very large genetic diversity and relatively little work on plant breeding to date suggest that substantial progress can be made in genetic improvement of grass biomass crops in the foreseeable future. In this regard, traits that are of particular interest include high yield potential with low input requirements, resistance to diseases, pests and drought, adaptation to a wide range of soils and climates, and biomass composition that is optimized for the intended conversion technology. Grasses can be used alone or in synergistic mixtures with wood and/or MSW to develop cellulosic biomass supply systems.

Acknowledgments The authors from Mendel acknowledge their Mendel and Tinplant colleagues for helpful discussions and insights, particularly Katrin Jakob, Erik Sacks, Fasong Zhou, Martin Deuter and Cora Münnich.

References

Alexander AG (1985) The energy cane alternative. Sugar series, vol 6. Elsevier, Amsterdam
Alexander AG (1991) High energy cane. In: Payne JH (ed) Cogeneration in the cane sugar industry. Elsevier, New York, pp 233–242

Alexandrova KS, Denchev PD, Conger BV (1996a) Micropropagation of switchgrass by node culture. Crop Sci 36:1709–1711

Alexandrova KS, Denchev PD, Conger BV (1996b) In vitro development of inflorescences from switchgrass nodal segments. Crop Sci 36:175–178

Amaducci S, Monti A, Venturi G (2004) Non-structural carbohydrates and fibre components in sweet and fibre sorghum as affected by low and normal input techniques. Ind Crops Prod 20:111–118

Amalraj VA, Balasundaram N (2006) On the taxonomy of the members of 'Saccharum complex'. Genet Resour Crop Evol 53:35–41

Andersson NJ (1855) Om de med Saccharum beslägtade genera. Öfvers Kungl Vet Akad Förn Stockholm 12:151–167

Anonymous (2007) Release of three high fiber sugarcane varieties: L 79-1002, Ho 00-961, and HoCP 91-552. Sugar Bull 85(10):21–26

Atienza G, Satovic Z, Petersen K, Dolstra O, Martin A (2002) Preliminary genetic linkage map of Miscanthus sinensis with RAPD markers. Theor Appl Genet 105:946–952

Atienza SG, Satovic Z, Petersen KK, Dolstra O, Martin A (2003a) Identification of QTLs influencing agronomic traits in Miscanthus sinensis Anderss. I. Total height, flag-leaf height and stem diameter. Theor Appl Genet 107:123–129

Atienza SG, Satovic Z, Petersen KK, Dolstra O, Martin A (2003b) Identification of QTLs influencing combustion quality in Miscanthus sinensis Anderss. II. Chlorine and potassium content. Theor Appl Genet 107:857–863

Atienza SG, Satovic Z, Petersen KK, Dolstra O, Martín A (2003c) Identification of QTLs associated with yield and its components in Miscanthus sinensis Anderss. Euphytica 132:353–361

Atienza SG, Satovic Z, Petersen KK, Dolstra O, Martin A (2003d) Influencing combustion quality in Miscanthus sinensis Anderss.: identification of QTLs for calcium, phosphorus and sulphur content. Plant Breed 122:141–145

Babu R, Nair SK, Prasanna BM, Gupta HS (2004) Integrating marker-assisted selection in crop breeding—prospects and challenges. Curr Sci 87:607–619

Baoder JD, Barrier JW (1990) Producing fuels and chemicals from cellulosic crops. In: Janick J, Simon JE (eds) Advances in new crops. Timber, Portland, pp 257–259

Basnayake J, Cooper M, Ludlow MM, Henzell RG, Snell PJ (1995) Inheritance of osmotic adjustment to water stress in three grain sorghum crosses. Theor Appl Genet 90:675–682

Beale CV, Long SP (1995) Can perennial C4 grasses attain high efficiencies of radiant energy conversion in cool climates? Plant Cell Environ 18:641–650

Billa E, Koullas DP, Monties B, Koukios EG (1997) Structure and composition of sweet sorghum stalk components. Ind Crops Prod 6:297–302

Blumenthal JB, Rooney WL, Wang D (2007) Yield and ethanol production in sorghum genotypes. In: Abstracts, Annual Meeting of ASA-CSSA-SSSA, New Orleans, 4–8 November 2007

Bourne JK Jr (2007) Green dreams. Natl Geogr October:38–59

Bouton JH (2007) Molecular breeding of switchgrass for use as a biofuel crop. Curr Opin Genet Dev 17:553–558

Bowers JE, Abbey C, Anderson S, Chang C, Draye X, Hoppe AH, Jessup R, Lemke C, Lennington J, Li ZK, Lin YR, Liu SC, Luo LJ, Marler BS, Ming R, Mitchell SE, Qiang D, Reischmann K, Schulze SR, Skinner DN, Wang YW, Kresovich S, Schertz KF, Paterson AH (2003) A high-density genetic recombination map of sequence-tagged sites for Sorghum, as a framework for comparative structural and evolutionary genomics of tropical grains and grasses. Genetics 165:367–386

Bransby DI, Smith HA, Taylor CR, Duffy PA (2005). An interactive budget model for producing and delivering switchgrass to a bioprocessing plant. Ind Biotechnol 1(2):122–125

Buchanan CD, Lim S, Salzman RA, Kagiampakis I, Morishige DT, Weers B, Klein RR, Pratt LH, Cordonnier-Pratt M-M, Klein PE, Mullet JE (2005) Sorghum bicolor's transcriptome response to dehydration, high salinity and ABA. Plant Mol Biol 58:699–720

Burton GW (1986) Biomass production from herbaceous plant. In: Smith WH (ed) Biomass energy development. Plenum, New York, pp 163–175

Cai Q, Aitken KS, Deng HH, Chen XW, Fu C, Jackson PA, McIntyre CL (2005) Verification of the introgression of *Erianthus arundinaceus* into sugarcane using molecular markers. Plant Breed 124:322–328

Chiang YC, Schaal BA, Chou CH, Huang S, Chiang TY (2003) Contrasting selection modes at the Adh1 locus in outcrossing *Miscanthus sinensis* vs. inbreeding *Miscanthus condensatus* (Poaceae). Am J Bot 90:561–570

Christian DG, Lamptey JNL, Forde SMD, Plumb RT (1994) First report of barley yellow dwarf luteovirus on *Miscanthus* in the United Kingdom. Eur J Plant Pathol 100:167–170

Christian DG, Riche AB, Yates NE (2008) Growth, yield and mineral content of *Miscanthus* × *giganteus* grown as a biofuel for 14 successive harvests. Ind Crops Prod 28:320–327

Clark J (1981) The inheritance of fermentable carbohydrates in stems of *Sorghum bicolor* (L.) Moench. PhD. Dissertation, Texas A&M University, College Station, Texas

Clifton-Brown J, Chiang YC, Hodkinson TR (2008) *Miscanthus*: genetic resources and breeding potential to enhance bioenergy production. In: Vermeris W (ed) Genetic improvement of bioenergy crops. Springer, New York, pp 273–294

Clifton-Brown JC, Lewandowski I (2000a) Water use efficiency and biomass partitioning of three different *Miscanthus* genotypes with limited and unlimited water supply. Ann Bot 86:191–200

Clifton-Brown JC, Lewandowski I (2000b) Overwintering problems of newly established *Miscanthus* plantations can be overcome by identifying genotypes with improved rhizome cold tolerance. New Phytol 148:287–294

Clifton-Brown JC, Lewandowski I, Andersson B, Basch G, Christian DG, Kjeldsen JB, Jorgensen U, Mortensen JV, Riche AB, Schwartz KU (2001) Performance of 15 *Miscanthus* genotypes at five sites in Europe. Agron J 93:1013–1019

Coombs J (1984) Sugar-cane as an energy crop. Biotechnol Genet Eng Rev 1:311–345

Cordeiro GM, Pan YB, Henry RJ (2003) Sugarcane microsatellites for the assessment of genetic diversity in sugarcane germplasm. Plant Sci 165:181–189

Corn RJ (2009) Heterosis and composition of sweet sorghum. PhD Dissertation, Texas A&M University, College Station, Texas

Das MK, Fuentes RG, Taliaferro CM (2004) Genetic variability and trait relationships in switchgrass. Crop Sci 44: 443–448

Dawson L, Boopathy R (2007) Use of post-harvest sugarcane residue for ethanol production. Bioresour Technol 98:1695–1699

Denchev PD, Conger BV (1994) Plant regeneration from callus cultures of switchgrass. Crop Sci 34:1623–1627

Denchev PD, Conger BV (1995) In vitro culture of switchgrass: influence of 2,4-D and picloram in combination with benzyladenine on callus initiation and regeneration. Plant Cell Tissue Organ Cult 40:43–48

Edme SJ, Miller JD, Glaz B, Tai PYP, Comstock JC (2005) Genetic contributions to yield gains in the Florida sugarcane industry across thirty-three years. Crop Sci 45:92–97

Farrell AD, Clifton-Brown JC, Lewandowski I, Jones MB (2006) Genotypic variation in cold tolerance influences the yield of *Miscanthus*. Ann Appl Biol 149:337–345

Florida Division of Agriculture and Consumer Services (2002) Division of Plant Industry, TRI-OLOGY, vol 41, no. 2, March–April 2002, http://www.doacs.state.fl.us/pi/enpp/02-mar-apr.html

Florida Division of Agriculture and Consumer Services (2005) Division of Plant Industry, TRI-OLOGY, vol 44, no. 5, September–October 2005

Gams W, Klamer M, O'Donnell K (1999) *Fusarium miscanthi* sp. nov. from *Miscanthus* litter. Mycol 91:263–268

Gonzalez-Hernandez J, Sarath G, Stein J, Owens V, Gedye K, Boe A (2009) A multiple species approach to biomass production from native herbaceous perennial feedstocks. In Vitro Cell Dev Biol Plant 45:267–281

Gossmann M (2000) Schadwirkung einer pilzparasitären Rhizombesiedlung und Maßnahmen zur Verbesserung der Austriebs- und Biomasseleistung bei *Miscanthus* x *giganteus* Greef et Deu. In: Pude R (ed) Miscanthus—Vom Anbau bis zur Verwertung Miscanthus—Symposium. Beiträge Agrarwissenschaften, Bonn, pp 26–31

Greef JM, Deuter M (1993) Syntaxonomy of *Miscanthus* x *giganteus* GREEF et DEU. Angew Bot 67:87–90

Gupta PK, Rustgi S, Mir RR (2008) Array-based high-throughput DNA markers for crop improvement. Heredity 101:5

Gupta SD, Conger BV (1998) In vitro differentiation of multiple shoot clumps from intact seedlings of switchgrass. In Vitro Cell Dev Biol Plant 34:196–202

Gupta SD, Conger BV (1999) Somatic embryogenesis and plant regeneration from suspension cultures of switchgrass. Crop Sci 39:243–247

Halbert SE, Remaudiere G (2000) A new oriental *Melanaphis* species recently introduced in North America [Hemiptera, Aphididae]. Rev Fr Entomol 22:109–117

Hallam AI, Anderson C, Buxton DR (2001) Comparative economic analysis of perennial, annual and intercrops for biomass production. Biomass Bioenergy 21:407–424

Heaton E, Voigt T, Long SP (2004) A quantitative review comparing the yields of two candidate C4 perennial biomass crops in relation to nitrogen, temperature and water. Biomass Bioenergy 27:21–30

Heaton EA, Dohleman FG, Long SP (2008) Meeting US Biofuel goals with less land: the potential of *Miscanthus*. Glob Change Biol 14:1–15

Heaton E, Dohleman FG, Long SP (2009) Seasonal nitrogen dynamics of *Miscanthus* x *giganteus* and *Panicum virgatum*. GCB Bioenergy 1:297–307

Heichel GH (1974) Comparative efficiency of energy use in crop production. Bull Conn Agric Exp Stn New Haven 739:1–26

Hernández P, Dorado G, Laurie DA, Martín A, Snape JW (2001) Microsatellites and RFLP probes from maize are efficient sources of molecular markers for the biomass energy crop Miscanthus. Theor Appl Genet 102:616–622

Hirata M, Hasegawa N, Nogami K, Sonoda T (1999) Use of a young tree plantation for grazing of cattle in southern Kyushu, Japan: 3. Non-destructive estimation of basal area and biomass of *Miscanthus sinensis* grass plants. Proc Int Rangeland Congr 480–481

Hodkinson TR, Renvoize SA, Chase MW (1997) Systematics of *Miscanthus*. Aspects Appl Biol 49:189–198

Hodkinson TR, Chase MW, Renvoize SA (2002) Characterization of a genetic resource collection for *Miscanthus* (Saccharinae, Andropogoneae, Poaceae) using AFLP and ISSR PCR. Ann Bot 89:627

Honda M (1930) Monographia Poacearum Japonicarum, *Bambusoides exclusis*. J Fac Sci Imperial U Tokyo III Bot HI:484

Huggett DAJ, Leather SR, Walters KFA (1999) Suitability of the biomass crop *Miscanthus sinensis* as a host for the aphids *Rhopalosiphum padi* (L.) and *Rhopalosiphum maidis* (F.), and its susceptibility to the plant luteovirus barley yellow dwarf virus. Agric For Entomol 1:143–149

Jakob K, Zhou F, Paterson AH (2009) Genetic improvement of C4 grasses as cellulosic biofuel feedstocks. In Vitro Cell Dev Biol Plant 45:291–305

Jessup R (2009) Development and status of dedicated energy crops in the United States. In Vitro Cell Dev Biol Plant 45:282–290

Jezowski S (2008) Yield traits of six clones of *Miscanthus* in the first 3 years following planting in Poland. Ind Crops Prod 27:65–68

Jordan WR, Miller FR (1980) Genetic variability in sorghum root systems: implications for drought tolerance. In: Turner PC, Kramer PJ (eds) Adaptation of plants to water and high temperature stress. Wiley, New York, pp 383–399

Keng YL (1932) The gross morphology of Andropogoneae. PhD Thesis. George Washington University

Kimber C (2000) Origins of domesticated sorghum and its early diffusion to India and China. In: Smith CW, Frederiksen RA (eds) Sorghum. Wiley, New York, pp 3–98

Klein PE, Klein RR, Cartinhour SW, Ulanch PE, Dong J, Obert JA, Morishige DT, Schlueter SD, Childs KL, Ale M, Mullet JE (2000) A high-throughput AFLP-based method for constructing integrated genetic and physical maps: progress toward a sorghum genome map. Genome Res 10:789–807

Klein RR, Klein PE, Mullet JE, Minx P, Rooney WL, Schertz KF (2005) Fertility restorer locus *Rf1* of sorghum (*Sorghum bicolor* L.) encodes a pentatricopeptide repeat protein not present in the colinear region of rice chromosome 12. Theor Appl Genet 111:994–1012

Konishi S, Izawa T, Lin S, Ebana K, Fukuta Y, Sasaki T, Yano M (2006) An SNP caused loss of seed shattering during rice domestication. Am Assoc Adv Sci 312:1392–1396

Lam E, Shine J Jr, daSilva J, Lawton M, Bonos S, Martin C, Carrer H, Silva-Filho MC, Glynn N, Helsel Z, Jiong M, Richard EP Jr, Souza G, Ming R (2009) Improving sugarcane for biofuel: engineering for an even better feedstock. GCB Bioenergy 1:251–255

Lee YN (1964) Taxonomic studies on the genus *Miscanthus*. 3. Relationship among the section, subsection and species. J Jpn Bot 38:197–205

Legendre BL, Burner DM (1995) Biomass production of sugarcane cultivars and early-generation hybrids. Biomass Bioenergy 8:55–61

Li C, Zhou A, Sang T (2006) Rice domestication by reducing shattering. Am Assoc Adv Sci 311:1936–1939

Lo CC, Chen YH, Huang YJ, Shih SC (1986) Recent progress in *Miscanthus* nobilization program. Proc Int Soc Sugar Cane Technol 19:514–521

Loomis RS, Williams WA (1963) Maximum crop productivity: an estimate. Crop Sci 3:67–72

Macedo IC, Leal MRLV, da Silva JEAR (2004) Assessment of greenhouse gas emissions in the production and use of fuel ethanol in Brazil. Sao Paulo. http://www.unica.com.br/i_pages/files/pdf_ingles.pdf

Maiti RK, Rao KE, Raju PS, House LR, Prasada-Rao KE (1984) The glossy trait in sorghum: its characteristics and significance in crop improvement. Field Crops Res 9:279–289

Martinez-Reyna JM, Vogel KP (2002) Incompatibility systems in switchgrass. Crop Sci 42:1800–1805

Martinez-Reyna JM, Vogel KP (2008) Heterosis in switchgrass: spaced plants. Crop Sci 48:1312–1320

McBee GG, Miller FR, Dominy RE, Monk RL (1987) Quality of sorghum biomass for methanogenesis. In: Klass DL (ed) Energy from biomass and waste. Elsevier, London, pp 251–260

McCollum FT III, McCuistion K, Bean B (2005) Brown Midrib and photoperiod-sensitive forage sorghums. In: Proceedings of the 2005 Plains Nutrition Council Spring Conference. Publication No. AREC 05-20. Texas A&M University Agricultural Research and Extension Center, Amarillo, TX, pp 36–46

McLaughlin SB, Kszos LA (2005) Development of switchgrass (*Panicum virgatum*) as a bioenergy feedstock in the United States. Biomass Bioenergy 28:515–535

McLaughlin SB, Kiniry JR, Taliaferro CM, de la Torre Ugarte D (2006) Projecting yield and utilization potential of switchgrass as an energy crop. Adv Agron 90:267–297

Menz MA, Klein RR, Mullet JE, Obert JA, Unruh NC, Klein PE (2002) A high-density genetic map of *Sorghum bicolor* (L.) Moench based on 2926 AFLP$^{(R)}$, RFLP and SSR markers. Plant Mol Biol 48:483–499

Ming R, Del Monte TA, Hernandez E, Moore PH, Irvine JE, Paterson AH (2002) Comparative analysis of QTLs affecting plant height and flowering among closely related diploid and polyploid genomes. Genome 45:794–803

Ming R, Moore PH, Wu KK, D'Hont A, Tew TL, Mirkov TE, da Silva J, Schnell RJ, Brumbley SM, Lakshmanan P, Jifon J, Rai M, Comstock JC, Glaszmann JC, Paterson AH (2006) Sugarcane improvement through breeding and biotechnology. Plant Breed Rev 27:17–118

Missaoui AM, Paterson AH, Bouton JH (2006) Molecular markers for the classification of switchgrass (*Panicum virgatum* L.) germplasm and to assess genetic diversity in three synthetic switchgrass populations. Genet Resour Crop Evol 53:1291–1302

Monk RL, Miller FR, McBee GG (1984) Sorghum improvement for energy production. Biomass 6:145–385

Moore KJ, Boote KJ, Sanderson MA (2004) Physiology and developmental morphology. In: Moser L, Burson B, Sollenberger L (eds) Warm-season (C4) grasses. American Society for Agronomy, Madison, pp 179–216

Muchow RC, Spilman MF, Wood WW, Thomas MR (1994) Radiation interception and biomass accumulation in a sugarcane crop under irrigated tropical conditions. Aust J Agric Res 45:3–49

Muchow RC, Cooper M, Hammer GL (1996) Characterizing environmental challenges using models. In: Cooper M, Hammer GL, Wallingford UK (eds) Plant adaptation and crop improvement. CAB International, Wallingford, pp 349–364

Mukherjee SK (1950) Search for wild relatives of sugarcane in India. Int Sugar J 52:261–262

Murray SC, Sharma A, Rooney WL, Klein PE, Mullet JE, Mitchell SE, Kresovich S (2008) Genetic improvement of sorghum as a biofuel feedstock: I. QTL for stem and grain nonstructural carbohydrates. Crop Sci 48:2165–2179

Naidu SL, Moose SP, Al-Shoaibi AK, Raines CA, Long SP (2003) Cold tolerance of C4 photosynthesis in *Miscanthus × giganteus*: adaptation in amounts and sequence of C4 photosynthetic enzymes. Plant Physiol 132:1688–1697

Nilsson-Leissner G (1942) A case of increased vitality in sibpollinated later generations of self-fertilized *Dactylis glomerata* strains. Hereditas 28:222–224

Ohwi J (1965) Flora of Japan. Smithsonian Institution, Washington, DC

O'Neill NR, Farr DF (1996) *Miscanthus* blight, a new foliar disease of ornamental grasses and sugarcane incited by *Leptosphaeria* sp. and its anamorphic state *Stagonospora* sp. Plant Dis 80:980–987

Panje RR (1972) The role of *Saccharum spontaneum* in sugarcane breeding. Proc Int Soc Sugar Cane Technol 14:217–223

Panje RR, Babu CN (1960) Studies in *Saccharum spontaneum*. Distribution and geographical association of chromosome numbers. Cytologia 25:152–172

Parrish DJ, Fike JH (2005) The biology and agronomy of switchgrass for biofuels. Crit Rev Plant Sci 24:423–459

Paterson AH, Bowers JE, Bruggmann R, Dubchak I, Grimwood J, Gundlach H, Haberer G, Hellsten U, Mitros T, Poliakov A, Schmutz J, Spannagl M, Tang H, Wang X, Wicker T, Bharti AK, Chapman J, Feltus FA, Gowik U, Grigoriev IV, Lyons E, Maher CA, Martis M, Narechania A, Otillar RP, Penning BW, Salamov AA, Wang Y, Zhang L, Carpita NC, Freeling M, Gingle AR, Hash CT, Keller B, Klein P, Kresovich S, McCann MC, Peterson DG, Rahman M, Ware D, Westhoff P, Mayer KFX, Messing J, Rokhsar DS (2009) The *Sorghum bicolor* genome and the diversification of grasses. Nature 457:551–556

Perlack RD, Wright LL, Turhollow AF, Graham RL, Stokes BJ, Erbach DC (2005) Biomass as feedstock for a bioenergy and bioproducts industry: the technical feasibility of a billion-ton annual supply. Oak Ridge National Laboratory, Oak Ridge, TN

Pratt LH, Liang C, Shah M, Sun F, Wang HM, St. Patrick R, Gingle AR, Paterson AH, Wing R, Dean R, Klein R, Nguyen HT, Ma HM, Zhao X, Morishige DT, Mullet JE, Cordonnier-Pratt MM (2005) Sorghum expressed sequence tags identify signature genes for drought, pathogenesis, and skotomorphogenesis from a milestone set of 16,801 unique transcripts, Plant Physiol 139:869–884

Pude R, Diepenbrock W, Franken H, Greef JM (1996) Impact and causes of winter kills of *Miscanthus*. Mitt Ges Pflanzenbauwiss (Germany) 9:61–62

Richard EP Jr (1999) Management of chopper harvester-generated green cane trash blankets: a new concern for Louisiana. Proc Int Soc Sugar Cane Technol 23:52–62

Roach BT (1978) Utilization of *Saccharum spontaneum* in sugarcane breeding. Proc Int Soc Sugar Cane Technol 16:43–58

Rooney WL (2004) Sorghum improvement—integrating traditional and new technology to produce improved genotypes. Adv Agron 83:37–109

Rooney WL, Aydin S (1999) The genetic control of a photoperiod sensitive response in *Sorghum bicolor* (L.) Moench. Crop Sci 39:397–400

Rosenow DT, Quisenberry JE, Wendt CW, Clark LE (1983) Drought tolerant sorghum and cotton germplasm. Agric Water Manag 7:207–222

Sanderson MA, Reed RL, McLaughlin SB, Wullschleger SD, Conger BV, Parrish DJ, Wolf DD, Taliaferro C, Hopkins AA, Ocumpaugh WR, Hussey MA, Read JC, Tischler CR (1996) Switchgrass as a sustainable bioenergy crop. Bioresour Technol 56:83–93

Sanderson MA, Adler PR, Boateng AA, Casler MD, Sarath G (2006) Switchgrass as a biofuels feedstock in the USA. Can J Plant Sci 86:1315–1325

Salvi S, Tuberosa R, Chiapparino E, Maccaferri M, Veillet S, van Beuningen L, Isaac P, Edwards K, Phillips RL (2002) Toward positional cloning of Vgt1, a QTL controlling the transition from the vegetative to the reproductive phase in maize. Plant Mol Biol 48:601–613

Smart AJ, Moser LE, Vogel KP (2003a) Establishment and seedling growth of big bluestem and switchgrass populations divergently selected for seedling tiller number. Crop Sci 43:1434–1440

Smart AJ, Vogel KP, Moser LE, Stroup WW (2003b) Divergent selection for seedling tiller number in big bluestem and switchgrass. Crop Sci 43:1427–1433

Smart AJ, Moser LE, Vogel KP (2004) Morphological characteristics of big bluestem and switchgrass plants divergently selected for seedling tiller number. Crop Sci 44:607–613

Somleva MN, Tomaszewski Z, Conger BV (2002) *Agrobacterium* mediated genetic transformation of switchgrass. Crop Sci 42:2080–2087

Song JZ, Soller M, Genizi A (1999) The full-sib intercross line (FSIL): a QTL mapping design for outcrossing species. Genet Res 73:61–73

Subbotin SA, Vierstraete A, De Ley P, Rowe J, Waeyenberge L, Moens M, Vanfleteren JR (2001) Phylogenetic relationships within the cyst-forming nematodes (Nematoda, Heteroderidae) based on analysis of sequences from the ITS regions of ribosomal DNA. Mol Phylogenet Evol 21:1–16

Taliaferro CM (2002) Breeding and selection of new switchgrass varieties for increased biomass production. Vol ORNL/SUB-02-19XSY162C/01. Oak Ridge National Laboratory, Oak Ridge, TN

Taliaferro CM, Vogel KP, Bouton JH, McLaughlin SB, Tuskan GA (1999) Reproductive characteristics and breeding improvement potential of switchgrass. In: Overend RP, Chonet E (eds) Biomass: a growth opportunity in green energy and value-added products, vol 1. Pergamon, Elsevier, Amsterdam, pp 147–153

Teakle DS, Shukla DD, Ford RE (1989) Sugarcane mosaic virus. AAB Descriptions Plant Viruses 5

Tew TL (2003) World sugarcane variety census—year 2000. Sugar Cane Int March/April:12–18

Tew TL, Cobill RM (2008) Genetic improvement of sugarcane (*Saccharum* spp.) as an energy crop. In: Vermerris W (ed) Genetic improvement of bioenergy crops. Springer, New York, pp 249–272

Thinggaard K (1997) Study of the role of *Fusarium* in the field establishment problem of *Miscanthus*. Acta Agric Scand B Plant Soil Sci 47:238–241

Tobias CM, Twigg P, Hayden DM, Vogel KP, Mitchell RM, Lazo GR, Chow EK, Sarath G (2005) Analysis of expressed sequence tags and the identification of associated short tandem repeats in switchgrass. Theor Appl Genet 111:956–964

Varshney RK, Hoisington DA, Tyagi AK (2006) Advances in cereal genomics and applications in crop breeding. Trends Biotechnol 24:490–499

Viator RP, Johnson RM, Grimm CC, Richard EP Jr (2006) Allelopathic, autotoxic, and hormetic effects of postharvest sugarcane residue. Agron J 98:1526–1531

Viator RP, Johnson RM, Richard EP Jr (2009a) Mechanical removal and incorporation of post-harvest residue effects on sugarcane ratoon yields. Sugar Cane Int 24:149–152

Viator RP, Johnson RM, Boykin DL, Richard EP Jr (2009b) Sugarcane post-harvest residue management in the temperate climate of Louisiana. Crop Sci 49:1023–1028

Vogel KP (2004) Switchgrass. In: Moser L, Burson B, Sollenberger L (eds) Warm-season (C4) grasses. Am Soc Agron, Madison, pp 561–588

Vogel KP, Mitchell RB (2008) Heterosis in switchgrass: biomass yield in swards. Crop Sci 48:2159–2164

West MAL, van Leeuwen H, Kozik A, Kliebenstein DJ, Doerge RW, St Clair DA, Michelmore RW (2006) High-density haplotyping with microarray-based expression and single feature polymorphism markers in Arabidopsis. Genome Res 16:787–795

Wu X, Zhao R, Bean SR, Seib PA, McLaren JS, Madl RL, Tuinstra MR, Lenz MC, Wang D (2007) Factors impacting ethanol production from grain sorghum in the dry-grind process. Cereal Chem 84:130–136

Chapter 7
Woody Biomass and Purpose-Grown Trees as Feedstocks for Renewable Energy

Maud A. W. Hinchee, Lauren N. Mullinax, and William H. Rottmann

7.1 The Forest Industry and Renewable Energy

The US government has committed to renewable energy and the recent American Recovery and Reinvestment Act of 2009 echoes the objectives of the 2007 Energy Independence and Security Act (EISA) and the Food, Conservation and Energy Act of 2008 (2008 Farm Bill). These bills have allocated more than US $16.8 billion in funding for renewable energy and energy efficiency projects over the next 10 years (ACORE 2009). Multiple Southeastern states have developed policy initiatives to encourage the development of the bioenergy industry, including consumption standards, tax incentives, subsidies and loans, as well as the identification of potential bioenergy crops including perennial grasses and trees.

Trees have been managed as a biomass crop for generations. Wood provides raw materials to a large wood products industry that includes sawmills, pulp and paper manufacturers, manufactured wood products facilities, and numerous niche applications. The US produces the largest amount of wood for industrial applications in the world, with the traditional forest products industry consuming approximately 600 million green short tons in 2006. In the Southeastern US, the total volume of hardwood and pine harvested annually is more than 365 million green short tons (RISI 2008). Much of this resource is renewed through re-planting, with approximately 950 million pine and 30 million hardwoods planted in the 2008/2009 planting year (McNabb and Enebak 2008). The majority of planted softwoods in the Southeastern US are *Pinus taeda* (loblolly pine) purpose grown on managed plantations, while the small number of planted hardwoods includes dozens of species intended primarily for non-industrial end uses such as recreation, conservation, restoration, reclamation and aesthetic values. Most harvested hardwoods for industrial uses come primarily from naturally regenerated stands.

M.A.W. Hinchee, L.N. Mullinax, and W.H. Rottmann
ArborGen, LLC, 840001 Summerville, SC 29484, USA
e-mail: mahinch@arborgen.com; lnmulli@arborgen.com; whrottm@arborgen.com

Wood production and wood harvests are dynamic in the US over time due to changing wood product demands and associated costs. The peak US roundwood harvest was in 1991 when industrial production was 18.8 billion cubic feet (ft^3 = 532.4 million m^3). Lumber and pulpwood-based products constitute the largest share (80%) of roundwood use. Since 1986, the harvest of wood for fuel and plywood has declined, with the largest increase in harvests directed toward pulpwood production (Howard 2005). However, the Southeastern US has increased timber production due to increased growth by non-industrial private landowners and fast growing pine tree plantations (Adams et al. 2006). Factors that are anticipated to affect the planting and harvest of trees in the future include emerging biopower and biofuels opportunities with new biopower and biofuel policies, loss of forest land to other uses such as development, and the emergence of carbon offset markets.

The use of trees for energy production is not new. Papermakers and lumber producers are the nation's largest renewable energy producers and consumers producing an estimated 28.5 million megawatts (MW) of electricity annually or enough to power approximately 2.7 million homes (AFPA 2009). The pulp and paper industry, as of 2005, contributed 1.22 quadrillion British thermal units (BTU) of energy to the nation's energy resources (one-fourth in the form of electricity and three-fourths in the form of useful thermal output), or slightly more than 1% of the nation's energy budget (Brown and Atamturk 2008). The industry itself provides for 60% of its own needs (Murray et al. 2006). This trend has been encouraged by rising energy prices for natural gas and petroleum, by environmental regulations, and by new and emerging technologies (Brown and Atamturk 2008). In the US in recent years, significant focus has been devoted to increasing energy independence and security, reducing dependence on non-renewable sources of energy, reducing our level of greenhouse gas emissions to mitigate climate change and fostering the development of rural economies. All of this has contributed to an increasing interest in the development of renewable energy technology from woody biomass.

Trees and wood were identified as part of the US bioenergy solution in the *Billion Ton Report* (Perlack et al. 2005). This report investigated the feasibility of producing the estimated 1 billion dry tons of biomass needed annually to meet the 30 x '30 goal of 30% replacement of the US petroleum consumption with biofuels by 2030. Two types for woody biomass were addressed: (1) wood residuals resulting from logging and mill/construction waste from the traditional forest products industry, and (2) wood produced specifically for energy as "perennial energy crops" that are grown more as an agricultural resource. Within the next 20 years, purpose-grown trees for energy are expected to account for 377 million dry short tons of the 1.37 billion dry ton total biomass resource potential at projected yields of 8 dry tons per acre per year (Perlack et al. 2005). The *Billion Ton Report* (Perlack et al. 2005) indicated that forest residues could supply approximately only 368 million dry tons biomass (Fig. 7.1).

There are multiple drivers for the use of trees for biopower and biofuels applications in the US. As mentioned earlier, there are substantial existing forest resources with an associated supply of unused or underused woody biomass that

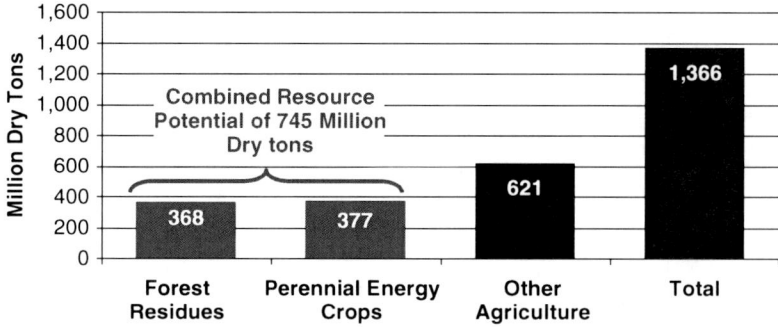

Fig. 7.1 Annual biomass resource potential (as million dry tons) in the US from wood and agricultural resources as presented in Perlack et al. (2005)

could be utilized for biopower and biofuels applications. Another driver for the utilization of woody biomass is the multiple energy applications associated with wood. Woody biomass, in particular wood waste, can generate power as a low cost boiler fuel, as a clean-coal alternative to coal fired utilities, as a source of gas that can be used in gas turbines to generate power, or as a substitute for natural gas used by certain industries. In addition, the technology to convert woody biomass cost-effectively into biofuels has been demonstrated on a pilot scale, with multiple commercial scale wood biomass biorefinery projects under way worldwide. Wood also can be converted into higher value fuels, such as clean diesel and jet fuel (Sklar 2009). Currently, the Southeastern US is a major supplier of renewable energy: its biodiesel plants produce 22% of the nation's biodiesel; its 12 operating ethanol plants produce 6.4% of the nation's ethanol; and its 534 biodiesel and E85 fueling stations represent 23% of the overall total (SAFER 2009). This presents an excellent opportunity for woody biomass utilization.

The Southern Alliance for Clean Energy (2009) estimates that biomass represents 60% of the near-term potential for expanding renewable energy. There has recently been a great deal of new development of cogeneration and wood pellet projects throughout the Southeastern US. The largest pellet mills ever built opened in this market during the past year and several additional projects have been announced. Wood demand from projects with announced start dates in the Southeastern US is expected to approach 50 million green tons by 2015 (Forisk Consulting 2009; L.M., unpublished results).

An independent study by the University of Tennessee estimates that the 25% replacement of petroleum-based fuel with renewable fuels by 2025 in the US equates to 86 billion gallons ethanol and 962 billion kWh energy by the year 2025 (English et al. 2006). Assuming trees represent 50% of the "energy crop" opportunity in the Southeastern US, this represents a total market of 258 million green short tons in the region by 2025. It is believed that wood residuals will be utilized first and that the expected woody biomass demand will quickly exceed residual supply (Fig. 7.2) (Abt et al. 2010). Multiple regional studies show dramatic increases in biomass prices with increasing demand when only forest residues wood are

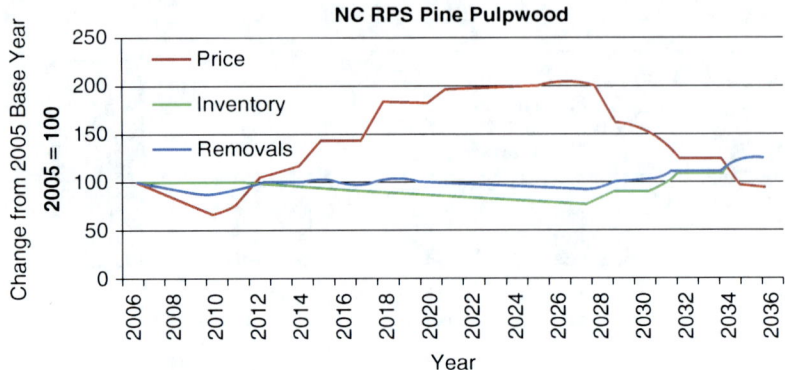

Fig. 7.2 Impact of North Carolina Renewable Energy Portfolio Standard on forest residual supply and price in the absence of purpose-grown energy crops (modified from Abt et al. 2010)

considered. It is therefore expected that future consumption will come from a mix of purpose-grown pine and hardwood trees. Using expected harvest yields and standard conversion rate assumptions this is the equivalent of 1,033 million and 574 million trees annually for pine and hardwoods, respectively. These numbers would more than double the current markets for woody biomass in the Southeastern US.

Woody biomass can provide renewable energy through two basic approaches: electricity generation through direct firing or co-firing of biomass with coal (biopower) and production of liquid fuels (biofuels). The technology and the fit of different species of woody biomass for these two renewable energy categories will be discussed in some detail in the following sections of this chapter. In addition, the improvements in tree genetics and silviculture that will be beneficial, improving the economic feasibility of woody feedstock for sustainable production of renewable energy will be addressed.

7.2 Biopower

The demand for renewable biopower is growing very quickly. Demand is influenced in large part by government mandates and incentives, and, to date, 28 states have implemented Renewable Portfolio Standard (RPS) legislation requiring that a portion of that state's electricity supply come from renewable sources (Pew Center on Global Climate Change 2008). This type of legislation has already been included in the American Clean Energy and Security (ACES) Act; H.R. 2454 was passed by the US House of Representatives in June 2009, and is currently being considered by the US Senate. The 25 x '25 alliance, comprised of a broad spectrum of participation from the agricultural and forestry sectors, has set a target of 962 billion kWh energy from biomass by 2025. Assuming trees represent 50% of the energy crop opportunity in the Southeastern United States, this represents a total market of 109 million green short tons by 2025.

The abundance of woody biomass resources in the Southeastern US have positioned this region to be a leader in bioenergy generation where 46% of the nation's biomass electricity was produced in 2007 (SAFER 2009). Wood and derived fuels are estimated to provide 184 trillion BTUs, with 50% of the feedstock coming from the Southeastern US (Energy Information Administration 2008). The vast majority of this can be attributed to the pulp and paper industry where by-products from production processes are routinely combusted to produce steam and electricity. The utilization of woody biomass for the production of electric power is already common but there is substantial potential for expansion. Biomass power plants exist outside of the pulp and paper industry that use forest-based biomass for power generation. In 2007, greater than 200 companies outside the wood products and food industries generated biomass power in the US to save fuel costs and earn emissions credits, particularly in regions where there is access to very low cost biomass supplies [US Department of Energy (DOE) 2007a].

Many coal-burning power plants in the US co-fire with wood, especially in the Southeastern US where highly productive forest plantations are prevalent. According to TimberMart (2009) a total of 32 biopower projects that would consume nearly 17 million green tons of woody biomass have been announced in the Southeastern US in the 21 months prior to the second quarter of 2009. For example, Southern Company has announced plans to convert six of its coal-fired power plants to co-fire with wood throughout the Southeastern US. Southern Company's Plant Mitchell and Southern Power's Nacogdoches Power Plant when combined are projected to consume more than 1 million green short tons of woody biomass annually (Forisk Consulting 2009). The wood consumption on an annual basis to generate biopower in the Southeastern US is projected to be nearly 28 million green short tons (Table 7.1).

7.2.1 Processes for Energy Production from Woody Biomass

The direct production of power from woody biomass involves the processes of co-firing and direct firing. Most of today's biomass power plants are direct-fired systems that are similar to most fossil fuel fired power plants (Bergman and Zerbe 2008).

7.2.1.1 Co-firing

Forest residues and low-value wood produced from traditional forest industries, agricultural crop residues, construction waste, municipal waste, storm debris, and dedicated energy crops, such as switchgrass, willow, and hybrid poplar, are typical biomass sources that can be used for simultaneous combustion with a base fuel such as coal to produce energy (typically electrical power). Wood can substitute for up to 20% of the coal in the boiler of coal-burning power plants. However power plants

Table 7.1 Wood biopower plants being established in the Southeastern US

Name	State	Electricity capacity (MW)	Start-up year	Estimated wood use (green tons)
Adage	FL	50	2012	500,000
Adage	FL	50	2012	500,000
Alabama Renewable Energy Alliance LLC	AL	13	2015	130,000
American Clean Energy	WV	28	2011	400,000
American Renewables LLC (Gainesville Renewable Energy Center)	FL	100	2013	1,000,000
American Renewables LLC (Hamilton County Renewable Energy Center)	FL	100	2013	1,000,000
Aspen Power	TX	50	2010	525,000
Bio-Gen	NC	6	2008	60,000
Bio-Gen	NC	4	2008	40,000
Bio-Gen	NC	2.5	2008	25,000
Bio-Gen	NC	5.3	2008	53,000
Biomass Gas & Electric	FL	75	2015	750,000
Biomass Gas & Electric	GA	30	2015	300,000
Biomass Gas & Electric – Northwest Florida Renewable Energy Center	FL	44	2011	410,625
Buckeye Cellulose – University of Florida	FL	12	2011	120,000
Coastal Carolina Clean Power	NC	25	2008	250,000
Decker Energy	NC	48	1990	550,000
Decker Energy	NC	50	2011	500,000
Decker Energy	TX	35–50	2014	500,000
Decker Energy (Fitzgerald Renewable Energy LLC)	GA	50	2011	600,000
Dominion Virginia City Hybrid Energy Center	VA	585	2012	585,000
Dominion/Pittsylvania Power Station	VA	80	2004	800,000
East Texas Electric Cooperative	TX	50	2014	500,000
Florida Biomass Energy (FB Energy)	FL	60	2012	600,000
Florida Crystals	FL	68	1995	900,000
Gainesville Regional Utilities/Gainesville Renewable Energy Center (GREC)	FL	100	2013	1,000,000
Global Energy Systems	GA	20	2013	200,000
Intrinergy – Coastal Paper	MS	30,000	2007	150,000
Milledgeville Central State Hospital	GA		1990	190,000
Multitrade Biomass Holdings	GA	18	2009	180,000
Novi Energy	VA	55	2012	1,000,000
NRG Energy	LA	1,730	2010	625,000
Oak Ridge National Laboratory/Nexterra Energy/Johnson Controls	TN		2011	70,000
Oglethorpe Power Corporation	GA	100	2014	1,000,000
Oglethorpe Power Corporation	GA	100	2015	16,000
Oglethorpe Power Corporation	GA	100	2015	32,000
Orangeburg County Biomass LLC	SC	35	2013	350,000
Peregrine Energy	SC	50	2012	200,000
Phoenix Renewable Energy	AR	20	2011	200,000
Plant Carl – Green Energy Partners, LLC	GA	25	2009	1,000,000
Pratt Industries	GA	9	2009	300,000

(*continued*)

Table 7.1 (continued)

Name	State	Electricity capacity (MW)	Start-up year	Estimated wood use (green tons)
Ridge Generating Station – Waste Management	FL	39.6	1994	500,000
Rollcast Energy	GA	50	2012	500,000
Rollcast Energy – Greenway Renewable Energy	GA	50	2010	400,000
Rollcast Energy – Loblolly Green Power	SC	50	2012	70,000
Rollcast Energy – Piedmont Green Power LLC	GA	50	2012	250,000
Savannah River Site D-Area/Ameresco	SC	20	2011	322,000
SI Group	FL	7	2006	27,000
Southern Company – Plant Mitchell	GA	96	2012	800,000
Southern Company – Plant Scholtz	FL	100	2013	1,000,000
Southern Company – Plant Sweatt	MS	80	2013	800,000
Southern Company (Nacogdoches Power)	TX	100	2012	700,000
Southern Company (Plant Gadsden)	AL	120	2013	1,200,000
SRS – A Area Steam Plant	SC		2008	450,000
SRS – K/L Area Heat	SC		2010	600,000
Sterling Planet at IP Riegelwood	NC	45	2013	450,000
SunMark Energy	TX	60	2012	2,500
Wiregrass Power LLC/Sterling Energy Assets	GA	40	2012	600,000
Yellow Pine Energy/Georgia Power	GA	110	2012	1,100,000
Electricity Total				27,883,125

Source: Wood Bioenergy South and ArborGen estimates

require some alterations in their process to utilize woody biomass, and this includes addressing the needs associated with woody biomass emissions procurement, handling and preprocessing (grinding to a fine size).

There are cost and environmental benefits to using woody biomass. It has been demonstrated that effective substitutions of biomass energy can make up 15% of the total energy input. Investments to modify coal-burning plants are expected to be US $100 to $700 per kW of bio-mass capacity, with the average ranging from $180 to $200 per kW. Co-firing results in a net reduction in emissions of sulfur dioxide (SO_2), nitrogen oxides (NO_x), and non-renewable carbon dioxide (CO_2); Bergman and Zerbe 2008). According to the National Renewable Energy Laboratory (1999), seven utilities burning at least 7% wood reduced their NOx emissions by 15% compared to burning 100% coal.

7.2.1.2 Direct Firing

In addition to the more common practice of co-firing, direct fire and gasification systems are also utilized. Most of today's biomass power plants that are direct-fired systems are similar to most fossil-fuel-fired power plants. In a direct-fired system, a large volume single-combustion chamber produces combustion gases that rise to the top of the boiler chamber to the heat exchange passages. Relative simplicity and low costs are features of direct-burn systems. Biomass power boilers are typically in the 20–50 MW range, compared with coal-fired plants in the 100–1,500 MW range (Bergman and Zerbe 2008). Use of wood pellets for direct fire energy production is becoming common in Europe in response to the European Union's ratification of the Kyoto Protocol (European Union 2002), and imported wood pellets from the Southeastern US are currently a primary source of woody biomass serving European markets. Several new pellet plants have been established to address this rapidly growing market (Table 7.2, Fig. 7.3).

Table 7.2 Wood pellet plants being established in the Southeastern US

Name	State	Pellet capacity (tons)	Start-up year	Estimated wood use (green tons)
American Green Holdings	GA	125,000	2010	300,000
Carolina Pacific Briquetting Co., LLC	SC	24,000 (2009) 300,000 (2010)	2009	600,000
Carolina Wood Pellets	NC	68,000	2009	136,000
FRAM / Appling County Pellets	GA	145,000	2007	140,000
Fulghum Fibers	GA	150000	2012	300,000
Green Circle Bio Energy	FL	500,000	2008	1,000,000
Green Circle Bio Energy	MS or GA	560,000	2011	1,250,000
Greenville Wood Products	FL	150,000	2009	300,000
Indeck Magnolia LLC	MS	90,000	2010	180,000
Integro EarthFuels (torrefied biomass)	NC	87,000 (2009), 350,000 (2012)	2009	700,000
Integro EarthFuels (torrefied biomass)	GA	168,000 (2012), 300,000 (2015)	2012	600,000
Lee Energy Solutions LLC	AL	75000	2010	150,000
Magnolia BioPower	GA	330,000 (2012), 660,000 (2015). 1,000,000 (2017)	2010	2,175,000
Nature's Earth Pellet	AL	90,000	1990	270,000
NexGen Biomass	AR	237,000 (2010), 395,000 (2011)	2010	790,000
Palmetto Renewable Energy	SC	16,000 (2008), 32,000 (2010)	2008	32,000
Phoenix Renewable Energy	AR	250,000	2011	500,000
Piney Woods Pellets	MS	50,000	2009	1,000,000
Point Bio Energy, LLC	LA	400,000	2010	540,000
Rockwood Pellets	GA	15,000	2006	500,000
Sparkman Wood Pellets	AR	18,000	1990	5,000
Woodfuels Virginia, LLC	VA	75,000	2008	300,000
Woodlands Alternative Fuels	GA	150,000 (2010), 300,000 (2011)	2011	600,000
Pellet Total				12,368,000

Source: Wood Bioenergy South and ArborGen estimates: Forisk Consulting 2009 and ArborGen estimates

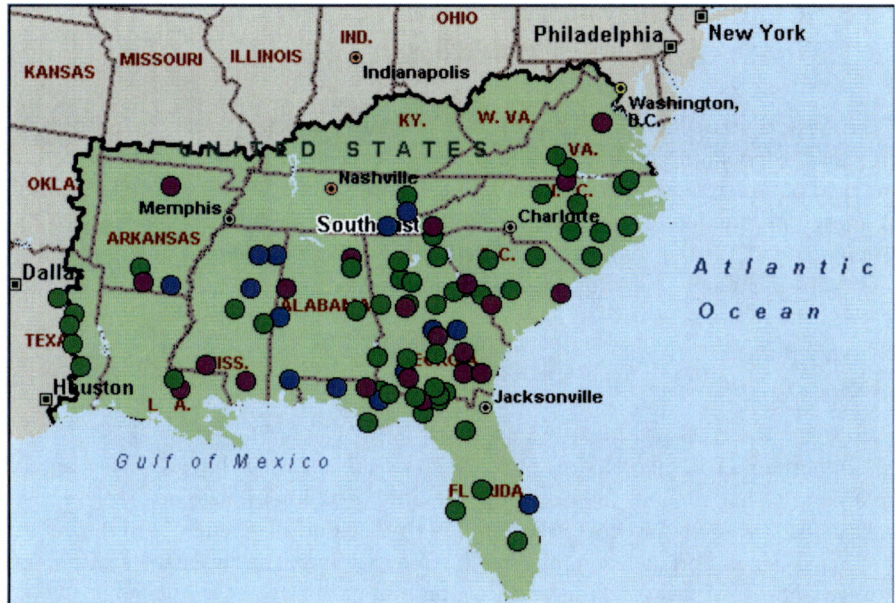

- Electricity
- Liquid Fuels
- Wood Pellets

Fig. 7.3 Map of Woody Biomass Projects in the Southeastern US
Sources: Forisk Consulting 2009 & ArborGen unpublished data). Green dots Plants generating primarily electricity, blue dots plants generating liquid fuels, purple dots plants generating pellets to be used as solid fuel

7.2.1.3 Gasification

Wood gasifiers can be more efficient than direct burning, although the gas may require cleaning to remove undesired chemical compounds. New low-energy, gas-producing gasifiers with better cleaning and control systems are being developed. Air-blown circulating fluidized bed appliances utilizing woody biomass have provided hot-fuel gas for lime kilns and boilers since the 1980s. The size and moisture content of the feedstock can vary when used in this type of combustion bed. Circulating fluidized bed gasifiers are currently being implemented in coal- and natural gas-fired utility boilers, and for integration with gas turbines. In the integrated gas turbine system, heat is recovered from gas turbine exhaust (flue gas) and used to generate power and heat in a steam turbine. This combined-cycle technology has a positive environmental impact because more energy can be produced per pound of CO_2 emitted than in simple-cycle technology. Currently, it is expected that biomass gasification power plants will be relatively small (30–40 MW) due to the conversion efficiency of the technology and the amount of biomass and the land base required to support this type of facility (Bridgwater 2003).

7.2.2 Characteristics of Wood Feedstock that Impact Bioenergy Production

The physical and chemical properties of woody biomass affect its processing efficiency (depending on the specific process being used) as well as the logistical and cost parameters associated with delivered cost of the biomass and overall cost of energy production. The characteristics that contribute most significantly to the efficiency and costs of biopower generation are described below. These are characteristics that can be targets of genetic improvement for bioenergy applications.

7.2.2.1 Density

An important characteristic of biomass materials is their bulk density and/or volume after harvest and subsequent processing. The importance of the harvested bulk density is related to cost associated with transport. The energy produced per volume is an important economic parameter that impacts the cost of shipment, storage and use. Hardwoods have an average bulk density of 0.23 tons per cubic meter (t/m^3), while softwood has a bulk density of 0.19 t/m^3, which is higher than the average value for baled straw (Biomass Energy Foundation 2009). When energy content is expressed relative to volume, there can be significant variation reported in this energy content metric. This is due mainly to genotype x environment interactions that affect bulk densities as well as the impact of harvesting techniques on water content.

7.2.2.2 Energy Content

The mean energy content associated with biomass dry matter is a stable feature for a particular biomass type, and is relatively independent of external factors such as environment. Biofuels have a higher proportion of oxygen and hydrogen relative to carbon as compared to fossil fuels, and this reduces the energy value of the fuel as the higher energy is contained in the carbon–carbon bonds. Woody biomass energy yield is determined mainly by the contents of energy-rich compounds such as lignin, resin or cellulose, and bark generally has a 2% higher energy content than wood on a dry weight basis (Kauter et al. 2003). Fuel analysis for solid fuels, such as coal, measures chemical energy content stored in two forms: fixed carbon and volatiles. The calorific value (CV) or heat value is defined as the heat released during combustion per mass unit fuel. Wood has an average calorific value of 18.6 megajoules per kilogram (MJ/kg), which is slightly higher than that of straw (approximately 16–17 MJ/kg) and more than 50% less than that of coal (34 MJ/kg; Kauter et al. 2003). More detailed energy values for different types of wood can be found in Gaur and Reed (1998).

Depending on the energy conversion process selected, particular material properties become important during subsequent processing. Some constituents, such as moisture, ash, and alkali content, in woody biomass used for biopower can have a negative impact on the cost effectiveness of the process. The principal targets for biomass genetic improvements for bioenergy generation include intrinsic properties such as bulk density and chemical composition and biomass yield per harvest cycle.

7.2.2.3 Ash and Alkali Metal Content

The chemical breakdown of a biomass fuel, by either thermo-chemical or biochemical processes, produces a solid residue. When produced by combustion in air, this solid residue is called "ash" and forms a standard measurement parameter for solid and liquid fuels. The ash content of biomass affects both the handling and processing costs of the overall biomass energy conversion cost (McKendry 2002). In a thermochemical conversion process, ash chemical composition can be a significant factor in the combustion process because the alkali metal content [potassium (K), sodium (Na), calcium (Ca), magnesium (Mg) and silicon oxide (SiO_2) content] of biomass can react with any silica present in the ash to produce "slag", which is a sticky, mobile liquid phase capable of blocking airways in the furnace and boiler. This can reduce plant efficiency and throughput, and result in increased operating costs. In addition, the chlorine (Cl) content of biomass contributes to corrosion and the silicate content can cause abrasion (Kauter et al. 2003). Silicate content can be a problem even if it is low in the intrinsic composition of the biomass. This is because, at harvest, contamination from soil can significantly increase the total silica content moisture content.

7.2.2.4 Moisture Content

Two forms of moisture content affect the thermochemical processing of woody biomass. Intrinsic moisture is the inherent moisture content of the material regardless of weather, and extrinsic moisture is the moisture content related to weather conditions during harvesting (McKendry 2002). Water content strongly influences the efficiency of energy production, as high water content reduces the net energy gain due to the thermal energy required for its vaporization.

An example of this is seen in poplar, a short rotation woody crop that is frequently used for the generation of biopower. The most problematic quality characteristic of this biomass is the elevated water content at harvest. The water content of poplar at harvest time typically varies between 54% and 61% of the fresh mass (Kauter et al. 2003). After felling, it changes depending on storage and processing. Agronomic approaches to control the drying process, such as leaving the felled biomass to dry prior to transport, are possible. The optimal moisture content for woody biomass transport and storage is 16% (Kauter et al 2003). The net

CV of wood with a water content of 16% is 15 MJ/kg (dry basis) while it is 6 MJ/kg (dry basis) for wood with a 60% moisture content.

One technology that can reduce the variability and negative consequences associated with moisture content is torrefaction. In torrefaction, green chipped wood waste is subjected to high temperatures under pressure, which drives off the moisture content and forms a high carbon content char. Torrefaction also produces gases from partial combustion and these gases can then be used to fuel the torrefaction process. The torrefied wood char is formed into torrefied pellets or briquettes that are low in sulfur. Torrefied wood has a BTU content that is slightly less than coal (Uslu et al. 2008) and can be substituted for coal in coal fired boilers without any boiler modifications required. Other advantages of torrefied pellets are that they do not collect moisture when stored and can be transported and stored at a lower cost than untorrefied wood. In the future, it may be possible for torrefied wood to be economically processed into a syngas that can then be liquefied into biofuel.

7.2.2.5 Yield

The quantity of woody biomass dry matter that can be harvested per unit land area determines the potential energy yield of the available land area. The biomass yield [dry metric tons (1,000 kg) of biomass per hectare or per acre (dmt/ha or dmt/ac)] combined with the high heating value (HHV) of the biomass determines the energy yield of a particular biomass crop. Experimental work on hybrid poplar species in the US Pacific Northwest, has produced yields of 43 dry metric tons per hectare per year (dmt ha^{-1} year^{-1}) compared to *Eucalyptus* yields in Brazil of 39 dmt ha^{-1} year^{-1} (Hislop and Hall 1996). Dedicated energy plantations to grow biomass will most likely address optimal yields as a function of species and available land. For example, a tree species that produces a high biomass yield, but requires intense management, will ideally be grown on good quality agricultural land to optimize productivity, while a tree species with a lower biomass yield but which is highly adaptive and does not require intense management will be grown on marginal land.

7.2.2.6 Energy Output

Energy output is the net energy balance for biomass after processing. For example, for every unit of fossil fuel energy used to grow, transport and convert willow biomass to electricity, 11–16 units of usable electrical energy are produced (Heller et al. 2003). At full generation rate, 1 kg woodchips converts to 1 kilowatt hours (kWh) via use in a gasifier/gas engine generator, giving an overall efficiency of conversion to electricity of about 20% without taking into account the potentially useful heat available from the gasifier/gas engine (Warren et al. 1995). The actual amount of energy recovered will vary with the conversion technology.

Independent of the process efficiency, the energy yield per hectare is affected by the biomass productivity of a specific energy crop. At yields of 15 dmt ha^{-1} year^{-1}, and with 1 dmt typically generating 1 megawatt hour (MWh), then 1 ha (based on a 3-year harvesting cycle) of short rotation woody biomass, such as poplar or willow, would provide 15 MWh/year (McKendry 2002). Therefore with the assumption of an annual operating time of 95%, a 1 MW gas engine/generator set would need approximately 550 ha to provide the necessary feedstock. This indicates that a significant amount of land (with the current productivity rates) is required to produce a relatively modest energy output of electricity due to the low efficiency of 20% in converting biomass into electricity.

However, combustion processes using high-efficiency, multi-pass, steam turbines to produce electricity can achieve an energy conversion efficiency of 35–40%, thereby reducing the required land to supply a 1 MW generator to between 270 and 310 ha (McKendry 2002). Integrated gasification combined cycle gas turbines can achieve energy conversion at about 60% efficiency, requiring approximately 110 ha to support feedstock production. However, the necessary land to grow a woody biomass crop to produce 1 MW is also dependent on the yield of that crop, indicating that efforts to increase yield through optimal genetics and silviculture is important.

7.2.3 Tree Species for Biopower

Many tree species already used for fiber and sawtimber applications are amenable to biopower and bioenergy production. Loblolly pine (Pinus taeda) and sweetgum (Liquidambar styraciflua) are highly adaptable trees commonly used in the Southeastern US for the production of pulp and paper, and are also potentially useful for bioenergy applications (Davis and Trettin 2006; Dickmann 2006), although their productivity per acre per year limits their potential to be used in short rotation energy plantations.

A sweetgum plantation grown for 7 years on old agricultural land has a productivity of only 1 oven-dry short ton per acre per year (odt ac^{-1} year^{-1}), although this productivity will increase with a longer rotation (Davis and Trettin 2006). Loblolly pine grown to a 20-year rotation can produce an average 4 odt ac^{-1} year^{-1} (Mercker 2007). Hardwood species of Populus, Salix, and Eucalyptus are suitable for short rotation woody crops due to their higher productivity and their ability to readily regenerate from the stump after harvest. Currently 12,000 acres in the US are planted as intensively managed short rotation hardwoods.

Research has been conducted on growing short rotation trees for fiber, biopower, and biofuels in the US and other countries (Short Rotation Forestry Handbook 1995). The Biofuels Feedstock Development Program of the US DOE has funded research on improved clonal planting stock and associated silvicultural systems for short rotation woody crops (SRWC; Tuskan 1998) in order to achieve target productivity rates of 8 odt ac^{-1} year^{-1} or greater (English et al. 2006).

Poplars and aspen (Populus sp. and their hybrids) as well as willows (Salix sp. and hybrids) are SRWC that can be successfully established in plantations. Populus is grown for bioenergy in Europe, the US Midwest and US Pacific Northwest. Salix is grown for energy in Europe, especially in England, and in the US Northeast. For example, short rotation willow crop yields range from 3 to 7 odt ac^{-1} year^{-1} on 1–4 year rotations (Mead 2005). Populus deltoides (Eastern Cottonwood) planted on good sites can produce an average yield of 5 odt ac^{-1} year^{-1} on 7–10 year rotations. For bioenergy applications the choice of species, the specific genotypes (clone or variety) and rotation length impact yield and quality. This is due to genotypic differences in growth pattern, nutrient use efficiency and relative proportions of different biomass fractions such as wood and bark.

Optimizing the management systems with respect to yield and fuel quality performance must be both genotype-and site-specific. Populus and Salix crops for bioenergy are typically managed as agriculture crops on arable soils. Short rotation coppicing, where new shoots and trees are regenerated from a cut stump, provides a cost effective option for increasing productivity per acre and decreasing feedstock costs. There is currently renewed interest in coppiced hardwoods for bioenergy and this is a subject of focused research (Andersson et al. 2002; Dickmann 2006).

7.2.3.1 Populus Species and Hybrids

Populus species fall into six morphologically and ecologically distinct sections, with the section *Aigeiros* containing cottonwood (*Populus deltoides* and *Populus nigra*), and the section *Populus*, which contains aspens (*Populus tremuloides*, *Populus tremula*, *Populus grandidentata*) and white poplars (*Populus alba*), representing the species most often used for bioenergy applications. However, the ease of hybridization allows for the development of many hybrids between these and other *Populus* species, such as *Populus maximowizcii*. Traditional breeding of poplars as single-trunk trees for wood production in short rotation forestry has been extremely successful. Hybrid poplars can be faster growing and more productive than parental species in short rotation silvicultural regimes. Sylleptic branching, in which lateral branches develop in the same year that their initial bud is formed, contribute positively to biomass in poplar by creating additional leaf area thereby increasing photosynthetic capacity and increasing volume (Ceulemans et al.1990; DeBell et al. 1996).

Poplars tend to be very site-sensitive and different genotypes may be limited to very specific sites in order to achieve adequate growth. Basic requirements for most *Populus* species are adequate water (minimum of 350 mm rainfall during the growing season) and nutrient supplies, deep soils and mild climatic conditions (average air temperature between June and September of at least 14°C). Biomass yield in poplar can be severely compromised when water is limited (Liang et al. 2006), and studies indicate that poplar tolerates water stress and low atmospheric moisture less than willow (Hinckley et al. 1994; Johnson et al. 2002). However, different poplar genotypes (clones) vary in their response to water stress indicating that genetic improvement for improved water stress response is possible (Karp and Shield 2008). In addition, for *Populus* to be grown as a SRWC, weed control is essential during plantation establishment (Kauter et al. 2003).

Aspen, unlike poplars and cottonwood, are more suited to poorer soils and can grow adequately in less favorable climatic conditions, such as those found in upland areas. Due to their slower typical growth rates, the silvicultural management system for aspen biomass involves longer rotations (approximately 10–12 years). These longer rotations enable aspen to attain mean annual growth increments of 10 dmt acre^{-1} year^{-1} as well as a higher proportion of high single shoot yields. Fuel quality improves with the single shoot mass due to an increased share of stem wood relative to bark (Kauter et al. 2003).

Two fundamentally different management systems are possible for SRWC. The traditional hybrid poplar system uses rotation cycles of 3–5 years, but this system is best for more fertile sites. As these fertile sites are also valuable for traditional agricultural use, the farmer may choose to harvest on longer rotation cycles of at least 7 years in order to achieve the yield and fuel quality to be economically competitive (Karp and Shield 2008). Poplars can be coppiced, initially producing many shoots that are subsequently thinned. However, in the 1st year after coppice, competition for light can result in very high shoot and stool mortality, so bioenergy poplars are often grown at densities of approximately 10,000 cuttings per hectare (Karp and Shield 2008). In comparative short rotation uncoppiced versus short rotation coppice trials, the coppiced systems yielded less, although coppicing enhanced intrinsic growth rates (Herve and Ceulemans 1996; Proe et al. 2002).

Biomass yields among poplars and aspen grown as SRWC differ considerably. Reported maximum yields can be 20–35 dmt ha^{-1} year^{-1} (Kauter et al. 2003). The highest yields were obtained from research trials with small plots that are actively managed and have significant edge effects while lower yields were reported for large demonstration plots. It is estimated that the average harvestable yields of poplars in the temperate regions of North America range between 10 and 12 dmt ha^{-1} year^{-1} (Kauter et al. 2003). This research has been validated by data presented in several reports. Poplar production in the US has been reported as ranging from approximately 8 t ha^{-1} year^{-1}) for cottonwood grown in flood plains on former agricultural land in Mississippi (Stanturf et al. 2003) to approximately 12 t ha^{-1} year^{-1} for hybrid poplar in the Southeastern US (de la Torre Ugarte et al. 2003) and approximately 13 t ha^{-1} year^{-1} for hybrid poplar in the Pacific Northwest (Stanton et al. 2002). These numbers reflect yields that are obtained reliably in large operational settings. To achieve biomass production at an overall rate of 11–12 t ha^{-1} year^{-1}, current poplar and aspen germplasm needs to be grown at 6- to 7-year rotations and 10- to 12-year rotations, respectively.

Increased silvicultural intensity (fertilization, irrigation, weed control) can improve yield, but at increased economic and energy costs. Breeding programs, such as described by GreenWood Resources (Stanton et al. 2002), can potentially increase yield by approximately 10% per rotation. In order to improve the sustainability and cost-effectiveness of *Populus* as a dedicated bioenergy crop, elite genotypes will need to be matched with the appropriate silvicultural management. For example, 20 Mt ha^{-1} year^{-1} yields of eastern cottonwood are possible on irrigated drylands in the Southeastern US. However, a yield improvement of 40% is considered necessary to achieve the desired cost-effectiveness, and this level of yield improvement will require genetic improvements and optimized silvicultural improvements (Gallagher et al. 2006).

The most problematic quality characteristic of *Populus* grown as a short rotation woody crop for biopower generation is its typically high water content (55–60%) at harvest time. This can be alleviated by storing unchipped material after felling on the field to lower moisture content.

7.2.3.2 Salix Species and Hybrids

Of the over 330 species of willows it is the shrub willows (*Salix viminalis* in Europe and *Salix eriocephala* in North America and Canada) that are most commonly used as bioenergy crops along with *S. dasyclados*, *S. schwerinii*, *S. triandra*, *S. caprea*, *S. daphnoides*, *S. purpurea*, and interspecific hybrids (Kuzovkina et al. 2008). Willow biomass plantations are easily and efficiently established from dormant stem cuttings using mechanical systems. The average planting density is 15,000–18,000 stools (individual plants or regenerating cut stumps) per hectare (Karp and Shield 2008). Coppicing is practiced routinely in willow SRWC production. Shrub willows respond to coppicing after the first growing season by prolific production of new stem growth in the second growing season. Coppiced willow is characterized by fast growth with many stems, followed by subsequent self-thinning. Above ground woody biomass is harvested during the dormant season on a 3- or 4-year rotation by a self-propelled forage harvester with a specialized cutter head, leaving the stool (remaining portion) behind. This cycle can be repeated for six to eight harvests before the stools need to be replaced (Karp and Shield 2008).

Genetic improvement in energy willow is focused on stem characteristics (height, diameter, straightness) and coppicing response (shoot number and vigor), as well as resistance to diseases, insects and frost damage (Karp and Shield 2008). However, different vegetative forms can achieve the desired biomass productivity. Some willows form a large number of thin stems (typically 11 per stool), relatively low leaf area index, and specific leaf area, while others produce fewer, larger-diameter stems (typically six per stool), and have a high leaf area index and specific leaf area. Both of these ideotypes can produce high yield (Tharakan et al. 2003). Willow genotypes that have early bud flush (even by a few days) may have a positive effect on total stem weight compared to genotypes with delayed growth cessation in the fall (Ronnberg-Wastljung and Gullberg 1999). Willow genotypes with higher water use efficiency and drought tolerance have been identified (Linderson et al. 2007). In the US, the most significant willow breeding and improvement program is based at The State University of New York College of Environmental Science and Forestry which conducts breeding and selection for high yielding clones that can sustainably produce high yields on a diversity of sites (Smart et al. 2005).

7.2.3.3 Eucalyptus

Eucalyptus is the most valuable and widely planted genus of hardwoods in the world and eucalypts demonstrate many agronomic traits, such as indeterminate shoot pattern, ability to coppice, resistance to drought, fire, and insects, and the ability to grow in acidic and low fertility soils, that enable them to be grown easily in large-scale plantings (Eldridge et al. 1994). In addition, many eucalypts display wood properties, such as high density, that make them very suitable for fuel and charcoal production, pulp and paper manufacturing, and sawn wood. In 2000, it was estimated that the area planted globally for Eucalyptus was 18 million ha in 90 countries (FAO 2000), with Brazil having the most intensively managed plantations with average productivities of 45–60m^3 ha^{-1} year^{-1} (Mora and Garcia 2000). *E. globulus* is the premier species for temperate zone plantations in Portugal, Spain, Chile and Australia. For pulp production and increasingly for solid wood, *E. grandis* and its hybrid with *E. urophylla* (*E. urograndis*) are grown in tropical and subtropical regions such as Brazil, South Africa, Congo, and China. The expansion of eucalypt plantations throughout the world can be attributed mostly to its superior fiber and pulping properties (Turnbull 1999). Genomics-aided improvement of Eucalyptus is underway (Grattapaglia 2004, Grattapaglia et al. 2004) and the Joint Genome Institute of the US DOE (2009) is nearing completion of the full genomic sequencing of *E. grandis*. It is anticipated that genetic improvement of Eucalyptus for industrial applications will increase dramatically after the publication of the Eucalyptus genome.

In the United States, some species of Eucalyptus are currently grown in Hawaii and warmer regions of California and Florida. In Florida, *E. grandis* operational plantings began in 1972 (Meskimen and Francis 1990). By the 1980s, it was planted on 5,650 ha (14,000 acres) in southwest Florida primarily as a source for landscape mulch. *E. grandis* grown in Florida has been tested successfully for pulpwood and fuel, and its wood has potential for poles, pallets, veneer, medium density fiberboard, and other products.

Eucalyptus is an ideal energy crop, with certain species and hybrids having excellent biomass productivity, relatively low lignin content, and a short rotation time. In Brazil, commonly planted Eucalyptushybrids such as *E. urograndis*, routinely yield 10–12 dry short tons acre^{-1} year^{-1}. Two Eucalyptus species, *E. grandis* and *E. amplifolia*, can be grown in Florida and are suitable for biofuel and biopower production. *E. grandis* grown in Florida has achieved a productivity exceeding 30 green short tons (approximately 15 dry short tons or 13.6 metric tons) acre^{-1} year^{-1}, with the potential to reach 55 green tons acre^{-1} year^{-1} (Stricker et al. 2000). This scale of productivity surpasses that seen with Populus and *Salix*, and addresses the biomass requirements for cost effective generation of biofuels and biopower from lignocellulosic feedstocks (Rockwood et al. 2008). In Florida, *Eucalyptus* utilization is likely to expand from the current mulchwood market to pulp and paper, biopower and biofuels (Rockwood et al. 2008).

7.2.4 Softwood Species for Bioenergy

Pinus taeda (Loblolly pine), also called Arkansas pine, North Carolina pine, and oldfield pine, is the most commercially important forest species in the Southeastern US. Loblolly pine plantations of the Southeastern US are prime candidates for the production of biomass for bioenergy because these plantations are already managed industrially to produce wood for pulp and paper production and for sawtimber. Loblolly pine occurs in subgenus *Pinus*, subsection *Australes* Loudon. This subsection is comprised largely of species found in the Southeastern US and the Caribbean, and most of the pines that co-occur naturally with this species in mixed stands (such as *P. echinata*, *P. elliottii*, *P. glabra*, *P. palustris*, and *P. serotina*) are in the same subsection (Kral 1993). The native range of loblolly pine extends through 14 States from southern New Jersey south to central Florida and west to eastern Texas, and it is the dominant tree on about 11.7 million ha (29 million acres) making up over one-half of the standing pine volume. It is a medium-lived tree with rapid juvenile growth. Loblolly pine is a highly adaptable species that responds well to silvicultural treatments and can be managed as either even-aged or uneven-aged natural stands, or can be regenerated artificially and managed in plantations.

Intensively managed pine plantations can be adapted easily to produce biomass for bioenergy applications. Loblolly pine biomass for bioenergy is currently being obtained from pre-commercial thinnings (see below) and from logging residues. As the pulpwood market in the US has decreased in recent years, there is a growing interest in providing wood to the energy industry. It will be possible for pine plantations to meet both the traditional forest industries and the bioenergy industry at the same time by using appropriate management practices. For example, agroforestry practices can be applied in plantation establishment in which double and triple rows of trees are planted, allowing some rows to be thinned at 12–15 years for bioenergy leaving the remaining trees to grow to a full 26-year sawtimber rotation (Foster and Mayfield 2007). It is anticipated that loblolly pine energy plantations will rapidly become a reality.

Yields of planted loblolly pine vary with plantation age, site quality, number of trees planted, and interactions of these variables. Yields generally increase with increasing age and site quality. Yields also increase with higher planting density or closer spacing; however, on some sites, moderately wide spacing of 2.4 x 2.4 m (8 x 8 feet) or 3.0 x 3.0 m (10 x 10 feet) can out-produce both wider and closer spacing. Mean annual increment in volume levels off at younger ages on better sites than on poorer ones. Better sites can carry more tree stocking densities than poor sites; consequently, initial spacing can be closer (Baker and Langdon 1990). Average total solid-wood yields of un-thinned loblolly pine planted at 1,730 seedlings/ha

(700/acre) on non-old-field sites at various locations within its range were predicted to increase from approximately 155 m^3/ha (2,200 cubic feet per acre) at age 15, to 300 m^3/ha (4,200 ft^3/acre) at age 30. Mean annual increment at age 30 was about 10 m^3/ha (145 ft^3/acre; Baker and Langdon 1990). Estimates are also available for a variety of site and stand conditions and geographic areas. If sawtimber is a primary management objective, wider tree spacing is used. However, thinning more densely planted stands can increase the diameter growth of the remaining trees, enabling multiple markets to be addressed where thinnings can be harvested at an intermediate time in the rotation for pulp or bioenergy with the full rotation trees being provided to the sawtimber market.

One current, but unfortunate, source of pine biomass for bioenergy applications are pine trees that have been killed by insect infestations in the western regions of North America, such as the mountain pine beetle infestations in Canada. It is estimated 10 million ha forest is already infected in Canada, potentially causing losses of up to 960 million m^3 wood. Bark beetles, such as the mountain pine beetle, are causing tree losses in the western areas of the US and Canada. Use of surplus wood from mountain pine beetle killed trees to generate power is being considered at several small size power plants (Kumar et al. 2008).

7.3 Liquid Biofuels

The established corn ethanol industry has laid the foundation for advanced biofuels in the US with ethanol facilities producing 9.2 billion gallons in 2008. Recent global food supply concerns as well as intrinsic limits to the sustainable production of corn have led to an increasing interest in the production of ethanol from non-food sources, including cellulose. The biofuels industry in the US is being driven in large part by government mandates for increased use of biofuels in the nation's transportation fuel supply. The Renewable Fuel Standard (RFS) included in the 2007 EISA mandates that 36 billion gallons of biofuels be included in the fuel supply by 2022 (US Environmental Protection Agency 2007).

7.3.1 Cellulosic Ethanol

Ethanol is the most common biofuel produced from wood, although wood can also be used to produce gasoline and diesel fuels. Ethanol is normally blended with gasoline to produce E10 (10% ethanol, 90% gasoline). In the US, the limitations on the blend ratios are currently driven by the automotive industry and the technologies used for automotive engines. In Brazil, E85 and E95 fuels are used and the "flex fuel" technology to allow the use of these fuels is readily available. The expansion of flex fuel vehicles (FFV) in the US will increase the consumption of ethanol by allowing the common use of fuels with higher ethanol contents. The domestic automobile manufacturers, seeing an opportunity to turn over the US fleet more rapidly, have committed to half of their vehicles sold being FFV by 2012 (Austin 2008).

The RFS mandates that 21 billion gallons of the 2022 biofuel supply will have to be "advanced biofuels" produced from non-corn feedstock (US Environmental Protection Agency 2008). Producing 21 billion gallons of cellulosic ethanol at a conversion rate of 100 gallons per dry short ton would require 210 million dry tons or roughly 420 million green tons if sourced entirely from woody biomass. Assuming trees represent 50% of the 'energy crop' opportunity in the Southeastern US, this equates to a total market of 150 million green tons by 2025 in the Southeastern US or 40% of current consumption for traditional forest products.

The US biofuel DOE has identified research to improve the cost-effectiveness of producing ethanol from lignocellulosic feedstock. Of significance is the US $375 million in funding the DOE has provided to three bioenergy research centers to develop advanced technologies to address current limitations in the production of bioethanol, including limitations that are related to feedstock inherent properties as well as productivity (US DOE 2007b). One of the centers, the BioEnergy Research Center led by Oak Ridge National Laboratory, is focusing its biomass improvement efforts on Populus in addition to switchgrass.

In the future, it is anticipated that ethanol production facilities will use forest-based biomass widely. Several commercial companies are developing pilot level or early commercial scale advanced biofuels production facilities using wood in the Southeastern US. According to *Wood Biomass South* published by Forisk Consulting (2009) and ArborGen unpublished data, cellulosic ethanol projects have been announced throughout the Southeastern US that would consume more than 9 million green short tons of biomass annually by 2015 (Table 7.3). For example, Range Fuels is constructing an ethanol plant in the state of Georgia that is expected to be operational in 2010. At this plant, wood waste is to be converted via a two-step, thermo-chemical conversion technology to a synthetic gas, to create power and ethanol using existing combined cycle generator technology.

Outside of the Southeastern US, another cellulosic ethanol plant, built by KL Process Design Group in Wyoming, became operational in 2008 and uses wood waste (KL Process Design Group 2008). The Flambeau River Biorefinery project will be the first modern US-based pulp mill biorefinery to produce cellulosic ethanol. It will be designed to produce 20 million gallons of cellulosic ethanol per year from spent pulping liquor. The new biorefinery, as designed, is expected to have a positive carbon impact of approximately 140,000 t/year. Project engineering has commenced with production of ethanol expected to begin as early as 2009 (PR Newswire 2009).

C2 Biofuels has a demonstration cellulosic ethanol plant that relies on Georgia pine trees as a feedstock and plans its first commercial plant in 2010. Their plant will use an enzyme-based system to convert the complex sugars in the wood chips into fermentable simple sugars. C2 Biofuels plans to have plants in various rural locations across Georgia, each capable of producing 50 million gallons ethanol a year (Center of Innovation for Agribusiness 2006).

7 Woody Biomass and Purpose-Grown Trees as Feedstocks for Renewable Energy

Table 7.3 Woody biomass liquid fuel projects in the Southeastern US excludes mill residuals

Name	State	Liquid fuel capacity (mmgal)	Start-up year	Estimated wood use (green tons)
Bleckley County Biorefinery	GA	20	2015	200,000
BlueFire Ethanol	MS	18	2015	180,000
Buckeye Cellulose – University of Florida	FL	0.14	2010	13,800
C2 Biofuels	GA	55	2011	550,000
Cello Energy LLC	AL	20	2008	120,000
Enerkem	MS	10 (2013), 20 (2015)	2015	200,000
Gulf Coast Energy	AL	0.23 (2008), 10 (2013), 100 (2016)	2009	1,000,000
Gulf Coast Energy	FL	45 (2010), 80 (2013)	2010	800,000
Gulf Coast Energy	TN	10 (2014), 100 (2017)	2014	1,000,000
Jerome Bio-Refinery LLC	AR	110	2015	1,100,000
Liberty Industries Inc.	FL	7 (2011), 77 (2020)	2011	770,000
Mascoma Corp.	TN	5	2010	180,000
New Planet Energy (INEOS)	FL	8 (2010), 21 (2011), 100 (2013)	2010	1,000,000
Range Fuels	GA	10 (2010), 100 (2013)	2010	1,000,000
Raven Biofuels	MS	11 (2015), 33 (2018)	2015	1,157,625
Liquid Fuels Total				9,271,425

Source: Wood Bioenergy South and ArborGen estimates

7.3.2 Conversion Processes

7.3.2.1 Thermal Conversion

Fuel gas can be produced from biomass and related materials by either partial oxidation or by steam or pyrolytic gasification. The process of gasification involves heating biomass with about one-third of the oxygen necessary for complete combustion, producing a syngas composed of a mixture of carbon monoxide and carbon dioxide that may then be fed into a special kind of fermenter. Instead of sugar fermentation with yeast, this process uses a microorganism named *Clostridium ljungdahlii* (Gaddy 2000). This microorganism will ingest carbon monoxide, carbon dioxide and hydrogen and produce ethanol and water.

Alternatively, the synthesis gas from gasification may be fed to a catalytic reactor where the synthesis gas is used to produce ethanol and other higher alcohols through a thermochemical process. This process can also generate other types of liquid fuels. Currently, the most promising technology for processing woody biomass to biofuel is "two-stage thermal conversion" or "gasification with gas-to-liquids conversion" process. The gasification with gas-to-liquids conversion process converts all of the woody biomass to syngas, leaving no lignin components behind.

A major upside of biofuels produced by thermal conversion is the cogeneration of heat and power using diesel engine and gas turbine technology to generate ancillary revenue from the sale of excess steam, waste gases and heat that they produce (Sklar 2009). In some configurations, the excess steam and syngas is used to co-generate green power for sale to the grid. These second generation biorefineries can become much more economical than stand-alone biorefineries, if they are partially integrated into strategically located host facilities such as pulp and paper mills, and if they are able to realize a fair return on the steam, power and heat they provide these host plants.

7.3.2.2 Biochemical Conversion

Lignocellulosic biomass is a non-food, inexpensive and abundantly available source for fermentation into transportation fuels. Biochemical conversion of lignocellulose through saccharification and fermentation is a major pathway for liquid fuel production from biomass (Zhu et al. 2008). In this approach, biomass cellulose is converted to glucose using microbial or enzymatic actions. The glucose is then converted to alcohols through fermentation. Improved and novel methods are currently in development. It is anticipated that genomics-enabled 'synthetic biology' approaches will be key in developing efficient, inexpensive biofuel production systems (Rubin 2008).

Bio-based processes used to produce biofuels are currently relatively inefficient and expensive. Currently, bioconversion into ethanol has four unit operations: pre-treatment, saccharification, fermentation, and distillation. It is the pre-treatment and saccharification steps that most affect the economics of bioconversion. One of the major limitations is the accessibility of the cellulose in the plant cell wall to hydrolysis and saccharification agents due to the tight bonding of lignin to cellulose. Pre-treatment is a very critical step in the conversion of woody biomass into ethanol because its cost, efficiency and potential to produce compounds negatively influencing downstream processes can determine the economic feasibility of bioethanol production. The goal of the pre-treatment process is to remove lignin and hemicellulose, reduce the crystallinity of cellulose, and increase the porosity of the lignocellulosic materials.

After pre-treatment, cellulases are then able to saccharify or degrade the cellulose to glucose which in turn is converted to ethanol by yeast or bacteria during fermentation. Lack of low-cost and high-activity cellulose hydrolytic enzymes is currently an economic barrier to cellulosic ethanol production. When lignocellulosic raw materials are used in ethanol production, the main by-product is lignin, which can be used as an ash-free solid fuel for production of heat and/or electricity.

Limited progress has been made in improvements of woody feedstock pre-treatment, despite efforts made in the last several decades (Kumar et al. 2009). Dilute acid pre-treatment developed over a century ago is still the dominant process for biomass pre-treatment today (Zhu et al. 2008). The current available technology, which is based on dilute acid hydrolysis, has about 35% efficiency from biomass to ethanol. The overall efficiency, with electricity co-produced from the non-fermentable lignin, is about 60%.

It is estimated that current investment costs for a biorefinery are US $3,000 per KWh (at a nominal 2,000 dry metric tons per day input; Hamelinck et al. 2005). Improvements in pre-treatment and advances in biotechnology, especially through process combinations, have the potential to bring the ethanol generation efficiency to 48%. If the byproduct lignin is used for energy generation then the overall process efficiency is expected to be 68%. The combination of improved hydrolysis-fermentation efficiency, lower required capital investments, increased scale and cheaper biomass could bring the ethanol production costs from an anticipated US $31 per Gigajoule (GJ) ($0.65/l ethanol) in the next 5 years to $18 per GJ ($0.38/l ethanol) in 10–15 years (Hamelinck et al. 2005).

Pre-treatment

Pre-treatment must improve the formation of sugars or the sugar production via hydrolysis, avoid the degradation or loss of carbohydrate, avoid the formation of by-products inhibitory to hydrolysis and fermentation, and be cost-effective. Pre-treatment methods are varied and include physical (milling and grinding), physicochemical (steam pre-treatment/autohydrolysis, hydrothermolysis, and wet oxidation), chemical (alkali, dilute acid, oxidizing agents, and organic solvents), biological, or electrical (Kumar et al. 2009). The conditions employed in the chosen pre-treatment method will affect various substrate characteristics, which in turn govern the susceptibility of the substrate to hydrolysis and the subsequent fermentation of the released sugars.

For woody biomass, physical pre-treatment through size reduction is critical, regardless of whether other forms of pre-treatment are used. Reduction of wood to the level of fibers or fiber bundles is necessary to increase microbial or enzymatic reaction surfaces in the breakdown of cellulose (Kumar et al. 2009) and this process can be very energy intensive. First logs are reduced to chips then the chips are reduced to fibers approximately 2 mm long. This process is relatively energy intensive, taking up about 10–30% of the calculated wood ethanol energy available (Kumar et al. 2009). Chemical pre-treatment, prior to size reduction, can reduce the energy requirements of the size reduction step.

Steam explosion is the most common and cost-effective pre-treatment process for wood. In this method, biomass is treated with high-pressure saturated steam, followed by explosive decompression upon the sudden release of pressure. Steam explosion is typically conducted at a temperature of 160–260°C, with a corresponding pressure ranging from 0.69 to 4.83 MPa. The high temperature of this process causes hemicellulose degradation and lignin release. An enzymatic hydrolysis efficiency of 90% has been reported for poplar chips pre-treated by steam explosion (Grous et al. 1986). Steam pre-treatment with the addition of a catalyst is proving useful with a variety of lignocellulosic feedstock including wood (Kumar et al. 2009). The technology has been scaled-up and operated at the pilot-plant scale at the Iogen demonstration plant in Canada (Jorgensen et al. 2007). Although steam explosion is viewed as a relatively cost-effective pre-treatment process for hardwoods and agricultural residues, it is not equally effective for softwoods.

Ammonia fiber explosion (AFEX) is a physicochemical pre-treatment process in which lignocellulosic biomass is exposed to liquid ammonia at high temperature and pressure for a period of time followed by a sudden pressure reduction (similar to steam explosion). However, the AFEX process was not very effective for biomass with higher lignin content such as newspaper and aspen chips (25% lignin). Hydrolysis yields of AFEX-pre-treated newspaper and aspen chips were reported to be only 40% and below 50%, respectively (Kumar et al. 2009).

Dilute-acid pre-treatment is not highly effective on wood. Pre-treatment with 0.25–1.0% weight-to-volume hydrogen sulfate (H_2SO_4, sulfuric acid) of four timber species, aspen, balsam fir, basswood, and red maple resulted in hydrolysis yields that were high only for xylose (70% for balsam fir and up to 85% for aspen, with 90% for switchgrass) while very low yields were obtained for glucose for all species (approximately 11–14%). Balsam fir, unlike the other species, also had high yields for mannose (approximately 60%) due to its higher inherent mannan content (Yat et al. 2008).

Alkali pre-treatments, using sodium hydroxide (NaOH) or calcium hydroxide (CaOH) can be beneficial because they cause cell wall swelling, which increases exposed internal surface area while decreasing cellulose polymerization and crystallinity, breaking linkages between lignin and carbohydrates, and disrupting lignin (Kumar et al. 2009; Fan et al. 1982). However, alkali pre-treatments yield better results when lignin contents are 20%, which makes most wood less suitable than herbaceous feedstock due to its higher lignin content. Hardwoods respond better to this pre-treatment than do softwoods.

An organosolv pre-treatment method is being commercialized by Lignol Innovations (Bevill 2008). This process uses an ethanol-based organosolv step to separate lignin, hemicellulose, and extractives from the cellulosic fraction of woody biomass. The resultant cellulosic fraction is highly amenable to enzymatic hydrolysis and capable of generating glucose yields over 90% in 48 h or less (Pan et al. 2005). The liquor from the organosolv step can be processed readily to recover lignin, furfural, xylose, acetic acid, and a lipophilic extractives fraction. Revenues generated from the these byproducts, in addition to the ethanol produced from the glucose, generates sufficient revenue to enable cost-effective operations for even small facilities that process only 100 metric tons per day (Arato et al. 2005), which is a scale similar to a large sawmill.

Biochemical Process Commercial Status

To date, no large scale bioethanol plants that rely on woody biomass are operating in the Southeastern US. However, such a plant that utilizes urban wood waste operates in Japan (Green Car Congress 2009). A new plant, under construction by Verenium in Louisiana, will use primarily wood, including southern pine, as a source of biomass (Verenium 2009). Zeachem, Inc. uses a thermochemical and biological process hybrid process that has demonstrated, at laboratory scale, production rates of 135 gallons of ethanol per short ton biomass (Eggeman 2009). In 2008, Zeachem announced plans to build a 1.5 million gallon per year production plant in Oregon that will be sourced with poplar trees grown by GreenWood Resources, Inc. on intensively managed tree farms (Zeachem 2008).

7.3.3 Other Cellulosic Liquid Fuels

Although the bioethanol industry is growing rapidly, with ethanol production facilities and distribution centers increasing rapidly in number, ethanol as a transportation fuel has several disadvantages. Disadvantages of ethanol as a biofuel include its high solubility in water, which requires an energy-intensive distillation step; lower energy content relative to petroleum; its hygroscopic nature, which creates difficulties in pipeline distribution; and incompatibilities with many current vehicles at higher blending volumes. These disadvantages make other potential liquid fuels, such as methanol and butanol, worthy of consideration.

7.3.3.1 Methanol

Methanol is a one-carbon alcohol most frequently used in the production of formaldehyde, but it is also used in the production of biodiesel. In Europe, 10–20% of the weight of biodiesel can be methanol, while the remaining 90% is oil processed from agricultural crops (Kavalov and Peteves 2005). It also could become a useful transportation fuel on its own as methanol-powered fuel cells and fuel-cell/electric vehicles are developed. Methanol is also known as "wood alcohol", as it was produced from wood in the early 1900s. With the development of a natural gas to methanol conversion process, the cost of producing methanol from wood was no longer cost competitive. However, technology exists today to produce methanol from wood, in a manner similar to methanol produced from natural gas, through gasification in which the syngas is converted methanol. Today, 1 short ton dry biomass can produce about 157 gallons methanol at a conversion efficiency of 50% (Vogt et al. 2009). This efficiency of conversion means that forest residuals and other wood sources now become an economically feasible feedstock.

7.3.3.2 Butanol

Butanol is a four-carbon alcohol commonly used in many industrial chemical manufacturing processes, although is can be used as a stand-alone transportation fuel or blended with gasoline or diesel. Its inherent chemical properties make it superior to ethanol for use in combustion engines due to its tolerance for water contamination and relatively low corrosive nature compared to ethanol (Vessia 2005), Each year 220 million gallons of butanol produced by petrochemical means are consumed in the United States (Young et al. 2007). Butanol can be produced from solventogenic bacteria, such as *Clostridium acetobutylicum*, which is used in the Acetone-Butanol-Ethanol (ABE) industrial fermentation process used widely in the early 1900s. This process utilizes a variety of feedstock, including ligncellulosic materials (Schuster 2000).

More advanced methods have made the process more cost-effective, and these are based on engineering microorganisms with butanol biosynthetic capabilities but also with increased tolerance to high levels of butanol thereby enabling greater butanol yields. For example, *Escherichia coli* has recently been engineered to produce isobutanol and other alcohols via a non-fermentative pathway that may be more readily adapted to large-scale production, using heterologous expression of *Lactococcus lactis* and *Bacillus subtilis* genes (Atsumi et al. 2007). Several companies are starting commercial-scale production of butanol with recently patented technologies. Gevo Inc. announced in 2009 the deployment of a 1 million gallon per year demonstration plant with engineered cellulosic yeast technology (Gevo 2009). ButylFuel, LLC, is developing a pilot fermentation plant in Ohio that uses patented technology demonstrated to produce butanol at a higher efficiency than the ABE process (Kiplinger Washington 2007). DuPont (E.I. du Pont de Nemours and Company) and British Petroleum Company plan to make biobutanol the first product of their joint effort, Butamax Advanced Biofuels, LLC, to develop, produce, and market advanced biofuels (Lane 2009). Their process also produces a recoverable amount of hydrogen, and other by-products such as acetic acid, lactic acid, propionic acid, acetone, isopropanol and ethanol.

7.3.4 Feedstock Characteristics Affecting Biofuel Production

Sections 7.2.3 and 7.2.4 describe characteristics of different tree species that can be used as energy crops. Features that affect delivered cost of the biomass to the processing plant are similarly important, whether or not the wood is undergoing a thermochemical or biochemical process, therefore many of the feedstock characteristics that affect yield and delivered cost are relevant for biofuel production. Wood chemical composition and moisture content have impacts that are specific for the particular process being used. In the biochemical processes, moisture content of the feedstock does not affect the energy required to generate liquid biofuels as it does in the thermochemical process.

Although both the biochemical and thermochemical processes are affected by wood chemical composition, the specific effects of chemical characteristics are not similar between the biochemical and thermochemical processes. Dry matter composition varies considerably among lignocellulosic energy crops affecting the ultimate yield and efficiencies in different bioconversion processes. Table 7.4 shows the percent dry weight of cellulose, hemicellulose, and lignin in softwoods and hardwoods as well for energy grasses. Since lignin is not easily biodegradable, the amount of lignin and the type of lignin can make a difference in the efficiency and costs of biochemical processing. In addition, the level of cellulose crystallinity affects the accessible cellulose surface area, and the type and amount of lignin and

Table 7.4 Consensus values for cell wall chemistry attributes as percentage of total dry weight of different energy crops. Values vary with site, season, cultivar and management. All values are percentages of mass at 100% dry matter

Feedstock	Lignin	Hemicellulose	Cellulose
Wheat straw[a]	15.1	24.6	33.2
Maize stover[a]	10.4	28	35
Switchgrass[a]	6.1	36	31.6
Miscanthus[a]	10.5	15.9	57.6
Willow[a]	19	14.0	55.9
Poplar[a]	20	23	40
Pine[b]	27–30	25–30	35–40
Eucalyptus[c]	34	13	38
Eucalyptus[d]	30	16	42

[a]Karp and Shield 2008
[b]McKendry 2002
[c]Rockwood et al. 2008, ArborGen unpublished data for 16
[d]*Eucalyptus urophylla* x *Eucalyptus urograndis* hybrid "EH"

hemicellulose can affect how cellulose is "protected" by lignin during the hydrolysis step. In the biochemical processes, the total amount of cellulose potentially affects the amount of glucose produced in hydrolysis and therefore ultimate biofuel yield, but this is dependent upon the pre-treatment method and the accessibility of cellulose to hydrolysis.

During hydrolysis of lignocellulosic materials a wide range of inhibitory compounds deleterious to fermenting microorganisms are formed or released. Hydrolysis procedures which involve treatment of lignocellulose at a high temperature under acidic conditions lead to the formation and liberation of a range of inhibitory compounds. These inhibitors are usually divided into three major groups: weak acids, furan derivatives, and phenolic compounds. These compounds limit efficient utilization of sugar hydrolysates during ethanol fermentation. Hardwoods and softwoods release different inhibitory compounds and this is somewhat dependent on the amount and types of lignin found in their wood. Hardwoods have syringyl and quaiacyl lignin while softwoods primarily have quaiacyl lignin. Hardwoods tend to have lower lignin levels than softwoods. For example, poplars contain much lower levels of fermentation-inhibiting extractives compared to softwood feedstock such as pine, thus the biochemical conversion efficiency of the biomass is correspondingly higher (Davis 2008; Palmqvist and Hahn-Hägerdal 2000). However, it is possible to generate ethanol through hydrolysis and fermentation of softwoods and it is projected that technology can be developed that will make the process cost-competitive with other feedstocks (Frederick et al. 2008).

7.4 Purpose-Grown Trees for Renewable Energy

Woody biomass from trees will compete with a variety of lignocellulosic energy crops, most of which are annually harvested (switchgrass, corn, sugarcane). Multiple approaches utilizing a variety of different energy crop species and production systems will be needed to meet the nation's renewable energy goals and it is clear that biomass from trees will play a significant role. The choice to use a specific energy crop must take into account regional conditions and needs, as well as logistical elements such as biomass transportation costs. Short rotation, purpose-grown trees, developed for high biomass productivity, have a variety of inherent logistical benefits and economic advantages relative to other lignocellulosic energy crops. It is expected that short rotation hardwood crops, such as fast growing species of Populus, Salix, and Eucalyptus and their respective hybrids, will be planted as purpose-grown wood on sites that enable high productivity and proximity to the processing plant, while existing pine plantations will be managed in a manner that enables harvests for energy end uses in addition to the traditional forest products industry.

In the Southeastern US, where the traditional forest products industry is an important economic driver, there already exists an established and accessible wood inventory and harvesting infrastructure. Current conventional forestry systems now provide biomass for energy applications as a by-product of timber production systems. Any harvesting operation, whether thinning in young stands, or cutting in older stands for timber or pulpwood, can yield tops and branches usable for bioenergy. Stands damaged by disease or fire are also sources of biomass. Forest residues can represent 25–45% of the total biomass at the time of harvest, and utilization of these residues can provide additional revenue from a stand while reducing the risks of possible forest fires, pests and disease.

It is anticipated that, in the future, forestry systems may also produce biomass for energy as a primary product. However, landowners who grow woody biomass have the advantage of having multiple markets for their wood, such as sawtimber and pulp and paper, in addition to biopower and biofuels. This provides the landowner an economic hedge when the market demands for a particular end-use might be lower than expected. The grower is provided a choice in harvest time and multiple end uses such as traditional forest products and energy products such as cellulosic ethanol and power generation through direct firing, co-firing, or wood pellet systems. The stand uniformity and improved genetics of plantations of purpose-grown trees, as compared to naturally regenerated forests, enables overall higher yields and improved management and harvesting efficiencies.

Trees have a variety of inherent logistical benefits and economic advantages relative to other biomass production systems. Many of these advantages are driven by the fact that trees can typically be harvested year-round, and provide a living inventory of available biomass as the trees continue to grow (Sims and Venturi 2004). The multi-year harvest cycle of trees serves to mitigate the risk of annual yield fluctuations experienced by other crops due to the acute effects drought, disease, pests and abiotic stresses. Multi-year harvest rotations also allow the biomass to be harvested in response to market demand, as well as provide a processor with multiple years to secure the acres needed to supply a given quantity of biomass, which improves deployment feasibility. Trees, in general, compare favorably with respect to storage-related costs, as annual herbaceous species typically must be harvested within a defined window of time and processed immediately to reduce the risk of decay in storage. In contrast, trees are stored essentially in the field. Wood can be stored easily dried at the roadside, at a distribution center, or at a conversion facility, and distributed during peak demand periods. Processors utilizing annually harvested feedstock must either secure more feedstock than is needed or run the risk of having to shut down the facility during a bad year. The standing inventory and ease of storage helps the processor avoid this by smoothing natural yield fluctuations allowing for supply to be more easily matched with demand.

The higher density of wood relative to cereals or grasses increases the cost-effectiveness of transport from the field to the processor, especially for logs. Transport costs are largely a function of the distance traveled and the energy density, e.g., energy content per unit volume (frequently measured as Megajoules per cubic meter, MJ/m^3), of the biomass being transported, and road transport accounts for about 70% of the total delivered biomass fuel cost (Transport Studies Group 1996). Forest residues normally have much lower densities and fuel values than logs, and transport cost effectiveness can be enhanced by increasing density close to the harvest site. This is done by size reduction via grinding or chipping, or by compaction into "compact residue logs" that can be handled efficiently (IEA BioEnergy 2002).

The multi-year harvest cycle for trees also provides some environmental advantages compared to crops that are harvested annually. For example, soil cultivation in a poplar energy plantation is only required twice for a 6- to 8-year harvest rotation, which reduces soil erosion associated with annual cultivation. While biomass yields and total land footprint to feed a processing plant may be similar between trees and annually harvested crops, trees only require that a fraction of the total land footprint would be planted or harvested in any given year (Table 7.5). This has a positive impact on the energy costs and CO_2 inputs required to supply an energy feedstock. Trees provide the added benefit of mitigating climate change through the sequestration of atmospheric carbon.

Many tree species already used for fiber and sawtimber applications are amenable to biopower and bioenergy production. Loblolly pine (*Pinus taeda*) and sweet-gum (*Liquidambar styraciflua*) are species commonly used for pulp and paper production in the Southeastern US, but their typical productivity and rotation age limits their potential for use as SRWC energy crops (Davis and Trettin 2006; Dickmann 2006), although these species, especially pine, will definitely have use

Table 7.5 Total and annual acreage needs for trees relative to annually-harvested energy crops, data (generated from an ArborGen financial model by L. Mullinax)

Feedstock	Biomass needed (MM green tons/year)	Productivity (green tons acre^{-1} year^{-1})	Rotation length (years)	Total acres needed (MM)	Acres planted annually (MM)
Trees	100	20	6	5	0.83
Annually harvested crop	100	20	1	5	5

as bioenergy feedstock through increased biomass yields per acre via altered management regimes such as increased planting density. Research has been conducted on growing short rotation trees for fiber, biopower, and biofuels in the US and other countries (Short Rotation Forestry Handbook 1995). Attempts at shorter rotation sweetgum (7-year-rotation on old agricultural land) produced only 1 odt acre^{-1} year^{-1} (Davis and Trettin 2006), and shorter rotation loblolly pine (20 year rotation on nonagricultural land) produced an average 4 dry tons acre^{-1} year^{-1} (Mercker 2007). The Biofuels Feedstock Development Program of the US DOE set their target productivity for short rotation woody crops as 8–10 odt acre^{-1} year^{-1} and have invested in developing improved clonal planting stock and associated silvicultural systems to achieve this productivity level (Tuskan 1998; English et al. 2006). Populus and Salix species are hardwoods that currently can be managed intensively in plantations, but these trees still fall short of achieving the desired productivity levels. Short rotation willow crop yields range from 3 to 7 dry short tons acre^{-1} year^{-1} (Mead 2005). *Populus deltoides* (Eastern Cottonwood) planted on good sites can produce an average yield of approximately 5 odt acre^{-1} year^{-1}.

Currently, the two main silvicultural systems for biomass hardwoods utilize the establishment of moderately dense stands of cottonwood (*Populus deltoides*) with 7- to10-year rotations, or the establishment of stands of willow (*Salix* species) harvested as a coppice on a 1- to 4-year rotation. Short rotation coppicing of hardwoods offers the promise to produce biomass for bioenergy. Coppicing is the process by which new shoots and trees are regenerated from a cut stump following harvest. The use of coppiced hardwoods for this purpose is not novel although it is the subject of renewed interest and focused research (Andersson et al. 2002; Dickmann 2006).

Tree genetic and silvicultural research is underway to provide the optimal productivity gains required for tree biofuel and biopower applications. Productivity improvements, such as those provided through advanced planting stock, optimal siliviculture, and biotech traits, will be important for the cost effective deployment of short rotation trees for biomass applications. The economics of a purpose-grown tree feedstock for energy may not be feasible without significant genetic improvement in the base growth rate. Other useful trait improvements will be dependent on the inherent biological and genetic limitations of the utilized genotype and its ability to respond to varying climate, soil and moisture conditions.

7.4.1 Genetic Improvement for Productivity

Domestication, breeding and selection of forest tree species has resulted in direct and indirect changes in the genetic make up of trees grown for industrial applications. Forest tree domestication was accelerated during the latter half of the twentieth century, with conventional breeding methods applied to forest tree populations to improve growth, volume and wood quality traits (Burdon and Libby 2006). The application of biotechnology to forest tree species is expected to further accelerate such improvement (Sedjo 2001), including new developments for bioenergy applications. Commercialization of improved planting stocks, based on new varieties generated through clonal propagation and advanced breeding programs, as well as further improvements with biotechnology with high value traits, will occur in the near future. These trees will further enhance the quality and productivity of plantation forests (Nehra et al. 2005) and will provide a renewable resource for industrial applications.

There are two major approaches to achieve dramatic productivity improvements in purpose-grown plantation trees. The first is to use genes to genetically improve the productivity of native trees, such as Populus species and pine. The second strategy is to genetically improve the adaptability of exceptionally productive non-native trees, such as eucalypts, to enable their operational use in the Southeastern US.

7.4.1.1 Native Tree Species—Populus

Various native Populus species and their hybrids (with native and introduced species) are among the most rapidly growing trees adapted to temperate climates. However, the high inherent growth potential of trees in this genus is often manifested only at the most favorable sites. *Populus* genotypes typically respond optimally to specific sites and cannot be grown across diverse sites. A research objective is to develop *Populus* genotypes that can demonstrate wide adaptability across a range of sites and environmental conditions. Unless strategies to increase productivity are employed together with tolerance to abiotic and biotic stresses, plantations of *Populus* species and hybrids will remain limited.

Several different strategies offer the potential to overcome these limitations and allow *Populus* to play an increasingly important role in bioenergy initiatives. The first and most straightforward of these strategies is through traditional breeding to generate hybrids and varieties that grow fast, have high volume increments, and can grow across a wide range of sites. For example, advanced Populus clones have been developed by companies such as MeadWestvaco (Robinson et al. 2006) and International Paper, with continued development through ArborGen, LLC, and GreenWood Resources, Inc., to have greater productivity and adaptability. These programs typically consist of breeding among selected genotypes within a species, or between species, and testing the seedling progeny in a series of field trials. The first tests are in a nursery at close spacing to evaluate the genotypes for broad adaptability and resistance to various pests, which are then followed by one or more series of vegetatively propagated field trials in which the varieties are further screened for suitability to diverse planting sites. Commercial candidates are typically selected based on projected yields and wood properties after as many as 10 years of field testing.

Genomic-aided improvement is now possible in *Populus* species. *Populus trichocarpa* was the first tree to have its genome sequenced, and genes related to traits that contribute to maximizing biomass yield per unit land area will be investigated

(Tuskan et al. 2006). The US DOE has described the optimal bioenergy tree attained through accelerated domestication enabled by genomics-aided breeding: short stature to increase light access and enable dense growth, large stem diameter, and reduced branch count to maximize energy density for transport and processing, harvest time of only a few years, and modification of the macromolecules that comprise the cell (Ragauskas et al. 2006; Tuskan 2007). Using a gene homology approach, in conjunction with gene function and expression analysis, candidate domestication genes for these traits have been identified. These include many genes involved in cellulose and hemicellulose synthesis as well as those believed to influence various morphological growth characteristics such as height, branch number and stem thickness (Kalluri et al. 2007; Busov et al. 2008; Ragauskas et al. 2006). In addition, the Oak Ridge National Laboratory, in conjunction with the Bioenergy Feedstock Development Program and Boise Cascade Corp., is developing methods to identify drought-tolerant genotypes based on the presence of certain leaf metabolites (Oak Ridge National Laboratory 2009). Functional validation of candidate genes will occur through high-throughput production and phenotypic analysis of transgenic *Populus* in which these candidate genes are up- or down-regulated.

Populus improvement via transgene insertion is a second viable approach to achieve productivity improvements. Introduced agronomic traits, such as herbicide tolerance and insect resistance, can result in significant biomass yield benefits (Meilan et al. 2000). Weeds have been shown to have a major negative impact on the growth and establishment of *Populus* grown as SRWC, with survival and height growth being significantly improved with the use of herbicide treatments (Hansen et al. 1984; Schuette 2000). Transgenic *Populus* resistant to glyphosate have demonstrated the ability to effectively control weeds by an over-the-top spray regime (Meilan et al. 2000; James et al. 2002; Donahue et al. 1994) and to potentially reduce the costs associated with alternative methods such as cultivation or directed herbicide spray to control weeds. *Populus* has been transformed with modified *Bacillus thuringiensis* (Bt) endotoxin genes that conferred resistance to lepidopteran pests, such as gypsy moth (*Lymantria dispar*), the forest tent caterpillar (*Malacosoma disstria*) and the chocolate tip moth (*Clostera anachoreta*; McCown et al. 1991; Robison et al. 1994; Wang et al. 1996) as well as to coleopteran insects such as the cottonwood leaf beetle (*Chrysomela scripta*; Meilan et al. 2000). The cottonwood leaf beetle resistant transgenic trees (Meilan et al. 2000) and chocolate tip moth (chocolate tip moth; Ewald et al. 2006) demonstrated reduced defoliation and increased growth in field trials in the US and China, respectively.

Another approach to increase total biomass production in trees is to improve the ability of trees to thrive at less than optimal nitrogen levels. For example, glutamine synthetase (GS) is responsible for ammonium assimilation by producing the amino acid glutamine from glutamic acid (Good et al. 2004). Kirby and co-workers at Rutgers University, who studied over-expression of a conifer cytosolic glutamine synthetase (GS1) in *Populus* (Fu et al. 2003; Man et al. 2005), showed that greenhouse-grown GS1 transgenic trees had a greater than 100% increase in leaf biomass relative to controls when grown under low nitrogen conditions (Man et al. 2005). The effect was less marked when more nitrogen was available to the trees.

Transgenic *Populus* characterized by over-expression of a pine cytosolic GS gene exhibits other beneficial phenotypes including enhanced tolerance to water stress (El-Khatib et al. 2004), and enhanced nitrogen use efficiency (Man et al. 2005). One other gene involved in nitrogen utilization is glutamate synthase (GOGAT), which is involved in recycling of ammonium ions in plant tissues. When NADH-GOGAT, found predominantly in non-photosynthetic cells, was overexpressed in tobacco, a 30% increase in foliar biomass was achieved although the nitrogen-to-carbon ratio remained unchanged (Chichkova et al. 2001).

Suboptimal nutrient and water availability limit *Populus* productivity on many sites. Genes and gene families have been identified that have the ability to alter plant responses to water and nutrient limitations (Tuskan et al. 2006). Introduced genes being tested in Populus include the *Populus tremula* and *Arabidopsis* stable protein 1 (SP-1) gene, as well as genes involved in metabolic processes responsive to drought, such as redox proteins, transporter proteins, signal transduction proteins and transcription factors (Polle et al. 2006). Drought and salt tolerance was observed in transgenic *Populus tomentosa* transformed with a constitutively expressed *Arabidopsis thaliana* phospholipase Dα (AtPLDα) gene and the level of tolerance was correlated to the level of expression of the introduced gene (Zhang et al. 2008). Various forms of phospholipase D have been implicated in signal transduction, membrane trafficking, and membrane degradation. Over-expression of At*PLD*α affected stomatal closure and plant water status via increased sensitivity to abscisic acid (Sang et al. 2001). General stress tolerance genes might also have a positive impact on productivity. It is also possible that genes that confer stress tolerance or delay senescence could also improve growth under environmental stress conditions or could extend the growing season, and it has been shown that suppression of a gene, deoxyhypusine synthase (DHS), which is part of the stress response pathway, increases vegetative and reproductive growth in the model plant *Arabidopsis* (Duguay et al. 2007).

Tree growth is also affected by genes involved in cell, tissue and organ development. An introduced β-1,4-endoglucanase (cel1) involved in cell wall synthesis during cell growth improved growth in transgenic Populus (Shani et al. 2004). Genes involved in leaf size and structure, stem development, timing of bud flush, and leaf senescence all may influence biomass accumulation, rotation time, and growth rate. For example, it has been shown that a gibberellin catabolism gene, GA 2-oxidase, can affect tree height (Busov et al. 2003). Over-expression of a Populus FT (flowering time) gene enabled transgenic *Populus* to continue growing despite exposure to short days that stimulate dormancy (Böhlenius 2006). This strategy could possibly be used to delay dormancy and extend the growing season by several days and enhance biomass yields over a multi-year rotation.

7.4.1.2 Native Tree Species—Pinus

Loblolly pine, unlike many Populus species, has the advantage of being widely adaptive across many sites in the Southeastern US below 2,000 feet in elevation.

Fig. 7.4 Three-year-old field trials of transgenic trees. **a** Control transgenic cottonwood (Populus deltoides) expressing the β-glucuronidase gene (GUS). **b** Transgenic cottonwood with ArborGen proprietary growth gene; **c** Control transgenic loblolly pine (Populus taeda) with GUS gene; **d** Transgenic loblolly pine with the same ArborGen proprietary growth gene as in **b**

However, loblolly pine's long rotation time (15 years for pulp wood applications and 23 years for sawtimber applications) is an economic limitation for its use in biofuel and biopower applications. To address this limitation, ArborGen, LLC, and other tree breeding organizations, are employing advanced breeding and crossing methods to develop high-performing traditional seedlings that have improved growth, disease resistance and form. ArborGen and two other tree improvement companies (CellFor, Inc. and Weyerhaeuser Co.) have utilized a tissue culture process called somatic embryogenesis to mass propagate selected elite loblolly genotypes (Grossnickle and Pait 2008). Improvements in traditional breeding and selection are predicted incremental to achieve 35% volume gains and sawtimber rotation times of approximately 20 years. Biotech gene insertion methods will be necessary to develop loblolly pine with the productivity levels required to enable the production of high value sawtimber in shorter rotation cycles. Early research results indicate that rotation times of 15 years may be possible. ArborGen has introduced candidate genes associated with improved growth into loblolly pine, and obtained results that indicate that 3-year-old transgenic trees in field trials can demonstrate nearly double the i biomass production relative to control trees (Fig. 7.4). Transgenic pine with reduced and altered lignin have also been produced (M.A.H., unpublished results).

7.4.1.3 Introduced Tree Species—Eucalyptus

The use of introduced trees is not new to forestry, with many of planted Populus trees in the US being hybrids with species brought to the US from Europe and Asia

(Stanton. 2003). Eucalyptus is not native to the US, but it is is grown widely for industrial forestry globally because of its good wood fiber properties, high pulp yields, excellent biomass productivity and ability to grow on upland sites. *Eucalyptus grandis* and its hybrids have the greatest biomass yield potential and are commonly used in the Brazilian pulp and paper industry. The positive attributes of Eucalyptus make it attractive for commercial use in the Southeastern US. However, Brazilian elite operational genotypes (or clonal varieties) are typically adapted to tropical climates and are sensitive to freezing temperatures. Efforts to introduce Eucalyptus to the Southeastern US began in 1971 as part of the North Carolina State University Hardwood Cooperative research program (Jahromi 1982). Later the International Paper Company in Bainbridge, Georgia conducted the first *Eucalyptus* screening trial in 1972 using *Eucalyptus viminalis*, a species with natural frost tolerance (Jahromi 1982). However, even these more tolerant trees demonstrated dieback and reduced height growth by the extreme fluctuations in temperatures in the Southeastern United States where temperatures can go from 21°C during the day to −1°C at night (Meskimen et al. 1987). Unfortunately, the species that survived the best in cold temperatures did not have the desired productivity.

This situation is changing. It is now possible to plant highly productive genotypes of Eucalyptus that have been modified via biotechnology to tolerate the freezing winter conditions in the Southeastern US. This advance was made possible by the discovery of the C-repeat/dehydration-responsive element binding (CBF/DREB) factor cold-response pathway (Jaglo-Ottosen et al. 1998; Liu et al. 1998). Over-expression of CBF genes conferred freezing tolerance, drought tolerance, as well as salt tolerance in *Arabidopsis* (Liu et al. 1998; Kasuga et al. 1999). Similarly, over-expression of *Arabidopsis* CBF genes in *Brassica napus*, *Nicotiana tabacum*, and *Populus* induced the expression of CBF-targeted genes and increased freezing and drought tolerance of transgenic plants (Jaglo-Ottosen et al. 2001; Kasuga et al. 2004; Benedict et al. 2006). Genes homologous to the *Arabidopsis* CBF genes have been demonstrated to be effective in inducing freeze tolerance. Over-expression of a pepper CBF homolog (CaPF1) in tobacco and *Arabidopsis* improved their freeze hardiness (Yi et al. 2004). The maize CBF homolog, ZmDREB1A, improved freezing and drought tolerance in *Arabidopsis* (Qin et al. 2004). However, not all CBF homologs are equally effective, differing in the interaction of the specific CBF homolog with the species in which it is being expressed.

Heterologous expression of the *Arabidopsis* CBF1 resulted in enhanced chilling tolerance but not freezing tolerance in transgenic tomato (Hsieh et al. 2002; Zhang et al. 2004). In rice, four CBF homologs have been identified, and over-expression of one of the homologs (OsDREB1A) in *Arabidopsis* conferred enhanced freezing tolerance and high-salinity tolerance (Dubouzet et al. 2003; Ito et al. 2006). One undesirable phenotype associated with strong constitutive over-expression of CBF genes is dwarfing (Zhang et al. 2004). To overcome this problem, stress-inducible plant promoters with a low background expression level have been used in conjunction with the cold tolerance genes (Yamaguchi-Shinozaki and Shinozaki 1994).

ArborGen has introduced the *Arabidopsis* CBF2 transcription factor driven by the *Arabidopsis* rd29a stress-inducible promoter (Yamaguchi-Shinozaki et al.

Fig. 7.5 ArborGen's Freeze Tolerant Eucalyptus field trials. **a** Aerial photograph of Freeze Tolerant Eucalyptus field trial planted in Alabama in a block plot design at age 2. EH Nontransgenic control trees; blocks 42, 43 two different transgenic lines of ArborGen Freeze Tolerant Eucalyptus containing the Arabidopsis CBF2 transcription factor driven by the Arabidopsis rd29a promoter. **b** Photograph of ArborGen Freeze Tolerant Eucalyptus trees at 3 years of age in Florida

1993) into variety EH1, a highly productive tropical *Eucalyptus* hybrid (*E. grandis* x *E. urophylla*). The new transgenic variety, "Freeze Tolerant Eucalyptus", has demonstrated tolerance to approximately 16°F (−9°C) across multiple years and multiple field trial locations (Fig. 7.5), while essentially maintaining its exceptional productivity (Cunningham et al. 2010; ArborGen, unpublished data).

The yields achievable with Freeze Tolerant Eucalyptus are predicted to meet or exceed those that have been defined by the DOE and others for the long-term feasibility of renewable energy production [i.e., 8–10 dry short tons acre^{-1} year^{-1} (English et al. 2006)]. The application of total biomass-driven management systems could further increase yields and reduce delivered costs. An added benefit of Freeze Tolerant Eucalyptus is its ability to coppice when managed appropriately. Coppicing allows for subsequent crops without the added costs of establishment (site preparation, seedling and planting costs), which can provide a higher return to landowners. Coppice crops can show increases in productivity relative to the initial single-stem harvest (Sims et al. 2001), but coppice yields will decline eventually over time. Re-planting will then become economically attractive as new varieties become available.

The types of genetic improvements discussed above will be important in optimizing the potential for purpose-grown trees to provide sustainable and economical feedstock options for the production of cellulosic ethanol and other forms of bioenergy. Table 7.6 summarizes the theoretical acreage needed to meet the "advanced biofuels" target in the 2007 RFS in the Southeastern US based on current productivity assumptions for loblolly pine and Eucalyptus under pulpwood and high-density coppicing scenarios. It is projected that it would require 118 million green short tons biomass to produce the 36 billion gallons ethanol mandated under the RFS. Productivity improvements with fast-growing species like Eucalyptuscan reduce the amount of land required to meet this target substantially (by a factor of three relative to traditionally managed open pollinated loblolly pine).

7 Woody Biomass and Purpose-Grown Trees as Feedstocks for Renewable Energy 191

Table 7.6 Approximate productivity and total planted acreage needed to meet the Renewable Fuel Standard[d] (RFS) in the southeastern US using purpose grown pine or Eucalyptus

	Pinus taeda	Eucalyptus urograndis	
		Pulpwood management	Total biomass management
Productivity(green tons acre^{-1} year^{-1})	10[a]	20[b]	30[c]
Planted acres (million) needed to meet target 118 million green tons year^{-1e}	12	6	4

[a]ArborGen, unpublished data, assumes a 10-year rotation with a planting density of 1,000 trees per acre

[b]ArborGen, unpublished data, assumes a 7-year rotation, with 450 trees per acre

[c]ArborGen unpublished data, assumes an average product of an initial harvest at 3 years, followed by three coppice rotations of approximately 3 years coppice rotation, using a similar coppicing regime as described in Sims et al. 2001)

[d] On 19 December 2007, the Energy Independence and Security Act of 2007 (H.R. 6) was signed into law. This comprehensive energy legislation amends the RFS signed into law in 2005, growing to 36 billion gallons biofuels available in 2022

[e]Assumes 21 billion gallon 'advanced biofuels' target in RFS, Southeastern share of 28%, 100 gallons ethanol produced per dry ton biomass and a 50% moisture content (wet basis)

7.4.2 Genetic Improvement for Wood Properties

Once the trees planted for woody biomass achieve productivity levels that enable economically sustainable delivered feedstock costs, improvements in the wood itself can provide additional efficiency and cost improvements in biofuel or biopower conversion. There are two general technical targets for modifying the wood to increase the yield of ethanol per ton of wood via biochemical/biological processes, thereby reducing the "recalcitrance" of wood in bioconversion to ethanol. The first target is to modify the accessibility and/or degradability of cell wall polysaccharides to the enzymes and/or microorganisms used to break them down. The second target is to reduce the concentrations of fermentation inhibitory compounds that reduce the efficiency of ethanol production from released sugars.

The DOE has funded several bioenergy science centers to address biological and technological barriers in the cost effective production of biofuels from lignocellusic feedstocks. One such center, the BioEnergy Science Center managed by Oak Ridge National Laboratory, has focused research on reducing factors contributing to the recalcitrance of wood in biofuels conversion (US DOE Office of Science 2009) In addition, quantitative trait analysis of natural variation coupled with phenotypic screens for recalcitrance traits such sugar release efficiency, acid digestibility and general cell wall composition are underway as part of the BioEnergy Science Center.

One approach that addresses both technical targets discussed above is reduction of lignin in plant cell walls, or potentially the modification of the chemical composition of lignin, making it easier to remove. Reduction of lignin has long been a target of interest in crop and forestry species, because lignin content negatively affects the feed efficiency of forage for farm animals and lignin removal is one of the more costly steps in the kraft pulping process used in the paper industry. Extensive research has been conducted to understand the lignin biosynthetic pathway and enable the transgenic manipulation of lignin quantity and composition over the past 15 years (as reviewed in Anterola and Lewis 2002; Boerjan et al. 2003; Li et al. 2008; Vanholme et al. 2008; Weng et al. 2008). In general, it has been shown that lignin content can be significantly reduced in trees and herbaceous plants and that this can improve the ability to remove lignin and access cellulose for bioconversion processes (Hu et al. 1999; Lapierre et al. 1999; Chen and Dixon 2007; Li et al. 2008). For example, several of the transgenic plants with down-regulated genes for lignin biosynthetic enzymes in the early steps of the pathway showed that syringyl lignin (S-lignin) content was affected more strongly than the guaiacyl lignin (G-lignin) content (Vanholme et al. 2008).

Lignin reduction is often associated with negative pleiotropic effects. Examples of pleiotropic effects from lignin reduction in Populus include the accumulation of sugars in leaves and concomitant reduction of photosynthetic capacity following down-regulation of p-coumaroyl shikimate 3δ-hydroxylase (Coleman et al. 2008), or the reduction of hemicellulose associated with the down-regulation of cinnamoyl-coenzyme A reductase (Leplé et al. 2007). Increased cavitation and vessel collapse have also been observed in some plants with reduced lignin (Coleman et al. 2008). Achievement of economically valuable levels of lignin reduction will require the use of carefully selected promoters and target genes. Using transcription factors involved in controlling elements of the lignin biosynthetic pathway as the target genes may minimize the impact of negative pleiotropic effects since these genes control a group of genes in the lignin biosynthesis pathway. For example, two plant-specific transcription factors, designated NAC secondary wall thickenings promoting factor 1 (NST1) and 3 (NST3), regulate the formation of secondary walls in woody tissues of *Arabidopsis* (Mitsuda et al. 2007).

An increased syringyl lignin to guaiacyl lignin (S/G) ratio is thought to be desirable, because the more oxygen-rich S-lignin is easier to remove through chemical treatment (Chiang and Funaoka 1990). This can be achieved by over-expression of ferulate 5-hydroxylase (F5H), also known as coniferaldehyde 5-hydroxylase (Cald5H; Huntley et al. 2003). Interestingly, it has been reported that the composition of lignin does not have a strong effect on biological degradation of cell walls (Grabber et al. 1997), but that lignin levels have a highly significant impact on cell wall recalcitrance to saccharification (Chen and Dixon 2007; Li et al. 2008).

The polysaccharides themselves contribute a large part to recalcitrance because of their insolubility. In conifer and hardwood (angiosperm) secondary cell walls, cellulose is approximately 45% of the dry weight of the wood while hemicellulose is approximately 20% (Table 7.4). Hemicellulose composition varies strongly depending upon the wood source. The hemicellulose of hardwoods such as Populus is comprised of about 80% xylans, with the remainder being mannans (10%), galactans (5%) and arabinans (<5%). In pine, mannans comprise about 50%, xylans 30%, galactans 10%, and arabinans 5% of the total hemicellulose (Rowell 2005).

The DOE has funded three collaborative centers for research focused on improving lignocellulosic feedstock and related downstream processes for improved conversion efficiency. It is anticipated that within the next 5 years, from the overall US government investment in these research centers, much more will be known about genes that can be used to address the recalcitrance of plant cell walls to enzymatic biofuels conversion methods.

If wood is to be used simply as a fuel via burning, the two obvious targets for modification are increasing density and increasing lignin content. Increasing wood density has the potential of increasing the yield of fuel per acre and decreasing transportation and storage costs (increasing the energy yield per truckload of wood or chips), although the amount of energy per unit weight of wood would remain the same. Increases in wood density would improve the energy yield of wood produced via gasification. It seems unlikely that it would be possible to insert significantly more material into the cell wall matrix, but decreasing the ratio of lumen volume to cell wall volume is one avenue to increase wood density. As yet there is not much research demonstrating biotech approaches to improve wood density, although it has been shown that over-expression of a *Eucalyptus gunnii* MYB transcription factor (EgMYB2) in tobacco led to significant thickening of xylem fiber cell walls (Goicoechea et al. 2005). It was theorized that the thickened cell wall phenotype is due to an overall rate increase in deposition of the S2 layer of the secondary cell wall.

Increasing lignin content would increase the thermal energy of wood. Pure cellulose has a calorific value of approximately 8,000 BTU per pound of dry biomass (BTU/lb), while pure lignin is approximately 11,000 BTU/lb (White 1987), so an increase of 10% in lignin content would increase the calorific value of wood by approximately 450 BTU/lb. Lignin up-regulation could potentially be achieved by over-expression of an enzyme that is a kinetic barrier in the lignin biosynthetic pathway or via over-expression of a transcription factor that controls multiple lignin biosynthetic genes. Increased lignin deposition has been observed with the over-expression of certain MYB transcription factors (Patzlaff et al. 2003; Goicoechea et al. 2005).

7.5 Sustainable Production of Purpose-Grown Trees

Sustainability of natural resources involves economic, environmental, and social considerations. There are environmental impacts arising from the production of biomass, as there are in managing many natural resources. Sustainable production takes into account conservation of biological diversity, maintaining production capacity and forest ecosystem health, enabling soil- and water resource-conservation, attaining a positive impact on global carbon cycles, and allowing associated long-term socio-economic benefits.

Sustainable production is an important practice in forestry, and core principles involve managing working forests as landscapes for the preservation of biodiversity and watersheds, while allowing a diversity of uses and benefits for current and future generations (Reynolds et al. 2003; Reynolds 2005). Sustainable production of wood grown for industrial applications has been a priority of the forestry industry, and it is expected that sustainable practices will continue to be applied for wood production for biopower and biofuels applications (Raison 2002). Appropriate silvicultural practices can ensure sustainable wood production. Science-based studies of site productivity and harvesting have allowed the development of operational forest management guidelines that adjust harvesting operations to the biological and physical requirements of the site.

Land management has intensified over the last 20 years, particularly in Southeastern and Pacific-Northwest regions of the US, in tree plantations grown for pulp and lumber end-uses. Sustainably high yields have been demonstrated on tree plantations (or farms) over cycles of planting improved tree stocks and through management for enhanced growth that involves tillage, fertilization, irrigation, herbicide use, clear-cutting, and replanting (US Department of Agriculture 2002).

There is concern that repeated rotations of tree plantation silviculture may degrade site fertility and thus not be sustainable. This has not been demonstrated to be the case. In pine plantations in the Southeastern US, 50 lbs phosphorous per acre at establishment may have a positive impact on biomass yield for 20 years or more, and this fertilization can improve growth by over 100% on deficient sites. In addition, at canopy closure an additional fertilization of nitrogen and phosphorous can be applied, which results in an average growth improvement of 30% (Fox et al. 2006). A study on two loblolly pine plantations evaluated nitrogen (N) and phosphorous (P) pools before harvest, and compared the value of these two nutrients after ten growing seasons of the second rotation following plantation re-establishment. Second rotation soil N pools ranged from 3,134 to 5,148 kg/ha, and soil P pools ranged from 578 to 767 kg/ha, and these were equal to or significantly greater than those found at the end of the first rotation. Both plantations accumulated N and P faster than predicted by their age, which indicates that the management practices even conserved and allowed for the accumulation of nutrients. Maintenance of soil nutrient levels can be considered a strong indicator of sustainable production (Gresham 2002).

Intensively managed tree farms have a high potential for storing carbon (US Department of Agriculture 2008). Irrigated Pacific Northwest poplar plantations produce 49 tons of trunk and branch wood and 2 tons leaves per acre by the 4th year of production. This level of productivity increases the total amount of carbon that can be captured. This capture is manifested in the short term as paper and wood products, and in the long term as soil carbon from roots, exudates, and decomposed litter. Evidence of improved carbon content in tree plantations relative to annual crop plantings was described by Mann and Tolbert (2000). After the initial 4- to 6-year establishment period for a hybrid poplar plantation, the measured amount of organic carbon stored in the soil was 191 Mg/ha, and this was greater than that for row crops or grass. The increased carbon storage was especially noticeable at soil depths below 30 cm, where most of the coarse root development occurred (Mann and Tolbert 2000).

Biomass crop systems that increase organic carbon generally yield positive changes in soil structure, water-holding capacity, and the storage and availability of nutrients, and this facilitates increased abundance and diversity of soil biota as well as increased resistance to compaction. The amount of equipment traffic on the land and the amount of soil disturbance is much less in tree plantations than in land with annually planted crops. This is due to multiyear harvest cycle possible with trees. A comparison of cottonwood planting to cotton in the Southeastern US revealed that the cottonwood plantation had better soil structure than the cotton field (decreased density and increase in water stable soil aggregates), which correlated with a five-fold reduction of surface sediment runoff over a 2-year period of time (Mann and Tolbert 2000). Minimum tillage, such as occurs in short rotation woody crops after establishment, in particular can increase the organic carbon content in soils. Erosion losses in SRWC range from 2 to 4 metric tons ha^{-1} $year^{-1}$ as compared to the 14–41 metric tons ha^{-1} $year^{-1}$ losses associated with annual row crops in the southeast (Mann and Tolbert 2000).

Biodiversity conservation is a central subject in forest management and associated public policy. In managing planted forests there is emphasis on retaining patches of natural vegetation or riparian or wildlife corridors, and in some cases re-establishing native vegetation as part of overall plantation development. Trees compare favorably when compared to annually harvested bioenergy crops with respect to biodiversity and wildlife since tree plantations will remain largely undisturbed for several years. SRWC such as willow that are harvested on a 3-year rotation have provided a variation in habitat for wildlife since approximately one-third of the land that is just starting to grow a new stand is a open habitat while more mature stands are more of a shrub habitat. Therefore, SRWC would provide a structurally diverse environment across their managed landscape that would support a wider range of biodiversity than an annually harvested energy crop (Volk et al. 2004). Careful forest management in natural and/or planted forests can contribute to the conservation of biodiversity and to water regulation, carbon sequestration and recreational benefits. Short rotation tree plantations have much higher productivity than unmanaged forests and it is expected that concerns over the compatibility of bioenergy and biodiversity can be met by keeping biodiversity as a key element in managing woody biomass for bioenergy as a component of the overall forest landscape.

Woody biomass as bioenergy feedstock offers overall energy balance improvements, reduced greenhouse gas emissions, and reversal of deforestation via afforestation. Well-managed short rotation forests are a sustainable and renewable resource that is CO_2 neutral. Typical energy balances for forestry and agriculture systems indicate that 25–50 units of bioenergy are produced for every unit of fossil energy consumed in production (IEA BioEnergy 2002). Woody biomass grown for bioenergy systems could enhance carbon sequestration since short rotation trees produce vegetation and roots that are strong carbon sinks, with carbon being sequestered over a tree's multi-year rotation and through the persistence of roots in the soil after harvest (IEA BioEnergy 2002; Satori et al. 2006). Another potential benefit to short rotation tree plantations is that they can be used not just for wood production but also for bioremediation. SRWC can be used in conjunction with sewage waste disposal (Smart et al. 2005) and for soil amelioration (Rockwood et al. 2004). Economic sustainability is also enabled, since tree bioenergy plantations can be planted on unused or non-productive agricultural lands, creating additional jobs in rural economies. Given supportive policies, woody biomass will provide a sustainable solution to future bioenergy demands.

Genetically improved trees through the insertion of biotech traits will contribute significantly to the US reaching its objectives of sustainable production of biopower and biofuels. Since the commercialization of the first biotech crops in 1996, the adoption rate of biotech crops has increased rapidly worldwide, with over 2 billion acres planted in 25 different countries (ISAAA 2008). Field tests of genetically modified trees are being conducted in several countries, with the majority of these field tests occurring in the US (van Frankenhuyzen and Beardmore 2004). To date, only two tree species, papaya and plum, have been granted non-regulated status for planting in the US. These fruit trees were made resistant to viruses that can have devastating impacts on fruit production and quality. Currently the only large-scale plantings of transgenic forest trees are insect-resistant poplars that are being grown in China (Hu et al. 2001). More recently, a petition was submitted to USDA APHIS Biotechnology Regulatory Services (BRS) requesting non-regulated status for the Freeze Tolerant *Eucalyptus* hybrid described in Sect. 7.5.1.3.

While it is recognized that the use of biotechnology for tree improvement can bring significant economic, social and environmental benefits, some concerns have been expressed (van Frankenhuyzen and Beardmore 2004), particularly associated with the potential dispersal of pollen, seeds, or vegetative propagules. Gene flow via pollen and seed dispersal is an important natural phenomenon for genetic improvement and evolution in plant species. The potential for gene flow from crops, be it from traditional breeding or developed through biotechnology, is considered to be less for non-native self-pollinated crops compared to native and wind pollinated species. Perennial wind pollinated species models predict that a small proportion of pollen and seed can travel considerable distances (Nathan et al. 2002; Williams 2005; Williams and Davis 2005; see Chaps. 9–11, this volume). However, for there to be any consequences of such dispersal, these models assume that the following: (1) viable pollen is able to fertilize receptive ovules of a related species resulting in viable seed production, followed by establishment of this seed in the environment; and (2) adverse consequences can occur only if the inserted genes are considered a significant risk to other organisms or can cause unintended effects on the fitness of the species. Traits including improved growth, wood quality and abiotic or biotic stress tolerance that are also being altered via traditional breeding might be considered as being inherently low risk, particularly when using genes from the tree itself, or other plant genes that are homologous to genes already present in the tree, and where no unintended phenotypes have been observed after extensive field testing. In some cases, especially for biomass production for bioenergy uses, short rotation trees may not even produce significant amounts of pollen or seed prior to harvest. In addition, there are proven and well tested technologies that selectively prevent or reduce pollen formation without affecting other functions of the plant species (Nasrallah et al. 1999; Gomez Jimenez et al. 2006; Yanofsky 2006; Chap. 10). ArborGen has adapted this technology for use in tree species and has demonstrated high levels of efficacy (Hinchee et al. 2009). Therefore, a number of tools exist that can minimize the potential risk of gene flow via pollen or seed dispersal from plantations. It is important however, that any risk assessments for trees take into account scientifically informed arguments for incorporating any such gene flow control mechanisms on a case-by-case basis.

7.6 Conclusion

The existing wood products industries have led the development of sustainable production, management, harvest, and transportation systems for the production of hundreds of millions of tons of wood from planted trees in the US annually. Existing infrastructure and know-how in intensively managed tree plantations for purpose-grown trees already exists, and this resource can be built upon to meet renewable energy goals and help mitigate climate change.

The high productivity of purpose-grown, short rotation trees is expected to improve the economic feasibility of biopower and biofuels production from woody biomass plantations. Woody biomass represents about two-thirds of the near-term potential of the Southeastern US for producing renewable energy, and bioenergy is already becoming a substantial market outlet for wood. Sixteen bioenergy projects have been announced throughout the Southeast that would consume nearly 50 million green short tons woody biomass by 2015. The development of a bioenergy sector in the Southeastern US holds great economic promise, and it is anticipated that purpose-grown, short rotation trees will be planted to address the 120 million green short tons of biomass that will be needed annually as a feedstock for advanced biofuels. At an estimated price of US $20 to US $30 per green short ton, this represents US $2–4 billion in economic opportunity associated with biomass production for the Southeastern US. Upside demand exists because of the suitability of wood for other bioenergy applications, such as the production of electricity through direct burning of wood or co-firing with coal.

To achieve lower production costs, the supply of cheap raw materials is thus a necessity. Accordingly, it will be possible to produce large amounts of low-cost ethanol only if lignocellulosic feedstock, such as fast-growing trees and grasses, are used along with waste products (including agricultural and forestry residues) and municipal and industrial wastes (Wheals et al. 1999). The inherent logistical benefits of trees, in combination with the high productivity of new varieties of short rotation trees such as Freeze Tolerant Eucalyptus, make it an ideal biomass for traditional industrial end uses such as pulp and paper, as well as for energy products such as cellulosic ethanol and electric power generation. Short rotation trees will generate more wood on less land, requiring a smaller plantation footprint to generate the necessary dry tons to feed industrial processing plants. This, in turn, will lessen pressure to harvest from native and old growth forests in order to meet society's demand for pulp, paper and energy. The addition biotech traits to elite varieties of purpose-grown trees will help achieve the short rotation times, high productivity, and target delivered costs of wood to make lignocellulosic feedstocks cost effective.

The choice of energy crops to plant must take into consideration regional conditions and needs, both in minimizing transportation costs as well as avoiding the current long distance distribution limitations of ethanol. In the Southeastern US, where accessible inventory and harvesting infrastructure for forestry operations are already well established, trees provide a clear advantage for biomass production compared to annual crops. Trees will play a significant role in helping to meet renewable energy standards, although it is recognized that multiple integrated approaches with a variety of different crop species and production systems will be required to meet our total renewable energy objectives.

References

Abt RC, Cubbage FW, Galik C, Henderson JD (2010) Effect of policy-based bioenergy demand on southern timber markets: a case study in North Carolina. Biomass Bioenergy (in press)
Adams DM, Haynes RW, Daigneault AJ (2006) Estimated timber harvest by US region and ownership, 1950-2002. United States Department of Agriculture Forest Service, Pacific Northwest Research Station, General Technical Report PNW-GTR-659, January 2006.http://www.fs.fed.us/pnw/pubs/pnw_gtr659.pdf
ACORE (2009) Overview of Renewable Energy Provisions in the American Recovery and Reinvestment Act of 2009. American Council on Renewable Energy, http://www.acore.org/files/images/email/acore_stimulus_overview.pdf
AFPA (2009) Alternative fuel mixture tax credit. 17 June 17 2009, American Forest and Paper Association,http://www.afandpa.org/whatwebelieve.aspx?id=917&terms=alternative+tax+credit. Cited 2 January 2010
Andersson G, Asikainen A, Bjorheden R, Hall PW, Hudson JB, Jirjis R, Mead DJ, Nurmi J, Weetman GF (2002) Production of forest energy. In: Richardson J, Bjorheden R, Hakkila P, Lowe AT, Smith CT (eds) Bioenergy from sustainable forestry: guiding principles and practice. Kluwer, Dordrecht, pp 49–123
Anterola AM, Lewis NG (2002) Trends in lignin modification: a comprehensive analysis of the effects of genetic manipulations/mutations on lignification and vascular integrity. Phytochemistry 61:221–294
Arato C, Pye EK, Gjennestad G (2005) The lignol approach to biorefining of woody biomass to produce ethanol and chemicals. J Appl Biochem Biotechnol 123:871–882. doi:10.1385/ABAB:123:1-3:0871
Atsumi S, Hanai T, Liao JC (2007) Non-fermentative pathways for synthesis of branched-chain higher alcohols as biofuels. Nature 451:86–89
Austin A (2008) The road ahead for FFVs. Ethanol Producer Magazine. Retrieved 16 October 2009. http://www.ethanol-producer.com/article.jsp?article_id=5010&q=&page=all
Baker JB, Langdon OG (1990) Loblolly pine. In: Burns RM, Honkala BH (eds) Silvics of North America, vol 1. Conifers. Agriculture Handbook. US Department of Agriculture. Forest Service, Washington DC 654:497–512
Benedict C, Skinner JS, Meng R, Chang Y, Bhalerao R, Huner NP, Finn CE, Chen TH, Hurry V (2006) The CBF1-dependent low temperature signalling pathway, regulon and increase in freeze tolerance are conserved in *Populus* spp. Plant Cell Environ 29:1259–1272
Bergman R, Zerbe J (2008) Primer on wood biomass for energy. USDA Forest Service State and Private Forestry Technology Marketing Unit Forest Products Laboratory, Madison, WI, http://128.104.77.228/documnts/tmu/biomass_energy/primer_on_wood_biomass_for_enegy.pdf
Bevill K (2008) Lignol receives $2 million for cellulosic plant. Biomass Magazine, 28 July 2008, http://www.biomassmagazine.com/article.jsp?article_id=1821

Biomass Energy Foundation (2009) Woodgas: fuel densitities. http://www.woodgas.com/fuel_densities.htm, cited 23 November 2009

Boerjan W, Ralph J, Baucher M (2003) Lignin biosynthesis. Annu Rev Plant Biol 54:519–546

Böhlenius H, Huang T, Charbonnel-Campaa L, Brunner AM, Jansson S, Strauss SH, Nilsson O (2006) CO/FT regulatory module controls timing of flowering and seasonal growth cessation in trees. Science 312:1040–1043

Bridgwater AV (2003) Renewable fuels and chemicals by thermal processing of biomass. Chem Eng J 91:87–102

Brown MA, Atamturk N (2008) Potential impacts of energy and climate policies on the US pulp and paper industry. Georgia Institute of Technology, School of Public Policy Working Paper No. 40, June 2008, http://www.spp.gatech.edu/faculty/workingpapers/wp40

Burdon RD, Libby WJ (2006) Genetically modified forests: from Stone age to modern biotechnology. Forest History Society. Durham, NC

Busov BV, Meilan R, Pearce DW, Ma C, Rood SB, Strauss SH (2003) Activation tagging of a dominant gibberellin catabolism gene (*GA-2 oxidase*) from poplar that regulates tree stature. Plant Physiol 132:1283–1291

Busov BV, Brunner AM, Strauss SH (2008) Genes for control of plant stature and form. New Phytol 177:589–607

Center of Innovation in Agribusiness (2009) Alternative Fuel Technology Company Locates in Georgia, 10 July 2006, http://agribusiness.georgiainnovation.org/highlights/details/3

Ceulemans R, Stettler RF, Hinckley TM, Isebrands JG, Heilman PE (1990) Crown architecture of *Populus* clones as determined by branch orientation and branch characteristics. Tree Physiol 7:157–167

Chen F, Dixon RA (2007) Lignin modification improves fermentable sugar yields for biofuel production. Nat Biotechnol 25:759–761, doi:10.1038/nbt1316

Chiang VL, Funaoka M (1990) The difference between guaiacyl and guaiacyl-syringyl lignins in their responses to Kraft delignification. Holzforschung 44:309–313

Chichkova S, Arellano J, Vance CP, Hernandex G (2001) Transgenic tobacco plants that overexpress alfalfa NADH-glutamate synthase have higher carbon and nitrogen content. J Exp Bot 52:2079–2084

Coleman HD, Samuels AL, Guy RD, Mansfield SD (2008) Perturbed lignification impacts tree growth in hybrid poplar—a function of sink strength, vascular integrity, and photosynthetic assimilation. Plant Physiol 148:1229–1237

Cunningham MW, Hinchee MA, Mullinax LN (2010) Freeze tolerant Eucalyptus: a biotech solution for purpose grown hardwoods. In: Proceedings of the Southern Forest Tree Improvement Conference, 2 June 2009, Blacksburg, VA (in press)

Davis AA, Trettin CC (2006) Sycamore and sweetgum plantation productivity on former agricultural land in South Carolina. Biomass Bioenergy 30:769–777

Davis JM (2008) Genetic improvement of Poplar (*Populus* spp.) as a bioenergy crop. In: Vermerris W (ed) Genetic improvement of bioenergy crops. Springer, New York, pp 397–419, doi: 10.1007/978-0-387-70805-8_14

DeBell DS, Clendenen GW, Harrington CA, Zasada JC (1996) Tree growth and stand development in short-rotation *Populus* plantings: 7-year results for two clones at three spacings. Biomass Bioenergy 11:253–269

De la Torre Ugarte DG, Walsh ME, Shapouri H, Slinsky SP (2003) The economic impacts of bioenergy crop production on US agriculture. Agricultural Economics Report No. 816, Washington DC, US Department of Agriculture, Office of the Chief Economist, Office of Energy Policy and New Uses, http://www.usda.gov/oce/reports/energy/AER816Bi.pdf

Dickmann DL (2006) Silviculture and biology of short rotation woody crops in temperate regions: then and now. Biomass Bioenergy 30:696–705

Doran-Peterson J, Cook DM, Brandon SK (2008) Microbial conversion of sugars from plant biomass to lactic acid or ethanol. Plant J 54:582–592

Donahue RA, Davis TD, Michler CH, Riemenschneider DE, Carter DR, Marquardt PE, Sankhla N, Hassig BE, Isebrands JG (1994) Growth, photosynthesis, and herbicide tolerance of genetically modified hybrid poplar. Can J For Res 24:2377–2383

Dubouzet JG, Sakuma Y, Ito Y, Kasuga M, Dubouzet EG, Miura S, Seki M, Shinozaki K, Yamaguchi-Shinozaki K (2003) OsDREB genes in rice, *Oryza sativa* L., encode transcription activators that function in drought-, high-salt- and cold-responsive gene expression. Plant J 33:751–763

Duguay J, Sadaf J, Liua Z, Wanga T-W, Thompson JE (2007) Leaf-specific suppression of deoxyhypusine synthase in*Arabidopsis thaliana* enhances growth without negative pleiotropic effects. J Plant Physiol 164:408–420

Eggeman T (2009) Third-generation cellulosic biofuels: sustainable, efficient, cost-effective. AiCHE 2009 Annual Conference, http://aiche.confex.com/aiche/icosse09/webprogram/Paper146972.html

Eldridge KG, Harwood C, van Wyk G, Davidson J (1994) Eucalypt domestication and breeding. Oxford University Press, Oxford, UK

El-Khatib R, Hamerlynck EP, Gallardo F, Kirby EG (2004) Transgenic poplar characterized by ectopic expression of a pine cytosolic glutamine synthetase gene exhibits enhanced tolerance to water stress. Tree Physiol 24:729–736

Energy Information Administration (2008) Renewable energy consumption and electricity. Preliminary 2007 statistics. United States Department of Energy, Washington, DC. Retrieved 16 November 2009 http://www.eia.doe.gov/cneaf/alternate/page/renew_energy_consump/rea_prereport.html

English BC, De La Torre Ugarte DG, Jensen K, Helwinckel C, Menard J, Wilson B, Roberts R, Walsh M (2006) 25% Renewable Energy for the United States by 2025: agricultural and economic impacts. The University of Tennessee, Institute of Agriculture, Department of Agricultural Economics, Biobased Energy Analysis Group http://www.25x25.org/storage/25x25/documents/RANDandUT/UT-EXECsummary25X25FINALFF.pdf, retrieved 26 October 2009

European Union (2002) European Union ratifies the Kyoto Protocol. Press release, 31 May 2002, http://europa.eu/rapid/pressReleasesAction.do?reference=IP/02/794&format=HTML&aged=0&language=EN&guiLanguage=en

Ewald D, Hu J, Yang M (2006) Transgenic forest trees in China. In: Fladung M, Ewald D (eds) Tree transgenesis recent developments. Springer, Berlin, pp 25–45 doi: 10.1007/3-540-32199-3_8

Fan LT, Le Y-H, Gharpuray MM (1982) The nature of lignocellulosics and their pretreatments for enzymatic hydrolysis. In: Fiechter A (ed) Advances in biochemical engineering/biotechnology, vol. 23. Microbial reactions. pp 157–187, doi: 10.1007/3-540-11698-2

FAO (2000) Global forest resources assessment 2000—main report. Food and Agriculture Organization of the United Nations Forestry paper, http://ftp://ftp.fao.org/docrep/fao/008/A0400E/A0400E00.pdf

Forisk Consulting (2009) Wood biomass south report. http://www.foriskstore.com/servlet/the-16/Wood-Bioenergy-South/Detail

Foster CD, Mayfield C (2007) Bioenergy production in planted pine forests. In: Hubbard W, Biles L, Mayfield C, Ashton S (eds) Sustainable forestry for bioenergy and bio-based products: trainers curriculum notebook. Southern Forest Research Partnership, Athens, GA, pp 117–120

Fox TR, Allen HL, Albaugh TJ, Rubilar R, Carlson CA (2006) Forest fertilization in southern pine plantations. Better Crops 90:12–15

Frederick WJ Jr, Lien SJ, Courchene CE, DeMartini NA, Ragauskas AJ, Lisa K (2008) Production of ethanol from carbohydrates from loblolly pine: a technical an economic assessment. Bioresour Technol 99:5051–5057

Fu J, Sampalo R, Gallardo F, Canavos FM, Kirby EG (2003) Assembly of a cytosolic pine glutamine synthetase holoenzyme in leaves of transgenic poplar leads to enhanced vegetative growth in young plants. Plant Cell Environ 26:411–418

Gaddy James L (2000) Biological production of ethanol from waste gases. US Patent number: 6 136-577, Filing date: 1 July 1996, Issue date: 24 October 2000

Gallagher T, Shaffer B, Rummer B (2006) An economic analysis of hardwood fiber production on dryland irrigated sites in the US Southeast. Biomass Bioenergy 30:794–802

Gaur S, Reed TB (1998) Thermal data for natural and synthetic fuels. 1998. Dekker, NY

Gevo (2009) Gevo wins US grant for bimass to butanol development. Press release, http://www.gevo.com/news_Gevo-Wins-US-Grant-pr_111309.php

Goicoechea M, Lacombe E, Legay S, Mihaljevic S, Rech P, Jauneau A, Lapierre C, Pollet B, Verhaegen D, Chaubet-Gigot N, Grima-Pettenati J (2005) EgMYB2, a new transcriptional activator from Eucalytpus xylem, regulates secondary cell wall formation and lignin biosynthesis. Plant J 43:553–567

Gomez Jimenez MD, Canas Clemente LA, Madueno Albi F, Beltran Porter JP (2006) Sequence regulating the anther-specific expression of a gene and its use in the production of androsterile plants and hybrid seeds. US Patent no. 7078593

Good AG, Swarat AK, Muench DG (2004) Can less yield more? Is reducing nutrient input into the environment compatible with maintaining crop production? Trends Plant Sci 9:597–605

Grabber JH, Ralph J, Hatfield RD, Quideau S (1997) p-Hydroxyphenyl, guaiacyl, and syringyl lignins have similar inhibitory effects on cell wall degradation. J Agric Food Chem 45:2530–2532

Grattapaglia D (2004) Integrating genomics into Eucalyptus breeding. Genet Mol Res 3:369–379

Grattapaglia D, Ribeiro D, Rezende VJ, Rezende GDSP (2004) Retrospective selection of elite parent trees using paternity testing with microsatellite markers: an alternative short term breeding tactic for Eucalyptus. Theor Appl Genet 109:192–199

Green Car Congress (2009) BioEthanol Japan begins production of cellulosic ethanol from wood scraps; uses Celunol Technology, 16 January 2007, http://www.greencarcongress.com/2007/01/bioethanol_japa.html, cited 1 December 2009

Gresham CA (2002) Sustainability of intensive loblolly pine plantation management in Carolina coastal plain, USA. For Ecol Manag 155:69–80

Grossnickle SC, Pait J (2008) Somatic embryogenesis tissue culture for applying varietal forestry to conifer species. USDA Forest Service Proceedings, RMRS-P-57:135–139, http://www.fs.fed.us/rm/pubs/rmrs_p057/rmrs_p057_135_139.pdf

Grous WR, Converse AO, Grethlein HE (1986) Effect of steam explosion pretreatment on pore size and enzymatic hydrolysis of poplar. Enzyme Microb Technol 8:274–280

Hamelinck CN, van Hooijdonk G, Faaij APC (2005) Ethanol from lignocellulosic biomass: techno-economic performance in short-, middle- and long-term. Biomass Bioenergy 28: 384–410

Hansen E, Netzer D, Reitveld WJ (1984) Weed control for establishing extensively managed hybrid poplar plantations, US Department of Agriculture, US Forest Service, North Central Forest Experiment Station, Research Note NC-317

Heller MC, Keoleian GA, Volk TA (2003) Life cycle assessment of a willow bioenergy cropping system. Biomass Bioenergy 25:147–165

Herve C, Ceulemans R (1996) Short-rotation coppiced vs non-coppiced poplar: a comparative study at two different field sites. Biomass Bioenergy11:139–150

Hinchee M, Rottmann W, Mullinax L, Zhang C, Chang S, Cunningham M, Pearson L, Nehra N (2009) Short-rotation woody crops for bioenergy and biofuels applications. In Vitro Cell Dev Biol Plant 45:619–629

Hinckley TM, Brooks JR, Cermak J, Ceulemans R, Kucera J, Meinzer FC, Roberts DA (1994) Water flux in a hybrid poplar stand. Tree Physiol 14:1005–1018

Hislop D, Hall DO (1996) Biomass resources for gasificiation power plant. A report for the IEA Bio-Energy Agreement, Thermal Gasfification Activity Task 33, London, IEA 28 ETSU B/M3/00388/31/REP, April 1995. http://media.godashboard.com/gti/IEA/IEABMFeedHall.pdf

Howard J (2005) Estimation of US timber harvest using roundwood equivalents. In: Proceedings of the Sixth Annual Forest Inventory and Analysis Symposium, 21–24 September 2004,

Denver, CO. General Technical Report WO-70. US Department of Agriculture Forest Service, Washington DC, http://www.treesearch.fs.fed.us/pubs/14240

Hsieh TH, Lee JT, Yang PT, Chiu LH, Charng YY, Wang YC, Chan MT (2002) Heterologous expression of the Arabidopsis C-repeat/dehydration response element binding factor 1 gene confers elevated tolerance to chilling and oxidative stresses in transgenic tomato. Plant Physiol 129:1086–1094

Hu JJ, Tian YC, Han YF, Li L, Zhang BE (2001) Field evaluation of insect resistant transgenic *Populus nigra* trees. Euphytica 121:123–127

Hu WJ, Harding SA, Lung J, Popko JL, Ralph J, Stokke DD, Tsai CJ, Chiang VL (1999) Repression of lignin biosynthesis promotes cellulose accumulation and growth in transgenic trees. Nat Biotechnol 17:808–812

Huntley SK, Ellis D, Gilbert M, Chapple C, Mansfield SD (2003) Significant increases in pulping efficiency in C4H-F5H-transformed poplars: improved chemical savings and reduced environmental toxins. J Agric Food Chem 51:6178–6183

IEA BioEnergy (2002) Sustainable production of woody biomass for energy. IEA BioEnergy Position Paper. http://www.ieabioenergy.com/LibItem.aspx?id=157

ISAAA (2008) The global status of commercialized biotech/GM crops: 2008. Retrieved 20 October 2009. International Service for the Acquisition of Agri-Biotech Applications, http://www.isaaa.org

Ito Y, Katsura K, Maruyama K, Taji T, Kobayashi M, Seki M, Shinozaki K, Yamaguchi-Shinozaki K (2006) Functional analysis of rice DREB1/CBF-type transcription factors involved in cold-responsive gene expression in transgenic rice. Plant Cell Physiol 47:141–153

Jaglo-Ottosen KR, Gilmour SJ, Zarka DG, Schabenberger O, Thomashow MF (1998) Arabidopsis CBF1 overexpression induces COR genes and enhances freezing tolerance. Science 280:104–106

Jaglo-Ottosen KR, Kleff S, Amundsen KL, Zhang X, Haake V, Zhang JZ, Deits T, Thomashow MF (2001) Components of the Arabidopsis C-repeat/dehydration-responsive element binding factor cold-response pathway are conserved in *Brassica napus* and other plant species. Plant Physiol 127:910–917

Jahromi ST (1982) Variation in cold resistance and growth in *Eucalyptus viminalis*. South J Appl For 6:221–225

James RR, Meilan R, Skinner JS, Han K-H, Jouanin L, Ma C, Pilate G, DiFaxio SP, Strauss SH, Eaton JA, Hoien EA, Stanton BJ, Crockett RP, Taylor ML (2002) The *CP4* transgene provides high levels of tolerance to Roundup® herbicide in field-grown hybrid poplars. Can J For Res 32:967–976

Johnson JD, Tognetti R, Paris P (2002) Water relations and gas exchange in poplar and willow under water stress and elevated atmospheric CO_2. Physiol Planta 115:93–100

Joint Genome Institute (2009) Eucalyptus genomic sequencing progress. United States Department of Energy, Office of Science, http://www.jgi.doe.gov/sequencing/statistics.html, cited 22 December 2009

Jorgensen H, Kristensen JB, Felby C (2007) Enzymatic conversion of lignocellulose into fermentable sugars: challenges and opportunities. Biofuels Bioprod Bioref 1:119–134

Kalluri UC, Difazio SP, Brunner AM, Tuskan GA (2007) Genome-wide analysis of Aux/IAA and ARF gene families in Populus trichocarpa. BMC Plant Biol 7:59, doi:10.1186/1471-2229-7-59, http://www.biomedcentral.com/1471-2229/7/59

Karp A, Shield I (2008) Bioenergy from plants and the sustainable yield challenge. New Phytol 179:15–32

Kasuga M, Liu Q, Miura S, Yamaguchi-Shinozaki K, Shinozaki K (1999) Improving plant drought, salt, and freezing tolerance by gene transfer of a single stress-inducible transcription factor. Nat Biotechnol 17:287–291

Kasuga M, Miura S, Shinozaki K, Yamaguchi-Shinozaki K (2004) A combination of the Arabidopsis DREB1A gene and stress-inducible rd29A promoter improved drought- and low-temperature stress tolerance in tobacco by gene transfer. Plant Cell Physiol 45:346-350

Kauter D, Lewandowski I, Claupein W (2003) Quantity and quality of harvestable biomass from Populus short rotation coppice for solid fuel use—a review of the physiological basis and management influences. Biomass Bioenergy 24:411–427

Kavalov B, Peteves SD (2005) Status and perspectives of biomass to liquid fuels in the European Union. European Commission, Directorate General Joint Research Center, Institute for Energy, Petten, The Netherlands. EUR21745. http://ie.jrc.ec.europa.eu/publications/scientific_publications/2005/EUR%2021745%20EN.pdf

Kiplinger Washington (ed) (2007) Kiplinger's biofuels market report vol 1, http://www.biontech.com/news/print-air/KiplingerBiofuels.070702.pdf

KL Process Design Group (2008) The 2nd generation cellulosic ethanol leader, becomes public and changes it's name to KL Energy Corp., October 2008, http://www.klprocess.com/BiomassProcess.htm

Kral R (1993) Pinus. In: Flora of North America Editorial Committee (eds) 1993+. Flora of North America North of Mexico. 15+ vols. New York and Oxford. Vol 3, pp 356–357

Kumar A, Flynn P, Sokhansanj S (2008) Biopower generation from mountain pine infested wood in Canada: an economical opportunity for greenhouse gas mitigation. Renew Energy 33:1354–1363

Kumar P, Barrett DM, Delwiche MJ, Stroeve P (2009) Methods for pretreatment of lignocellulosic biomass for efficient hydrolysis and biofuel production. Ind Eng Chem Res 48: 3713–3729

Kuzovkina YA, Weih M, Romero MA, Charles J, Hurst S, McIvor I, Karp A, Trybush S, Labrecque M, Teodorescu TI, Singh NB, Smart LB, Volk TA (2008) Salix: botany and global horticulture. Hortic Rev 34:447–489

Lane J (2009) BP, DuPont butanol JV, Butamax, heads for commercialization; BP stands for "butanol play"? How will Gevo, Cobalt counter? Biofuels Digest, 14 July 2009. http://www.biofuelsdigest.com/blog2/2009/07/14/bp-dupont-butanol-jv-is-named-butamax-heads-for-commercialization-bp-stands-for-butanol-play-how-will-gevo-cobalt-counter/. Retrieved 6 November 2009

Lapierre C, Pollet B, Petit-Conil M, Toval G, Romero J, Pilate G, Leplé J-C, Boerjan W, Ferret V, De Nadai V, Jouanin L (1999) Structural alterations of lignins in transgenic poplars with depressed cinnamyl alcohol dehydrogenase or caffeic acid O-methyltransferase activity have an opposite impact on the efficiency of industrial kraft pulping. Plant Physiol 119:153–164

Leplé JC, Dauwe R, Morreel K, Storme V, Lapierre C, Pollet B, Naumann A, Kang KY, Kim H, Ruel K, Lefèbvre A, Joseleau JP, Grima-Pettenati J, De Rycke R, Andersson-Gunnerås S, Erban A, Fehrle I, Petit-Conil M, Kopka J, Polle A, Messens E, Sundberg B, Mansfield SD, Ralph J, Pilate G, Boerjan W (2007) Downregulation of cinnamoyl-coenzyme A reductase in poplar: multiple-level phenotyping reveals effects on cell wall polymer metabolism and structure. Plant Cell 19:3669–3691

Li X, Weng JK, Chapple C (2008) Improvement of biomass through lignin modification. Plant J. 54:569–581

Liang ZS, Yang HW, Shao HB, Han RL (2006) Investigation on water consumption characteristics and water use efficiency of poplar under soil water deficits on the loess plateau. Colloids Surf B 53 23–28

Linderson ML, Iritz Z, Lindroth A (2007) The effect of water availability on stand-level productivity, transpiration, water use efficiency and radiation use efficiency of field-grown willow clones. Biomass Bioenergy 31:460–468

Liu Q, Kasuga M, Sakuma Y, Abe H, Miura S, Yamaguchi-Shinozaki K, Shinozaki K (1998) Two transcription factors, DREB1 and DREB2, with an EREBP/AP2 DNA binding domain separate two cellular signal transduction pathways in drought- and low-temperature-responsive gene expression, respectively, in Arabidopsis. Plant Cell 10:1391–1406

Man Hui-min R, Boriel R, El-Khatib Kirby EG (2005) Characterization of transgenic poplar with ectopic expression of pine cyotsolic glutamine synthetase under conditions of varying nitrogen availability. New Phytol 167:31–39

Mann L, Tolbert V (2000) Soil sustainability in renewable biomass plantations. Ambio 29: 492–498

McCown BH, McCabe DE, Russell DR, Robison DJ, Barton KA, Raffa KF (1991) Stable transformation of Populus and incorporation of pest resistance by electric discharge particle acceleration. Plant Cell Rep 9:590–594

McKendry P (2002) Energy production from biomass (part 1): overview of biomass. Bioresour Technol 83:37–46

McNabb K, Enebak S (2008) Forest tree seedling production in the Southern United States: the 2005–2006 planting season. Tree Planters Notes 53:47–56 http://forestry.about.com/gi/o.htm?zi=1/XJ&zTi=1&sdn=forestry&cdn=education&tm=23&gps=245_302_1020_503&f=00&su=p897.6.336.ip_&tt=2&bt=0&bts=0&zu=http://%3A//www.rngr.net/Publications/tpn/tpn_back"

Mead DJ (2005) Forests for energy and the role of planted trees. Crit Rev Plant Sci 24:407–421

Meilan R, Ma C, Cheng S, Eaton J, Miller LK, Crockett RP, DiFazio SP, Strauss SH (2000) High levels of Roundup® and leaf-beetle resistance in genetically engineered hybrid cottonwoods. In: Blatner KA, Johnson JD, Baumgartner DM (eds) Hybrid poplars in the Pacific Northwest: culture, commerce and capability. Washington State University Cooperative Extension Bulletin MISC0272, Pullman, WA, pp 29–38

Mercker D (2007) Short rotation woody crops for biofuels. University of Tennessee Agricultural Experiment Station. Retrieved 3 November 2009. http://www.utextension.utk.edu/publications/spfiles/SP702-C.pdf

Meskimen G, Francis JK (1990) Rose gum Eucalyptus. In: Burns RM, BH Honkala (eds) Silvics of North America, vol 2. Hardwoods. Agriculture Handbook 654. US Department of Agriculture, Forest Service, Washington, DC, http://www.na.fs.fed.us/spfo/pubs/silvics_manual/volume_2/eucalyptus/grandis.htm

Meskimen GF, Rockwood DL, Reddy KV (1987) Development of Eucalyptus clones for a summer rainfall environment with periodic severe frosts. New For 1:197–205

Mitsuda N, Iwase A, Yamamoto H, Yoshida M, Seki M, Shinozaki K, Ohme-Takagi M (2007) NAC transcription factors, NST1 and NST3, are key regulators of the formation of secondary walls in woody tissues of Arabidopsis. Plant Cell 19:270–280

Mora AL, Garcia CH (2000) Eucalypt cultivation in Brazil. Brazilian Society of Silviculture (SBS), Sao Paulo, Brazil

Murray B, Nicholson R, Ross M, Holloway T, Patil S (2006) Biomass energy consumption in the forest products industry. RTI Project Number 0209217.002, RTI International, Research Triangle Park, NC, tonto.eia.doe.gov/FTPROOT/renewables/rti_biomass_report_2006.pdf

Nasrallah ME, Nasrallah JB, Thorsness MK (1999) Isolated DNA elements that direct pistil-specific and anther-specific gene expression and methods of using same. United States Patent No. 5,859,328

Nathan R, Katul GG, Horn HS, Thomas SM, Oren R, Avissar R, Pacala SW, Levin SA (2002) Mechanisms of long-distance dispersal of seeds by wind. Nature 418:409–413

Nehra NS, Becwar MR, Rottmann WH, Pearson L, Chowdhury K, Chang S, Wilde HD, Kodrzycki RJ, Zhang C, Gause KC, Parks DW, Hinchee MA (2005) Forest biotechnology: innovative methods, emerging opportunities. In Vitro Cell Dev Biol Plant 41:701–717

National Renewable Energy Laboraory (1999) A renewable alternative for utlities and their customers. US Department of Energy, Biomass Power, Washington DC, DOE/GO-200099-759, May 1999, http://www.nrel.gov/docs/fy99osti/24933.pdf

Oak Ridge National Laboratory (2009) ORNL/BFDP Poplar Drought Tolerance Research, http://bioenergy.ornl.gov/papers/misc/drotpopl.html; accessed 10 January 2009

Palmqvist E, Hahn-Hägerdal B (2000) Fermentation of lignocellulosic hydrolysates. II: Inhibitors and mechanisms of inhibition. Bioresour Technol 74:25–33

Pan X, Arato C, Gilkes N, Gregg, Mabee W, Pye K, Xiao Z, Zhang X, Saddler J (2005) Biorefining of softwoods using ethanol organosolv pulping: preliminary evaluation of process streams for manufacture of fuel-grade ethanol and co-products. Biotechnol Bioeng 90:473–481

Patzlaff A, McInnis S, Courtenay A, Surman C, Newman LJ, Smith C, Bevan MW, Mansfield S, Whetten RW, Sederoff RR, Campbell MM (2003) Characterisation of a pine MYB that regulates lignification. Plant J 46:743–754

Perlack RD, Wright LL Turhollow AF, Graham RL, Stokes BJ, Erbach DC (2005) Biomass as a feedstock for a bioenergy and bioproducts industry: the technical feasibility of a billion-ton annual supply. US Department of Energy, Oak Ridge National Laboratory, Oak Ridge, TN

Pew Center on Global Climate Change (2008) Renewable Portfolio Standards (RPS), http://www.pewclimate.org/what_s_being_done/in_the_states/rps.cfm. cited August 19, 2008

Polle A, Altman A, Jiang X (2006) Towards genetic engineering for drought tolerance in trees. In: Fladung M, Ewald D (eds) Tree transgenesis: recent developments. Springer, Berlin, pp 275–297

PR NewsWire (2009) Cellulosic ethanol takes off! 23 January 2009, http://www.prnewswire.com/news-releases/cellulosic-ethanol-takes-off-53732747.html

Proe MF, Griffiths JH, Craig J (2002) Effects of spacing, species and coppicing on leaf area, light interception and photosynthesis in short rotation forestry. Biomass Bioenergy 23:315–326

Qin F, Sakuma Y, Li J, Liu Q, Li YQ, Shinozaki K, Yamaguchi-Shinozaki K (2004) Cloning and functional analysis of a novel DREB1/CBF transcription factor involved in cold-responsive gene expression in Zea mays L. Plant Cell Phys 45:1042–1052

Ragauskas AJ, Williams CK, Davison BH, Britovsek G, Cairney J, Eckert CJ, Frederick WJ, Hallett JP, Leak DJ, Liotta CL, Mielenz JR, Murphy R, Templer R, Tschaplinski T (2006) The path forward for biofuels and biomaterials. Science 311:484–489, doi: 10.1126/science.1114736

Raison RJ (2002) Environmental sustainability of forest energy production. In: Richardson J, Boerjhedan J, Hakkila P, Lowe AT, Smith CT (eds) Bioenergy from sustainable forestry: guiding principles and practice. Kluwer, Dordrecht, pp 159–262

Reynolds KM (2005) Integrated decision support for sustainable forest management in the United States: fact or fiction? Comput Electron Agric 49:6–23

Reynolds KM, Johnson KN, Gordon SN (2003) The science/policy interface in logic-based evaluation of forest ecosystem sustainability. For Policy Econ 5:433–446

RISI (2008) North American Timber Annual Historical Data, 2007. RISI, http://www.risiinfo.com/risi-store/do/product/detail/north-american-timber-annual-historical-data.html;jsessionid=3F8266CC8AEBE4090F798CF085D85324

Robinson TL, Rousseau RJ, Zhang J (2006) Biomass productivity improvement for eastern cottonwood. Biomass Bioenergy 30:735–739

Robison DJ, McCown BH, Raffa KF (1994) Responses of gypsy moth (Lepidoptera: Lymantriidae) and forest tent caterpillar (Lepidoptera: Lasiocampidae) to transgenic poplar, Populus spp., containing a Bacillus thuringiensis gene. Environ Entomol 23:1030–1041

Rockwood DL, Naidu CV, Carter DR, Rahmani M, Spriggs TA, Lin C, Alker GR, Isebrands JG, Segrest SA (2004) Short-rotation woody crops and phytoremediation: opportunities for agroforestry? J Agrofor Syst 61–62:51–63, doi: 10.1023/B:AGFO.0000028989.72186.e6

Rockwood DL, Rudie AW, Ralph SA, Zhu JY, Winandy JE (2008) Energy product options for Eucalyptus species grown as short rotation woody crops. Int J Mol Sci 9:1361–1378

Ronnberg-Wastljung AC, Gullberg U (1999) Genetics of breeding characters with possible effects on biomass production in Salix viminalis (L.). Theor Appl Genet 98:531–540

Rowell RM (2005) Handbook of wood chemistry and wood composites. Taylor & Francis/CRC, Boca Raton

Rubin EM (2008) Genomics of cellulosic biofuels. Nature 454:841–845

Sang Y, Zheng S, Li W, Huang B, Wang X (2001) Regulation of plant water loss by manipulating the expression of phospholipas Dα. Plant J 23:135–144

Satori F, Lal R, Ebinger MH, Parrish DJ (2006) Potential soil carbon sequestration and CO_2 offset by dedicated energy crops in the USA. Crit Rev Plant Sci 25:441–472

Schuette B (2000) Weed management strategies for hybrid poplar plantings. In: Blatner KA, Johnson JD, Baumbartner DM (eds) Hybrid poplars in the Pacific Northwest: culture, commerce, and capability. In: Symposium Proceedings, 7–9 April 1999, Pasco, WA. Washington State University Cooperative Extension Publication MISC0272, Pullman, WA, pp 83–86

Schuster KC (2000) Applied acetone-butanol fermentation. J Mol Microbiol Biotechnol 2:3–4

Sedjo RA (2001) Biotechnology in forestry: considering the costs and benefits. Resoures 145:10–12, http://rff.org/RFF/Documents/RFF-Resources-145-biotech.pdf

Shani Z, Dekel M, Tsabary G, Goren R, Shoseyov O (2004) Growth enhancement of transgenic poplar plants by over expression of Arabidopsis thaliana endo-1,4-β-glucanase (cel1). Mol Breed 14:321–330

Short Rotation Forestry Handbook (1995) University of Aberdeen. Retrieved 20 October 2009, http://www.abdn.ac.uk/wsrg/srfhbook

Sims RH, Venturi P (2004) All year-round harvesting of short rotation coppice Eucalyptus compared with the delivered costs of biomass from more conventional short season, harvesting systems. Biomass Bioenergy 26:27–37

Sims RH, Maiava TG, Bullock BT (2001) Short rotation coppice tree species selection for woody biomass production in New Zealand. Biomass Bioenergy 20:329–335

Sklar T (2009) Making the case for woody biomass based biofuels and bioenergy. Biofuels Digest, 19 February 2009, http://www.biofuelsdigest.com/blog2/2009/02/19/making-the-case-for-woody-biomass-based-biofuels-and-bioenergy-a-special-biofuels-digest-report/

Smart LB, Volk TA, Lin J, Kopp RF, Phillips IS, Cameron KD, White EH, Abrahamson LP (2005) Genetic improvement of shrub willow (Salix spp.) crops for bioenergy and environmental applications in the United States. Unasylva 56:51–55

SAFER (2009) Southern Bioenergy Roadmap. Southern Agriculture & Forestry Resource Alliance, http://www.saferalliance.net/projects/roadmap.html

Southern Alliance for Clean Energy (2009) Yes we can: Southern solutions for a national renewable energy standard, http://www.cleanenergy.org/images/stories/serenewables022309rev.pdf posted 23 February 2009

Stanton BJ (2003) Poplars. In: Burley J, Evans J, Younquist JA (eds) Encyclopedia of forest sciences. Elsevier, Oxford, pp 1441–1449

Stanton B, Eaton J, Johnson J, Rice D, Schuette B, Moser B (2002) Hybrid poplar in the Pacific Northwest: the effects of market-driven management. J For 100:28–33

Stanturf J, Gardiner E, Schoenholtz S (2003) Interplanting for bioenergy and riparian restoration in the southeastern USA. In: Nicholas ID (ed) Proceedings, Short Rotation Crops for Bioenergy. International Energy Agency Task 30 Conferences 1–5 December 2003, Tauranga, New Zealand, pp 241–250

Stricker J, Rockwood DL, Segrest SA, Alker GR, Prine RM, Carter DR (2000) Short rotation woody crops for Florida. In: Proceedings of 3rd Biennial Short Rotation Woody Crops Operations Working Group Conference, Syracuse, NY, pp 15–23, http://www.woodycrops.org/reports/2000%20Proceedings.pdf

Tharakan PJ, Volk TA, Abrahamson LP, White EH (2003) Energy feedstock characteristics of willow and hybrid poplar clones at harvest age. Biomass Bioenergy 25:571–580

TimberMart (2008) South Market News Quarterly 13:1 p 28, Retrieved 3 Nov. 2009 http://www.tmart-south.com/tmart/pdf/Qtr_01Q08news.pdf

Transport Studies Group, University of Westminster (1996) Supply chain options for biomass fuels. In: Transport and supply logistics of biomass fuels, vol. 1. ETSU Report B/W2/00399/Rep/2

Turnbull J (1999) Eucalyptus plantations. New For 17:37–52

Tuskan GA (1998) Short-rotation woody crop supply systems in the United States: what do we know and what do we need to know? Biomass Bioenergy 14:307–315

Tuskan GA (2007) Bioenergy, genomics, and accelerated domestication: a US example. In: FAO, Papers and Presentations from The Role of Agricultural Biotechnologies for Production of Bioenergy in Developing Countries, http://www.fao.org/biotech/seminaroct2007.htm

Tuskan G, DiFazio S, Hellsten U, Jansson S, Rombauts S, Putnam N, Sterck L, Bohlmann J, Schein J, Bhalerao RR, Bhalerao RP, Blaudez D, Boerjan W, Brun A, Brunner A, Busov V,

Campbell M, Carlson J, Chalot M, Chapman J, Chen G, Cooper D, Coutinho PM, Couturier J, Covert S, Cunningham R, Davis J, Degroeve S, dePamphilis C, Detter J, Dirks B, Dubchak I, Duplessis S, Ehlting J, Ellis B, Gendler K, Goodstein D, Gribskov M, Grigoriev I, Groover A, Gunter L, Hamberger B, Heinze B, Helariutta Y, Henrissat B, Holligan D, Islam-Faridi N, Jones-Rhoades M, Jorgensen R, Joshi C, Kangasjärvi J, Karlsson J, Kelleher C, Kirkpatrick R, Kirst M, Kohler A, Kalluri U, Larimer F, Leebens-Mack J, Leplé JC, Déjardin A, Pilate G, Locascio P, Lucas S, Martin F, Montanini B, Napoli C, Nelson DR, Nelson CD, Nieminen KM, Nilsson O, Peter G, Philippe R, Poliakov A, Ralph S, Richardson P, Rinaldi C, Ritland K, Rouzé P, Ryaboy D, Salamov A, Schrader J, Segerman B, Sterky F, Souza C, Tsai C, Unneberg P, Wall K, Wessler S, Yang G, Yin T, Douglas C, Sandberg G, Van de Peer Y, Rokhsar D (2006) The genome of black cottonwood, Populus trichocarpa. Science 313:1596–1604

US DOE (2007a) DOE selects six cellulosic ethanol plants for up to $385 million in Federal funding. Press release 28 February 2007. Retrieved 26 Oct. 2009. http://www.energy.gov/news/4827.htm

US DOE (2007b) DOE Bioenergy Research Centers, http://genomicscience.energy.gov/centers/ Retrieved 31 October 2009

US DOE Office of Science (2009) DOE BioEnergy Science Center (BESC), http://genomicsgtl.energy.gov/centers/center_ORNL.shtml accessed 10 December 2009

US Department of Agriculture (2002) Agricultural Research Service Global Change National Program. http://www.ars.usda.gov/SP2UserFiles/Program/204/NP204ActionPlan.pdf

US Department of Agriculture (2008) Agricultural Research Service Global Change National Program, Research: Plantation Tree Farming, http://www.ars.usda.gov/research/programs/programs.htm?np_code=204&docid=308&page=5 last modified 28 October 2008

US EPA (2007) Renewable Fuel Standard Program, http://www.epa.gov/OMS/renewablefuels/ Retrieved 23 November 2009

US EPA (2008) US Climate Policy and Actions. http://www.epa.gov/climatechange/policy/index.html Accessed 19 August 2008

Uslu A, Faaij APC, Bergman PCA (2008) Pre-treatment technologies, and their effect on international bioenergy supply chain logistics. Techno-economic evaluation of torrefaction, fast pyrolysis and pelletisation. Energy 33:1206–1223

Van Frankenhuyzen K, Beardmore T (2004) Current status and environmental impact of transgenic forest trees. Can J For Res 34:1163–1180

Vanholme R, Morreel K, Ralph J, Boerjan W (2008) Lignin engineering. Curr Opin Plant Biol 11:278–285

Verenium Corporation (2009) Verenium Corporation: combining world-class enzyme science with expertise in complex, large-plant infrastructure, Verenium Corporation Fact Sheet. http://www.verenium.com/pdf/Veren_BBU.pdf cited 11 November 2009

Vessia Ø (2005) Biofuels from lignocellulosic material—in the Norwegian context 2010—technology, potential and costs. NTNU, Norwegian University of Science and Technology, Trondheim, http://www.zero.no/transport/bio/vessia_version3-20-12-05.pdf

Vogt DJ, Vogt KA, Gordon JC, Miller ML, Mukumoto C, Upadhye R, Miller MH (2009) Wood methanol as a renewable energy source in the western United States. In: Solomon B, Luzadis VA (eds) Renewable energy from forest resources in the United States. Routledge, New York, pp 299–322

Volk TA, Verwijist T, Therakan PJ, Abrahamson LP, White EH (2004) Growing fuel: a sustainability assessment for willow biomass crops. Frontiers Ecol Environ 2:411–418

Wang G, Castiglione S, Chen Y, Li L, Han Y, Tian Y, Gabriel DW, Han Y, Mang K, Sala F (1996) Poplar (Populus nigra L.) plants transformed with a Bacillus thuringiensis toxin gene: insecticidal activity and genomic analysis. Transgenic Res 5:289–301

Warren TJ, Poulter R, Parfitt R (1995) Converting biomass to electricity on a farm-sized scale using downdraft gasification and a spark-ignition engine. Bioresour Technol 52:95–98

Weng J-K, Li X, Bonawitz N, Chapple C (2008) Emerging strategies of lignin engineering and degradation for cellulosic biofuel production. Curr Opin Biotechnol 19:166–172

Wheals AE, Basso LC, Denise M, Alves G, Amorim H (1999) Fuel ethanol after 25 years. Trends Biotechnol 17:482–487

White RH (1987) Effect of lignin content and extractives on the higher heating value of wood. Wood Fiber Sci 19:446–452

Williams CG (2005) Framing the issues on transgenic forests. Nat Biotechnol 23:530–532

Williams CG, Davis B (2005) Rate of transgene spread via long-distance seed dispersal in Pinus taeda. For Ecol Manage 217:95–102

Yamaguchi-Shinozaki K, Shinozaki K (1993) Characterization of the expression of a desiccation-responsive rd29 gene of Arabidopsis thaliana and analysis of its promoter in transgenic plants. Mol Gen Genet 236:331–340

Yamaguchi-Shinozaki K, Shinozaki K (1994) A novel cis-acting element in an Arabidopsis gene is involved in responsiveness to drought, low temperature, or high salt-stress. Plant Cell 6:251–264

Yanofsky MF (2006) Methods of suppressing flowering in transgenic plants. United States Patent No. 6,987,214 B1

Yat SC, Berger A, Shonnard DR (2008) Kinetic characterization of dilute surface acid hydrolysis of timber varieties and switchgrass. Bioresour Technol 99:3855–3863

Yi SY, Kim JH, Joung YH, Lee S, Kim WT, Yu SH, Choi D (2004) The pepper transcription factor CaPF1 confers pathogen and freezing tolerance in Arabidopsis. Plant Physiol 136:2862–2874

Young M, Boland M, Hofstrand D (2007) Current issues in ethanol production. Agricultural Marketing Resource Center, Iowa State University, http://www.agmrc.org/media/cms/EthanolPaperRev_049574599F3A9.pdf Retrieved 7 November 2009

Zeachem (2008) Zeachem, Inc announces long-term poplar tree feedstocksupply agreement with Greenwood Resources, Inc for a biorefinery (press release 11 February 2008), http://www.zeachem.com/press/pressrelease01.php Retrieved 10 November 2009

Zhang TT, Song YZ, Liu YD, Guo XQ, Zhu CX, Wen FJ (2008) Overexpression of phospholipase Dα gene enhances drought and salt tolerance of Populus tomentosa. J Chin Sci Bull 53L:3658–3665, doi:10.1007/s11434-008-0476-1

Zhang X, Fowler SG, Cheng H, Lou Y, Rhee SY, Stockinger EJ, Thomashow MF (2004) Freezing-sensitive tomato has a functional CBF cold response pathway, but a CBF regulon that differs from that of freezing-tolerant Arabidopsis. Plant J 39:905–919

Zhu, JY, Wang GS, Pan XJ, Gleisner (2008) The status of and key barriers in lignocellulosic ethanol production: a technological perspective. In: Wu C, Yuan Z, Ma L, Zhuang X (eds) International Conference on Biomass Energy Technologies, 3–5 December 2008, vol 1. The Guangzhou Institute of Energy Conversion, Chinese Academy of Science, Guangzhou, China. pp 1–12

Chapter 8
Engineering Status, Challenges and Advantages of Oil Crops

Richard F. Wilson and David F. Hildebrand

8.1 Global Trends in Supply and Demand for Edible Oils

8.1.1 Constraints on the Use of Edible Crop Products for Biofuel

Near the end of the nineteenth century, vegetable oil gained prominence as the fuel for pressure-ignited internal combustion engines. The first engine designed by Rudolf Diesel (Fig. 8.1) ran on peanut oil (Quick 1989). However, discovery of vast petroleum deposits and the evolution of that industry provided the world with a more convenient, and seemingly ample, supply of petroleum-based diesel fuel. In 2008, the US consumed a total of 7.14 billion barrels of crude petroleum—about 23% of total world consumption (USDOE 2009a). Assuming 10 gallons diesel fuel per barrel, that volume produced about 71 billion gallons of diesel, 57% of which was derived from imported crude oil. Suffice to say, vegetable oil-based or any other form of biofuel has received serious attention only in times of energy crisis, such as the shortages experienced during World War II and more recent market disorders leading to record prices for crude petroleum. Although current fuel prices have subsided substantially since those historic peaks, price plus other social factors once again have brought biofuel to the forefront, prompting governmental energy policies to reduce dependence on petroleum and fossil-fuels. For example, the *US Energy Policy Act of 2005* mandated that renewable fuel use in gasoline and diesel reach 7.5 billion gallons by 2012 (Westcott 2007). The *US Energy Independence & Security Act of 2007* Public Law 110-140 also provided for a renewable fuel

R.F. Wilson
Oilseeds & Biosciences LLC, 5517 Hickory Leaf Drive, Raleigh, NC 27606-9502, USA
e-mail: rfwilson@mindspring.com

D.F. Hildebrand
University of Kentucky, 403 Plant Science Building, 1405 Veterans Drive, Lexington KY 40546-0312, USA
e-mail: dhild@uky.edu

Fig. 8.1 The first diesel engine (photo taken in the German Museum, Munich)

standard to increase the supply of alternative fuel sources, and increased the use mandate to about 36 billion gallons of biofuel (15 billion gallons from ethanol) by 2022 (Malcolm and Allery 2009).

The share of US corn production that supplies the ethanol market essentially was non-measurable prior to 2000 (Fig. 8.2), but accounted for about 20% of US corn production in 2009 (USDA 2009a). However, expansion of ethanol production has not come at the expense of domestic feed or food applications for corn. Indeed, there has been little growth or decline in these food and feed market sectors over the past 15 years. The same is generally true for US corn exports. There also is little evidence to support prognostications that US corn production will cut deeply into the planted acreage for other crops in future years. Breeding and other supporting sciences will continue to increase yields while constrained by a relatively constant land area for renewable oil production (Egli 2008a, b). The area for US corn harvest averaged 30.2 ± 2.4 million hectares (Mha), while US harvested area for soybean averaged 29.0 ± 1.2 Mha from 1997 to 2009 (USDA 2009a). Therefore, the rise in domestic corn consumption from 7.4 to 10.9 million bushels (Mbu) during that period is attributed almost exclusively to ethanol production. Current estimates indicate 197 US ethanol plants are in operation, with a capacity of 12.7 billion gallons per year; the 20 more plants under construction should add another 1.8 billion gallons of ethanol per year. The capacity of these facilities suggests that federal mandates for ethanol production can be met in a timely manner. In addition, each bushel of corn produces equal mass (18 lbs each) of ethanol and distillers' dried grains (DDGS)—the solid co-product of ethanol production from corn

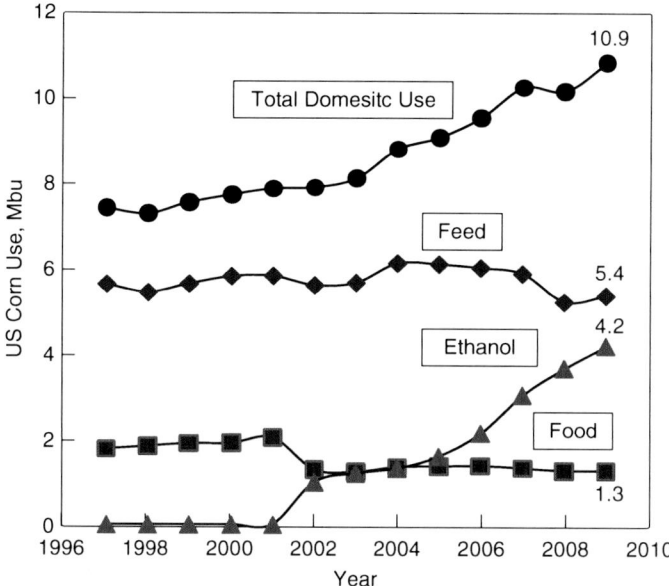

Fig. 8.2 Trends in US domestic consumption of corn and corn products (USDA 2009a)

(Rosentrater 2009). The main outlet for DDGS is livestock feed. DDGS typically contain about 27% protein, 9% oil, 47% fiber, 9% starch and 8% ash. Although nutritional value is comparatively low, DDGS supplemented with synthetic amino acids could have significant economic consequences on future inclusion of corn and soybean in feed for pork and poultry production. With US corn yields escalating to an estimated average of 186 bushels per acre by 2020, the US corn industry may become quite dependent upon demand for ethanol and DDGS production to balance the expected supply of corn and corn products.

Another constraint that should be factored into the equation is the situation between food and industrial use of the major edible oils (USDA 2009b). Prior to 1997, about 10% of the world edible oil supply was consumed in non-fuel industrial uses of oil-based esters such as plasticizers, solvents, adhesives, surfactants, adjuvants, pharmaceuticals, cosmetics, printing inks, plastics and lubricants (Fig. 8.3). Since then there has been a marked rise in the market share taken by industrial applications. Industrial applications accounted for 27.7 million metric tons (MMT) or 20.6% of world edible oil consumption in 2009. Although this trend appears to be slowing, it is conceivable that 25% of the world edible oil supply may be used in biofuel plus oleochemical applications by 2020. The escalation of industrial uses for edible oils may be attributed in part to governmental legislation that provides economic incentives for biofuel production in the US and Europe. Currently, the European Union produces about 65% of the world's total biodiesel output which is estimated to be 3.9 billion gallons (Licht 2009). However, at B100, the total volume of biodiesel equates to 14.9 MMT vegetable oil—or only 10% of world supply.

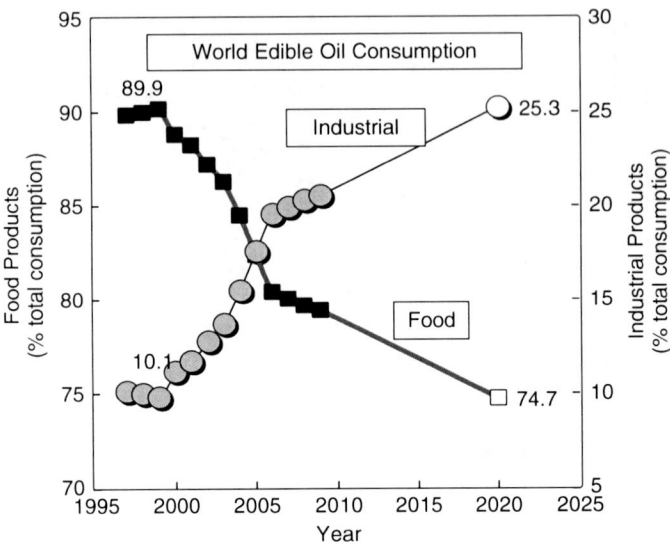

Fig. 8.3 World trends in food and industrial applications of edible oils (USDA 2009b)

Likewise, the US produced 700 million gallons of B100 in 2008 from about 2.6 MMT vegetable oil (USDOE 2009b). Although that amount accounted for about 23% of the total US supply of edible crop oils, US vegetable oil endstocks that year were about 1.7 MMT. Thus, speculations that biodiesel production may cause severe shortfalls in edible oil food products (Ash et al. 2006) appear to be overstated. However, at present, only a small percentage of the world's supply of diesel fuel is biodiesel. With respect to precautionary opinions, events likely will prove that a robust biodiesel/biofuel market will be quite welcome in future years if edible oil and corn production continues to grow at current rates. However, demand always will be tempered by product cost, and supply will remain a function of product availability.

8.1.2 Availability and Cost of Biodiesel Feedstocks

The list of plants that can produce edible oil is quite extensive, and includes almond, apricot, avocado, cocoa butter, corn, grapeseed, hazelnut, mustard, nutmeg, sesame, sheanut, and walnut oils (Salunkhe 1992). A longer list of lesser known plant oils might be added to this discussion. However, USDA statistics generally track commodities that account for the majority of world production, such as: coconut, cottonseed, olive, palm, palm kernel, peanut, rapeseed (canola), safflower, soybean, and sunflower[1]. In 2009, the USDA estimated world edible oil

---

[1] Note: corn is a major crop; but the amount of corn oil available for dietary and energy applications is very low, typically less than 15,000 MT per year (USDA 1989, 2009a)

supply (production plus end-stocks) for those 'major' commodities at 148.2 MMT; and world edible oil demand (domestic consumption plus exports) at 134.8 MMT (USDA 2009b). These agricultural statistics indicate an ample supply of edible oil, but what levels of edible oil supply and demand might be anticipated in the future?

Based on regression analyses, world edible oil supply increased at 5.57 MMT/year from 1997 to 2009. Domestic consumption of edible oil on a global basis increased at 5.22 MMT/year during the same period. These rates project edible oil supply at 209.4 MMT, and demand at 192.2 MMT in 2020. If these rates persist, the gap between supply and demand should increase, with advantage to supply; a condition that could presumably favor production of biofuel. Other features of the trend also become more pronounced. For example, production of major commodity oils such as peanut, olive, coconut and palm kernel appear to have limited capacity for substantial growth. Although all will make important contributions to total supply, long-term positive gains in future world edible oil production will be attributed to a diminishing number of commodities. Going forward, the edible oil market will be dominated by those commodities with the production systems capable of sustaining the observed linear growth rate (Fig. 8.4). Palm, soybean, canola and sunflower currently account for about 88% of world edible oil production, but in future years this analysis suggests that the portion contributed by peanut and canola may not advance at as great a rate as palm or soybean oil production, due primarily to slow expansion of arable land where those crops are adapted. Although geographic regions may tend to utilize domestic product, e.g., canola in Europe, the

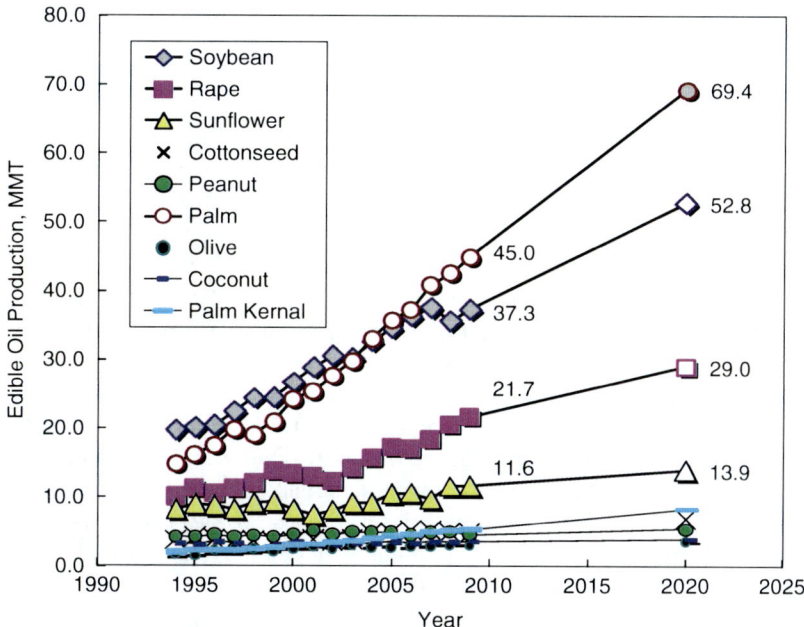

Fig. 8.4 Trends in global production of major edible oils (USDA 2009b)

Table 8.1 Trends in commodity prices for major edible oils

Year	Soybean	Peanut	Sunflower	Rapeseed	Cottonseed	Palm	Corn	Coconut
				US $ per pound oil				
1996	$0.22	$0.44	$0.23	$0.24	$0.26	$0.24	$0.24	$0.31
1997	$0.26	$0.49	$0.27	$0.29	$0.29	$0.27	$0.29	$0.28
1998	$0.20	$0.40	$0.20	$0.22	$0.27	$0.22	$0.25	$0.34
1999	$0.16	$0.35	$0.17	$0.16	$0.22	$0.14	$0.18	$0.24
2000	$0.14	$0.35	$0.16	$0.17	$0.16	$0.11	$0.14	$0.15
2001	$0.16	$0.32	$0.23	$0.20	$0.18	$0.15	$0.19	$0.18
2002	$0.22	$0.47	$0.33	$0.27	$0.38	$0.19	$0.28	$0.20
2003	$0.30	$0.60	$0.33	$0.30	$0.31	$0.22	$0.28	$0.29
2004	$0.23	$0.53	$0.44	$0.30	$0.28	$0.18	$0.28	$0.29
2005	$0.23	$0.44	$0.41	$0.35	$0.29	$0.19	$0.25	$0.26
2006	$0.31	$0.57	$0.58	$0.39	$0.36	$0.30	$0.32	$0.37
2007	$0.52	$1.01	$0.91	$0.64	$0.74	$0.48	$0.69	$0.59
2008	$0.32	$0.70	$0.50	$0.39	$0.37	$0.29	$0.33	$0.33

Source: USDA 2009b

overall situation will be dominated by growing dependence on palm and soybean to meet future consumer demand for renewable oils in food and fuel applications.

Sustainable biofuel production is also a function of the market price for edible oil commodities (USDA 2009b). With the exception of peanut oil and sunflower oil, the respective prices of major edible oils have co-meandered within a relatively narrow trading range throughout the decade ending in 2006. However, energy-related speculation within the edible oil trade in 2007 exposed the world market to aggressive price volatility. Nearly every major edible oil recorded historic high prices. Since then, USDA statistics now show a decline in edible oil prices (Table 8.1). The apparent downward price inflection toward the historical average (1996–2008) for each commodity suggests that the global vegetable oil market can adapt to an expanding market share for industrial products. This should be welcome news to consumers, and especially to the US biodiesel industry in view of the failure of the 111th Congress to extend the biodiesel tax credit before 31 December 2009[2]. Although governmental subsidies in any form for agricultural crops often are berated as unfair competitive or non-economical practices, tax incentives that encourage innovation and development of the biodiesel industry help ensure competition and sustainability, especially during the formative years of the industry. The retail price of B20 diesel fuel obviously will vary depending on the price of a feedstock and how feedstocks are blended. Yet, as the retail price for diesel fuel continues to rise, price differentials between B20 and retail diesel fuel should narrow, thus strengthening the emergence of a more diverse and competitive energy market (CAST 2008). This economic dynamic already appears to be in motion, at least in the US, as evidenced by recent price estimates for B20 from soybean, palm and corn oil (Table 8.2). Furthermore, it is apparent that the retail price of petroleum

---

[2]Note: this incentive may be reinstated in the next session of the 111th Congress (American Soybean Association, personal communication)

8 Engineering Status, Challenges and Advantages of Oil Crops

Table 8.2 Average US retail price for diesel and proximate cost of B20 biofuel

Year	Diesel	Soybean	Peanut	Sunflower	Canola	Cotton	Palm	Corn
				US $/gallon				
2003	$1.51	$1.87	$2.37	$1.93	$1.88	$1.89	$1.73	$1.84
2004	$1.81	$2.01	$2.52	$2.36	$2.13	$2.26	$1.93	$2.10
2005	$2.40	$2.55	$2.90	$2.84	$2.74	$2.65	$2.47	$2.58
2006	$2.71	$2.94	$3.37	$3.39	$3.06	$3.01	$2.91	$2.95
2007	$2.88	$3.42	$4.24	$4.08	$3.62	$3.78	$3.36	$3.71
2008	$3.81	$3.90	$4.52	$4.20	$4.02	$3.98	$3.84	$3.90

Sources: CAST 2008; US DOE 2009c

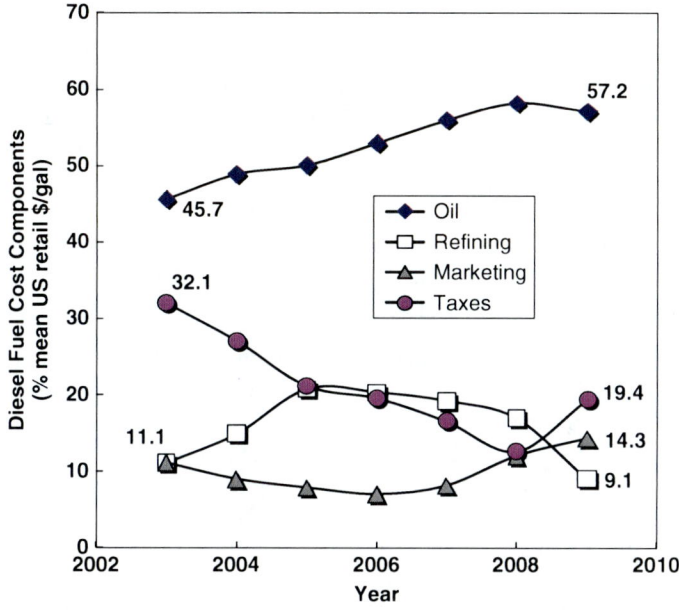

Fig. 8.5 Recent history of the component costs of petroleum diesel fuel (USDOE 2009c)

diesel also has been modulated through tax adjustments (Fig. 8.5). Taxes accounted for about 32% of the average annual retail price for diesel fuel in 2003, but declined to about 13% in 2007 (USDOE 2009c). If the proportion attributed to taxes had remained relatively constant, the current average retail price of diesel fuel in the US would be between US $4.00–4.50/gallon. Therefore, perceptions that biodiesel is not price competitive with petroleum diesel may require adjustment. Perhaps a more appropriate question might be: Is petroleum diesel *sans* price-reducing taxation policies economically competitive with biodiesel?

In addition, it is worth considering the Brazilian ethanol experience. In 1980, the cost of ethanol production from sugarcane was about three-fold greater than the cost of gasoline. The Brazilian government provided subsidies to help the fledgling sugarcane-ethanol industry remain cost competitive in the marketplace. It is estimated that these subsidies amounted to about US $30 billion during the period they were in effect. However, that investment produced enough ethanol to offset US $50 billion in petroleum imports. Since 2004, Brazil has become the world leader in renewable fuel production (Goldemberg 2007).

8.1.3 Sustainability

Feedstock availability and cost are the traditional factors that determine whether a market for a given product can be sustained. However, concern for 'global climate change' has introduced a new category of concern that will have increasingly important implications for biofuel. Briefly, the issue involves the impact that biofuel production may have on biodiversity and atmospheric 'greenhouse-gas (GHG)' emissions. For example, expansion of oil palm plantations in Indonesia and Malaysia at the expense of tropical rainforests or use of organic soils has received considerable public attention, especially when the habitat of rare species such as orangutan, sun bear, and Borneo gibbon are endangered. As a result, many companies and governments, primarily in Europe, are requiring that the palm oil industry meet standards developed by conventions such as the *International Roundtable for Sustainable Biofuels* (IRSB; Schill 2008). The standards established by the IRSB are embodied by 12 principles:

(1) Biofuel production shall follow international treaties and national laws regarding such things as air quality, water resources, agricultural practices, labor conditions, and more.
(2) Biofuel projects shall be designed and operated in participatory processes that involve all relevant stakeholders in planning and monitoring.
(3) Biofuel shall significantly reduce GHG emissions as compared to fossil fuels. The principle seeks to establish a standard methodology for comparing GHG benefits.
(4) Biofuel production shall not violate human rights or labor rights, and shall ensure decent work and the well-being of workers.
(5) Biofuel production shall contribute to the social and economic development of local, rural and indigenous peoples and communities.
(6) Biofuel production shall not impair food security.
(7) Biofuel production shall avoid negative impacts on biodiversity, ecosystems and areas of high conservation value.
(8) Biofuel production shall promote practices that improve soil health and minimize degradation.
(9) Surface and groundwater use will be optimized and contamination or depletion of water resources minimized.
(10) Air pollution shall be minimized along the supply chain.
(11) Biofuel shall be produced in the most cost-effective way, with a commitment to improve production efficiency and social and environmental performance in all stages of the biofuel value chain.
(12) Biofuel production shall not violate land rights

This standard remains controversial. Nevertheless, in Europe, governmental policies that require suppliers of palm oil to certify that their product contributes to a specified level of GHG emission reduction, and was not produced in areas of high biodiversity or carbon-enriched soils, are already being implemented (Block 2009a, b). These certification mandates are supported further by the actions and public statements of companies such as Unilever, the world's largest buyer of palm oil (Stareborn 2009).

Soybeans produced in countries like Brazil may also be subjected to sustainability certification in view of aggressive claims that soybean production is driving deforestation of the Amazon biome (Vera-Diaz et al. 2009; Fernandes 2009). However, other sources estimate that about 9% of the 420 Mha Amazon biome has been deforested since 1988, while current soybean production in the Amazon biome per se may total less than 1.5 Mha (Butler 2009). Butler (2009) also notes major soybean producing states such as Mato Grosso are included in the area known as 'Amazon Legal', but most soybean production in those Brazilian States occurs outside of the rainforests of the Amazon biome, on cerrado grassland or previously cleared land. Total harvested area for 2009 soybean production in Brazil is estimated at 22.7 Mha (USDA 2009b).

Unlike oil palm, soybean is geographically adapted to an extremely broad range of latitudes across Asia and the Americas. There are 13 recognized maturity classes for soybeans. These range from maturity groups 000, 00, 0, and I through X. Those varieties with the lowest number designation (000 to IV) are adapted to regions with shorter growing seasons, while soybeans in maturity groups V through X are grown at latitudes that are progressively closer to the equator. A majority of the soybeans produced in the US have maturities from I to IV (Wilcox 2004). However, soybean as well as other oilseed crops do not escape the sustainability inquisition based upon the environmental impact of biodiesel use on GHG emissions. Although additional research is needed, USDA estimates suggest that soy-biodiesel has a negative carbon profile, and is capable of reducing CO_2 emissions by 78% compared to fossil fuels. The sustainability of soy-biodiesel also may be affirmed by the 'fossil energy ratio (FER)', the amount of energy derived from biodiesel compared to the amount of energy that was required to produce biodiesel (assume B100). A fuel that has a high FER suggests a positive energy balance that should mediate a reduced contribution to atmospheric CO_2. In that regard, the most recent and most comprehensive energy life-cycle assessment of soybean biodiesel to date (USDA 2009c) has determined the FER of soybean oil to be 4.56; the FER of petroleum diesel is about 0.87. Moreover, the FER for soybean oil is expected to increase by 0.45 for every 1-bushel/acre increase in future US soybean yields. Among all of the major oilseeds, soybean alone exhibits the greatest long-term capacity for enhanced yielding ability compared to canola, sunflower, peanut, and cotton (Fig. 8.6). Advances in soybean genomics and biotechnology will ensure the continuation of a high-rate of improvement in yielding ability and product quality, to a substantially greater degree than any major agricultural crop in the foreseeable future. These attributes add to the list of strong positive indicators that attest to the sustainability of soy biodiesel.

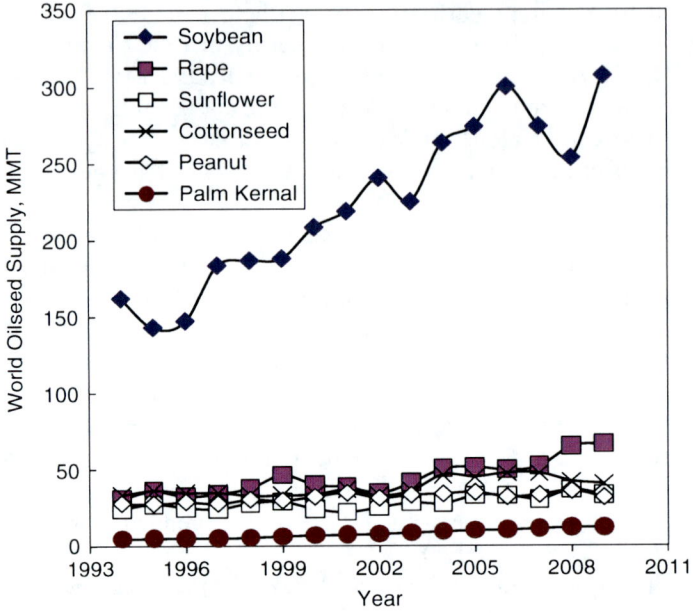

Fig. 8.6 Trends in world supply of major oilseeds (USDA 2009b)

8.2 Technology Trends to Further Enhance the Sustainability of Edible Oils for Biofuel

8.2.1 Physical Properties of Edible Oils

Several more factors contribute to the sustainability of biodiesel fuel based on comparison of the inherent physical properties of edible oils with those of diesel fuel. It is generally agreed that biodiesel conveys: (1) greater lubricity, which reduces engine wear; (2) greater heat of combustion, which not only indicates higher energy value but also more complete combustion; (3) more complete combustion, which helps reduce carbon monoxide, particulate matter, hydrocarbons and sulfur oxides in engine exhaust emissions, and (4) reduced engine noise pollution. However, the medium is not perfect. Problems may arise from higher viscosity in cool weather, poor oxidative stability and nitrous oxides (NO_x) in exhaust emissions (Knothe 2008). Low temperature properties may be measured as a function of melting point, cloud point, pour point, or cold-filter plugging point; the temperatures at which biodiesel transitions from a liquid to a solid state. Oxidative stability may be measured as a function of oil degradation, which has negative impact on energy value and exhaust emissions. Higher combustion temperatures enhance the formation to NO_x from atmospheric N_2 and O_2 during the operation of diesel engines, and contributes to the formation of smog and ozone (Demirbas 2008). The degree to which each of these three problems may affect the sustainability of biodiesel is relative to the fatty acid composition of the biodiesel feedstock. Fortunately, this relation suggests that all three problems may be mitigated by a common tactic.

Natural variation in the traditional or conventional fatty acid composition among oil crops listed in Table 8.3 is quite remarkable (Reeves and Weihrauch 1979). For example, oil (glycerides) from nutmeg butter, coconut, palm kernel, cocoa butter, palm and sheanut are endowed with high levels of saturated fatty acids, predominately myristic acid (14:0), palmitic acid (16:0) and stearic acid (18:0). Crop oils

Table 8.3 Traditional fatty acid distribution among plant sources of edible oils. *CN* Cetaine number (Bamgboye and Hansen 2008), CN for D1 diesel fuel = 47

Commodity (*Genus species*)	Fatty acid (% oil)[a]									CN[c]
	14:0	16:0	18:0	16:1	18:1	18:2	18:3	Other	Sats[b]	
Nutmeg butter (*Myristica fragrans*)	87	5	0	0	5	0	0	3	95	69
Coconut (*Cocos nucifera*)	18	9	3	0	6	2	0	62	92	64
Palm kernal (*Elaeis guineensis*)	17	9	3	0	12	2	0	57	86	63
Cocoa butter (*Theobroma cacao*)	0	27	35	0	35	3	0	0	62	68
Palm (*Elaeis guineensis*)	1	46	4	0	38	10	0	1	51	64
Sheanut (*Butyrospermum paradoxum*)	0	5	41	0	46	5	0	2	48	65
Hazelnut (*Corylus avellana*)	0	5	2	0	82	11	0	0	8	56
Olive (*Olea europaea*)	0	11	2	1	76	8	1	1	14	58
Avacado (*Persea americana* mill.)	0	11	1	3	71	13	1	0	12	56
Almond (*Prunus dulcis*)	0	7	2	1	73	18	0	0	9	55
Mustard (*Sinapis alba*)	2	4	1	0	13	18	7	55	7	54
Apricot kernal (*Purnus armeniacia*)	0	6	1	2	61	31	0	0	7	52
Canola (*Brassica napus*)	0	4	2	0	59	21	10	3	7	51
Peanut (*Arachis hypogea*)	0	10	2	0	48	34	0	5	17	53
Sesame (*Sesamum* spp)	0	9	5	0	41	43	0	0	14	51
Corn (*Zea mays*)	0	11	2	0	25	61	1	0	13	47
Soybean (*Glycine max*)	0	11	4	0	24	54	7	0	15	46
Sunflower (*Helianthus annuus*)	0	7	4	0	21	69	0	0	11	50
Walnut (*Juglans regia*)	0	7	2	0	23	56	11	0	9	44
Cottonseed (*Gossypium* spp.)	1	24	2	1	18	54	0	0	37	51
Grapeseed (*Vitis vinifera*)	0	7	3	0	17	73	0	0	10	44
Safflower (*Carthamus tinctorius*)	0	7	2	0	12	78	0	0	9	43

[a] *14:0* myristic, *16:0* palmitic, *18:0* stearic, *16:1* palmitoleic, *18:1* oleic, *18:2* linoleic, *18:3* linolenic acid
[b] Total saturated fatty acid concentration
[c] $CN = 61.1 + 0.088\ (\%14{:}0) + 0.133\ (\%16{:}0) + 0.152\ (\%18{:}0) - 0.101\ (\%16{:}1) - 0.039\ (\%18{:}1) - 0.243\ (\%18{:}2) - 0.395\ (\%18{:}3)$
Source (fatty acid composition): Reeves and Weihrauch (1979)

that exhibit comparatively high levels of monounsaturated fatty acids include: hazelnut, olive, avocado, almond, mustard, apricot kernel, canola, peanut, macadamia, and sesame. The emphasized characteristic of those oils is elevated oleic acid (18:1) concentration. A third grouping of glycerides exhibits relatively high levels of polyunsaturated fatty acids, primarily linoleic acid (18:2) and linolenic acid (18:3). These oils include: corn, soybean, walnut, sunflower, cottonseed, grapeseed, linseed, Perilla, chia and safflower. This wide range of natural genetic diversity in fatty acid composition among crop genera consequently imparts differences in functional properties that are important in the manufacture of food products as well as biodiesel. For example, the melting point of palm oil typically occurs at about 35°C. Although there are exceptions, oils that exhibit elevated oleic acid concentration, such as olive and peanut, melt between –6°C and 3°C (Weiss 1983). The melting point of oils with higher levels of polyunsaturated fatty acids, such as corn, soybean, sunflower and safflower, is usually between –17°C and –11°C (Table 8.4). These and other estimates of the functional properties of edible oils may vary with the type and degree of oil processing technology (Tan et al. 2002). However, the general association between melting point and level of saturation in edible oils prevails. A strong positive correlation also exists between oxidative stability and level of saturation. Oils naturally high in polyunsaturated fatty acids are more prone to degradation and off-odor formation than oils that are rich in saturated fatty acids. In addition, a strong positive correlation is found between the level of fatty acid saturation and ignition quality of biodiesel. Ignition quality in diesel fuel is measured by the 'cetane number' (CN), which is derived empirically from engine performance tests. Representative estimates of CN also may be calculated from the fatty acid composition of oil (Bamgboye and Hansen 2008). A high CN is associated with easy starting, low ignition pressure, smooth operation, lower engine noise or 'knocking', and decreased NO_x emissions (Knothe 2008). The calculated CN for various crop oils is listed in Table 8.3. In general, there is a strong positive correlation between CN and increased saturation and/or the number, type and position of carbon–carbon double bonds within unsaturated fatty acids. For example, the CN for palm oil (enriched in 16:0) is greater than olive

Table 8.4 Characteristic physical properties of major edible oils. *HHV* Higher-heating-value of heat of combustion (Demirbas 2008)

Commodity (Genus species)	Melting point (°C; Wiess 1983)	HHV (MJ kg^{-1})	Stability (min)[a]
Palm (*Elaeis guineensis*)	35	39.7	>1,000
Olive (*Olea europaea*)	–6	39.5	291
Canola (*Brassica napus*)	–10	39.5	366
Peanut (*Arachis hypogea*)	3	39.9	304
Corn (*Zea mays*)	–17	39.7	195
Soybean (*Glycine max*)	–11	39.4	146
Sunflower (*Helianthus annuus*)	–16	39.6	133
Cottonseed (*Gossypium* spp.)	–1	40.4	255
Safflower (*Carthamus tinctorius*)	–17	39.8	137

[a]Oxidative stability index (OSI) of refined oils at 110°C (Tan et al. 2002)

(high levels of 18:1), which is greater than safflower (high levels of 18:2). These associations represent the natural diversity among functional and physical properties of vegetable oils relative to biodiesel quality; and further suggest genetic modification of fatty acid composition as a feasible strategy for mitigating problems with cold flow, oxidative stability and NO_x emissions that mediate the utility and sustainability of edible oils in biodiesel.

8.2.2 Genetic Modification of the Physical Properties of Edible Oils

Considering the fatty acid composition of traditional soybean oil, the CN of individual fatty acid methyl esters were estimated by Knothe (2008) to be: 85.9 (16:0), 101 (18:0), 56.5 (18:1), 38.2 (18:2) and 22.7 (18:3). Respective melting points for these methyl esters are: 30°C (16:0), 39°C (18:0), −19°C (18:1), −35°C (18:2) and −52°C (18:3). Hence, a soybean oil feedstock with reduced 16:0 and elevated 18:1 concentration should enhance the sustainability of soy-biodiesel by balancing the tradeoff between cold flow and oxidative stability. This hypothesis is supported by experimental evidence with soybean lines that exhibit a variety of genetic modifications in fatty acid composition. For example, soybean oil with a low 16:0 plus high-18:1 phenotype exhibit the highest calculated CN (Table 8.5). Corroborating evidence has shown that (1) a low-16:0 trait in soybean oil reduces crystallization of biodiesel blends at cold operating temperatures (Johnson and Hammond 1996); (2) a high-18:1 trait in soybean oil confers superior oxidative stability (Knowlton 1999); and (3) a high-18:1 trait in soybean oil significantly reduced NO_x emissions compared to conventional soy-biodiesel and commercial diesel fuel (Tat et al. 2007). Therefore, these remaining questions concerning the sustainability of an edible oil feedstock for biodiesel may be addressed through genomic and biotechnical approaches. Indeed, Table 8.6 shows that experimental germplasm with 18:1 concentrations that greatly exceed the conventional or normal

Table 8.5 Fatty acid composition of genetically modified soybean oils

Soybean (*Glycine max*)	Fatty acid (% oil)						CN[a]
Modified trait oil	16:0	18:0	18:1	18:2	18:3	Sats[b]	
High oleic	4	3	85	6	2	7	57
Mid-oleic	8	4	62	24	2	12	54
High-stearic	10	25	15	42	8	35	52
Low-linolenic	12	4	24	59	1	16	48
High-palmitic	18	4	17	53	8	22	47
Conventional	11	4	24	54	7	15	46
High-linolenic	12	4	11	58	15	16	43
Low-palmitic	3	2	25	62	8	5	43

[a]See Table 8.4
[b]Total saturated fatty acid concentration

Table 8.6 Commodities with high-oleic acid trait options

Commodity	2009 World oil consumption[a]		Trait option	
			Normal	High
	MMT	%	% Oleic acid	
Palm	45.0	33	36	59
Soybean	37.3	27	24	84
Rapeseed	21.3	16	57	89
Sunflower	11.1	8	29	84
Peanut	4.7	3	55	76
Cottonseed	4.7	3	13	78
Olive	3.0	2	76	76
Safflower	1.1	1	10	81

[a]Total 2009 World Oil Consumption, 135.9 MMT (USDA 2009b)
Sources: Trait option (Liu et al. 2002b). *MMT* Million metric tons

option have been developed for each of the major oilseed crops (Liu et al. 2002b). Some commodities, like sunflower, have moved forward aggressively to convert all varieties to mid-oleic (NuSun®) or high-oleic types. Genomic and biotechnologies will help expedite varietal development with this modified trait as the other commodities move forward to capitalize on this market opportunity.

8.2.3 Development of Markets for Edible Oils with Modified Traits

Regardless of how compelling the need to develop alternatives to petroleum diesel, it is unreasonable to assume that genetically embellished traits might be implemented commercially for the sole reason of enhancing the sustainability of biodiesel. However, these benefits might accrue indirectly through strengthened consumer demand for more healthful and nutritious food products. Throughout the past 50 years, USDA and other nutritionists have conditioned public opinion on the dietary health benefits of fats and oils. The association of dietary saturated fatty acids with incidence of arteriosclerosis has now been extended to unsaturated fatty acids with process-induced rearrangement of molecular structure. The implication of trans-fatty acids that are formed in partially hydrogenated oils with an elevation of serum levels of low-density lipoproteins (LDL or 'bad' cholesterol) eventually led to the FDA Final Rule: *Trans Fatty Acids in Nutrition Labeling, Nutrient Content Claims, and Health Claims* in 2003 (USFDA 2003). The Final Rule became effective in January 2006. Public disclosure of trans-fat levels in vegetable shortenings, margarines, crackers, candies, cookies, snack foods, fried foods, baked goods, and other processed foods led to over reaction in the public good. Certain cities and state governments imposed bans on restaurant foods that contained trans-fat, without informed thinking on how menus might be depleted. Turmoil ensued as food manufacturers reformulated product lines with alternatives to partially hydrogenated oil while trying to preserve desired characteristics that consumers associated with their products (Kodali and List 2005). As this scenario played on, the outcome had a very unsettling impact on commodity markets for vegetable oils. In effect, enactment of the 'Final Rule' precipitated about a 2.2 billion pound net loss in US domestic soybean oil consumption during a time when food manufacturers sought alternatives to partially hydrogenated oils (USDA 2009b). Oilseed processing technologies at the time offered no way to mitigate this trend.

A bold but effective countermeasure taken by the US soybean industry was the introduction of soybean oil with genetically modified linolenic acid concentration by Pioneer/DuPont/Bunge (Treus^{TM}), Monsanto/Cargill (Vistive^{TM}), and Iowa State University (Asoyia UltraLow-Lin^{TM}). Soybean oils with 1.0–3.5% linolenic acid—about the same level as partially hydrogenated soybean—had been shown to produce better tasting fried foods with less room odor than partially hydrogenated soybean oil (Mounts et al. 1994). Thus these low-linolenic soybean oils provided the US industry with much needed flexibility to deal with the *trans*-fat issue. They also helped the industry to develop a production and market infrastructure that could accommodate new raw commodity streams for additional edible oils with other genetically modified quality traits.

Next generation soybean oils that feature higher 18:1 levels and lower levels of 16:0, 18:2 and 18:3 are now entering the vegetable oil market with the assistance of organizations such as *Qualisoy*. These new products include: Plenish^{TM}, genetically engineered high oleic acid oil (85% 18:1) from Pioneer/DuPont; and Vistive-Gold^{TM} with about 75% oleic acid, 15% linoleic acid, and less than 3% linolenic acid, from Monsanto. Qualisoy is a unique coalition of 22 members representing all parts of the soybean value chain: research, farmers, seed companies, processors, food and feed manufacturers, and trade associations (http://www.qualisoy.com/). This coalition promotes awareness of advances in relevant technologies to encourage commercial production and industry adoption of soybeans with genetically modified composition. Even at this early stage of product availability, USDA statistics show that soybean oils with genetically modified traits that do not require hydrogenation have blunted the recent decline in US domestic soybean oil consumption. Collectively, these oils are projected to claim a 40+% share of the US edible market for soybean oil, and to help restore growth in domestic soybean oil consumption to the 'pre-Final Rule' rate by 2020 (Fig. 8.7).

8.3 Advances in Genetically Modified Oil Trait Technology in Major Oilseed Crops

8.3.1 Biological Basis for Trait Modified Oils

Molecular genetic technologies have opened new insights into the biological mechanisms that govern fatty acid composition in many plants and oilseed crops. Genes that encode nearly every enzyme in the fatty acid and glycerolipid synthetic pathway of plants (Fig. 8.8) have been cloned and sequenced (Ohlrogge and

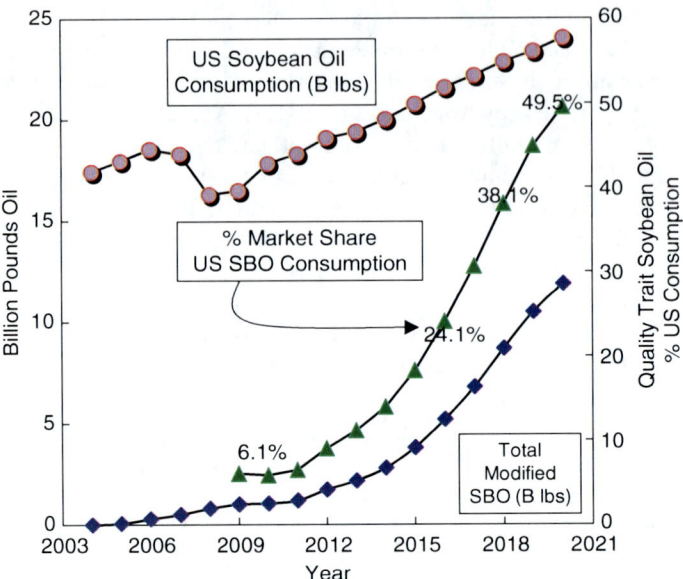

Fig. 8.7 Anticipated impact of modified trait soybean oils on US domestic use (USDA 2009b; Qualisoy Value Chain Analysis, personal communication)

Jaworski 1997). Transgenic modification of these genes have revealed underlying biological mechanisms, such as confirmation of the pathway for linoleic and linolenic acid synthesis in soybean (Hitz et al. 1995), and advances in protein chemistry have helped envision the functional structure of enzymes, such as acyl-desaturases (Shanklin et al. 1997). Recently, a roadmap to all genes in the soybean genome became available at http://soybeanphysicalmap.org/ (Shoemaker et al. 2008). Since sequencing and reassembly of other crop genomes may be some years away, a reference soybean genome sequence stands as the most significant scientific breakthrough achieved thus far in the advancement of genetic science in oilseed crop species. The high-resolution physical map of the soybean genome enables the identification of all the copies and structural variations of a gene for any trait of interest. In that regard, plant genomes usually contain multiple copies of a given gene, and each of those copies may be distinguished by a mutation or variation in the nucleotide alignment at a point within a gene sequence. These mutations can be manifested as point-mutations (single nucleotide polymorphisms, SNP), sequence deletions, inversions or additions. All plants contain genes with natural mutations in one form or another. When mutations occur within a gene sequence at a position that affects the function or activity of the transcribed gene product (usually an enzyme) in a manner that produces a heritable modification of a phenotypic trait, the modified gene sequence is called a recessive allele. Alleles give rise to genetic diversity and confer phenotypic differences among the germplasm accessions that constitute a plant species. Study of the inheritance of alleles

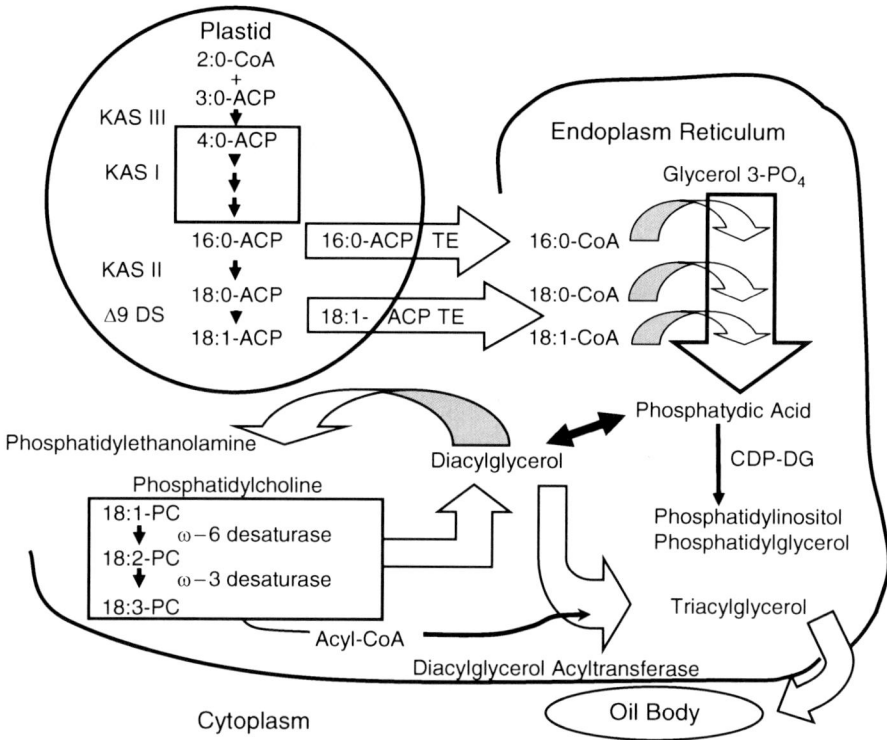

Fig. 8.8 Diagram of the fatty acid and glycerolipid synthetic pathway in plants

that mediate phenotypic traits is the historical basis of plant breeding, especially with respect to genetic modification of oil composition.

Modern innovations in crop genomics now make it possible to associate an allele unequivocally with an enzymatic step within a metabolic pathway and develop DNA markers that can identify the presence or absence of the allele among progeny in a breeding population that is segregating for the trait (Wilson 2004). For example, in soybean, additive genetic effects of four recessive alleles of the FAT B gene (which encodes the 16:0 ACP thioesterase, the enzyme that converts 16:0-ACP to 16:0-CoA) can reduce palmitic acid concentration from about 11% to less than 4% (Cardinal et al. 2007). Two recessive alleles of the gene that encodes the 3-ketoacyl ACP synthase (KAS II), the enzyme that converts 16:0-ACP to 18:0-ACP, mediates a high-16:0 phenotype, which may range from 17% to 40% of soybean oil (Aghoram et al. 2006). Three alleles of the gene that encodes the Δ-9 18:0-ACP desaturase, which mediates the level of 18:1-ACP, usually contribute to a high-stearic phenotype ranging from 8% to 24% of soybean oil (Zhang et al. 2008).

As mentioned earlier, a transgenic approach was used to confirm the mechanism of oleic acid synthesis in soybean oil (Hitz et al. 1995). Co-suppression of the FAD2 gene, which encodes the ω-6 desaturase (Δ-12 oleyl-phosphatidylcholine desaturase), limited gene expression and significantly increased oleic concentration up to about 85% of soybean oil at the expense of linoleic acid. However, it also is possible to achieve higher levels of oleic acid via a genomic approach. Inspection of the soybean genome has revealed at least three different FAD2 genes in soybean, each with one to three recessive alleles [*Glycine max* (Gm)FAD2-1A, GmFAD2-1B, GmFAD2-2A, GmFAD2-2B, GmFAD2-2C, GmFAD2-3A]. Appropriate combinations of these recessive alleles will give a phenotype of between 40% and 70% 18:1 (Schlueter et al. 2007). Finally, three different FAD3 genes encode the ω-3 desaturase in soybean. Recessive alleles of these genes (GmFAD3A-1, GmFAD3A-2, GmFAD3B-1, GmFAD3C-1, GmFAD3C-2) mediate lower levels of linolenic acid from 1% to 3.5% of soybean oil (Flores et al. 2008). By the same token, multiple genes and alleles for fatty acid traits also are being discovered in canola (Health Canada 1999), peanut (Chu et al. 2009), and cottonseed (Liu et al. 2002a).

Obviously, the genetic regulation of these traits is more complex than could be previously imagined. This knowledge helps explain why it has been relatively difficult to breed crops with genetically modified oil composition using conventional methods. For example, if a trait required the combination of only three recessive alleles, a breeder relying on random selection would have less than a 1% chance of finding the desired genotype in a F3 population. Those odds may be improved dramatically if the breeder has access to and uses DNA markers for each of the segregating alleles. Allelic markers also are needed to maintain traits during the subsequent development of new elite cultivars. The ability to track specific genes with DNA markers can significantly reduce the time it takes to develop a new crop variety for commercial production. For example, the addition of a high oleic trait to an established peanut cultivar by conventional breeding usually requires about 10–15 years, whereas with marker-assisted selection the new high-oleic peanut variety can be delivered in about 2 years. Such progress is timely considering current trends toward strong consumer demand for oils with higher oleic acid concentration—a trait that is perceived to convey nutritional and human health benefits. As mentioned previously, commodities that account for about 94% of world consumption of edible oils have developed high-oleic options (Liu et al. 2002b), many of which are entering commercial production.

8.3.2 Modified Oil Traits in the Commercial Pipeline

The material presented in the following discussion is not intended to be inclusive of all possible efforts to develop and market commodity oils with genetically modified traits. No preference is intended among proprietary products that might be mentioned. No guarantee or absolute claim is implied by listing a concentration reported for a trait. The length of time required for product testing, regulatory approvals and variety development of a trait for commercial production may be longer than anticipated. Some of these products may never achieve commercial production or sustainable market penetration. However, these data should project a reasonable estimation of the scope and objectives of industry and research efforts involving the genetic improvement of oil traits in major crop species.

8.3.2.1 Sunflower

Discovery of genes for a high oleic acid trait in the open pollinated cultivar 'Pervenets' enabled the development of sunflower oil that did not require hydrogenation to maintain oxidative stability (Dow AgroSciences 2007). High oleic sunflower oil is usually defined as having a minimum 80% oleic acid. Production and merchandizing of this oil is handled primarily on a contracted acre basis, with customer needs driving total acres. Other sources of high oleic genes have been used to develop NuSun® oil, a mid-oleic sunflower oil with less than 10% saturated fat, 55–75% oleic acid, and 15–35% linoleic acid. In 1995, the National Sunflower Association, with encouragement from Archer Daniels Midland Co. and Frito-Lay Inc. and the assistance of the USDA Agricultural Research Service, launched an initiative to promote the development of commercial varieties that met NuSun® specifications. The first commercial volumes were available in 1998 (Gupta 1998). In 2009, the National Sunflower Association reported that NuSun® varieties accounted for 90+% of total sunflower acres planted in the US. The latest innovation in sunflower oil composition is Nutrisun®, which features about 18% stearic acid, 70% oleic acid, and 4% linoleic acid. This type of sunflower oil may be appropriate for applications that typically use cocoa butter and sheanut butter (Martinez-Force et al. 2006). All mid- and high-oleic sunflower hybrids in commercial production today have been developed through conventional breeding.

8.3.2.2 Canola

Canola varieties typically have about 10% linolenic acid, but this trait has been reduced to about 3% linolenic in cultivars released in the 1980s (Prévôt 1990). High oleic, low linolenic canola varieties that feature induced gene mutations or transgenic events are becoming available from a number of sources, such as Natreon™ and XCEED™ canola (Watkins 2009a).

8.3.2.3 Cottonseed

Genetic modification of cottonseed oil composition has been achieved by seed specific silencing of ghSAD-1, the gene that encodes the Δ-9 18:0-ACP desaturase, and ghFAD2-1, the gene that encodes the Δ-12 oleoyl-phosphatidylcholine desaturase or ω-6 desaturase (Liu et al. 2002b). RNAi-based silencing of ghSAD-1 increased stearic acid concentration from 2% to 40% of cottonseed oil; silencing of ghFAD2-1 elevated oleic acid concentration to 78% of total oil. The combination of these traits produced oil with 40% stearic plus 37% oleic acid. These innovations were reported by CSIRO about 10 years ago, but are still undergoing product evaluation (Watkins 2009b).

8.3.2.4 Palm and Coconut

The Malaysian Palm Oil Board reports development of experimental germplasm with transgenically modified FAD2-1 and Δ-9 18:0-ACP desaturase genes. These modifications are expected to produce high-oleic palm oil with reduced palmitic and elevated stearic acid concentrations. Commercial planting is expected by 2025. Conventional breeding methods also are being used to enhance oleic acid concentration, reduce lipase activity, and increase antioxidants such as β-carotene and vitamin E (Watkins 2009b).

8.3.2.5 Safflower

California high oleic safflower oil is available with about 77% oleic acid, 15% linoleic acid, and 7% total saturates. Apparently, few attempts have been made to commercialize other genetically modified oil traits in safflower (Watkins 2009c).

8.3.2.6 Peanut

Numerous varieties such as SunOleic 97R, Tamrun OL01, Olin, Georgia 04S, and Florida-07 have been derived from F435 (high 18:1 germplasm) by conventional breeding (Knauft et al. 1999). Two genes, ahFAD2A and ahFAD2B mediate oleic acid synthesis in cultivated (allotetraploid) peanut. F435 contains a natural recessive mutation in ahFAD2A. Two different recessive mutations in ahFAD2B have been identified thus far among US cultivars. The combination of the recessive allele for ahFAD2A with either of the recessive alleles for ahFAD2B, conditions the expression of 74% to 84% oleic acid and 2% to 8% linoleic acid (Chu et al. 2009). DNA markers that hybridize with the specific nucleotide sequence for each mutation in these alleles are now being used by peanut breeders to expedite the industry strategy to incorporate the high-oleic trait into all future US peanut varieties.

8.3.2.7 Soybean

Soybean oil with 1% to 3.5% linolenic acid is available as Treus^{TM} from Pioneer/DuPont/Bunge, Vistive^{TM} from Monsanto/Cargill, or Asoyia UltraLow-Lin^{TM} from Iowa State University. The next modified trait to emerge from the Qualisoy pipeline is Plenish^{TM}, a genetically engineered high oleic acid oil (85% 18:1) from Pioneer/DuPont. Initial commercial production of Plenish^{TM} is expected in 2010. In following years, Monsanto's Vistive-Gold^{TM} with about 75% oleic acid, 15%

linoleic acid, and less than 3% linolenic acid should debut in 2013 to 2015. Various universities in the US and Canada also may be planning public releases of soybean varieties exhibiting some of these modified traits (Watkins 2009c).

8.4 Advances in Genetically Modified Oil Trait Technology in New or Underdeveloped Oilseed Crops

Prior to the twentieth century, animal fats, principally lard, tallow and butter were the main available source of edible fats and oils, at least in the western world. For reasons already mentioned, liquid oils were preferable to fats as a fuel for the first diesel engine. However, peanut and cotton probably were the only edible vegetable oils available to Rudolph Diesel. At the time, limited palm oil production was sequestered in Western Africa. Sunflower oil was unknown in Europe. Soybean was not a commercial crop in the US until the early 1950s. Canola was not developed until the late 1960s in Canada. Obviously, there has been tremendous change within the fats and oils industry in a relatively short period of time. However, it also is apparent that the major sources of edible oil today were considered 'new' crops just a few decades ago. Understandably, the dynamics of the oilseed situation invites speculation of what may lay ahead in terms of crops that may emerge as significant contributors to global edible oil supply in the future.

8.4.1 New Crop Oils for Industrial Chemicals

8.4.1.1 Flax

Flax, *Linum usitatissimum* L., is in the *Linaceae* or flax family. This self-pollinated crop has a chromosome number of $2n = 30$. Flax or flaxseed is also called linseed. Seed color ranges from a deep brown to a light yellow (golden) determined by the amount of pigment in the outer seed coat. Yellow-seeded flax has a higher seed weight and oil concentration than brown-seeded flax (Diederichsen and Raney 2006). Golden flaxseed is preferred in Europe and Asia as well as in the US food market as it blends well in food ingredients (Berglund 2002).

Flax is an annual plant. It is usually sown in the same type of land and climate required for wheat and barley production. Flax can fit into a small-grain rotation. It requires a 50-day vegetative period, a 25-day flowering period and about 35 days to mature. Frost seldom kills flax seedlings. They tolerate temperatures down to –2°C for a few hours. Flax is grown commercially mainly in Canada, China, India, the US, Ethiopia, Bangladesh, Russia, Ukraine, France and Argentina (Wittkop et al. 2009). In the US, flax is grown mostly in the states of North Dakota, South Dakota, Minnesota and Montana. In 2000, oilseed flax production in Canada was reported to be 707,000 t (Berglund 2002).

Flax seed generally contains ~40% oil, 20% protein, 28% dietary fiber and 3.4% ash (http://www.flaxcouncil.ca 2009). Flax oil has approximately 46% linolenic acid (omega-3 fatty acid; Wittkop et al. 2009). It has long been known that flax seed hulls contain mucilaginous substances. Flax is a good source of fiber, both soluble and insoluble. The mucilage on flax hull comprises about 8% of the seed weight. It is primarily a mixture of polysaccharides composed of D-galacturonic acid, L-rhamnose, L-galactose and D-xylose units (Anderson and Lowe 1974; Erskine and Jones 1957; Fedeniuk and Biliaderis 1994). Flax is increasingly being recommended as a component of a healthy diets (Anonymous 2002; Oliff 2004). Flax is high in lignans, which appear to have anti-carcinogenic properties (Thompson et al. 1996). Evidence is accumulating that flaxseed consumption can have remarkable benefits to human health, leading a significant reduction in the incidence of heart disease and a number of types of cancer (Bloedon and Szapary 2004; Hutchins et al. 2001; Ogborn et al. 2002; Prasad 2009; Rafter 2002; Spencer et al. 2003; Yang et al. 2001). Flax is also reported to improve mental health. There is evidence that it can improve cognitive ability in mammals (Hartvigsen et al. 2004). Joshi et al. (2006) reported that supplementation with flax oil improves the outcome of attention deficit hyperactivity disorder (ADHD). An increasing number of foods containing flaxseed are being developed (Lee et al. 2004; Shearer and Davies 2005; Warrand et al. 2005).

Flax has been used in food in Europe, Africa and Asia since 5000–8000 BC, and flax fiber has been used for linen cloth and many other uses. Flax was first brought to North America for its stem fiber for use in making linen and paper. It has been grown in the US and Canada as a commercial oilseed crop since early colonial times and was a major crop and source of fiber for clothing for settlers from the old world (Berglund 2002; Thomas Jefferson Agricultural Institute 2007). Currently, flax seed is used as human food and animal feed and as a source of flax oil. Ground or whole flax seed can be added to almost any baked product and adds a nutty flavor to bread, waffles, cookies or other products. The addition of whole flax seed or linseed oil into animal diets increases the nutritional value of the resulting eggs (Bean and Leeson 2003; Grobas et al. 2001; Pekel et al. 2009), pork (Hoz et al. 2003) and milk (Gulati et al. 2002; Petit 2002; Petit et al. 2002; Ward et al. 2002). Flax is increasingly being recognized as an outstanding major dietary component for healthy fish (high in ω-3 fatty acids) production (Bell et al. 2004; Bendiksen et al. 2003; Izquierdo et al. 2003; Kiron et al. 2004; Montero et al. 2002; Regost et al. 2003; Rollin et al. 2003; Zheng et al. 2004). The health of consumers of animal products such as meat, eggs and butter can be improved when such animals are fed significant amounts of linseed (Weill et al. 2002).

Industrial uses of linseed are expanding (Rakotonirainy and Padua 2001). Linseed oil has been used as a drying oil in paints (Lazzari and Chiantore 1999). It is also epoxidized and used as reactive diluents for coatings (Ashby et al. 2000; Muturi et al. 1994). Flax is also a source of biodegradable plastics (Wrobel et al. 2004) because natural fibers have the advantage of low density, low cost and biodegradability (Li et al. 2007). The main disadvantage is poor compatibility between the flax fiber and the matrix. There are studies on preparations and modifications of flax fiber to improve composite properties (Arbelaiz et al. 2005; Liu et al. 2006). Flax fiber is processed and used for other products, for example, cigarette paper, pulp and paper, and erosion control mats (Berglund 2002).

8.4.1.2 Camelina

The crucifer oilseed plant camelina, *Camelina sativa*, also known as false flax or "gold of pleasure" is a member of the Brassicaceae family but has similar applications as flax (Wittkop et al. 2009). It may have been among the earliest cultivated plants, and Hovsepyan and Willcox (2008) found evidence of its production as early as the sixth millennium BC in the Ararat valley in Armenia, or at least *Camelina microcarpa* also known as false flax. Camelina is a self-pollinating spring-sown crop that can be attractive as an alternative crop for tight crop rotations. Seed yields of camelina varieties can reach ~ 2 Mt ha^{-1} and an oil content of 44–46% (Wittkop et al. 2009). Camelina is relatively rich in α-linolenic acid, with genotypes ranging from 25% to 42% C18:3, whereas the erucic acid concentration is low, ranging from 2% to 6% (Vollmann et al. 2007). It is reported to have better yield stability than canola in areas with limited water or nutrients (Wittkop et al. 2009) but otherwise can be grown in areas canola does well (Carlsson 2009). Camelina meal contains the secondary plant metabolites, glucosinolates common in *Brassica* crops including canola. These metabolites can considerably reduce feed performance in animals (Pekel et al. 2009). Liu et al. (2008) recently report on a simple floral dip transformation system for Camelina that can facilitate the genetic improvement of this oilseed.

8.4.1.3 Crambe

There is interest in Crambe (*Crambe abyssinica*, L.; $n = 45$) as an industrial oil as its seed oil contains up to 60% erucic acid, which has useful chemical properties (Carlsson 2009). With 34 species, the *Crambe* genus is second only to *Brassica*, which includes canola in numbers of species within the tribe Brassiceae of the Brassicaceae family (Prina and Martinez-Laborde 2008). *Crambe abyssinica* is thought to have originated in the Abyssinian highlands of Ethiopia as its Latin name implies (Warwick and Gugel 2003). Crambe oil is some 10% higher in erucic acid than high erucic acid rapeseed oil. This is useful as a lubricant or formulations in the textile, steel and related metal industries, in drilling oils and marine lubricants. Erucic acid can be oxidatively cleaved into brassylic acid, which can be converted into polyesters (Warwick and Gugel 2003). However the purity of erucic acid in such applications is important and obtaining good quantities of relatively pure erucic acid is challenging.

Crambe is adapted to cool temperate areas. Commercial production began in the northern US Great Plains in 1990, and production reached 24,000 ha in North Dakota by 1994. It has greater drought tolerance than canola, reaching almost that of cereal grains (Warwick and Gugel 2003). Crambe seeds have oil contents up to 38% and oil yields of up to 1,000 kg ha^{-1} have been reported. In addition to erucic acid, crambe oil contains 16% oleic, 9% linoleic, 5% stearic and 3% palmitic acids. Major research needs are focusing on increased seed yield, winter hardiness and development of transformation and metabolic engineering protocols.

8.4.1.4 Pennycress

Field pennycress, *Thlaspi arvense* L., also known as stinkweed or French-weed, is yet another member of the Brassicaceae family with oilseed potential. It is a winter annual native to Eurasia but is now naturalized throughout temperate regions of the world including North America. It is a pioneer plant of disturbed soils and a common agricultural weed (Moser et al. 2009). Pennycress seeds and leaves contain high glucosinolate levels that can be toxic to animals and the meal therefore cannot be used in animal feed. The seeds have oil contents of 20–36% containing ∼33% erucic acid. The moderately high erucic acid content make it unsuitable as an edible oil but this adds to its potential value as a biodiesel source. Biodiesel (methyl esters) prepared from pennycress oil is reported to have a CN value of 59.8, i.e. well above that of regular canola oil (Table 8.3).

Pennycress is an important weed of grain, canola and forage fields in cool temperate areas of the world such as the northern prairies in North America (Moser et al. 2009). Fields heavily infested with pennycress have been reported to produce seed yields as much as 1,345 kg/ha with few inputs and wild population yields are reported to produce 1,120–2,240 kg/ha. Application of modern breeding to pennycress improvement should increase seed yields readily. This weed might be able to be turned into industrial oilseed crop and grown as a winter annual double cropped with corn or soybeans.

8.4.1.5 Lesquerella

Lesquerella fendleri (A. Gray) S. Wats, sometimes known as bladder-pod, is another oilseed in the Brassicaceae family. The main interest in lesquerella is its high content of the C20:1 hydroxy fatty acid, lesquerolic acid (14-hydroxy-cis-11-eicosenoic acid; Blackmer and Byers 2009; Jenderek et al. 2009; Thompson and Dierig 1994). The genus *Lesquerella* contains more than 90 species but only *L. fendleri* is currently being developed into a commercial crop (Blackmer and Byers 2009; Jenderek et al. 2009; Thompson and Dierig 1994). Many of the species in this genus, including *L. fendleri*, are native to the southwest US and northern Mexico. *L. fendleri* seeds contain ∼20–30% oil and ∼55–60% lesquerolic acid. The relatively high long-chain hydroxy fatty-acid content of lesquerella oil makes it useful for a number of industrial products, including lubricants, heavy duty detergents, soaps, grease thickeners, inks, coatings, plastics, resins, adhesives, waxes and cosmetics. These uses are similar to those of castor oil, which contains 90% or more of an 18 carbon hydroxy fatty acid. Lesquerella meal contains 30–35% protein that is rich in lysine and, unlike castor seed meal, appears promising as a protein supplement for animal feed (Dierig 1995). Many *Lesquerella* species accumulate various hydroxy fatty acids in their seed oil and *L. pallida* and *L. lindheimeri* average >80% lesquerolic acid (C20:1 OH), *L. perforata*, *L. stonensis*, *L. densipila*, *L. lyrata* and *L. lescurii* average >40% densipolic acid (C18:2 OH) and *L. auriculata* and *L. densiflora* average >30% auricolic acid (C20:2 OH; Dierig 1995).

In the wild, Lesquerella occurs mainly on well-drained calcareous soils derived from limestone 600–1,800 m above sea level. It grows best in areas with 250–400 mm rainfall. Most current production is in Arizona, where it is planted in late autumn, flowers in late winter and is harvested in late spring (Dierig 1995). Seed yields of 950–1,120 kg/ha are reported, with an oil content of 21%. As a newly domesticated crop, it is anticipated it can be improved by breeding. Efficient protocols for *Lesquerella* transformation and regeneration have been developed (Wang et al. 2008) that should further facilitate improvement of this new crop.

8.4.1.6 Jojoba

Jojoba, *Simmondsia chinensis*, is a particularly interesting "oilseed" native to the Sonoran desert of southwestern US and northern Mexico (Ash et al. 2005). Jojoba isthe only plant known to accumulate wax esters, although a number of prokaryotic organisms are known to store hydrocarbon in wax esters (Waltermann and Steinbuchel 2005). Jojoba has recently been placed in a new monotypic family: *Simmondsiaceae*. Like many desert plants, it is a true xerophyte and can store substantial amounts of water in swollen stems and leaves. Roots are reported to grow to depths of 9 m. It can withstand severe droughts and very high temperatures. However, unlike many xerophytes, jojoba does not have crassulacean metabolism but utilizes the C3 pathway instead (Ash et al. 2005). The plant is a slow-growing, evergreen, woody perennial shrub with a life span of 100–200 years with mature plants growing from 0.6 to ~4.6 m in height. It is dioecious and often dimorphic.

The seeds contain ~50% of a light yellow liquid wax known as jojoba oil that consists of straight-chain esters of long-chain monounsaturated fatty acids and alcohols with esters containing 40 and 42 carbons comprising 80% of the wax (Le Dréau et al. 2009). Its excellent emolliency, moisturization, and oxidative stability provide unique properties making it an outstanding natural cosmetic lipid (Ash et al. 2005). Jojoba oil is similar to sperm whale oil and is also used in pharmaceuticals, dietetic foods, animal feed, lubricants and polishes. It can provide excellent biodiesel but not at competitive costs to date (Le Dréau et al. 2009).

Only about 7,000–10,000 ha jojoba, mainly in Argentina and the US, are reported to be in cultivation (Ash et al. 2005; Benzioni 2006). Promising breeding lines have been reported to yield ~5,260 kg/ha seed in Israel but yields can vary greatly from year to year, and good yields occur only after at least 4 years from planting (Ash et al. 2005; Benzioni 2006). Jojoba seeds can sell for US $2–$3/kg. The meal remaining after oil extraction contains 20–25% protein and another

10–20% of compounds known as simmondsins. Simmondsins are unique to jojoba and are potent food/feed-intake inhibitors that cause starvation in animals fed jojoba meal, thus preventing such traditional uses of jojoba meal (Benzioni 2006). Perhaps in carefully controlled doses, jojoba meal or simmondsin extracts might be used to help treat obesity in humans.

8.4.1.7 Perilla

Perilla frutescens (L.) Britton, is a member of the Lamiaceae or mint family. Perilla is thought to have been first domesticated in China and has been grown in Asia since ancient times. It exists as two distinct varieties, *P. frutescens* var. *crispa* and *P. frutescens* var. *frutescens* (Pandey and Bhatt 2008). These might be considered subspecies but both have $2n = 40$ and readily cross hybridize. The plants exist in purple and green leaf forms. *P. frutescens* var. *crispa* is grown as an ornamental and spice and has edible leaves known as *shiso* in Japan. However, it has been reported to be toxic to horses. *P. frutescens* var. *frutescens* has much larger seeds (~2 mg/seed), and has been grown as an oilseed in East Asia and the Himalayas. This oilseed type contains ~30–50% oil with a linolenic acid content of 60% (Park et al. 2000; Rao et al. 2008). Seeds of some species of the Lamiaceae or mint family are particularly high in α-linolenic acid (Rao et al. 2008). Perilla oil has been used as a drying oil similar to linseed and tung oils in paints, varnishes, linoleum, coatings, printing inks and for waterproofing cloth (Brenner 1995).

Because there have been very few efforts to improve *Perilla* through modern breeding, genotypes of *Perilla* crop still occur as landraces in farmer's fields in several areas of South Korea (Park et al. 2008). Nevertheless, it is one of the most important oil crops in Korea and in need of genetic improvement. Perilla breeding is underway in Korea (Park et al. 2002) and a genetic transformation system has been developed for Perilla improvement (Lee et al. 2005).

8.4.1.8 Chia

Salvia hispanica L. is a member of the Labiateae or Lamiaceae or mint family. A center of genetic diversity of chia is in the highlands of western Mexico (Cahill 2004). This annual seed crop has a chromosome number of $2n = 12$ (Estilai et al. 1990). Mean seed mass of domesticated chia is about 1.2 mg/seed (Cahill and Ehdaie 2005). Currently, chia is grown commercially in tropical and subtropical areas. Ayerza and Coates (2005) reported that chia is cultivated in Argentina, Bolivia, Colombia, Mexico and Peru, where latitudes range from 20°55′N to 25°05′S, and it is very frost sensitive. Chia seeds require wet soil to germinate, but once the seedlings are established, chia grows well with limited water. Chia grows well on soils containing widely varying levels of nutrients. Commercial chia seed yields of 500–600 kg/ha are reported (Coates and Ayerza 1996); however, some growers have obtained up to 1,260 kg/ha. Some experimental plots yield

2,500 kg/ha when irrigation and nitrogen fertilizer were applied. New lines for production of chia in the US have been developed at the University of Kentucky.

Chia seed is a good source of natural lipid antioxidants. Flavonol glycosides, chlorogenic acid and caffeic acid are found in chia extracts (Taga et al. 1984). The antioxidant activity of the fiber-rich fraction of chia flour was found to be higher than many cereals and similar to drinks such as wine, tea, coffee and orange juice (Vazquez et al. 2009). Chia seed coat is high in a fiber that becomes mucilagenous and expands considerably when soaked in water. The fiber consists of xylose, glucose and glucuronic acid monomers (Lin et al. 1994). Vasquez et al. (2009) reported that the fiber-rich fraction of chia flour has 56.5 g/100 g total dietary fiber content. The fiber-rich fraction water-holding capacity is 15.4 g/g.

Chia seed contains about 20% protein. Chia oil content ranges from 28.5% to 32.7% (Ayerza and Coates 2004, 2007). Chia is high in the $\omega 3$ fatty acid, α-linolenic acid, 18:3, at $\sim 60\%$. Chia diets dramatically decreased triacylglycerol levels and increased high density lipoprotein cholesterol and ω-3 fatty acid content in rat serum (Ayerza and Coates 2005). Dietary chia seed also improves adiposity and insulin resistance in dyslipeamic rats (Chicco et al. 2009). Diets supplemented with chia have been found to decrease risks from some types of cardiovascular diseases, cancers and diabetes. It has been reported that chia decreases tumor weight and metastasis number and also inhibits growth and metastasis in murine mammary gland adenocarcinoma (Espada et al. 2007). Long-term supplementation with chia attenuated a major cardiovascular risk factor and emerging factors safely beyond conventional therapy, while maintaining good glycemic and lipid control in people with well-controlled type 2 diabetes (Kreiter 2005; Vuksan et al. 2007, 2009).

Chia was used widely in pre-Columbian Mesoamerica for different purposes such as medicinal, culinary, artistic and religious use. The ethnobotany of chia in the sixteenth century is described extensively by Cahill (2003). Cultivation of chia was reduced drastically after Spanish colonization due to its religious significance in Mesoamerican cultures. Chia and chia oil is used as human food, animal feed, drying oil in paints, and as an ingredient in cosmetics. Chia leaf oil may be useful in flavorings or fragrances, and possibly pesticides since white flies and other insects avoid the plant (Ahmed et al. 1994). Broiler feed supplemented with chia seeds was shown to significantly lower saturated fatty acid content in white and dark meats (Ayerza et al. 2002). There have been studies on hens for the potential of chia diets as a source of ω-3 fatty acids in egg production (Ayerza and Coates 1999, 2000, 2001, 2002). A recent study was done to determine fatty acid composition and nutritive value of chia seed and vegetative parts as a possible source of polyunsaturated fatty acids for ruminants (Peiretti and Gai 2009).

8.4.1.9 Castor

Ricinus communis in the *Euphorbiaceae* family has a chromosome number of $2n = 20$ (Perry 1943). Castor is widespread as a wild plant through East and North Africa.

The diversity for many plant, fruit and seed characters is enormous and this has led to the assumption that cultivated castor is of African origin. As a perennial plant in the tropics, it can grow to 10–13 m tall with a stem diameter of 7.5–15 cm. As an annual plant in the temperate regions, it is usually ~1–3 m tall.

Castor seeds contain a potent toxin: ricin. Ricin is found only in seeds. Accumulation starts days after pollination and increases quickly until seed maturation. Ricin is degraded to below the level of detection 6 days after germination (Barnes et al. 2009). Ricin is a 6.6 kDa glycoprotein lectin that blocks protein synthesis via its ribosome inactivating properties, thus killing cells. Ricin toxicity and its mechanism of action are described extensively by Bagchi et al. (2009). The lethal oral dose in humans has been estimated to be 1–20 mg ricin/kg body weight (approximately eight beans). Castor also contains another glycoprotein lectin, the *R. communis* agglutinin. It is not directly cytotoxic but has an affinity to red blood cells, leading to agglutination and subsequent hemolysis. Ricinine is an alkaloidal toxin also found in the leaves and pericarp of the castor plant. In experimental mice models, ricinine causes convulsions and subsequent death. However, there are no reports of human ricinine poisoning (Audi et al. 2005).

Castor was in production as early as the mid-1850s in the central part of the United States. Yields of irrigated castor range from 2,242 kg/ha to 3,363 kg/ha, and some fields have produced 3,811 kg/ha to 4,035 kg/ha (Brigham 1993). Castor beans are processed throughout the world to make castor oil. India currently produces the largest amount of castor oil, followed by China and Brazil (Bagchi et al. 2009). India has 0.8 million ha of area under castor and 1.0 million t of production (Anjani et al. 2010). Li et al. (2010a) reported that there is a tremendous demand for castor in China. There are large areas of 3 million ha along coastal saline land that could be targeted for castor plantations in the future. Castor offers immense potential for improvement for water and salt stress tolerance, disease and pest resistance and toxin-free varieties through genetic engineering. Genetic engineering technique requires a good in vitro system. Tissue culture of castor has been reported (Reddy and Badahur 1989) but was not very efficient until more recent developments (Ahn et al. 2007; Ahn and Chen 2008; Kumari et al. 2008). The first successful castor transformation using *Agrobacterium* was reported in 2005 (Sujatha et al. 2008; Sujatha and Sailaja 2005). Semilooper (*Achoea janata* L.)-resistant transgenic castor has been developed (Malathi et al. 2006).

Even though castor grows naturally in the wild in Africa, these wild plants are usually too tall to harvest and seed capsules shatter when mature and dry. Improved commercial castor cultivars from America, Israel and South Africa were found to give good yield and suitable for production in Zimbabwe (Tongoona 1992).

Castor oil has a high level of ricinoleic (12-hydroxy-9-octadecenoic) acid (~900 g/kg) and a low level of oleic acid (~30 g/kg). However, through extensive screening of a world germplasm collection, the natural mutant line OLE-1 was found to have approximately 780 g/kg oleic acid while the ricinoleic acid content is as low as 140 g/kg (Rojas-Barros et al. 2004). A study was conducted to determine the inheritance of the high oleic/low ricinoleic acid content by crossing the OLE-1 line and a standard oleic/ricinoleic content line. F_2 segregation was consistent with the action of two independent major genes, which are designated *ol* and *Ml* (Rojas-Barros et al. 2005).

Castor seeds contain about 46–55% oil by weight. Castor oil is a viscous, pale yellow non-volatile and non-drying oil with a bland taste containing ~90% of the fatty acid ricinoleic acid (Yamamoto et al. 2008). The other major fatty acids are linoleic (4.2%), oleic (3.0%), stearic (1%), palmitic (1%) and linolenic acid (0.3%). The high ricinoleic acid content makes castor oil suitable for many chemical reactions. Suthar et al. (1991) and Ogunniyi (2006) reported that castor oil is used in isocyanate reactions to make polyurethane elastomers, polyurethane millable, castables, adhesives and coatings, interpenetrating polymer network from castor oil-based polyurethane and polyurethane foam. The oil is useful as a component in blending lubricants because it has high viscosity. Lipstick contains up to 80% castor oil.

Despite its high protein content, castor seed meal is not used as livestock feed due to the presence of toxic factors, ricin, ricinine and its properties as an allergen. A study was conducted to detoxify ricin in castor cake. It has been reported that autoclaving and lime treatment can completely destroy the toxins (Anandan et al. 2005). Advantages of castor oil biodiesel are the high caloric value, high CN, low contents of phosphorous and carbon residues. Disadvantages are that it has significantly higher viscosity at temperatures under 50°C, and possibly higher compressibility than the main biodiesel sources such as soy or canola. Its hygroscopicity causes relatively a high water content and thereby possible algae growth, filtration and corrosion problems. These properties complicate the use of castor oil as a fuel in engines (Scholza and da Silva 2008). There are research and development efforts to improve castor oil as fuel such as biodiesel (de Oliveira et al. 2004; Albuquergue et al. 2009; Jeong and Park 2009).

8.4.1.10 Jatropha

Jatropha curcas (L.), is a pan-tropical perennial shrub or small tree in the spurge or Euphorbiaceae family like castor (Divakara et al. 2010; Jongschaap et al. 2007; King et al. 2009; Makkar and Klaus 2009). The genus *Jatropha* contains 165–175 species in two subgenera, ten sections and ten subsections. The centers of diversity are dry areas of Mexico and NE Brazil. *J. curcas* is thought to have originated in Mexico and central America but has become naturalized in Africa and Asia. It is considered to be a diploid with $2n = 22$. Jatropha has generated considerable interest as a source of oil from land too dry or otherwise unsuitable for other oilseed production providing fuel and income in semi-arid areas. It is reported to have high resource use efficiency and productivity with low water input and is useful for controlling soil erosion (Kheira and Atta 2009; Makkar and Klaus 2009). Jatropha plants can reportedly survive up to 3 years with no rainfall and revive when rains resume. It is adapted mainly to tropical and subtropical regions but can reportedly withstand light frosts (Kheira and Atta 2009). For most accessions the entire plant, including the seed meal and oil, is toxic and is not eaten by livestock, and therefore Jatropha has been useful for fencing in livestock (Jongschaap et al. 2007). Such use for fencing was the main use farmers have made for Jatropha until the recent interest in additional renewable oil sources. Now, several million hectares of Jatropha have been planted in India and China for production of this plant for oil, mainly for biodiesel (King et al. 2009).

The main toxins in Jatropha seeds are curcin and phorbol-esters (King et al. 2009; Makkar and Klaus 2009). Curcin is a ribosome-inactivating lectin like ricin but curcin is not nearly as potent or toxic as ricin. The phorbol-esters, which are a group of diterpenes, are highly toxic and may also contribute to cancer formation. Non-toxic genotypes of *J. curcas* are known in Mexico but so far only toxic lines are being cultivated. It is estimated that mature Jatropha plantations can produce seed yields up to 4–5 t/ha resulting in ~1.5 t oil/ha. This is despite the fact that improved lines from sustained modern breeding are not yet available for Jatropha (King et al. 2009). Genetic transformation systems building on reliable organogenic and embryogenic regeneration systems are being established, which could facilitate Jatropha improvement (Divakara et al. 2010; Sujatha et al. 2008).

One report (Gerbens-Leenes et al. 2009a) suggests that, for entire plant biomass conversion to electricity, Jatropha and rapeseed have the highest (use the most water) water footprints (WFs), with sugar beet, corn and sugar cane having the lowest WFs of the crops evaluated (dedicated biomass crops such as Miscanthus or switch grass were not evaluated in their report). Several studies conclude that insufficient data is available at this time to draw such a conclusion (Gerbens-Leenes et al. 2009b; Jongschaap et al. 2009; Maes et al. 2009). The WFs for oil production among other oilseed crops and algae also remains to be determined.

8.4.1.11 Algae

The organisms known as algae include eukaryotic and prokaryotic groups. Eukaryotic algae account for more than half of the primary basis of the global food chain and are found from polar regions to deserts (Harwood and Guschina 2009). The majority of lichens involve a symbiotic relationship of algae with fungi. Nine divisions of eukaryotic algae are known, with the classification based mainly on pigmentation patterns. The groups with the most known species are the Chlorophyceae (green algae), Phaeophyceae (brown algae), Pyrrophyceae (dinoflagellates), Chrysophyceae (golden-brown algae), Bacillariophyceae (diatoms) and Rhodophyceae (red algae). Eukaryotic algae are particularly diverse in fatty acid compositions and many contain unusually high amounts of very polyunsaturated fatty acids (Harwood and Guschina 2009). Many marine algae contain high concentrations of the very long chain polyunsaturated fatty acids, arachidonic, eicosapentaenoic (EPA) and docosahexanoic (DHA) acids which are of keen interest in the health of humans and other animals. EPA and DHA are ω-3 fatty acids that are very important and often deficient in human diets. Fish, which accumulate high levels of

EPA and DHA ω-3 fatty acids, do not actually synthesize these molecules but accumulate them from the algae or other plants that make them at the base of the food chain. Since brain tissue is particularly high in DHA, and with the very large brain to total body mass ratio in humans, we benefit from high DHA levels in our diets, especially as infants. Human breast milk is correspondingly much higher in DHA than common milk sources such as cow's milk. Some marine algae, most notably the Thraustochytrids, accumulate unusually high levels of DHA, and biotechnology companies have been producing DHA from such algae for fortification of infant formula and other foods. Due to the high cost of DHA production in algae, oilseeds are being genetically engineered for production of this valuable fatty acid (Napier 2007). The main monounsaturated fatty acid in algae is palmitoleic acid (16:1 Δ9) rather than oleic acid (18:1 Δ9), which is the main monounsaturated fatty acid in higher plants. Palmitoleic acid is also of great interest in human health. Algae have also been used successfully to produce commercially other high value lipids such as arachidonic acid, which is also sometimes added to infant formula.

Algae have recently garnered considerable interest as a renewable oil source. This is based on estimates of potential oil yields per unit land area such as shown in Table 8.7 (Mata et al. 2010). For production of oil that can be achieved in the 1st year after planting, castor has the highest oil yields/ha, at about twice that of soybeans. Once the plantation trees reach maximum production, an oil palm can produce over eight-fold as much oil per hectare as soybeans. Some algal strains grown under certain condition can accumulate greater than 75% oil/dry weight (Mata et al. 2010). It has been estimated that high-oil algal strains can produce some 200-times the amount of oil vs soybeans per unit land area! This has spurred new investments in algal research since 2009. These estimates for algae are based on extrapolations that, to date, have not been shown to be scalable to any appreciable acreage.

In the 1960s, NASA invested in considerable research into algae as a food and protein source but this finally was never used. As a reaction to the oil price spikes in the 1970s, the US Department of Energy (DOE) invested US $25 million in developing algae as a source of liquid fuels. This included an outdoor test facility

Table 8.7 Comparison of potential yields among oil sources

Oil source	Oil production capacity (L ha^{-1} year^{-1})
Soybean	636
Jatropha	741
Camelina	915
Canola	974
Sunflower	1,070
Castor	1,307
Palm	5,366
Algae (low)	58,700
Algae (high)	136,900

in New Mexico consisting of two 1,000 m^2 high-flow-rate ponds. However, the program was terminated in 1996 as it was concluded that, although it might be technically feasible to produce cost-competitive biodiesel from algae, considerable long-term research and development is needed before this can be achieved (Liang et al. 2009; Mata et al. 2010). To date, a scalable and economically feasible system for algal biomass production has yet to be developed (Carlsson 2009; Liang et al. 2009; Philip and Al 2009). Among the technical hurdles to be overcome is light penetration in high density cultures. In fact, currently, compared with oil produced from algae grown photoautotrophically, the most economic algal oil production systems involve heterotrophic algal growth with plants providing the hydrocarbon source in the form of corn powder hydrolysate or glycerol from oilseed biodiesel production. Such heterotrophic production can increase oil yield by greater than 14-fold (Liang et al. 2009). As of 2009, global algal biomass production for high-value, low volume nutraceuticals and food supplements was only 5,000–10,000 t (Philip and Al 2009). Researchers are currently investigating the possibility of using algae to scrub CO_2 from coal burning in power plants to reduce the CO_2 emissions from this still abundant energy source. If a carbon tax is imposed on such coal burning, algal biomass production via this route might prove more economical than previously developed production systems.

8.4.1.12 Other species

Other industrial oil crops are being developed and grown, but only two additional plant groups will be briefly mentioned in this chapter. Meadowfoam, *Limnanthes alba*, has attracted attention because the seeds contain 20–30% oil consisting of over 98% long chain fatty acids, making the oil highly stable for various industrial applications such as in cosmetics, lubricants, rubber and plastics (Jenderek and Hannan 2009).

The genus *Cuphea* contains many species that accumulate high levels of medium-chain fatty acids in seed oil that are useful in detergents (Berti and Johnson 2008). *C. painteri* oil, for example, is ~75% caprylic acid; *C. carthagenensis* oil is ~80% lauric acid and *C. koehneana* oil is ~95% capric acid.

Tung, *Vernicia fordii*—a species in the spurge family—is native to southern China, Burma, and northern Vietnam. Tung is a small- to medium-sized deciduous tree growing to 20 m tall that bleeds latex if cut. Tung oil is derived from the seeds of the tree. Tung oil, also called China wood oil or nut oil, has been used traditionally in lamps in China. In modern times, it is used as an ingredient in paint and varnish, and also can be used as a motor oil. Tung has been introduced to Argentina, Paraguay, Thailand, and the US for oil production. Several cultivars have been released including 'Folsom', 'Cahl', 'Isabel', 'La Crosser', and 'Lampton'. The principle fatty acids of tung oil are: a-eleostearic acid (D9c, 11t, 13t linolenic acid), which accounts for about 72% of the oil and lignoceric acid (24:0) at about 10% of the total fatty acid composition (Sanford et al. 2009).

8 Engineering Status, Challenges and Advantages of Oil Crops 241

8.4.2 Biological Basis for Industrial Oil Traits

8.4.2.1 Triacylglcerol Biosynthesis

The hydrocarbon fueling triacylglycerol (TAG) accumulation in most oilseeds is sucrose. For many Rosaceae seeds, such as apple seeds, sorbitol is the main assimilate translocated from leaves (Nosarzewski and Archbold 2007). The main steps from delivery of assimilates from leaves to developing seeds to final seed oil, TAG, (and protein) accumulation is given in Fig. 8.9 (Bates et al. 2009; Bewley and Black 1994; Lonien and Schwender 2009; Ohlrogge and Browse 1995; Weselake et al. 2009). Assimilates move through the vascular connection to the mother plant through the funiculus, and move through vascular connections in the integuments and then into developing embryo tissue across the apoplastic space. The amino acids for seed protein synthesis, including storage proteins, arrive from leaves mainly in

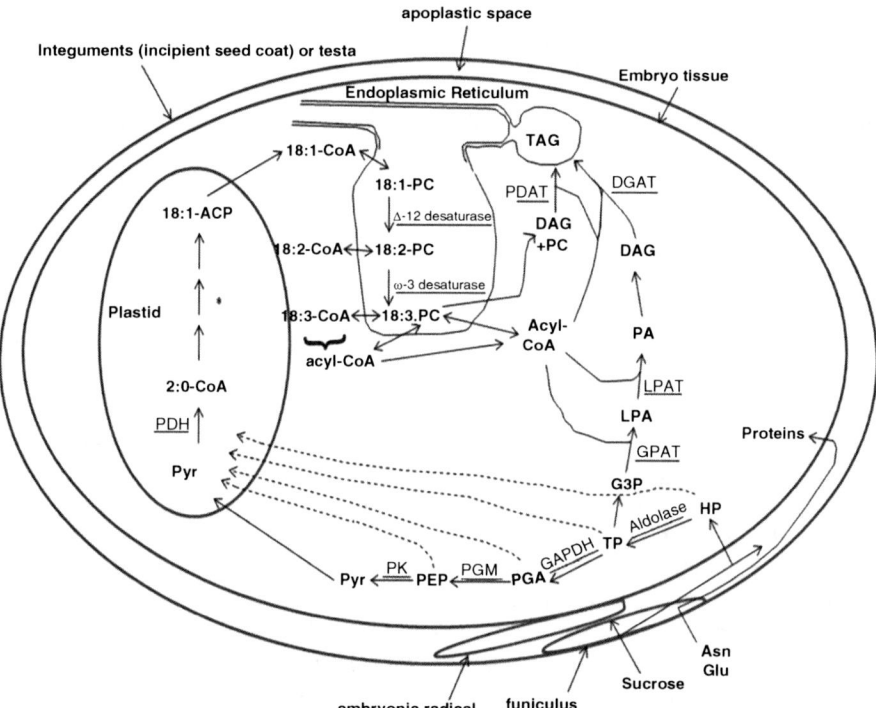

Fig. 8.9 Overall scheme from seed assimilates to final seed oil (and protein) accumulation. * See Fig. 8.8. *Underlined* Enzymes shown. *DGAT* Acyl-CoA:diacylglycerol acyltransferase; *GAPDH* glyceraldehyde 3-phosphate dehydrogenase; *G3P* glycerol-3 phosphate; *GPAT* glycerol-3 phosphate acyltransferase; *HP* hexose phosphate; *LPA* lysophosphatidic acid; *LPAT* lysophosphatidic acid acyltransferase; *PDAT* phosphatidylcholine: diacylglycerol acyltransferase; *PDH* Pyruvate dehydrogenase; *PEP* phosphoenolpyruvate; *PGA* 3-phosphoglycerate; *PGM* phosphoglycerate mutase; *PK* pyruvate kinase; Pyr Pyruvate; *TAG* triacylglycerol; *TP* triose phosphate

the form of the N-rich amino acids asparagine and glutamine. In nitrogen-fixing oilseeds such as soybeans and peanuts, fixed nitrogen is exported from nodules as ureides, which are reconverted into amino acids in leaves; ureides do not directly enter developing seeds. Sucrose is cleaved into hexose sugar monomers, and hexose phosphates can be cleaved into triose phosphates. Triose phosphates can be reduced to glycerol-3 phosphate, the backbone for glycerolipids, or oxidized to 3-phosphoglycerate. 3-Phosphoglycerate can be rearranged to phosphoenolpyruvate, which can be dephosphorylated to pyruvate. Intermediates from hexose phosphates to pyruvate and pyruvate itself can be translocated into plastids of developing seeds and pyruvate converted (decarboxylated) into acetyl-CoA (2:0-CoA). The acetyl-CoA then provides hydrocarbon for fatty acid biosynthesis to 18:1 in plastids but acetyl-CoA itself is not translocated into plastids from other compartments. Acetyl-CoA is a precursor to many molecules in plants and other organisms in multiple organelles. The first committed step of fatty acid biosynthesis is the conversion of acetyl-CoA into malonyl-CoA by acetyl-CoA carboxylase and then to malonyl-ACP by a transacylase. In most tissues of most eukaryotic organisms, including plants, malonyl-ACP is elongated in eight cycles, two carbon units at a time, via the fatty acid synthase complex, to palmitoyl (16:0)-ACP (or -CoA). The fatty acid synthase complex involves four different enzymatic reactions with each cycle starting with a condensation, followed by a reduction, a dehydration and a second reduction (Ohlrogge and Jaworski 1997). The condensation reactions are catalyzed by enzymes known as 3-ketoacyl-ACP synthases (KASs). The first condensation reaction going from acetyl-CoA to 3-ketobutyrate is catalyzed by KAS III, the reaction from butyryl-ACP (C4) to palmitoyl-ACP (C16) by KAS I and from palmitoyl-ACP to stearoyl-ACP (C18) by KAS II. The reaction stops at C16 and C18 fatty acids not only by virtue of the specificity of the KAS enzymes but also by the action of thioesterases (TEs), which hydrolyze the acyl-S-ACP thioester bonds. Some plants, such as coconuts, palm kernel and Cuphea, have unusual TEs, known as medium chain TEs, which stop the reaction at C8, C10, C12 or C14 fatty acid chain lengths, and these plants can accumulate medium chain fatty acids in their seed oil (Budziszewski et al. 1996; Voelker et al. 1996, 1992; Yuan et al. 1995).

Oil is biosynthesized during the second main stage of seed maturation (Goldberg et al. 1989; Harwood and Page 1994; Le et al. 2007), at which time the relevant biosynthetic enzymes are highly expressed. The major fatty acids of plants (and most other eukaryotic organisms) have a chain length of 16 or 18 carbons and contain from zero to three cis-double bonds. Five fatty acids (18:1, 18:2, 18:3, 16:0 and, in some species, 16:3) make up over 90% of acyl chains of structural glycerolipids of almost all plant membranes (Ohlrogge and Browse 1995). The nature of the acyl composition of the TAG is dependent on the availability of the fatty acids from the acyl-CoA substrate pool as well as the selectivity of the acyltransferases of the Kennedy pathway (Harwood 1997; Harwood and Page 1994) and possibly transacylases. These same five fatty acids are the main fatty acids present in plant oils in proportions generally quite different than that of membrane lipids.

A starting point of both membrane and TAG synthesis is the acylation of sn-glycerol-3-phosphate producing lysophosphatidic acid, catalyzed by glycerol-3-phosphate acyltransferase (GPAT). A second acylation of lysophosphatidic acid catalyzed by lysophatidic acid acyl transferase (LPAT) produces phosphatidic acid (PA; Fig. 8.9). The PA formed can be subsequently de-phosphorylated to diacylglycerol (DAG). The DAG then serves as a precursor for TAG. The third acylation step is catalyzed by transfer of acyl groups to DAG forming TAG. Phosphatidyl choline (PC) is a key intermediate in oil biosynthesis and plays a central role in the production of polyunsaturated fatty acids by serving as a substrate for Δ-6, Δ-9, Δ-12, and Δ-15 desaturases (Jackson et al. 1998). LPAT are known to display specificity for substrates with certain fatty acids and can be important in determining final TAG composition (Franzosi et al. 1998).

Another reaction that appears to be involved in TAG accumulation is the reversible conversion of PC into DAG in presence of cytidine diphosphate (CDP) choline transferase. Slack et al.(1985) gave indirect evidence for the reversibility of PC by labeling studies in vivo with linseed cotyledons and in vitro with safflower cotyledons. When sunflower microsomes were incubated with radiolabeled PC, the radioactivity was incorporated progressively into DAG. When the concentration of the microsomal protein was increased, the activity also increased, indicating the reversible reaction of choline transferase in sunflower (Triki et al. 1998). A soybean cDNA encoding an aminoalcoholphosphotransferase (AAPTase) that demonstrates high levels of CDP-choline:sn-1,2-diacylglycerol cholinephosphotransferase activity was isolated by complementation of a yeast strain deficient in this function (Dewey et al. 1994). AAPTases utilize diacylglycerols and CDP-aminoalcohols as substrates in the synthesis of the main membrane lipids phosphatidylcholine and phosphatidylethanolamine, and can possibly affect DAG pools for TAG synthesis. In animals, at least TAG can be synthesized by two major pathways: the glycerol 3-phosphate pathway and the monoacylglycerol pathway (Hiramine et al. 2010).

Acyl-CoA: diacylglycerol (DAG) acyltransferase (DGAT; EC 2.3.1.20) activity has long been detected in various animal and plant tissues active in TAG synthesis. DGAT catalyzes the reaction:
DGAT

$$Acyl - CoA + DAG \rightarrow TAG + CoASH \qquad (8.1)$$

As expected, this enzyme is membrane-bound or -associated and difficult to work with biochemically. As such, the first DGAT gene was not cloned until 1998 from mice (Cases et al. 1998). The groups of Hobbs (Hobbs and Hills 2000; Hobbs et al. 1999) and Zou et al. (1999) reported the cloning of a DGAT from Arabidopsis the next year. Lardizabal et al. reported the cloning of a second class of DGAT—DGAT2—from the oleagineous fungus *Mortierella ramanniana* (Lardizabal et al. 2001) that has no homology to the earlier identified DGAT sequences now known as DGAT1s. Cases et al. (2001) also cloned a mammalian DGAT2, and it is now known that humans have seven DGAT2s (Turkish et al. 2005). Only a single

DGAT1 gene (At2g19450) and a single DGAT2 gene (At3g51520) are present in the Arabidopsis genome (Beisson et al. 2003; Mhaske et al. 2005). The genomics of TAG biosynthesis in Arabidopsis has been studied rather well (although much remains to be elucidated) by Ohlrogge and colleagues (Beisson et al. 2003; Ruuska et al. 2004). A database for Arabidopsis lipid genes is available at: http://www.plantbiology.msu.edu/lipids/genesurvey/index.htm Soybeans have at least two DGAT1s (Hildebrand et al. 2008). A draft of the soybean genome has recently been reported (Shoemaker et al. 2008).

A second mechanism for the biosynthesis of TAG in yeast and plants was discovered and reported in 2000 (Dahlqvist et al. 2000; Oelkers et al. 2000) that has homology to lecithin cholesterol acyltransferases (LCATs). This reaction is catalyzed by an enzyme known as phospholipid: diacylglycerol acyltransferase (PDAT, EC 2.3.1.158) that transfers an acyl group (fatty acid) from a phospholipid (PL) to DAG forming TAG and a lysophospholipid (LPL):
PDAT

$$PL + DAG \rightarrow TAG + LPL \qquad (8.2)$$

Arabidopsis has six PDAT/LCAT homologs (Ståhl et al. 2004), of which (At5g13640) is most closely related to the PDAT identified in yeast. Ståhl et al. (2004) demonstrated that this gene is expressed widely in different Arabidopsis tissues and has PDAT activity. In humans, most TAG is synthesized by DGAT1 and DGAT2 and the only human gene similar to PDAT has phospholipase A_2 and phospholipid:ceramide transacylase activities (Hiraoka et al. 2002). Mhaske et al. (2005) generated a knockout for At5g13640, and their studies plus those of Ståhl et al. (2004) fairly well rule out a role for this gene in TAG synthesis in Arabidopsis seeds. A second arabidopsis PDAT/LCAT homolog most related to At5g13640 (57% identical) is At5g44830. This PDAT/LCAT-like gene was found by Ståhl et al. (2004) to be expressed mainly in developing seeds, and they speculate that it might have a role in seed oil biosynthesis. However this role has not yet been addressed directly nor has its activity been assessed.

A third TAG biosynthetic activity involving a DAG/DAG transacylase (DGTA) has been reported in animals (Lehner and Kuksis 1993) and plants (Stobart et al. 1997), and Ståhl et al. (2004) detect such an activity in Arabidopsis. Although the enzyme has not been characterized biochemically and the corresponding genes have not been cloned, DGTA is known to catalyze the following reaction:
DGTA

$$DAG + DAG \rightarrow TAG + MAG \text{ (monoacylglycerol)} \qquad (8.3)$$

A number of mutants with reduced seed oil contents have been reported in Arabidopsis and have been found to be due to defects in DGAT1 (Focks and

Benning 1998; Katavic et al. 1995; Lu and Hills 2002; Routaboul et al. 1999; Zou et al. 1999) or an impairment in transfer of carbon from sucrose and glucose to TAG, possibly due to impaired hexokinase and pyrophosphate-dependent phosphofructokinase (Focks and Benning 1998). Arabidopsis DGAT1 mutants have ~25% to 50% reductions in seed oil contents. DGAT1 is reported to be expressed maximally in developing seeds at a stage of high oil synthesis (Lu et al. 2003). Silencing of DGAT1 in tobacco has also been reported to reduce seed oil content (Zhang et al. 2005). Our preliminary data (Hildebrand et al. 2008) indicates a role for DGAT1(s) in soybean oil synthesis but this has not yet been directly addressed. The role of DGAT2 in oil accumulation in Arabidopsis and common oilseeds such as soybeans has not yet been investigated.

Several reports indicate a role for DGAT in oil accumulation in developing soybean seeds (Kwanyuen and Wilson 1986, 1990; Kwanyuen et al. 1988; Settlage et al. 1998). No mutants with large changes in oil levels or defects in DGAT have yet been reported in soybeans. It is not yet clear what contributions DGAT1, DGAT2 or other possible DGATs play in soybean oil biosynthesis. We detect transcripts for DGAT1, DGAT2 and PDAT in soybean tissues including developing seeds (Hildebrand et al. 2008; our unpublished results). Developing soybean seeds accumulate TAG after most cell division has ceased and cotyledons have been formed and cell expansion initiated (Dahmer et al. 1991; Le et al. 2007). Like most green tissues linolenate (18:3) is the most abundant fatty acid of soybean oil early in seed development. The 18:3 levels of soybean oil continue to decline throughout seed development, with linoleate (18:2) and oleate (18:1) becoming the predominate fatty acids of soybean oil as seeds mature (Dahmer et al. 1991). DGAT levels correlate with oil accumulation. We find that expression of a DGAT in leaves can cause accumulation of TAG in leaf tissue (Li et al. 2010). Baud and co-workers (Baud et al. 2009; Baud and Lepiniec 2009) report that expression of the B3 family domain transcription factor LEC2 can cause accumulation of TAG in Arabidopsis leaves. LEC2 induces other transcription factors, including FUS3, that can increase lipid biosynthesis leading to TAG accumulation in plant tissues including leaves (Vyacheslav et al. 2009; H.Y. Wang et al. 2007). Over-expression of the related transcription factor LEC1 (Baud and Lepiniec 2009) also causes lipid accumulation in Arabidopsis plants (Mu et al. 2008). Another transcription factor, WR1, can also increase oil accumulation in plants (Cernac and Benning 2004). Soybean Dof-type transcription factor genes are reported to increase lipid levels in transgenic Arabidopsis seeds (H.W. Wang et al. 2007).

Oilseeds including soybeans accumulate TAG in special organelles known as oil bodies. There is strong evidence that oil bodies form with the accumulation of TAG inside the phospholipid bilayer in specialized regions of the ER ballooning out from the accumulating TAG, and the remaining phospholipid forming a monolayer surrounding the growing lipid body. Concurrent with this, the oil body-specific protein oleosin is co-translationally inserted into the phospholipid monolayer of the oil bodies (Kalinski et al. 1991; Loer and Herman 1993; Sarmiento et al. 1997; Siloto et al. 2006; Tzen et al. 1990).

8.4.2.2 Genetic Engineering of Industrial Oil Targets

Plant oils are known to accumulate more than 300 unusual fatty acids (UFA) in addition to the five fatty acids common in most all plant tissues and the main components of most commodity oils (van de Loo et al. 1993). However, in most cases, sources of these UFA cannot be grown economically on a commercial scale. Hydroxy, epoxy, conjugated, acetylenic, very long chain and branched chain fatty acids and liquid waxes are among the industrial targets of greatest interest. Oils high in hydroxy fatty acids can be produced from castor and Lesquerella but it could be produced more economically on a large scale currently with canola or soybeans engineered with genes for such metabolism due to, among other reasons, the value of the meal co-product and better developed agronomic properties. Genes for most of these UFA have been cloned and good reviews have been written on this subject (e.g., Napier 2007). It is easy to produce such UFA in transgenic oilseeds with genes encoding enzymes for UFA biosynthesis but it has been very difficult to achieve accumulation of UFA to more than ~10% of total lipids. This is in contrast to the accumulation of as much of 95% of the seed oil TAG being composed of a single UFA such as the hydroxy fatty acid ricinoleate in castor oil, the epoxy fatty acid vernoleate in *Bernardia pulchella* oil, and the short-chain fatty acid caproate in *Cuphea koehneana* oil. In cases where details of the biosynthesis of the UFA are known, they are made on phosphatidyl choline (PC) in the ER but then accumulate selectively in seed oil TAG. They do not accumulate in membrane lipids such as the starting PC in plants that accumulate high levels in TAG, and neither do they show such selective distribution in transgenic oilseeds with UFA biosynthetic genes alone. This has led to studies on whether TAG biosynthetic enzymes might have selectivity for fatty acids that accumulate in the TAG.

Studies on the expression profiles of genes encoding TAG biosynthetic enzymes have found that only DGAT1, and not DGAT2 or PDAT, has an expression profile in different tissues of soybeans and Arabidopsis consistent with a role in seed oil synthesis. In contrast, DGAT1 and DGAT2 display expression consistent with a role in seed oil synthesis in high epoxy and hydroxy fatty acid accumulating plants, implicating DGATs, particularly DGAT2, as playing an important role in the selective accumulation of UFA in TAG (Li et al. 2005, 2010b). Jay Shockey and colleagues (Shockey et al. 2006a, b) presented good evidence that DGAT2 from tung trees, *Vernicia fordii*, has specificity for the conjugated fatty acid that accumulates in tung oil—eleostearic acid. Co-expression of hydroxylase and DGAT2 from castor and epoxygenase plus DGAT2 from *Bernardia* and *Vernonia* can increase the accumulation of hydroxy and epoxy fatty acids in seed oil up to five times over expression of the hydroxylase and epoxygenase genes alone (Burgal et al. 2008; Li et al. 2010a; Zhou et al. 2008). Li et al. (2010a) were the first to demonstrate this in a commercial oilseed, going from about 5% epoxy fatty acid in seed lipids of soybeans expressing an epoxygenase, increasing to >10% in soybeans expressing an epoxygenase plus a DGAT1 from a high epoxy fatty acid accumulating plant, and to >25% in soybeans expressing an epoxygenase plus a DGAT2 from the same high epoxy fatty acid accumulating plant (Fig. 8.10).

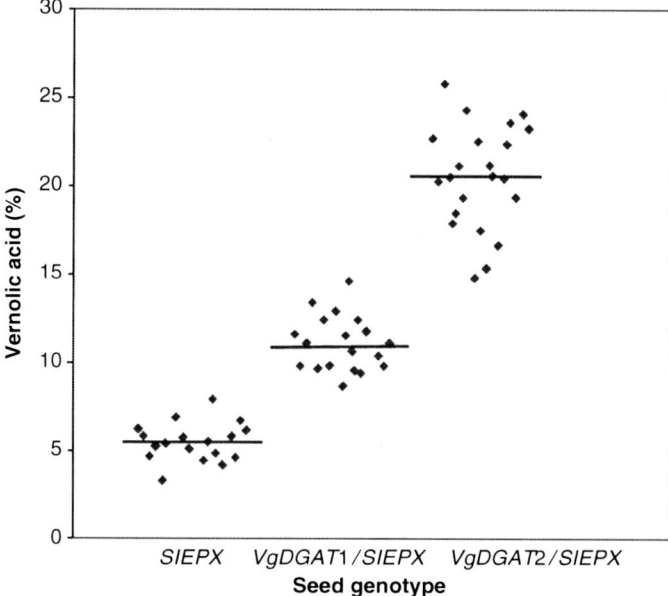

Fig. 8.10 Vernolic acid contents of transgenic soybean seeds from regenerated plants. Seed-chips of each mature seed of *SlEPX*-, *SlEPX/VgDGAT*- and *SlEPX/VgDGAT2*- transgenic soybeans were sampled for fatty acid analysis by gas chromatography (GC) and genotyping by polymerase chain reaction (PCR). Each *data point* represents the vernolic acid content in a seed sample. *Horizontal bars* indicate the mean for each data set

8.5 Conclusions

The outlook for adequate global supply of edible oils during the next decade is encouraging. Current rates of oil supply should be maintained. Demand is expected to be strong, but equilibrium will be reached between food and industrial applications. Indeed, a robust market for biofuel may be necessary to balance gains in productivity. However, total supply will become more dependent on crops that have sufficient production systems to sustain continued growth. Impact on the environment will be a major factor in determining the sustainability of biodiesel. The main problems with oxidative stability, poor cold-flow properties and NO_x emissions can be mitigated through genetic enhancement of oleic acid concentration. All edible oil crops now have a high-oleic option; many are entering commercial production. Looking forward research must continue to develop new sources of edible and industrial oils as a necessary step in meeting the consumer driven demands of twenty-first century markets. Many new plant and algal oil sources are being developed for biodiesel and specific industrial applications. Genetic engineering commercial oilseeds for biosynthesis and selective accumulation of specific fatty acids with unique chemical properties will enhance renewable chemical developments.

Acknowledgements The authors wish to thank the United Soybean Board for providing partial support for the preparation of this manuscript under Projects 9276 & 9306.

References

Aghoram K, Wilson RF, Burton JW, Dewey RE (2006) A mutation in a 3-keto-acyl ACP synthase II gene is associated with elevated palmitic acid levels in soybean seeds. Crop Sci 46:2453–2459

Ahmed M, Ting I, Scora RW (1994) Leaf oil composition of *Salvia hispanica* L. from three geographical areas. J Essent Oil Res 6:223–228

Ahn YJ, Chen GQ (2008) In vitro regeneration of castor (*Ricinus communis* L.) using cotyledon explants. Hortscience 43:215–219

Ahn Y-J, Vang L, McKeon T, Chen G (2007) High-frequency plant regeneration through adventitious shoot formation in castor (*Ricinus communis* L.). In Vitro Cell Dev Biol Plant 43:9–15

Albuquerque MCG, Machado YL, Torres AEB, Azevedo DCS, Cavalcante CL Jr, Firmiano LR, Parente EJS Jr (2009) Properties of biodiesel oils formulated using different biomass sources and their blends. Renew Energy 857–859

Anandan S, Kumar GKA, Ghosh J, Ramachandra KS (2005) Effect of different physical and chemical treatments on detoxification of ricin in castor cake. Anim Feed Sci Technol 120:159–168

Anderson E, Lowe HJ (1974) The composition of flaxseed mucilage. J Biol Chem 168:289–297

Anjani K, Pallavi M, Babu SNS (2010) Biochemical basis of resistance to leafminer in castor (*Ricinus communis* L.). Ind Crops Products 31:192–196

Anonymous (2002) Is there flaxseed in your fridge yet. Tufts Univ Health Nutr Lett 20:3

Arbelaiz A, Cantero G, Fernandez B, Mondragon I, Ganan P, Kenny JM (2005) Flax fiber surface modifications: effects on fiber physico mechanical and flax/polypropylene interface properties. Polym Compos 26:324–332

Ash GJ, Albiston A, Cother EJ (2005) Aspects of jojoba agronomy and management. Advances in Agronomy, vol 85. Elsevier, San Diego, pp 409–437

Ash M, Livezey J, Dohlman E (2006) Soybean backgrounder. OCS-2006-01. US Department of Agriculture, Economic Research Service, Washington DC

Ashby RD, Foglia TA, Solaiman DKY, Liu CK, Nunez A, Eggink G (2000) Viscoelastic properties of linseed oil-based medium chain length poly(hydroxyalkanoate) films: effects of epoxidation and curing. Int J Biol Macromol 27:355–361

Audi J, Belson M, Patel M, Schier J, Osterloh J (2005) Ricin poisoning—a comprehensive review. JAMA 294:2342–2351

Ayerza R, Coates W (1999) An omega-3 fatty acid enriched chia diet: influence on egg fatty acid composition, cholesterol and oil content. Can J Anim Sci 79:53–58

Ayerza R, Coates W (2000) Dietary levels of chia: influence on yolk cholesterol, lipid content and fatty acid composition for two strains of hens. Poult Sci 79:724–739

Ayerza R, Coates W (2001) Omega-3 enriched eggs: the influence of dietary alpha-linolenic fatty acid source on egg production and composition. Can J Anim Sci 81:355–362

Ayerza R, Coates W (2002) Dietary levels of chia: influence on hen weight, egg production and sensory quality, for two strains of hens. Br Poult Sci 43:283–290

Ayerza R, Coates W (2004) Composition of chia (*Salvia hispanica*) grown in six tropical and subtropical ecosystems of South America. Trop Sci 44:131–135

Ayerza R, Coates W (2005) Ground chia seed and chia oil effects on plasma lipids and fatty acids in the rat. Nutr Res 25:995–1003

Ayerza R Jr, Coates W (2007) Effect of dietary alpha-linolenic fatty acid derived from chia when fed as ground seed, whole seed and oil on lipid content and fatty acid composition of rat plasma. Ann Nutr Metabol 51:27–34

Ayerza R, Coates W, Lauria M (2002) Chia seed (*Salvia hispanica* L.) as an omega-3 fatty acid source for broilers: influence on fatty acid composition, cholesterol and fat content of white and dark meats, growth performance, and sensory characteristics. Poult Sci 81:826–837

Bagchi M, Zafra-Stone S, Lau FC, Bagchi D (2009) Ricin and abrin. In: Gupta RC (ed) Handbook of toxicology of chemical warfare agents. Academic, New York

Bamgboye AO, Hansen AC (2008) Prediction of cetane number of biodiesel fuel from the fatty acid methyl ester (FAME) composition. Int Agrophys 22:21–29

Barnes DJ, Baldwin BS, Braasch DA (2009) Ricin accumulation and degradation during castor seed development and late germination. Ind Crops Prod 30:254–258

Bates PD, Durrett TP, Ohlrogge JB, Pollard M (2009) Analysis of acyl fluxes through multiple pathways of triacylglycerol synthesis in developing soybean embryos. Plant Physiol 150:55–72

Baud S, Lepiniec L (2009) Regulation of de novo fatty acid synthesis in maturing oilseeds of Arabidopsis. Plant Physiol Biochem 47:448–455

Baud S, Wuilleme S, To A, Rochat C, Lepiniec L (2009) Role of WRINKLED1 in the transcriptional regulation of glycolytic and fatty acid biosynthetic genes in Arabidopsis. Plant J 60:933–947

Bean LD, Leeson S (2003) Long-term effects of feeding flaxseed on performance and egg fatty acid composition of brown and white hens. Poult Sci 82:388–394

Beisson F, Koo AJ, Ruuska S, Schwender J, Pollard M, Thelen JJ, Paddock T, Salas JJ, Savage L, Milcamps A, Mhaske VB, Cho Y, Ohlrogge JB (2003) Arabidopsis genes involved in acyl lipid metabolism. A 2003 census of the candidates, a study of the distribution of expressed sequence tags in organs, and a web-based database. Plant Physiol 132:681–697

Bell JG, Henderson RJ, Tocher DR, Sargent JR (2004) Replacement of dietary fish oil with increasing levels of linseed oil: modification of flesh fatty acid compositions in Atlantic salmon (*Salmo salar*) using a fish oil finishing diet. Lipids 39:223–232

Bendiksen EA, Berg OK, Jobling M, Arnesen AM, Masoval K (2003) Digestibility, growth and nutrient utilisation of Atlantic salmon parr (*Salmo salar* L.) in relation to temperature, feed fat content and oil source. Aquaculture 224:283–299

Benzioni A (2006) Jojoba research as basis for domestication of jojoba in Israel. Isr J Plant Sci 54:157–167

Berglund DR (2002) Flax: new uses and demands. In: Janick J, Whipkey A (eds) New crops. ASHS, Alexandria, VA, pp 358–360

Berti MT, Johnson BL (2008) Growth and development of cuphea. Ind Crops Prod 27:265–271

Bewley J, Black M (1994) Seeds: physiology of development and germination, 2nd edn. Plenum, New York

Blackmer JL, Byers JA (2009) *Lugus* spp. (Heteroptera: Miridae). Host-plant interactions with *Lesquerella fendleri*, (Brassicaceae), a new crop in the arid southwest. Environ Entomol 38:159–167

Block B (2009a) Global palm oil demand fueling deforestation. Available via World Watch Institute: www.worldwatch.org/node/6059

Block B (2009b) Can "sustainable" palm oil slow deforestation? Global palm oil demand fueling deforestation. Available via World Watch Institute: www.worldwatch.org/node/6082

Bloedon LT, Szapary PO (2004) Flaxseed and cardiovascular risk. Nutr Rev 62:18–27

Brenner D (1995) Perilla. Purdue University NewCrop Fact Sheet, wwwhortpurdueedu/newcrop/cropfactsheets/perillapdf.

Brigham RD (1993) Castor: return of an old crop. In: Janick, Simon JE (eds) New crops. Wiley, New York

Budziszewski GJ, Croft KPC, Hildebrand DF (1996) Uses of biotechnology in modifying plant lipids. Lipids 31:557–569

Burgal J, Shockey J, Lu CF, Dyer J, Larson T, Graham I, Browse J (2008) Metabolic engineering of hydroxy fatty acid production in plants: RcDGAT2 drives dramatic increases in ricinoleate levels in seed oil. Plant Biotechnol J 6:819–831

Butler RA (2009) Deforestation in the Amazon. Available via Mongabay, www.mongabay.com/brazil.html

Cahill JP (2003) Ethnobotany of Chia, *Salvia hispanica* L. (Lamiaceae). Econ Bot 57:604–618

Cahill JP (2004) Genetic diversity among varieties of Chia (*Salvia hispanica* L.). Genet Resour Crop Evol 51:773–781

Cahill JP, Ehdaie B (2005) Variation and heritability of seed mass in chia (*Salvia hispanica* L.). Genet Resour Crop Evolut 52:201–207

Cardinal AJ, Burton JW, Camacho-Roger AM, Yang JH, Wilson RF, Dewey RE (2007) Molecular analysis of soybean lines with low palmitic acid content in the seed oil. Crop Sci 47:304–310

Carlsson AS (2009) Plant oils as feedstock alternatives to petroleum—a short survey of potential oil crop platforms. Biochimie 91:665–670

Cases S, Smith SJ, Zheng Y, Myers HM, Lear SR, Sande E, Novak S, Collins C, Welch CB, Lusis AJ, Erickson SK, Farese RV Jr (1998) Identification of a gene encoding an acyl CoA: diacylglycerol acyltransferase, a key enzyme in triacylglycerol synthesis. Proc Natl Acad Sci USA 95:13018–13023

Cases S, Stone SJ, Zhou P, Yen E, Tow B, Lardizabal KD, Voelker T, Farese RV Jr (2001) Cloning of DGAT2, a second mammalian diacylglycerol acyltransferase, and related family members. J Biol Chem 276:38870–38876

CAST (2008) Convergence of Agriculture and Energy: III. Considerations in Biodiesel Production. CAST Commentary QTA2008-2. Council for Agricultural Science and Technology, Ames, IA

Cernac A, Benning C (2004) WRINKLED1 encodes an AP2/EREB domain protein involved in the control of storage compound biosynthesis in Arabidopsis. Plant J 40:575–585

Chicco AG, D'Alessandro ME, Hein GJ, Oliva ME, Lombardo YB (2009) Dietary chia seed (*Salvia hispanica* L.) rich in alpha-linolenic acid improves adiposity and normalises hypertriacylglycerolaemia and insulin resistance in dyslipaemic rats. Br J Nutr 101:41–50

Chu Y, Holbrook CC, Ozias-Akins P (2009) Two alleles of ahFAD2B control the high oleic trait in cultivated peanut. Crop Sci 49:2029–2036

Coates W, Ayerza R (1996) Production potential of chia in northwestern Argentina. Ind Crop Prod 5:229–233

Dahlqvist A, Ståhl U, Lenman M, Banas A, Lee M, Sandager L, Ronne H, Stymne S (2000) Phospholipid:diacylglycerol acyltransferase: an enzyme that catalyzes the acyl-CoA-independent formation of triacylglycerol in yeast and plants. Proc Natl Acad Sci USA 97:6487–6492

Dahmer ML, Collins GB, Hildebrand DF (1991) Lipid concentration and composition of soybean zygotic embryos maturing in vitro and in planta. Crop Sci 31:735–740

Demirbas A (2008) Relationships derived from physical properties of vegetable oil and biodiesel fuels. Fuel 87:1743–1748

De Oliveira D, Di Luccio M, Faccio C, Dalla Rosa C, Bender JP, Lipke N, Menoncin S, Amroginski C, De Oliveira JV (2004) Optimization of enzymatic production of biodiesel from castor oil in organic solvent medium. Appl Biochem Biotechnol 113:771–780

Dewey RE, Wilson RF, Novitzky WP, Goode JH (1994) The AAPT1 gene of soybean complements a cholinephosphotransferase-deficient mutant of yeast. Plant Cell 6:1495–1507

Diederichsen A, Raney JP (2006) Seed color, seed weight and seed oil content in *Linum usitatissimum* accessions held by Plant Gene Resources of Canada. Plant Breed 120:360–362

Dierig DA (1995) Lesquerella. New Crop FactSheet. (www.hort.purdue.edu/newcrop/cropfactsheets/Lesquerella.html)

Divakara BN, Upadhyaya HD, Wani SP, Gowda CLL (2010) Biology and genetic improvement of *Jatropha curcas* L.: a review. Appl Energy 87:732–742

Dow AgroSciences (2007) High oleic imidazolinone resistant sunflower. World Intellectual Property Organization, WO/2007/038738

Egli DB (2008a) Comparison of corn and soybean yields in the United States: historical trends and future prospects. Agron J 100:S79–S80

Egli DB (2008b) Soybean yield trends from 1972 to 2003 in mid-western USA. Field Crops Res 106:53–59

Erskine AJ, Jones JKN (1957) The structure of linseed mucilage. Can J Chem 35:1174–1182

Espada CE, Berra MA, Martinez MJ, Eynard AR, Pasqualini ME (2007) Effect of Chia oil (*Salvia hispanica*) rich in omega-3 fatty acids on the eicosanoid release, apoptosis and T-lymphocyte tumor infiltration in a murine mammary gland adenocarcinoma. Prostaglandins Leukot Essent Fatty Acids 77:21–28

Estilai A, Hashemi A, Truman K (1990) Chromosome number and meiotic behavior of cultivated chia, *Salvia hispanic* (Lamiaceae). Hortscience 25:1646–1647

Fedeniuk RW, Biliaderis CG (1994) Composition and physiochemical properties of linseed (*Linum usitatissimum* L.) mucilage. J Agric Food Chem 42: 240–247

Fernandes AT (2009) The social and environmental impacts of industrial agriculture in the Legal Amazon. Anais XIV Simposio Brasilerio de Sensoriamento Remoto, Natal Brazil, INPE pp 159–165. Available via marte.dpi.inpe.br/col/dpi.inpe.br/sbsr@80/2008/11.16.14.56/doc/159-165.pdf

Flores T, Karpova O, Su X, Zeng P, Bilyeu KD, Sleper DA, Nguyen HT, Zhang ZJ (2008) Silencing of GmFAD3 gene by siRNA leads to low a-linolenic acids (18:3) of a fad3-mutant phenotype in soybean [(*Glycine max* (Merr.)]. Transgenic Res 17:839–850

Focks N, Benning C (1998) wrinkled1: a novel, low-seed-oil mutant of arabidopsis with a deficiency in the seed-specific regulation of carbohydrate metabolism. Plant Physiol 118:91–101

Franzosi G, Battistel E, Santoro M, Iannacone R (1998) LPAAT and DAGAT activity and specificity in rapeseed (*Brassica napus* L. var. Canola) and sunflower (*Helianthus annuus*) developing seeds. In: Sánchez J, Cerdá-Olmedo E, Martínez-Force E (eds) Advances in plant lipid research. Universidad de Sevilla, Spain, pp 679–682

Gerbens-Leenes W, Hoekstra AY, van der Meer TH (2009a) The water footprint of bioenergy. Proc Natl Acad Sci USA 106:10219–10223

Gerbens-Leenes W, Hoekstra AY, van der Meer TH (2009b) Reply to Maes et al.: A global estimate of the water footprint of *Jatropha curcas* under limited data availability. Proc Natl Acad Sci USA 106:E113–E113

Goldberg RB, Barker SJ, Perez-Grau L (1989) Regulation of gene expression during plant embryogenesis. Cell 56:149–160

Goldemberg J (2007) Ethanol for a sustainable energy future. Science 315:808–810

Grobas S, Mendez J, Lazaro R, Blas Cd, Mateos GG (2001) Influence of source and percentage of fat added to diet on performance and fatty acid composition of egg yolks of two strains of laying hens. Poult Sci 80:1171–1179

Gulati SK, May C, Wynn PC, Scott TW (2002) Milk fat enriched in n-3 fatty acids. Anim Feed Sci Technol 98:143–152

Gupta MK (1998) NuSun—the future generation of oils. Inform 9:1150

Hartvigsen MS, Mu H, Hougaard KS, Lund SP, Xu X, Hoy CE (2004) Influence of dietary triacylglycerol structure and level of n-3 fatty acids administered during development on brain phospholipids and memory and learning ability of rats. Ann Nutr Metab 48:16–27

Harwood JL (1997) Plant lipid metabolism. In: Dey PM, Harborne JB (eds) Plant biochemistry. Academic, London, pp 237–272

Harwood JL, Guschina IA (2009) The versatility of algae and their lipid metabolism. Biochimie 91:679–684

Harwood JL, Page RA (1994) Biochemistry of oil synthesis. In: Murphy DJ (ed) Designer oil crops. VCH, Weinheim, pp 165–194

Health Canada (1999) High oleic acid/low linolenic acid canola lines 45A37, 46A40. Avaialble via: http://www.hc-sc.gc.ca/fn-an/gmf-agm/appro/ofb-096-228-a-eng.php

Hildebrand DF, Li R, Hatanaka T (2008) Genomics of soybean oil traits. In: Stacey G (ed) Genetics and genomics of soybean. Springer, New York, pp 185–210

Hiramine Y, Emoto H, Takasuga S, Hiramatsu R (2010) Novel acyl-coenzyme A:monoacylglycerol acyltransferase (MGAT) plays an important role in hepatic triacylglycerol secretion. J Lipid Res doi: 10.1194/jlr.M002584

Hiraoka M, Abe A, Shayman JA (2002) Cloning and characterization of a lysosomal phospholipase A2, 1-O-acylceramide synthase. J Biol Chem 277:10090–10099

Hitz WD, Yadav NS, Reiter RS, Mauvais CJ, Kinney AJ (1995) Reducing polyunsaturation in oils of transgenic canola and soybean. In: Kader JC (ed) Plant lipid metabolism. Kluwer, Dordrecht, pp 506–508

Hobbs DH, Hills MJ (2000) Expression and characterization of diacylglycerol acyltransferase from *Arabidopsis thaliana* in insect cell cultures. Biochem Soc Trans 28:687–689

Hobbs DH, Lu C, Hills MJ (1999) Cloning of a cDNA encoding diacylglycerol acyltransferase from *Arabidopsis thaliana* and its functional expression. FEBS Lett 452:145–149

Hovsepyan R, Willcox G (2008) The earliest finds of cultivated plants in Armenia: evidence from charred remains and crop processing residues in pise from the Neolithic settlements of Aratashen and Aknashen. Veg Hist Archaeobot 17:S63–S71

Hoz L, Lopez Bote CJ, Cambero MI, D'Arrigo M, Pin C, Santos C, Ordonez JA (2003) Effect of dietary linseed oil and alpha-tocopherol on pork tenderloin (*Psoas major*) muscle. Meat Sci 65:1039–1044

Hutchins AM, Martini MC, Olson BA, Thomas W, Slavin JL (2001) Flaxseed consumption influences endogenous hormone concentrations in postmenopausal women. Nutr Cancer 39:58–65

Izquierdo MS, Obach A, Arantzamendi L, Montero D, Robaina L, Rosenlund G (2003) Dietary lipid sources for seabream and seabass: growth performance, tissue composition and flesh quality. Aquacult Nutr 9:397–407

Jackson FM, Michaelson L, Fraser TCM, Stobart AK, Griffiths G (1998) Biosynthesis of triacylglycerol in the filamentous fungus *Mucor circinelloides*. Microbiology 144:2639–2645

Jenderek MM, Hannan RM (2009) Diversity in seed production characteristics within the USDA–ARS *Limnanthes alba* germplasm collection. Crop Sci 49:1387–1394

Jenderek MM, Dierig DA, Isbell TA (2009) Fatty-acid profile of Lesquerella germplasm in the National Plant Germplasm System collection. Ind Crops Prod 29:154–164

Jeong GT, Park DH (2009) Optimization of biodiesel production from castor oil using response surface methodology. Appl Biochem Biotechnol 156:431–441

Johnson LA, Hammond EG (1996) Soybean oil ester fuel blends. US Patent 5520708, Washington DC

Jongschaap R, Corré W, Bindraban P, Brandenburg W (2007) Claims and facts on *Jatropha curcas* L. Available at http://library.wur.nl/way/bestanden/clc/1858843.pd In: Global *Jatropha curcas* Evaluation, Breeding and Propagation Programme Report 158

Jongschaap REE, Blesgraaf RAR, Bogaard TA, van Loo EN, Savenije HHG (2009) The water footprint of bioenergy from *Jatropha curcas* L. Proc Natl Acad Sci USA 106:E92–E92

Joshi K, Lad S, Kale M, Patwardhan B, Mahadik SP, Patni B, Chaudhary A, Bhave S, Pandit A (2006) Supplementation with flax oil and vitamin C improves the outcome of attention deficit hyperactivity disorder. Prostaglandins Leukot Essent Fatty Acids 74:17–21

Kalinski A, Loer DS, Weisemann JM, Matthews BF, Herman EM (1991) Isoforms of soybean seed oil body membrane protein 24 kDa oleosin are encoded by closely related cDNAs. Plant Mol Biol 17:1095–1098

Katavic V, Reed DW, Taylor DC, Giblin EM, Barton DL, Zou J, Mackenzie SL, Covello PS, Kunst L (1995) Alteration of seed fatty acid composition by an ethyl methanesulfonate-induced mutation in *Arabidopsis thaliana* affecting diacylglycerol acyltransferase activity. Plant Physiol 108:399–409

Kheira AAA, Atta NMM (2009) Response of *Jatropha curcas* L. to water deficits: yield, water use efficiency and oilseed characteristics. Biomass Bioenergy 33:1343–1350

King AJ, He W, Cuevas JA, Freudenberger M, Ramiaramanana D, Graham IA (2009) Potential of *Jatropha curcas* as a source of renewable oil and animal feed. J Exp Bot 60:2897–2905

Kiron V, Puangkaew J, Ishizaka K, Satoh S, Watanabe T (2004) Antioxidant status and nonspecific immune responses in rainbow trout (*Oncorhynchus mykiss*) fed two levels of vitamin E along with three lipid sources. Aquaculture 234:361–379

Knauft DA, Gorbet DW, Norden AJ, Norden CK (1999) Peanut oil from enhanced peanut products. US Patent 5922390, Washington DC

Knothe G (2008) Designer biodiesel: optimizing fatty ester composition to improve fuel properties. Energy Fuels 22:1358–1364

Knowlton S (1999) Soybean oil having high oxidative stability. US Patent 5981781. Washington DC

Kodali DR, List GR (2005) Trans fats alternatives. AOCS, Champaign IL

Kreiter T (2005) Seeds of wellness: return of a supergrain. In: The Saturday Evening Post Nov/Dec

Kumari KG, Ganesan M, Jayabalan N (2008) Somatic organogenesis and plant regeneration in *Ricinus communis*. Biol Planta 52:17–25

Kwanyuen P, Wilson RF (1986) Isolation and purification of diacylglycerol acyltransferase from germinating soybean cotyledons. Biochim Biophys Acta 877:238–245

Kwanyuen P, Wilson RF (1990) Subunit and amino acid composition of diacylglycerol acyltransferase from germinating soybean cotyledons. Biochim Biophys Acta 1039:67–72

Kwanyuen P, Wilson RF, Burton JW (1988) Substrate specificity of diacylglycerol acyltransferase purified from soybean. In: Applewhite TH (ed) Proceedings of the World Conference on Biotechnology for the Fats and Oils Industry. American Oil Chemists' Society, Champaign, IL, pp 294–297

Lardizabal KD, Mai JT, Wagner NW, Wyrick A, Voelker T, Hawkins DJ (2001) DGAT2 is a new diacylglycerol acyltransferase gene family. Purification, cloning, and expression in insect sells of two polypeptides from *Mortierella ramanniana* with diacylglycerol acyltransferase activity. J Biol Chem 276:38862–38869

Lazzari M, Chiantore O (1999) Drying and oxidative degradation of linseed oil. Polym Degrad Stabil 65:303–313

Le BH, Wagmaister JA, Kawashima T, Bui AQ, Harada JJ, Goldberg RB (2007) Using genomics to study legume seed development. Plant Physiol 144:562–574

Le Dréau Y, Dupuy N, Gaydou V, Joachim J, Kister J (2009) Study of jojoba oil aging by FTIR. Anal Chim Acta 642:163–170

Lee B-K, Yu S-H, Kim Y-H, Ahn B-O, Hur H-S, Lee S-C, Zhang Z, Lee J-Y (2005) *Agrobacterium*-mediated transformation of Perilla (*Perilla frutescens*). Plant Cell Tissue Organ Cult 83:51–58

Lee RE, Manthey FA, Hall CA III (2004) Content and stability of hexane extractable lipid at various steps of producing macaroni containing ground flaxseed. J Food Process Preserv 28:133–144

Lehner R, Kuksis A (1993) Triacylglycerol synthesis by an sn-1,2(2,3)-diacylglycerol transacylase from rat intestinal microsomes. J Biol Chem 268:8781–8786

Li R, Yu K, Hatanaka T, Hildebrand DF (2005) Characterization of cDNAs involved in seed oil synthesis in some high epoxy and hydroxy fatty acid accumulators. In: Proceedings of XIII Plant & Animal Genomes Conference, San Diego, CA

Li R, Yu K, Hatanaka T, Hildebrand DF (2010a) Vernonia DGATs increase accumulation of epoxy fatty acids in oil. Plant Biotechnol J 8:184–195

Li R, Yu K, Hildebrand DF (2010b) DGAT1, DGAT2 and PDAT expression in seeds and other tissues of epoxy and hydroxy fatty acid accumulating plants. Lipids 45:145–157

Li X, Tabil LG, Panigrahi S (2007) Chemical treatments of natural fiber for use in natural fiber-reinforced composites: a review. J Polym Environ 15:25–33

Liang YN, Sarkany N, Cui Y (2009) Biomass and lipid productivities of *Chlorella vulgaris* under autotrophic, heterotrophic and mixotrophic growth conditions. Biotechnol Lett 31:1043–1049

Licht FO (2009) World Ethanol and Biofuels Report 7(14):288

Lin KY, Daniel JR, Whistler RL (1994) Structure of chia seed polysaccharide exudate. Carbohydr Polym 23:13–18

Liu Q, Surinder P, Green AG (2002a) High-stearic and high-oleic cottonseed oils produced by hairpin RNA-mediated post-transcriptional gene silencing. Plant Physiol 129:1732–1743

Liu Q, Surinder P, Green AG (2002b) High-oleic and high-stearic cottonseed oils: nutritionally improved cooking oils developed using gene silencing. J Am Coll Nutr 21(3):205S–211S

Liu XJ, Leung S, Brost J, Rooke S, Nguyen T (2008) *Camelina sativa* transformation by floral dip and simple large-scale screening of markerless transformants. In Vitro Cell Dev Biol Anim 44: S40–S41

Liu ZS, Erhan SZ, Akin DE, Barton FE (2006) "Green" composites from renewable resources: preparation of epoxidized soybean oil and flax fiber composites. J Agric Food Chem 54:2134–2137

Loer DS, Herman EM (1993) Cotranslational integration of soybean (*Glycine max*) oil body membrane protein oleosin into microsomal membranes. Plant Physiol 101:993–998

Lonien J, Schwender J (2009) Analysis of metabolic flux phenotypes for two Arabidopsis mutants with severe impairment in seed storage lipid synthesis. Plant Physiol 151:1617–1634

Lu C, Hills MJ (2002) Arabidopsis mutants deficient in diacylglycerol acyltransferase display increased sensitivity to abscisic acid, sugars, and osmotic stress during germination and seedling development. Plant Physiol 129:1352–1358

Lu CL, de Noyer SB, Hobbs DH, Kang J, Wen Y, Krachtus D, Hills MJ (2003) Expression pattern of diacylglycerol acyltransferase-1, an enzyme involved in triacylglycerol biosynthesis in *Arabidopsis thaliana*. Plant Mol Biol 52:31–41

Maes WH, Achten WMJ, Muys B (2009) Use of inadequate data and methodological errors lead to an overestimation of the water footprint of *Jatropha curcas*. Proc Natl Acad Sci USA 106: E91–E91

Makkar HPS, Klaus B (2009) *Jatropha curcas*, a promising crop for the generation of biodiesel and value-added coproducts. Eur J Lipid Sci Technol 111:773–787

Malathi B, Ramesh S, Rao K, Reddy V (2006) *Agrobacterium*-mediated genetic transformation and production of semilooper resistant transgenic castor (*Ricinus communis* L.). Euphytica 147:441–449

Malcolm, S, Allery M (2009) Growing crops for biofuels has spillover effects. Amber Waves, US Department of Agriculture, Economic Research Service, Washington DC. Available via www. ers.usda.gov/AmberWaves/March09/Features/Biofuels.htm

Martinez-Force E, Munoz-Ruz J, Fernandez-Martinez JM, Garces R (2006) High oleic/high stearic sunflower oils, US Patent 7141267, Washington DC

Mata TM, Martins AA, Caetano NS (2010) Microalgae for biodiesel production and other applications: a review. Renew Sustain Energy Rev 14:217–232

Mhaske V, Beldjilali K, Ohlrogge J, Pollard M (2005) Isolation and characterization of an *Arabidopsis thaliana* knockout line for phospholipid: diacylglycerol transacylase gene (At5g13640). Plant Physiol Biochem 43:413–417

Montero D, Kalinowski T, Obach A, Robaina L, Tort L, Caballero MJ, Izquierdo MS (2002) Vegetable lipid sources for gilthead seabream (*Sparus aurata*): effects on fish health. Aquaculture 225:353–370

Moser BR, Knothe G, Vaughn SF, Isbell TA (2009) Production and evaluation of biodiesel from field pennycress (*Thlaspi arvense* L.) oil. Energy Fuels 23:4149–4155

Mounts TL, Warner K, List GR, Neff WE, Wilson RF (1994) Low-linolenic acid soybean oils—alternatives to frying oils. J Am Oil Chem Soc 71:495–499

Mu JY, Tan HL, Zheng Q, Fu FY, Liang Y, Zhang JA, Yang XH, Wang T, Chong K, Wang XJ, Zuo JR (2008) LEAFY COTYLEDON1 is a key regulator of fatty acid biosynthesis in Arabidopsis. Plant Physiol 148:1042–1054

Muturi P, Wang DQ, Dirlikov S (1994) Epoxidized vegetable-oils as reactive diluents. 1. Comparison of vernonia, epoxidized soybean and epoxidized linseed oils. Prog Org Coat 25:85–94

Napier JA (2007) The production of unusual fatty acids in transgenic plants. Annu Rev Plant Biol 58:295–319

Nosarzewski M, Archbold DD (2007) Tissue-specific expression of SORBITOL DEHYDROGENASE in apple fruit during early development. J Exp Bot 58:1863–1872

Oelkers P, Tinkelenberg A, Erdeniz N, Cromley D, Billheimer JT, Sturley SL (2000) A lecithin cholesterol acyltransferase-like gene mediates diacylglycerol esterification in yeast. J Biol Chem 275:15609–15612

Ogborn MR, Nitschmann E, Bankovic Calic N, Weiler HA, Aukema H (2002) Dietary flax oil reduces renal injury, oxidized LDL content, and tissue n-6/n-3 FA ratio in experimental polycystic kidney disease. Lipids 37:1059–1065

Ohlrogge JB, Browse J (1995) Lipid biosynthesis. Plant Cell 7:957–970

Ohlrogge JB, Jaworski JG (1997) Regulation of fatty acid synthesis. Annu Rev Plant Physiol Plant Mol Biol 48:109–136

Ogunniyi DS (2006) Castor oil: vital industrial raw material. Bioresour Technol 97:1086–1091

Oliff HS (2004) Effects of flaxseed on lipids and bone metabolism in postmenopause. HerbalGram 24

Pandey A, Bhatt KC (2008) Diversity distribution and collection of genetic resources of cultivated and weedy type in *Perilla frutescens* (L.) Britton var. *frutescens* and their uses in Indian Himalaya. Genet Resour Crop Evol 55:883–892

Park C, Bang J, Lee B, Kim J, Lee B (2000) Research activity and achievement in mutation breeding of perilla in Korea. Korean J Int Agric 12:108–115

Park C, Bang J, Lee B, Kim J, Lee B, Chung M (2002) A high-oil and high-yielding perilla variety "Daesildeulkkae"' with dark brown seed coat. Korean J Breed 34:78–79

Park YJ, Dixit A, Ma KH, Lee JK, Lee MH, Chung CS, Nitta M, Okuno K, Kim TS, Cho EG, Rao VR (2008) Evaluation of genetic diversity and relationships within an on-farm collection of *Perilla frutescens* (L.) Britt. using microsatellite markers. Genet Resour Crop Evol 55:523–535

Peiretti PG, Gai F (2009) Fatty acid and nutritive quality of chia (*Salvia hispanica* L.) seeds and plant during growth. Anim Feed Sci Technol 148:267–275

Pekel AY, Patterson PH, Hulet RM, Acar N, Cravener TL, Dowler DB, Hunter JM (2009) Dietary camelina meal versus flaxseed with and without supplemental copper for broiler chickens: live performance and processing yield. Poult Sci 88:2392–2398

Perry BA (1943) Chromosome number and phylogenetic relationships in the Euphorbiaceae. Am J Bot 30:527–543

Petit HV (2002) Digestion, milk production, milk composition, and blood composition of dairy cows fed whole flaxseed. J Dairy Sci 85:1482–1490

Petit HV, Dewhurst RJ, Scollan ND, Proulx JG, Khalid M, Haresign W, Twagiramungu H, Mann GE (2002) Milk production and composition, ovarian function, and prostaglandin secretion of dairy cows fed omega-3 fats. J Dairy Sci 85:889–899

Philip TP, Al D (2009) The promise and challenges of microalgal-derived biofuels. Biofuels Bioprod Biorefin 3:431–440

Prasad K (2009) Flaxseed and cardiovascular health. J Cardiovasc Pharmacol 54:369–377

Prévôt A, Perrin JL, Laciaverie G, Auge P, Coustille JL (1990) A new variety of low-linolenic rapeseed oil: characteristics and room-odor tests. J Am Oil Chem Soc 67(3):161–164

Prina AO, Martinez-Laborde JB (2008) A taxonomic revision of *Crambe* section *Dendrocrambe* (Brassicaceae). Bot J Linn Soc 156:291–304

Quick GR (1989) Oilseed as energy crops. In: Robbelen G, Downey RK, Ashri A (eds) Oil crops of the world. McGraw-Hill, New York, pp 118–131

Rafter JJ (2002) Scientific basis of biomarkers and benefits of functional foods for reduction of disease risk: cancer. Br J Nutr 88:S219–S224

Rakotonirainy AM, Padua GW (2001) Effects of lamination and coating with drying oils on tensile and barrier properties of zein films. J Agric Food Chem 49:2860–2863

Rao S, Abdel-Reheem M, Bhella R, McCracken C, Hildebrand D (2008) Characteristics of high α-linolenic acid accumulation in seed oils. Lipids 43:749–755

Reddy KRK, Bahadur B (1989) Adventitious bud formation from leaf cultures of castor (*Ricinus communis* L). Curr Sci 58:152–154

Reeves JB III, Weihrauch JL (1979) Composition of foods: fats and oils. Agriculture Handbook 8-4, United States Department of Agriculture, Human Nutrition Information Service, Washington DC

Regost C, Arzel J, Cardinal M, Rosenlund G, Kaushik SJ (2003) Total replacement of fish oil by soybean or linseed oil with a return to fish oil in turbot (*Psetta maxima*). 2. Flesh quality properties. Aquaculture 14:737–747

Rojas-Barros P, Haro Ad, Munoz J, Fernandez-Martinez JM (2004) Isolation of a natural mutant in castor with high oleic/low ricinoleic acid content in the oil. Crop Sci 44:76–80

Rojas-Barros P, de Haro A, Fernandez-Martinez JM (2005) Inheritance of high oleic/low ricinoleic acid content in the seed oil of castor mutant *OLE-1*. Crop Sci 45:157–162

Rollin X, Peng J, Pham D, Ackman RG, Larondelle Y (2003) The effects of dietary lipid and strain difference on polyunsaturated fatty acid composition and conversion in anadromous and landlocked salmon (*Salmo salar* L.) parr. Comp Biochem Physiol 134B:349–366

Rosentrater KA (2009) Distillers' dried grains with solubles (DDGS): a key to the fuel ethanol industry. Inform 20(12):789–800

Routaboul J-M, Benning C, Bechtold N, Caboche M, Lepiniec L (1999) The TAG1 locus of Arabidopsis encodes for a diacylglycerol acyltransferase. Plant Physiol Biochem 37:831–840

Ruuska SA, Schwender J, Ohlrogge JB (2004) The capacity of green oilseeds to utilize photosynthesis to drive biosynthetic processes. Plant Physiol 136:2700–2709

Salunkhe DK, Chavan JK, Adsule RN, Kadam SS (1992) World oilseeds: chemistry, technology, and utilization. AVI, New York

Sanford SD, White JM, Shah PS, Wee C, Valverde MA, Meier GR (2009) Feedstock and biodiesel characteristics report. Renewable Energy Group, Ames, IA. Available at: www.regfuel.com

Sarmiento C, Ross JH, Herman E, Murphy DJ (1997) Expression and subcellular targeting of a soybean oleosin in transgenic rapeseed. Implications for the mechanism of oil-body formation in seeds. Plant J 11:783–796

Schill SR (2008) Roundtable for sustainable biofuel releases proposed standards for review, Biomass Magazine. Available via http://www.biomassmagazine.com/article.jsp?article_id=1914

Schlueter JA, Vasylenko-Sanders IF, Deshpande S, Yi J, Siegfried M, Roe BA, Schlueter SD, Scheffler BE, Shoemaker RC (2007) The FAD2 gene family of soybean: insights into the structural and functional divergence of a paleopolyploid genome. Crop Sci 47(S1):S14–S26

Scholza V, da Silva JN (2008) Prospects and risks of the use of castor oil as a fuel. Biomass Bioenergy 32:95–100

Settlage SB, Kwanyuen P, Wilson RF (1998) Relation between diacylglycerol acyltransferase activity and oil concentration in soybean. J Am Oil Chem Soc 75:775–781

Shanklin J, Cahoon EB, Whittle E, Lindqvist Y, Huang W, Schneider G, Schmidt H (1997) Structure–function studies on desaturases and related hydrocarbon hydroxylases. In: Williams JP (ed) Physiology, biochemistry, and molecular biology of plant lipids. Kluwer, Dordrecht, pp 6–10

Shearer AEH, Davies CGA (2005) Physicochemical properties of freshly baked and stored wholewheat muffins with and without flaxseed meal. J Food Qual 28:137–153

Shockey J, Gidda S, Burgal J, Chapital D, Kuan J-C, Rothstein S, Mullen R, Browse J, Dyer J (2006a) A new magic bullet? Type-2 diacylglycerol acyltransferases are key components to novel fatty acid accumulation in transgenic systems. In: Proceedings of 17th International Symposium on Plant Lipids, East Lansing, MI

Shockey JM, Gidda SK, Chapital DC, Kuan J-C, Dhanoa PK, Bland JM, Rothstein SJ, Mullen RT, Dyer JM (2006b) Tung tree DGAT1 and DGAT2 have nonredundant functions in triacylglycerol biosynthesis and are localized to different subdomains of the endoplasmic reticulum. Plant Cell 18:2294–2313

Shoemaker R, Grant D, Olson T, Warren WC, Wing R, Cregan P, Joseph B, Futrell-Griggs M, Nelson W, Davito J, Walker J, Wallis J, Kremitski C, Scheer D, Clifton S, Graves T, Nguyuen H, Wu X, Luo M, Dvorak J, Cannon S, Thomkins J, Schmutz J, Stacey G, Jackson S (2008) Microsatellite discovery from BAC end sequences and genetic mapping to anchor the soybean physical and genetic maps. Genome 51:294–302

Siloto RMP, Findlay K, Lopez-Villalobos A, Yeung EC, Nykiforuk CL, Moloney MM (2006) The accumulation of oleosins determines the size of seed oilbodies in Arabidopsis. Plant Cell 18:1961–1974

Slack CR, Roughan PG, Browse JA, Gardiner SE (1985) Some properties of choline phosphotransferase from developing safflower cotyledons. Biochim Biophys Acta 833:438–448

Spencer JD, Thornton T, Muir AD, Westcott ND (2003) The effect of flax seed cultivars with differing content of alpha-linolenic acid and lignans on responses to mental stress. J Am Coll Nutr 22:494–501

Ståhl U, Carlsson A, Lenman M, Dahlqvist A, Huang B, Bana W, Bana A, Stymne S (2004) Cloning and functional characterization of a phospholipid:diacylglycerol acyltransferase from Arabidopsis. Plant Physiol 135:1324–1335

Stareborn M (2009) Unilever calls for moratorium on deforestation tropical rainforest. Available via http://www.unilever.com/mediacentre/pressreleases/2009/Unilevercallsformoratoriumondeforestationoftropicalrainforest.aspx

Stobart K, Mancha M, Lenman M, Dahlqvist A, Stymne S (1997) Triacylglycerols are synthesised and utilized by transacylation reactions in microsomal preparations of developing safflower (*Carthamus tinctorius* L.) seeds. Planta 203:58–66

Sujatha M, Sailaja M (2005) Stable genetic transformation of castor (*Ricinus communis* L.) via *Agrobacterium tumefaciens*-mediated gene transfer using embryo axes from mature seeds. Plant Cell Rep 23:803–810

Sujatha M, Reddy TP, Mahasi MJ (2008) Role of biotechnological interventions in the improvement of castor (*Ricinus communis* L.) and *Jatropha curcas* L. Biotechnol Adv 26:424–435

Suthar B, Parikh N, Patel N (1991) Interpenetrating polymer networks from castor-oil based polyurethanes and poly(methyl methacrylate) XX. Int J Polym Mater 15:85–91

Taga MS, Miller EE, Pratt DE (1984) Chia seeds as a source of natural lipid antioxidants (*Salvia hispanica*, potential as food antioxidants). J Am Oil Chem Soc 61:928–931

Tan CP, Che Man YB, Selanmat J, Yusoff MSA (2002) Comparative studies of oxidative stability of edible oils by differential scanning calorimetry and oxidative stability index methods. Food Chem 76:385–389

Tat ME, Wang PS, Van Gerpen JH, Clemente TE (2007) Exhaust emissions from an engine fueled with biodiesel from high-oleic soybeans. J Am Oil Chem Soc 84:865–869

Thomas Jefferson Agricultural Institute (2007) http://www.jeffersoninstitute.org/flax.php

Thompson AE, Dierig DA (1994) Initial selection and breeding of *Lesquerella fendleri*, a new industrial oilseed. Ind Crops Prod 2:97–106

Thompson LU, Rickard SE, Orcheson LJ, Seidl MM (1996) Flaxseed and its lignin and oil components reduce mammary tumor growth at a late stage of carcinogenesis. Carcinogenesis 17:1373–1376

Tongoona P (1992) Castor (*Ricinus communis* L.) research and production prospects in Zimbabwe. Ind Crops Products 1:235–239

Triki S, Ben Hamida J, Mazliak P (1998) About the reversibility of the cholinephosphotransferase in developing sunflower seed microsomes. In: Sanchez J, Cerda-Olmedo E, Martinze-Force E (eds) Advances in plant lipid research. University of Sevilla, pp 236–239

Turkish AR, Henneberry AL, Cromley D, Padamsee M, Oelkers P, Bazzi H, Christiano AM, Billheimer JT, Sturley SL (2005) Identification of two novel human acyl-CoA wax alcohol acyltransferases: members of the diacylglycerol acyltransferase 2 (DGAT2) gene superfamily. J Biol Chem 280:14755–14764

Tzen JTC, Lai YK, Chan KL, Huang AHC (1990) Oleosin isoforms of high and low molecular weights are present in the oil bodies of diverse seed species. Plant Physiol 94:1282–1289

USDA (1989) Composition of foods: cereal grains and pasta. Human Nutrition Information Service. Agricultural Handbook 8–20. US Department of Agriculture, Washington, DC

USDA (2009a) World Agricultural Supply and Demand Estimates, WASDE–477, Economic Research Service. US Department of Agriculture, Washington, DC

USDA (2009b) Oilseeds: world markets and trade, FOP 10–09, Foreign Agricultural Service, US Department of Agriculture, Washington, DC

USDA (2009c) Energy life-cycle assessment of soybean biodiesel, Agricultural Economic Report 845, Economic Research Service, US Department of Agriculture, Washington DC

US DOE (2009a) How dependent is the US on foreign oil? Energy Information Administration, US Department of Energy, Washington DC Available via: tonto.eia.doe.gov/ask/crudeoil_faqs.asp#foreign_oil

US DOE (2009b) US crude oil supply and disposition, Energy Information Administration, US Department of Energy, Washington DC. Available via: tontoeia.doe.gov/oog/info/gdu/dieselpump.html

US DOE (2009c) Diesel fuel component history. Energy Information Administration, US Department of Energy, Washington DC. Available via: tonto.eia.doe.gov/oog/info/gdu/dieselpump.html

USFDA (2003) Final Rule: trans fatty acids in nutrition labeling, nutrient content claims and health claims, 21 CFR Part 101A. US Food and Drug Administration, Washington DC

Van de Loo FJ, Fox BG, Somerville C (1993) Unusual fatty acids. In: Moore JTS (ed) Lipid metabolism in plants. CRC, Boca Raton, pp 91–126

Vasquez OA, Rosado RG, Chel GL, Betancur AD (2009) Physicochemical properties of a fibrous fraction from chia (*Salvia hispanica* L.). LWT Food Science Technol 42:168–173

Vera-Diaz MC, Kaufmann RK, Nepstad DC (2009) The environmental impacts of soybean expansion and infrastructure development in Brazil's Amazon basin. Global Development and Environmental Institute Working Paper 09–05. Tufts University, Medford MA

Voelker TA, Worell AC, Anderson L, Bleiaum J, Fan C, Hawkins DJ, Radke SE, Davies HM (1992) Fatty acid biosynthesis redirected to medium chains in transgenic oilseed plants. Science 257:72–74

Voelker TA, Hayes TR, Cranmer AM, Turner JC, Davies HM (1996) Genetic engineering of a quantitative trait: metabolic and genetic parameters influencing the accumulation of laurate in rapeseed. Plant J 9:229–241

Vollmann J, Moritz T, Kargl C, Baumgartner S, Wagentristl H (2007) Agronomic evaluation of camelina genotypes selected for seed quality characteristics. Ind Crops Prod 26:270–277

Vuksan V, Whitham D, Sievenpiper J-L, Jenkins A-L, Rogovik A-L, Bazinet R-P, Vidgen E, Hanna A (2007) Supplementation of conventional therapy with the novel grain salba (*Salvia hispanica* L.) improves major and emerging cardiovascular risk factors in type 2 diabetes: results of a randomized controlled trial. Diabetes Care 30:2804–2810

Vuksan V, Jovanovski E, Dias A, Lee A, Rogovik A, Jenkins A (2009) Comparable dose-response glucose-lowering effect with whole versus finely ground novel omega-3-rich grain salba (*Salvia hispanica* L.) baked into white bread. Pharm Biol 47:S13

Vyacheslav A, Nikolai B, Natalia B, Anita B, Joseph D, Sergei S, John F, Paulina M, Karolina A, Marilyn L, Maxim G, Hilary K (2009) Tobacco as a production platform for biofuel: over-expression of *Arabidopsis DGAT* and *LEC2* genes increases accumulation and shifts the composition of lipids in green biomass. Plant Biotechnol J 8:1–11

Waltermann M, Steinbuchel A (2005) Neutral lipid bodies in prokaryotes: recent insights into structure, formation, and relationship to eukaryotic lipid depots. J Bacteriol 187:3607–3619

Wang HW, Zhang B, Hao YJ, Huang J, Tian AG, Liao Y, Zhang JS, Chen SY (2007) The soybean Dof-type transcription factor genes, *GmDof4* and *GmDof11*, enhance lipid content in the seeds of transgenic Arabidopsis plants. Plant J 52:716–729

Wang HY, Guo JH, Lambert KN, Lin Y (2007) Developmental control of Arabidopsis seed oil biosynthesis. Planta 226:773–783

Wang WY, Wang CG, Huang BL, Huang BQ (2008) *Agrobacterium tumefaciens*-mediated transformation of *Lesquerella fendleri* L., a potential new oil crop with rich lesquerolic acid. Plant Cell Tissue Organ Cult 92:165–171

Ward AT, Wittenberg KM, Przybylski R (2002) Bovine milk fatty acid profiles produced by feeding diets containing solin, flax and canola. J Dairy Sci 85:1191–1196

Warrand J, Michaud P, Picton L, Muller G, Courtois B, Ralainirina R, Courtois J (2005) Flax (*Linum usitatissimum*) seed cake: a potential source of high molecular weight arabinoxylans. J Agric Food Chem 53:1449–1452

Warwick SI, Gugel RK (2003) Genetic variation in the *Crambe abyssinica–C. hispanica–C. glabrata* complex. Genet Resour Crop Evol 50:291–305
Watkins C (2009a) Oilseeds of the future: Part 1. Inform 20(5):276–279
Watkins C (2009b) Oilseeds of the future: Part 2. Inform 20(6):342–347
Watkins C (2009c) Oilseeds of the future: Part 3. Inform 20(7):408–410
Weill P, Schmitt B, Chesneau G, Daniel N, Safraou F, Legrand P (2002) Effects of introducing linseed in livestock diet on blood fatty acid composition of consumers of animal products. Ann Nutr Metab 46:182–191
Weiss TJ (1983) Food oils and their uses, 2nd edn. AVI, Westport CN
Weselake RJ, Taylor DC, Rahman MH, Shah S, Laroche A, McVetty PBE, Harwood JL (2009) Increasing the flow of carbon into seed oil. Biotechnol Adv 27:866–878
Westcott P (2007) Agricultural projections to 2016, OCE 2007-01. United States Department of Agriculture, Washington, DC
Wilcox JR (2004) World distribution and trade of soybean. In: Boerma HR, Specht JE (eds) Soybeans: improvement, production and uses, 3rd edn. American Society of Agronomy, Madison, pp 1–14
Wilson RF (2004) Seed composition. In: Boerma HR, Specht JE (eds) Soybeans: improvement, production and uses, 3rd edn. American Society of Agronomy, Madison, Wisconsin, pp 621–678
Wittkop B, Snowdon R, Friedt W (2009) Status and perspectives of breeding for enhanced yield and quality of oilseed crops for Europe. Euphytica 170:131–140
Wrobel M, Zebrowski J, Szopa J (2004) Polyhydroxybutyrate synthesis in transgenic flax. J Biotechnol 41–54
Yamamoto K, Kinoshita A, Shibahara A (2008) Ricinoleic acid in common vegetable oils and oil seeds. Lipids 43:457–460
Yang CS, Landau JM, Huang MT, Newmark HL (2001) Inhibition of carcinogenesis by dietary polyphenolic compounds. Annu Rev Nutr 21:381–406
Yuan L, Voelker TA, Hawkins DJ (1995) Modification of the substrate specificity of an acyl-acyl carrier protein thioesterase by protein engineering. Proc Natl Acad Sci USA 92:10639–10643
Zhang F-Y, Yang M-F, Xu Y-N (2005) Silencing of *DGAT1* in tobacco causes a reduction in seed oil content. Plant Sci 169:689–694
Zhang P, Burton JW, Upchurch RG, Whittle E, Shanklin J, Dewey RE (2008) Mutations in a D-9 stearoyl-ACP-desaturase gene are associated with enhanced stearic acid levels in soybean seeds. Crop Sci 48:2305–2313
Zheng X, Tocher DR, Dickson CA, Bell JG, Teale AJ (2004) Effects of diets containing vegetable oil on expression of genes involved in highly unsaturated fatty acid biosynthesis in liver of Atlantic salmon (*Salmo salar*). Aquaculture 236:467–483
Zhou X-R, Singh S, Green A (2008) Increased accumulation of epoxy fatty acids in Arabidopsis by transgenic expression of TAG assembly genes from *Bernardia pulchella*. In: Proceedings of 18th International Symposium on Plant Lipids, Bordeaux, France
Zou J, Wei Y, Taylor DC (1999) The *Arabidopsis thaliana* TAG1 mutant has a mutation in a diacylglycerol acyl transferase gene. Plant J 19:645–654

Part C
Mitigating Invasiveness

Chapter 9
Invasive Species Biology, Ecology, Management and Risk Assessment: Evaluating and Mitigating the Invasion Risk of Biofuel Crops

Jacob N. Barney and Joseph M. DiTomaso

9.1 Biofuel Crops and Invasive Species

The global demand for biomass-based renewable energy continues to grow in an effort to reduce petroleum product dependence, stimulate rural economies, and stabilize national security. Many countries are mandating increasing amounts of liquid transportation fuels be biomass-based over the coming century. For example, in 2007 the US passed the Energy Independence and Security Act, which sets a 136 billion L goal for renewable liquid fuels by 2022, with 44% being derived from cellulose. Estimates vary, but approximately 60 million additional hectares of land will need to be cultivated to meet this mandate (Robertson et al. 2008), with projections of 1.5 billion ha required globally (Field et al. 2008). As an incentive to stimulate grower adoption of these novel crops, the US included a subsidy of US $45 per ton of eligible crops as part of the 2008 Farm Bill. Therefore, research effort is focused on identifying crops that will maximize yield while allowing cultivation on less productive, marginal lands.

Unlike traditional food, feed, and fiber crops, biofuel crops are being selected to be maximally productive on marginal land, which requires they be easy to establish, highly competitive, and thrive with minimal human intervention (Meyerson 2008). The most promising crops are perennial rhizomatous grasses that exhibit rapid growth rates, possess broad climatic tolerance, tolerate poor growing conditions, harbor few pests, and require minimal inputs (Table 9.1; Lewandowski et al. 2003). However, many of these agronomically desirable traits are shared by many of our worst invasive species (Raghu et al. 2006). Additionally, several candidate crop species are known invasive or noxious species, for which risk analyses indicate a high likelihood of invasiveness in target cropping regions (Barney and DiTomaso 2008). The majority of our worst invasive species were introduced intentionally

---

J.N. Barney and J.M. DiTomaso
Department of Plant Sciences, University of California, Davis, CA 95616, USA
e-mail: jbarney@ucdavis.edu; jmditomaso@ucdavis.edu

Table 9.1 Plant traits under selection for improved biofuel crop performance and economic suitability that overlap with characters of many invasive species. Comparison among traditional field crops, potential biofuel crops, and known invasive species that were introduced for agronomic purposes. Adapted from (DiTomaso et al. 2007)

	Agronomic crops		Biofuel crops			Invasive species with agronomic origin	
	Corn	Soybean	Switchgrass	*Miscanthus × giganteus*	Giant reed	Johnsongrass	Kudzu
Perennial	–	–	×	×	×	×	×
C_4 photosynthesis	×	–	×	×	–	×	–
Rapid establishment	×	×	–	×	×	×	×
Highly competitive	–	–	×/–	×/–	×	×	×
Drought tolerant	×	×	×	×	×	×	–
Salt tolerant	–	–	–	–	–	–	–
Reallocation of nutrients to roots	–	–	×	×	×	×	×
No major pests/diseases	–	–	×	×	×	×	×
Disperses readily from aboveground vegetative fragments	–	–	–	×	×	–	×
Prolific viable seed production	×	×	×	–	–	×	×

(Simberloff 2008), with some invasive species receiving government support for widespread adoption (Barney and DiTomaso 2008).

As noted by Raghu et al. (2006), the US may be unintentionally putting policy mandates at variance by creating a bioeconomy that relies on novel crops (EISA 2007) while preventing introduction and dissemination of invasive species (Executive Order 13112). We would be remiss to assume that widespread adoption of novel species is inherently safe, or dangerous, on the basis that they are crops (Barney and DiTomaso 2008). Rather, an empirical science-based evaluation of the invasive potential of each candidate biofuel crop is necessary, and should be conducted within each target cropping region (DiTomaso et al. 2007), not for broad climatically variable geographies. The invasion risk, along with other potential negative impacts (e.g., water use, biodiversity loss), should be contrasted with the economic, environmental, and other benefits before widespread adoption to ensure sustainable development of these potentially important crops (Robertson et al. 2008).

9.2 Invasive Species Biology and Ecology

> "For the most part, successful invasion is forever"—Daniel Simberloff (2005)

The US government defines an invasive species as an "alien (non-native) species whose introduction does or is likely to cause economic or environmental harm or harm to human health" (Executive Order 13112). In contrast, ecologists define a species as invasive when naturalized plants produce reproductive offspring, often in very large numbers, at considerable distances from parent plants, that are able to spread over broad areas (Richardson et al. 2000). Conceptual details aside, invasive species cost the global economy trillions in lost revenue and management costs annually (DiTomaso 2000; Pimentel et al. 2000), and are cited as the second greatest threat to biodiversity (Mack et al. 2000). Invasive species are known to reduce native species biodiversity (Gurevitch and Padilla 2004), alter biogeochemical cycles (Vitousek et al. 1987), modify food webs (Savidge 1987), transform natural disturbance regimes (D'Antonio and Hobbie 2005), increase fire frequency and intensity (Brooks et al. 2004), and reduce deep water storage (Sala et al. 1996).

Despite the dire ecological consequences some non-native species inflict, most introduced plants have neutral consequences while many provide tremendous benefits to society. Most of our food crops, fiber crops, and construction materials are non-native species that humans have domesticated to meet our needs (Sax et al. 2005). Domestication does not guarantee safe introduction of non-native species, as most of our worst invasive species were introduced intentionally (Simberloff 2008), many of which were provided environmental subsidization from their human introducers (Mack 2000; Reichard and White 2001). For example, kudzu (*Pueraria montana* [Lour.] Merr. var. *lobata* [Wild.] Maesen and S. Almeida) was promoted by the federal government as a forage crop, and later as a soil stabilizer, only to quickly escape anthropogenically imposed boundaries and dominate ecosystems regionally (Forseth Jr and Innis 2004). Japanese knotweed (*Polygonum cuspidatum* Sieb. & Zucc.)—a robust rhizomatous perennial with 'clouds of bloom'—was favored by the eminent horticulturalist Liberty Hyde Bailey in the early twentieth century, only to recant that endorsement a few years later when the species began to take over gardens and surrounding natural areas (Bailey 1916). Therefore, parsing the beneficial non-native species from the potentially harmful species is challenging at best, with enormous economic and environmental consequences for making an erroneous decision.

Despite numerous attempts, no checklist or protocol exists to assess the invasive potential of an introduced species (Pyšek and Richardson 2007). Frequent attempts have been made to find commonalities among introduced invasive species (Hayes and Barry 2008), or discover unique characteristics that separate them from introduced non-weedy species and native species (Sutherland 2004). In the middle of the twentieth century, Herbert Baker attempted to parse weeds—plants growing "entirely or predominantly in situations markedly disturbed by man (without being deliberately cultivated plants)"—from non-weeds by surveying plant traits (Baker 1965, 1974). However, this list of "Baker traits" has since been proven too simplistic (Pyšek and Richardson 2007). Despite the inconsistency in trait-based determinations of invasiveness, ecologists continue to search for commonalities. In an extensive survey of the North American flora, Sutherland (2004) found that invasive species were more likely to be perennial, monoecious, and self-incompatible compared to non-weedy introduced and native species. These characteristics statistically distinguish the chosen categories of plants, though the predictive capacity remains limited.

In spite of the limitations of trait-based assessments in segregating invasive from benign species, this methodology has become the foundation for pre-introduction risk assessments (see Sect. 9.3.1). Several protocols have been developed that are in various stages of implementation across the globe (Reichard and Hamilton 1997; Weber et al. 2009). The protocols are highly accurate in identifying invasive species (Gordon et al. 2008), but are severely limited for species with little published information, and have not been tested with crops or genetically modified plants. The necessity for basic biological and ecological information on introduced species becomes evident when invasions are viewed through the complex spatiotemporal lens that begins with an introduction and concludes with naturalized populations spreading beyond their local foci and causing ecological damage (Theoharides and Dukes 2007). There are numerous points along this continuum that the introduced species must pass through, including abiotic (e.g., climate, edaphic, disturbance) and biotic (e.g., herbivores, competitors) filters (Richardson et al. 2000; Theoharides and Dukes 2007). Additionally, habitats vary in their susceptibility to invasion, which is a function of colonization pressure (i.e., the number of species introduced to a location) and propagule pressure (i.e., the number of individuals and events introduced to a single location) (Lockwood et al. 2009) in light of the characteristics of that habitat (Barney and Whitlow 2008).

It is well recognized that invasions are not merely the result of stochastic events among randomly selected species. Additionally, a non-native species that is invasive in one region would not necessarily be invasive in all introduced regions—hence, the term invasive should not be applied globally to any species. Therefore, there are several critical components of the invasion process that need to be identified: (1) species characteristics, (2) receiving environment characteristics, and (3) the propagule pressure of each species to each habitat (Barney and Whitlow 2008). Complicating matters is the temporal aspect of invasions—introduced species do not manifest invasions immediately, rather they are protracted over periods of time: typically termed lag phases (Kowarik 1995). This has forced invasion ecology to become largely a post hoc science, because a species is not invasive until it is invasive. Despite these limitations, invasion ecology has developed into its own field of inquiry incorporating aspects of plant ecology, biogeochemistry, ecosystem succession, propagule biology (e.g., seed science), soil science, evolution, and plant–plant interactions (Sax et al. 2005).

One aspect of integrating biomass-based crops into our energy portfolio sustainably will be to utilize crops that present an acceptably low risk of becoming invasive. All introduced species present some risk of becoming invasive. The key is to identify that risk, and mitigate against known hazards. Ultimately, we will be required to balance the risks associated with biofuel use against the economic and national security benefits of the nascent bioeconomy (Meyerson 2008). Here we outline the potential risks posed by biofuel crops, methods for identifying those risks, and recommendations for mitigating against the introduction of future invasive species.

9.3 Assessing the Invasive Risk of Biofuel Crops

The combined factors of the scales of cultivation, the traits for which biofuel crops are being selected, the fact that many of the species are known to be invasive elsewhere (or at a minimum, non-native in the target cropping region), and the likelihood of unintentional propagule dissemination during harvest, transport, and storage result in a non-trivial probability that biomass crops may escape the cultivated environment and become invasive pests. The typical crop development, introduction, and commercialization timeframe will be highly compressed to meet the 2022 EISA mandate of 61 billion L cellulosic-based transportation fuels. This may lead to rash adoption of novel crops in the name of expedited production. However, we believe that with appropriate screening and cogent application of mitigation strategies at various points along the biofuel supply chain (see Sect. 9.4), expedient biofuel crop implementation can be efficient and environmentally safe.

The projected scales of cultivation for dedicated energy crops is astounding—60 million ha in the US (Robertson et al. 2008), and 1.5 billion ha across the globe by 2050 (Field et al. 2008). The FAO (2007) estimates that 1.25 billion ha were harvested in 2007 globally, which with future biofuel production would more than double the global cultivated land. This extraordinary demand will require crops that are high-yielding on historically less productive lands—so-called marginal land. Therefore, there exists a need to evaluate the invasive potential of candidate biofuel crops in each target cropping region.

We have outlined a series of studies that can be performed in parallel with agronomic studies that are aimed at identifying the invasive potential of candidate biofuel crops (DiTomaso et al. 2007). Below we detail each step in the overall assessment (see Sects. 9.3.1–9.3.7), which is then followed by mitigation recommendations at each point along the biofuel supply chain based on the proposed invasive assessment (see Sect. 9.4).

9.3.1 Risk Assessment

Risk assessment tools have been in development for decades to aid in decision making for proposed introductions of novel species for horticultural, agronomic, and other purposes. Several federal governments have instituted risk assessment protocols for pre-introduction evaluation with billions saved in economic benefits (Keller et al. 2007). Risk assessment should serve as a basic first step in evaluating the invasive potential of biofuel crops whether they are exotic species, native species, novel constructs (e.g., hybrids), or genetically modified species (Cousens 2008).

Australia has been overwhelmed with introduced species that bring about dramatic environmental damage—costing $4(Aus) billion per annum (Sinden et al. 2004). The Weed Risk Assessment (WRA) system was developed to allow entry of non-native species into Australia, while preventing the introduction of potentially harmful invasive species, and has been in official use since 1997 (Weber et al. 2009). This assessment is a simple protocol of 49 questions answered "yes", "no", or "unknown", which result in a final score that categorizes a species' invasiveness risk (Pheloung et al. 1999). The categories "accept", "reject", and "evaluate further", based on the final score, were validated on known datasets. The WRA has since been tested in seven regions globally, and identifies major invasive species with >90% accuracy (Gordon et al. 2008).

The WRA has been used to screen potential biofuel crops in several regions of the US (Barney and DiTomaso 2008; Buddenhagen et al. 2009; D.R. Gordon, K.J. Tancig, D.A. Onderdonk, and C.A. Gantz, unpublished data). Barney and DiTomaso (2008) screened three of the leading biofuel crops in the US: switchgrass (*Panicum virgatum* L.), giant miscanthus (*Miscanthus* × *giganteus*), and giant reed (*Arundo donax* L.). Switchgrass is native to most of North America east of the Rocky Mountains and has been identified by the US Department of Energy as a model biofuel crop (Parrish and Fike 2005). In the non-native range of California, switchgrass was found to be potentially invasive (Barney and DiTomaso 2008). However, a hypothetical sterile cultivar (i.e., no seed production) was found to have a low likelihood of invasiveness, and could be safely introduced to California. In contrast, the sterile perennial giant reed, which is one of the worst invasive species of California, was found to have a very high invasive potential in Florida where a large plantation is planned (Mack 2008). *Miscanthus* × *giganteus*, which is a naturally occurring sterile hybrid of *M. sinensis* and *M. sacchariflorus*, was found to have a low invasive potential in the US, despite sharing similar life history characters and growth habit with the known invasive giant reed (Barney and DiTomaso 2008).

In a comparison of 40 proposed biofuel crops and 40 introduced non-biofuel species, Buddenhagen et al. (2009) found that 70% of regionally suitable biofuel crops have a high risk of becoming invasive in Hawaii, and are two- to four-times more likely to establish wild populations. Gordon and colleagues (D.R. Gordon, K.J. Tancig, D.A. Onderdonk, and C.A. Gantz, unpublished data) found 7 of 12 proposed biofuel crops have a high probability of becoming invasive pests in Florida. A qualitative assessment of the invasive potential of biofuel crops in Australia deemed many potential crops as high risk (Low and Booth 2007). Therefore, the WRA suggests that many of the highly touted biofuel crops have a high risk of becoming invasive pests in their target cropping regions. It should be noted that the WRA was not designed to evaluate species intended for biofuel production, and may not be ideally suited to identifying invasive risk (Cousens 2008). A more targeted risk assessment protocol designed specifically for biofuel crops may lead to a more robust prediction of risk.

The Australian WRA serves as an important first step in assessing the invasive risk of biofuel crops, but should not serve as a sole source for policy recommendations due to the limitations of the model (i.e., WRA was designed primarily for ornamentals for pre-introduction assessment). Further studies should be conducted following initial qualitative risk assessments to begin quantifying the invasive potential in each target region (see Sect. 9.3.2–9.3.7).

9.3.2 Species Biology

Many of the species touted as potential biofuel crops are relatively undomesticated, and are only a few generations from wild-types (Lewandowski et al. 2003). Therefore, very little information exists on the basic biology and ecology of these species. The agronomic potential for many of these crops is largely untested, and the potential to escape the cultivated environment and become invasive pests is wholly unknown (Barney and DiTomaso 2008). Economics will likely dictate that biofuel crops be relegated to less productive regions where competition with food, fuel, and livestock is minimized (Royal Society 2008). Therefore, developers are generating crops that require minimal inputs of nutrients, water, and pesticides. In many cases, however, we do not have baseline information on the basic nutrient or water requirements, which reduces the efficacy of agronomic modeling and makes basic risk assessments difficult. Therefore, there exists the need to characterize the environmental tolerances of each biofuel crop, regardless of species' nativity, and to identify ecosystems most susceptible to invasion. Once described, these factors can be integrated into risk analysis and bioclimatic and agronomic models to estimate, and subsequently mitigate, the likelihood of invasion (Barney and DiTomaso 2008), thus leading to safer and more sustainable use of these important potentially important crops (Robertson et al. 2008).

Basic biological studies that relate to environmental tolerance to various abiotic factors will facilitate agronomic development, as well as assisting in climate-match modeling (see Sect. 9.3.3), determining habitat susceptibility (see Sect. 9.3.5), and risk assessment (see Sect. 9.3.1). Abiotic factors of interest include, but are not limited to, soil moisture, cold tolerance, heat tolerance, shade tolerance, salt tolerance, and nutrient requirements. Studies should begin by focusing on the extremes of each abiotic variable compared to "stress-free" controls, as the environments most likely to be cultivated will be marginal in their productivity.

We have begun to quantify the abiotic stress bandwidth (or environmental tolerance) of two of the leading biofuel crops: switchgrass and *Miscanthus × giganteus*. In the western US, water is a limiting resource for agricultural development, with the additional acreage of biomass crops adding strain to this limited resource (Schnoor et al. 2008). Therefore, our preliminary studies have focused on defining the soil moisture stress tolerance of switchgrass and *M. × giganteus* as they relate to known invasive species. Four common switchgrass cultivars belonging to upland (Blackwell, Cave-In-Rock) and lowland (Alamo, Kanlow) ecotypes were found to be tolerant of water deficit conditions (–4 MPa) and flooded conditions (Barney et al. 2009). Seeds of all accessions were capable of germinating and establishing in soils ranging from –0.3 MPa (10% soil moisture) to under water in flooded soils. Lowland ecotypes display higher fitness under all conditions, and are the primary target of germplasm improvement (Parrish and Fike 2005). Tolerance to a breadth of soil moisture conditions greatly increases the likely cultivatable range of switchgrass, but also increases the number of potentially susceptible environments to invasion.

We have also compared the soil moisture stress tolerance of *M. × giganteus* to giant reed (*A. donax*) due to the similarities in habit and life history traits (Mann et al. 2009). Giant reed is one of the worst invasive species in California, causing tremendous environmental damage resulting from a single clone—similar to the single *M. × giganteus* genotype. Both giant reed and *M. × giganteus* had reduced biomass at –0.5 and –4.0 MPa whether established for 8 weeks or newly planted. However, under flooded conditions both species performed as well as controls. Survival was high in both species at both establishment levels, and suggests that *M. × giganteus* has similar soil moisture stress tolerance as the invasive giant reed. This study alone does not allow conclusions to be made regarding invasiveness, but does provide one aspect of the invasiveness matrix.

Due to the hurried pace of the nascent bioeconomy, genetic modification will likely play a starring role in bringing these largely undomesticated crops to commercial status (Yuan et al. 2008). Gressel (2008) remarked that it is "naïve, ignorant of history, or conceited to think that one can efficiently grow species as biofuel crops that have not been domesticated for that purpose". He goes on to state that biotechnology will be necessary to provide efficient and sustainable biomass production (Gressel 2008). Therefore, basic biological studies will serve as baselines for genetic improvement. Additionally, engineering of crop species will necessitate unique assessments for invasiveness, even of native species like switchgrass. Modified switchgrass may not require a wholesale novel assessment, but sufficient study is necessary to characterize the effects of the transgenes on ecological interactions with abiotic and biotic components of the target environment.

9.3.3 Niche Modeling

It is well known that the natural distribution of a species is controlled largely by climate (Pearson and Dawson 2003), with precipitation and temperature playing the dominant roles (Sutherst 2003). One method for estimating a species' range in a novel region is to model the bioclimatic envelope, which is the relationship between climate variables and the current distribution (Sutherst 2003). This can be accomplished via correlation analyses, or empirically by relating physiological ecology to specific climate variables, or by a combination of the two methods (Franklin 1995). Bioclimatic envelopes, or "climate match", can be modeled easily to various resolutions, which provides an estimate of range suitability for the species outside cultivation, and also the agronomic potential of the biofuel crop in the target region.

Most weed risk assessments require an evaluation of the climatic suitability of the introduced region—does the target region's climate overlap sufficiently with that of the native range to afford establishment of the species under consideration? Despite this requirement of risk assessment, it is rarely performed due to lack of sufficient information to adequately address climatic suitability (Weber et al. 2009). Therefore, a robust risk assessment would include an empirically derived estimate of climate suitability in the target region.

Bioclimatic envelope models are designed to model the realized climate niche under the assumption that climatic requirements largely dictate distributions, that the taxa is in equilibrium with the current climate in its native range, and that this relationship is conserved across space (Beaumont et al. 2009). There are numerous methods for estimating the bioclimatic envelope: CLIMEX, Maxent, GARP, BIOCLIM, classification and regression tree, and simple logistic regression. Each method has benefits and drawbacks, with some methods capable of utilizing presence-only data (e.g., CLIMEX, Maxent), which is very beneficial when modeling a species' potential distribution. CLIMEX has been used to model the distribution of biocontrol agents (Poutsma et al. 2008), poikilothermic animals (Sutherst et al. 2007), and many invasive species (Holt and Boose 2000; Kriticos et al. 2005; Pattison and Mack 2008). CLIMEX is used widely to model the climatic suitability of an introduced species due to the ability to parameterize the model with empirical data, apply climate change parameters, use climate data at various resolutions, and evaluate parameter sensitivity (Sutherst et al. 2000, 2007). The strength of climate-matching analyses, especially using CLIMEX, is the ability to base a predictive model on the established range (e.g., from herbarium specimen data) and to supplement the model with empirically derived biological and physiological data (Sutherst et al. 1999). CLIMEX is also the suggested model to evaluate the climatic suitability of the target species in the Australian WRA (Pheloung et al. 1999).

As part of our evaluation of potential invasiveness of switchgrass in the non-native range of the western US, especially California, we performed a CLIMEX analysis. Most of the west is unsuitable climatically for switchgrass—due primarily to prolonged dry periods in the summer months (Barney and DiTomaso 2010). We ran a subsequent analysis that included an "irrigation" function that simulates the availability of water all year (*sensu* Pattison and Mack 2008). This "permanent water" scenario, which simulates both an irrigated system, as well as natural systems (e.g., riparian areas), demonstrates that nearly the entire west is climatically suitable (Barney and DiTomaso 2010). Greenhouse studies have confirmed that switchgrass is tolerant of up to 10 weeks of dry conditions, but the Mediterranean summer is rain-free for up to 9 months. This analysis will be used to target specific habitats in the non-native range of the western US, where field studies should be conducted to evaluate the susceptibility to switchgrass invasion. Additional analyses that incorporate climate change scenarios would assist in evaluating the impacts on agronomic production.

9.3.4 Propagule Biology

An invasion is not possible without the introduction of propagules into a novel environment. The number of species introduced to an environment is termed colonization pressure, while the number of introduction events and the number of propagules per event is termed propagule pressure (Lockwood et al. 2009). The probability of a non-native species successfully naturalizing in an introduced environment is proportional to the propagule pressure to that environment (Barney and DiTomaso 2008). In the case of biofuels, propagules (seeds, stem nodes, rhizome fragments) may be unintentionally dispersed during planting, field management, harvesting, shipping, and feedstock storage. With an estimated 60 million ha of land under biofuel cultivation, combined with the tens of thousands of transport kilometers for feedstock shipment, the potential propagule pressure to sensitive habitat is substantial.

To characterize the probability of biofuel crop propagules escaping the cultivated environment, we must understand the basic biology of each propagule type. Many biofuel crops will be perennial rhizomatous grasses, which typically possess three propagule sources: seeds, stem nodes, and rhizome fragments (Lewandowski et al. 2003). Seeds and stem fragments will be the disseminules most likely to be spread unintentionally, and should be the primary focus of propagule biology studies. Rhizome fragments may also serve as potential propagules under large disturbance events (i.e., floods, hurricanes).

Our initial weed risk assessment for the invasive potential of switchgrass in California determined that seed production and potential dissemination was the primary factor influencing its high invasiveness (Barney and DiTomaso 2008). Therefore, we conducted a germination and establishment study of four switchgrass accessions in various soil moisture conditions to evaluate the types of environments switchgrass may establish under (Barney et al. 2009). Some switchgrass seeds germinated and emerged from −0.3 MPa (10% moisture) to under water (flooded). In our study, 55% of the emerged seedlings survived at −0.3 MPa, which is 25% of all seeds and 50% of germinable seeds (55% were dormant or dead). Switchgrass seed production in biofuel crop field trials has been estimated between 300 and 900 kg ha^{-1}, with a mean seed weight of 100 mg per 100 seeds (Boe 2007; Kassel et al. 1985; Sanderson et al. 2004), resulting in 300–900 million seeds ha^{-1}. A conservative estimate of 300 million seeds ha^{-1} and 60% dormancy results in 75 million seeds ha^{-1} able to germinate in mesic soils ($\geq$ −0.3 MPa), and 18 million seeds ha^{-1} able to establish in flooded soils (Barney et al. 2009). Without mitigation practices to reduce the risk of spread, such as the use of sterile cultivars or closed transport systems, propagule pressure to adjacent sensitive ecosystems could be very high.

Unlike switchgrass, giant miscanthus and giant reed both reproduce exclusively via vegetative fragments in the US, and both share similar growth habits. In the invaded range of California and Texas, giant reed disperses along riparian corridors primarily through stem and rhizome fragments that are carried downstream (Khudamrongsawat et al. 2004). Therefore, studies should be conducted to determine the minimum dispersible stem fragment size for giant *Miscanthus* and giant reed and the timing of stem node viability. This will be particularly important if these grass crops are harvested while green and actively growing, and transported in open trucks. The viability of stem nodes is often very high and serve as suitable propagules to colonize new habitat. Harvest, transport, and storage practices (see Sect. 9.4.4) may be varied depending on the results of propagule biology studies.

9.3.5 Habitat Susceptibility

Not all habitats are equally susceptible to invasion by a non-native species, which is a result of the complex biotic and abiotic factors that influence invasion success (Barney and Whitlow 2008). Similarly, not all invasive species are invasive in all environments. Despite this lack of generality, certain patterns have emerged that suggest some habitats may be more susceptible to invasion than others, which may be attributed to the diversity (or lack thereof) of native species (Levine et al. 2004), resource availability (Davis et al. 2000), or a combination of as yet undetermined factors. Climate-matching analyses and biological and physiological studies will aid in identifying habitats most susceptible to invasion by biofuel crops, which can then be tested empirically. Introducing propagules into potentially susceptible habitats under controlled conditions will allow determination of survival and establishment potential under field conditions. Target plants should be monitored throughout all life stages over multiple years to determine if survival, sexual reproduction, and local population growth can be achieved.

Identification of potentially susceptible habitats will be different for each biofuel crop under consideration for each target region. For example, riparian areas of the west are more highly invaded than other habitat types due to the availability of water throughout the year (Levine 2000). Therefore, field studies on habitat invasibility should be focused on the most susceptible ecosystems first and progress from this baseline. If the most susceptible habitat appears to be resistant to invasion by a particular crop than habitats with fewer resources available are even less likely to sustain invasion. All habitats should be considered that not only border the cultivated regions, but also those that will be traversed during shipment, and those that surround storage sites and conversion facilities. Thus, a combination of climate modeling, physiological studies, and field experiments will elucidate the habitats most vulnerable to invasion.

9.3.6 Hybridization Potential

Large-scale introduction of biofuel crops will bring novel species and genotypes into the landscape (Chapotin and Wolt 2007; Cousens 2008). As with genetically modified food and feed crops, screening for possible hybridization with related species should be obligatory to reduce genetic contamination or creation of novel hybrids (DiTomaso et al. 2007). Crop adaptation to various abiotic stressors (e.g., drought, salt, temperature) and yield improvements will likely be necessary to meet many of the mandated biofuel targets (Gressel 2008). Despite many of the crops being non-native to the US and Canada (e.g., *Sorghum bicolor* and *Miscanthus* spp.), close relatives exist that should be screened for possible hybridization, introgression or gene flow. Similarly, crops that are native to North America (e.g., switchgrass) should be held to the same standards as non-native crops where novel genotypes are introduced. Switchgrass comprises a dominant species in many relict prairie stands across North America and serves as the remaining genetic repository, which could be swamped with pollen from thousands of cultivated hectares of modified switchgrass. Hybridization studies would confirm genetic safety of introduced biofuel crops.

9.3.7 Competitive Interactions

Once a potentially susceptible habitat likely to incur propagule pressure from biofuel crops is identified, the competitive ability of each biofuel crop should be evaluated against desirable native species and dominant non-native species within that habitat (Fig. 9.1). The majority of biofuel species will be non-native to the region where they are cultivated and, thus, present unknown threats to native

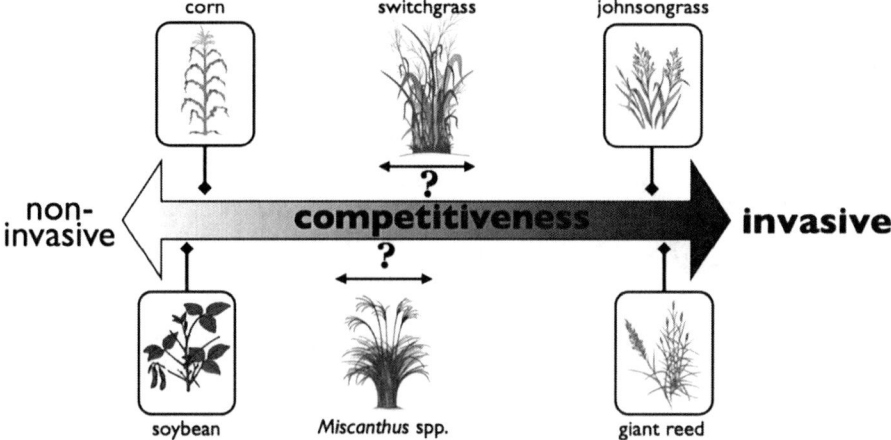

Fig. 9.1 Hypothetical competitiveness continuum. Traditional field crops corn and soybean are poorly competitive, while known invasive species johnsongrass and giant reed are highly competitive. The proposed biofuel crops switchgrass and *Miscanthus* have unknown competitive abilities. Despite switchgrass growing in the US for millennia, few empirical studies on the competitive ability have been conducted

communities and managed crop production. An additional complicating factor is the likely genetic modification of these crops (Gressel 2008), which further reduces the predictability of ecological interactions (DiTomaso et al. 2007). Therefore, quantitative studies should be performed that elucidate the competitive interactions with desirable native species and crops. To provide useful reference, these studies should also include known invasives within the same habit. Ideally, these comparisons would be conducted over a range of resource availabilities, which typically vary non-linearly under competitive conditions (Stachowicz and Tilman 2005). This relative comparison with known invasives serves as positive controls and will inform potential ecological interactions should the biofuel crops escape cultivation into natural and managed landscapes. Target species of conservation or agronomic value can be identified via the studies mentioned above.

9.4 Mitigating the Invasion Risk Along the Biofuel Chain

Of the four recognized stages of invasion—transport, colonization, establishment, and landscape spread (Theoharides and Dukes 2007)—plants may circumvent the first three via human cultivation. Typically, a lag phase punctuates the stages before landscape spread, thus protracting invasions over decades or centuries (Kowarik 1995). However, the likely scale of biofuel cultivation (1.5 billion ha globally) combined with propagule pressure from production fields, storage sites, field equipment, and transportation vehicles, may accelerate the latter stages of invasion

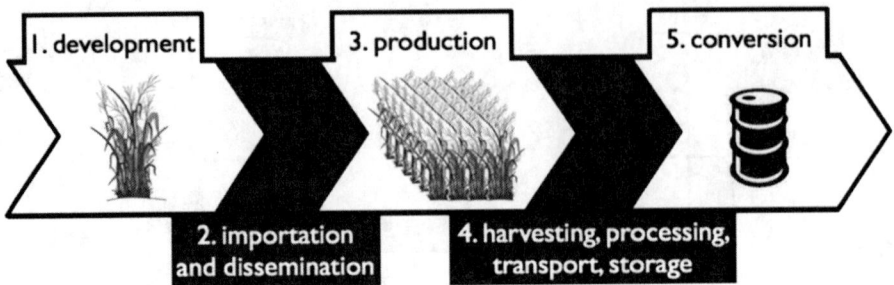

Fig. 9.2 Simplified biofuel supply chain showing *1* crop development; *2* crop importation and dissemination; *3* crop production, feedstock harvesting; *4* processing, transport, and storage; *5* feedstock conversion. *Black arrows* indicate links where propagule or feedstock movement is involved

(Barney and DiTomaso 2008). In addition to the introduced propagule load, the probability of establishing outside cultivation will be proportional to the environmental tolerance of the species, especially at the seedling stage (Barney and Whitlow 2008). The studies outlined above will allow quantification of the invasive potential of each biofuel crop in specified target cropping regions. This information will need to be incorporated into various points along the biofuel supply chain to minimize this risk and allow safe cultivation of these important crops. We have identified five stages along the biofuel supply chain, or biofuel pathway, that spans the seed-to-fuel lifecycle of a biofuel crop (Fig. 9.2). Below we outline a framework for the mitigating steps that should be taken at each point along the biofuel pathway, from crop development to feedstock conversion.

9.4.1 Crop Development

Biofuel crop developers should make every effort to not utilize known invasive species, while reducing the invasive potential of each chosen crop.

For example, the Biomass Crop Assistance Program of the 2008 US Farm Bill states that known invasive or noxious, or potentially invasive or noxious species are not eligible for crop subsidies designed to encourage biofuel adoption by growers. Noxious and invasive species lists are maintained nationally, regionally, and by state agencies and should be consulted by crop developers. State and regional lists are much more comprehensive (e.g., California Invasive Plant Council (Cal-IPC), Mid-Atlantic and Florida Exotic Pest Plant Councils), and are typically the results of science-based assessments, and should serve as primary sources of known invasive species to be avoided. However, it should be noted that inclusion on one list (e.g., Cal-IPC) should not be viewed as invasive everywhere. Despite invasiveness elsewhere being the most robust predictor of invasive potential in new ranges (Reichard and Hamilton 1997), this should be viewed at a regional level, because all known invasive species are not invasive everywhere within an introduced range. If a target biofuel crop is not known to be established in the target region then follow-up studies should be performed to quantify the invasive potential as outlined above.

The studies above will aid in identifying traits that contribute to invasiveness, or alternatively, reduce potential invasiveness, which should be incorporated into crop development programs. Crop developers should partner with plant scientists, ecologists, and agronomists to conduct the studies to identify invasive traits for each crop in each target cropping region. If a crop has a large and variable cropping region (e.g., switchgrass), the necessary studies will be broad in geographic scope, and should be conducted within the range of climates and habitats likely to be encountered. Each crop should be screened through qualitative risk assessment, followed by biological, ecological, and ecophysiological studies of the target crop combined with identification of susceptible habitat within the target region. The species-/genotype-specific studies will identify specific autecological traits that contribute to potential invasiveness, which can then be targeted for mitigation via breeding.

9.4.2 Crop Importation and Dissemination

Biofuel crop propagule movement prior to cultivation should minimize unintentional dissemination to non-target systems.

Not all biofuel crops will necessarily be developed domestically. Prudent introductions via quarantine facilities should be required prior to invasiveness assessment. The US government requires only that plant introductions not include federally noxious weed species, and does not mandate a weed risk assessment as is required in Australia (Weber et al. 2009). Therefore, plants introduced for biomass production should be screened for potential invasiveness in parallel with crop development. This would minimize the economic burden of screening while providing crops that present a minimal threat of escaping cultivation.

Secondly, following crop selection and development, care should be taken in transporting biofuel production propagules—most likely seeds—from "seed fields" to distributors and growers. Invasiveness assessments will identify sensitive habitat (e.g., wetlands), which should be avoided while moving "seed" material.

9.4.3 Crop Production

Growers should make every effort to follow crop developer guidelines, eliminate unintentional dispersal of biofuel crop propagules, maintain biofuel crops within production boundaries, and eradicate escapes.

As with genetically modified food, feed, and fiber crops, growers will play an important role in minimizing unintentional harm, as well as meeting sustainability goals (Robertson et al. 2008). Similar to genetically modified crops, restrictions and guidelines to growing dedicated energy crops, genetically modified or not, should not be overly costly to the grower (Bradford et al. 2005), but should be practical in mitigating invasion risk. Information regarding the risks of cultivating dedicated energy crops will be known before grower adoption, and can be integrated into their practices.

Biofuel feedstock crop selection will be chosen by the growers or dictated by conversion facilities. Production field site selection will be determined by growers based on current crop rotations, land relief and erodibility, accessibility to irrigation, and economics. Growers should also site biomass production fields away from known dispersal corridors (e.g., streams), while creating buffers larger than minimum dispersal distances when biofuel fields must be located near such corridors. Regardless of seed production capacity of the selected crop, scouting of field margins and along sensitive habitat should be compulsory to identify escapes.

9.4.4 Feedstock Harvesting, Processing, Transport, and Storage

Biofuel feedstock harvesting, processing, transport, and storage should minimize unintentional dispersal of viable propagules to non-target ecosystems.

Harvest practices will be defined by the selected crop, the geographic location of the production field, as well as the demands of the contracted conversion facility. The selected harvest technique should consider reducing viable propagule loads. For example, if seed is determined to be the primary source of invasiveness, effort should be made to harvest before flowering or seed set. Alternatively if stem nodes are the primary dispersal propagule (e.g., giant reed, *Arundo donax*) shredding or chipping harvesters should be used to eliminate viable stem node production.

Processing and transport to storage loci or conversion facilities should be performed to minimize viable propagule dispersal. When applicable, transport from fields should be done with closed trucks or shipping containers, and flat-bed trucks with open bales should be avoided, especially when viable seed is contained within the feedstock bales. Open shipping of alfalfa hay has led to the widespread introduction of feral alfalfa populations throughout much of the US (DiTomaso and Healy 2007).

Biomass refineries will likely operate year-round converting feedstocks into primary and secondary products. However, feedstocks will likely be harvested once, or at most three to four times a year—primarily in mid-summer through late fall. Therefore, feedstocks will be baled and stored for most of the year on either grower property, or more likely, refinery property near the conversion facility. Depending on the feedstock crop and the method and timing of harvest, storage sites may serve as propagule reservoirs if not managed properly. As with cultivation fields, storage sites would be ideally located away from dispersal corridors.

9.4.5 Feedstock Conversion

Conversion facilities should require non-invasive feedstocks that have been managed under contracted guidelines that minimize unintentional ecological harm.

In most cases, conversion facilities will be the direct consumers of biofuel feedstocks. Refineries will play a role in dictating crop selection, harvest timing, and storage loci. These companies should work with plant scientists, agronomists, and crop developers in selecting feedstock crops that present a negligible threat of becoming invasive, while encouraging production, harvest, and transport practices that minimize unintentional propagules dispersal.

9.5 Response to Biofuel Crop Escapes

Prior to commercialization and wide-scale cultivation of biofuel crops, eradication protocols need to be developed to rapidly respond to escaped plants or populations, reclaim abandoned production fields, or facilitate rotation to other crops. Such protocols will need to be crop specific and should be made readily available to growers and others involved along the biofuel processing pathway in each region of consideration. If an escaped population is found or unwanted population is identified (i.e., abandoned field), eradication techniques should be implemented with the goal of rapid removal and eventual elimination of all propagules. To effectively accomplish this, a suite of potential mechanical or chemical management practices should be evaluated alongside agronomic field trials during the crop development stage.

Because plant invasions are generally unpredictable and can spread initially without detection, escapes can quickly overwhelm local resources. To effectively respond to incipient populations of escaped biofuel plants, an early detection and rapid response (EDRR) system needs to be in place prior to the release of biofuel crops. An EDRR program should have a readily accessible multi-year funding source established, which is particularly critical when escapes occur on public lands or in proximity to federal lands.

9.5.1 Eradication Techniques

The mitigation strategies previously discussed are all part of an invasion prevention program. However, if a biofuel plant escapes cultivation, either an eradication or containment program should be established. Containment programs are typically employed for species that have widespread distribution and where eradication is economically infeasible (DiTomaso 2000). In most situations with biofuel plants, this will not be the case. In contrast, when invasions are relatively new and cover very small areas, successful eradication is possible and is the most cost-effective pest control approach (Rejmánek and Pitcairn 2002).

The key components of an effective eradication program are early recognition of the escaped plant or population and rapid, intensive and aggressive response to prevent reproduction through the development of a soil seedbank or a well established vegetative reproductive system (Zamora et al. 1989). The principle behind eradication is not to manage the population, but rather to completely eliminate the species from that site including all regenerative plant parts (e.g., seeds, rhizomes, tubers, root crowns, reproductive stem fragments). This generally requires multiple years.

While weed and invasive plant management strategies in agricultural and non-crop areas often incorporate mechanical, cultural, biological, and chemical control techniques, eradication programs are typically limited to mechanical removal or herbicide treatment. Biological control is restricted to widespread weeds or invasive plants and this approach is not desirable for invasives that also serve as important crops. Cultural control options in non-crop systems typically focus on prescribed burning, timely grazing, or revegetation efforts (DiTomaso 2000), all of which are impractical or ineffective for the management of small infestations.

Many mechanical control techniques are available for the eradication of small infestations. These can range from hand labor methods, such as weed whips, sling blades, clippers, shovels, hoes, mattocks, and weed wrenches, to mowing and tillage. These techniques are all impractical and not cost effective for large infestations, but can be selective and successful for small populations. The use of a particular technique depends on the invasive species, but regardless of the method used, it is necessary to make repeated treatments to ensure that the reproductive structures are completely eliminated.

Chemical control is the most economical and commonly used technique in eradication programs (DiTomaso 2000). Like other control methods, it also requires repeat applications to eliminate reproductive propagules. For escaped populations in non-crop and wildland areas, the most commonly used herbicides include the (1) auxin-like growth regulators that selectively control broadleaf species through soil or foliar application, (2) glyphosate, a non-selective foliar applied herbicide that has no soil activity, and (3) the imidazolinone and sulfonylurea herbicides that disrupt the synthesis of amino acids essential for plant growth and have a range of selectivity spectrums through soil or foliar applications. These compounds can also be used for reclaiming abandoned production fields. For rotational systems, however, the specific control options will depend upon the proceeding crop and can include an entirely different choice of mechanical or chemical tools.

9.6 Conclusion

The combined factors of the scales of cultivation, the traits for which biofuel crops are being selected, the fact that many of the species are known to be invasive elsewhere (or at a minimum, non-native in the target cropping region), and the likelihood of unintentional propagule dissemination during harvest, transport, and storage result in a non-trivial probability that biomass crops may escape the cultivated environment and become invasive pests. However, this risk can be quantified and subsequently reduced via a series of studies combining risk assessment, biological and ecological studies, niche modeling, and crop management in each target cropping region. Targeted studies conducted in parallel with agronomic trials should not be overly burdensome to developers, growers, refiners, or regulators, but will require their collaboration to ensure that biofuel crops present an acceptably low risk of invasiveness. If dedicated energy crops are developed, grown, harvested, transported, and stored responsibly the unintentional ecological risk may be acceptably low, and their cultivation will promote sustainable energy production.

References

Bailey LH (1916) Standard cyclopedia of horticulture. MacMillan, New York
Baker HG (1965) Characteristics and modes of origin of weeds. In: Baker HG, Stebbins GL (eds) The genetics of colonizing species. Academic, New York, pp 147–168
Baker HG (1974) The evolution of weeds. Annu Rev Ecol Syst 5:1–24
Barney JN, DiTomaso JM (2008) Nonnative species and bioenergy: are we cultivating the next invader? BioScience 58:64–70
Barney JN, DiTomaso JM (2010) Bioclimatic predictions of habitat suitability for the biofuel switchgrass in North America under current and future climate scenarios. Biomass Bioenergy 34:124–133
Barney JN, Whitlow TH (2008) A unifying framework for biological invasions: the state factor model. Biol Invasions 10:259–272
Barney JN, Mann JJ, Kyser GB, Blumwald E, Van Deynze A, DiTomaso JM (2009) Tolerance of switchgrass to extreme soil moisture stress: ecological implications. Plant Sci 177:724–732
Beaumont LJ, Gallagher RV, Thuiller W, Downey PO, Leishman MR, Hughes L (2009) Different climatic envelopes among invasive populations may lead to underestimations of current and future biological invasions. Divers Distrib 15:409–420
Boe A (2007) Variation between two switchgrass cultivars for components of vegetative and seed biomass. Crop Sci 47:636–642
Bradford KJ, Van Deynze AE, Gutterson N, Parrott W, Strauss SH (2005) Regulating transgenic crops sensibly: lessons from plant breeding, biotechnology and genomics. Nature 23:439–444
Brooks ML, D'Antonio CM, Richardson DM, Grace JB, Keeley JE, DiTomaso JM, Hobbs RJ, Pellant M, Pyke D (2004) Effects of invasive alien plants on fire regimes. BioScience 54:677–688
Buddenhagen CE, Chimera C, Clifford P (2009) Assessing biofuel crop invasiveness: a case study. PLoS ONE 4(4): e5261, doi:10.1371/ journal.pone.0005261
Chapotin SM, Wolt JD (2007) Genetically modified crops for the bioeconomy: meeting public and regulatory expectations. Transgenic Res 16:675–688
Cousens R (2008) Risk assessment of potential biofuel species: an application for trait-based models for predicting weediness? Weed Sci 56:873–882
D'Antonio CM, Hobbie SE (2005) Plant species effects on ecosystem processes. In: Sax DF, Stachowicz JJ, Gaines SD (eds) Species invasions: insights from ecology, evolution and biogeography. Sinauer, Sunderland, MA, pp 65–84

Davis MA, Grime JP, Thompson K (2000) Fluctuating resources in plant communities: a general theory of invasibility. J Ecol 88:528–534

DiTomaso JM (2000) Invasive weeds in rangelands: species, impacts, and management. Weed Sci 48:255–265

DiTomaso JM, Healy EA (2007) Weeds of California and other Western states. University of California, Agricultural and Natural Resources, Oakland, CA

DiTomaso JM, Barney JN, Fox A (2007) Biofuel feedstocks: the risk of future invasions. Council for Agricultural Science and Technology Commentary QTA 2007-1

Energy Independence and Security Act (2007) Public Law 110-140. H.R. 6

FAO (2007) Global crop area harvested—all crops. http://faostat.fao.org. Cited Aug 31 2009

Field CB, Campbell JE, Lobell DB (2008) Biomass energy: the scale of the potential resource. Trends Ecol Evolut 23:65–72

Forseth IN Jr, Innis AF (2004) Kudzu (*Pueraria montana*): history, physiology, and ecology combine to make a major ecosystem threat. Crit Rev Plant Sci 23:401–413

Franklin J (1995) Predictive vegetation mapping: geographic modeling of biospatial patterns in relation to environmental gradients. Prog Phys Geogr 19:474–499

Gordon DR, Onderdonk DA, Fox AM, Stocker RK (2008) Consistent accuracy of the Australian weed risk assessment system across varied geographies. Divers Distrib 14:234–242

Gressel J (2008) Transgenics are imperative for biofuel crops. Plant Sci 174:246–263

Gurevitch J, Padilla DK (2004) Are invasive species a major cause of extinctions? Trends Ecol Evolut 19:470–474

Hayes KR, Barry SC (2008) Are there any consistent predictors of invasion success? Biol Invasions 10:483–506

Holt JS, Boose AB (2000) Potential for spread of *Abutilon theophrasti* in California. Weed Sci 48:43–52

Kassel PC, Mullen RE, Bailey TB (1985) Seed yield of three switchgrass cultivars for different management practices. Agron J 77:214–218

Keller RP, Lodge DM, Finnoff DC (2007) Risk assessment for invasive species produces net bioeconomic benefits. Proc Natl Acad Sci USA 104:203–207

Khudamrongsawat J, Tayyar R, Holt JS (2004) Genetic diversity of giant reed (*Arundo donax*) in the Santa Ana River, California. Weed Sci 52:395–405

Kowarik I (1995) Time lags in biological invasions with regard to the success and failure of alien species. In: Pyšek P, Prach K, Rejmánek M, Wade M (eds) Plant invasions—general aspects and special problems. Academic, Amsterdam, pp 15–38

Kriticos D, Yonow T, McFadyen RE (2005) The potential distribution of *Chromolaena odorata* (Siam weed) in relation to climate. Weed Res 45:246–254

Levine JM (2000) Species diversity and biological invasions: relating local process to community pattern. Science 288:852–854

Levine JM, Adler PB, Yelenik SG (2004) A meta-analysis of biotic resistance to exotic plant invasions. Ecol Lett 7:975–989

Lewandowski I, Scurlock JMO, Lindvall E, Chistou M (2003) The development and current status of perennial rhizomatous grasses as energy crops in the US and Europe. Biomass Bioenergy 25:335–361

Lockwood JL, Cassey P, Blackburn TM (2009) The more you introduce the more you get: the role of colonization pressure and propagule pressure in invasion ecology. Divers Distrib 15:904–910

Low T, Booth C (2007) The weedy truth about biofuels. Invasive Species Council, Melbourne

Mack RN (2000) Cultivation fosters plant naturalization by reducing environmental stochasticity. Biol Invasions 2:111–122

Mack RN (2008) Evaluating the credits and debits of a proposed biofuel species: giant reed (*Arundo donax*). Weed Sci 56:883–888

Mack RN, Simberloff D, Lonsdale WM, Evans H, Clout M, Bazzaz FA (2000) Biotic invasions: causes, epidemiology, global consequences, and control. Ecol Appl 10:689–710

Mann JJ, Barney JN, Kyser GB, DiTomaso JM (2009) Pre-commercial screening of the leading biofuel crop *Miscanthus* x *giganteus* for invasive plant traits. Calif Weed Sci Soc 61:59

Meyerson LA (2008) Biosecurity, biofuels, and biodiversity. Front Ecol Environ 6:291

Parrish DJ, Fike JH (2005) The biology and agronomy of switchgrass for biofuels. Crit Rev Plant Sci 24:423–459

Pattison RR, Mack RN (2008) Potential distribution of the invasive tree *Triadica sebifera* (Euphorbiaceae) in the United States: evaluating CLIMEX predictions with field trials. Glob Change Biol 14:813–826

Pearson RG, Dawson TP (2003) Predicting the impacts of climate change on the distribution of species: are bioclimatic envelop models useful? Glob Ecol Biogeogr 12:361–371

Pheloung PC, Williams PA, Halloy SR (1999) A weed risk assessment model for use as a biosecurity tool evaluating plant introductions. J Environ Manage 57:239–251

Pimentel D, Lach L, Zuniga R, Morrison D (2000) Environmental and economic costs of nonindigenous species in the United States. BioScience 50:53–65

Poutsma J, Loomans AJM, Aukema B, Heijerman T (2008) Predicting the potential geographical distribution of the harlequin ladybird, *Harmonia axyridis*, using the CLIMEX model. BioControl 53:103–125

Pyšek P, Richardson DM (2007) Traits associated with invasiveness in alien plants: where do we stand? In: Nentwig W (ed) Biological invasions. Springer, Berlin, pp 97–125

Raghu S, Anderson RC, Daehler CC, Davis AS, Wiedenmann RN, Simberloff D, Mack RN (2006) Adding biofuels to the invasive species fire? Science 313:1742

Reichard SH, Hamilton CW (1997) Predicting invasions of woody plants introduced into North America. Conserv Biol 11:193–203

Reichard SH, White P (2001) Horticulture as a pathway of invasive plant introductions in the United States. BioScience 51:103–113

Rejmánek M, Pitcairn MJ (2002) When is eradication of exotic pest plants a realistic goal? In: Veitch CR, Clout MN (eds) Turning the tide: the eradication of invasive species. IUCN SSC Invasive Species Specialist Group, Cambridge, pp 249–253

Richardson DM, Pyšek P, Rejmanek M, Barbour MG, Panetta FD, West CJ (2000) Naturalization and invasion of alien plants: concepts and definitions. Diver Distrib 6:93–107

Robertson GP, Dale VH, Doering OC, Hamburg SP, Melillo JM, Wander MM, Parton WJ, Adler PR, Barney JN, Cruse RM, Duke CS, Fearnside PM, Follett RF, Gibbs HK, Goldemberg J, Mladenoff DJ, Ojima D, Palmer MW, Sharpley A, Wallace L, Weathers KC, Wiens JA, Wilhelm WW (2008) Sustainable biofuels redux. Science 322:49–50

Royal Society (2008) Sustainable biofuels: prospects and challenges. Royal Society, London, p 90

Sala A, Smith SD, Devitt DA (1996) Water use by *Tamarix ramosissima* and associated phreatophytes in a Mojave desert floodplain. Ecol Appl 63:888–898

Sanderson MA, Schnabel RR, Curran WS, Stout WL, Genito D, Tracy BF (2004) Switchgrass and big bluestem hay, biomass, and seed yield response to fire and glyphosate treatment. Agron J 96:1688–1692

Savidge JA (1987) Extinction of an island forest avifauna by an introduced snake. Ecology 68:660–668

Sax DF, Stachowicz JJ, Gaines SD (eds) (2005) Species invasions: insights into ecology, evolution, and biogeography. Sinauer, Sunderland, MA

Schnoor JL, Doering OC, Entekhabi D, Hiler EA, Hullar TL, Tilman D (2008) Water implications of biofuels production in the United States. National Academies Press, Washington, DC

Simberloff D (2005) The politics of assessing risk for biological invasions: the USA as a case study. Trends Ecol Evolut 20:216–222

Simberloff D (2008) Invasion biologists and the biofuels boom: Cassandras or colleagues? Weed Sci 56:867–872

Sinden J, Jones R, Hester S, Odom D, Kalisch C, James R, Cacho O (2004) The economic impact of weeds in Australia, CRC for Australian Weed Management Technical Series, p 55

Stachowicz JJ, Tilman D (2005) Species invasions and the relationships between species diversity, community saturation, and ecosystem functioning. In: Sax DF, Stachowicz JJ, Gaines SD (eds) Species invasions: insights into ecology, evolution, and biogeography. Sinauer, Sunderland, MA, pp 41–64

Sutherland S (2004) What makes a weed a weed: life history traits of native and exotic plants in the USA. Oecologia 141:24–39

Sutherst RW (2003) Prediction of species geographical ranges. J Biogeogr 30:805–816

Sutherst RW, Maywald GF, Yonow T, Stevens PM (1999). CLIMEX: predicting the effects of climate on plants and animals. CSIRO, Victoria

Sutherst RW, Maywald GF, Russell BL (2000) Estimating vulnerability under global change: modular modelling of pests. Agric Ecosyst Environ 82:303–319

Sutherst RW, Maywald GF, Bourne AS (2007) Including species interactions in risk assessments for global change. Glob Change Biol 13:1843–1859

Theoharides KA, Dukes JS (2007) Plant invasion across space and time: factors affecting nonindigenous species success during four stages of invasion. New Phytol 176:256–273

Vitousek PM, Walker LR, Whiteaker LD, Mueller-Dombois D, Matson P (1987) Biological invasion by *Myrica faya* alters ecosystem development in Hawaii. Science 238:802–804

Weber J, Panetta FD, Virtue JG, Pheloung PC (2009) An analysis of assessment outcomes from eight years' operation of the Australian border weed risk assessment system. J Environ Manage 90:798–807

Yuan JS, Tiller KH, Al-Ahmad H, Stewart NR, Stewart CN Jr (2008) Plants to power: bioenergy to fuel the future. Trends Plant Sci 13:421–429

Zamora DL, Thill DC, Eplee RE (1989) An eradication plan for plant invasions. Weed Technol 3:2–12

Chapter 10
Gene Flow in Genetically Engineered Perennial Grasses: Lessons for Modification of Dedicated Bioenergy Crops

Albert P. Kausch, Joel Hague, Melvin Oliver, Lidia S. Watrud, Carol Mallory-Smith, Virgil Meier, and C. Neal Stewart Jr

10.1 Introduction

The potential ecological consequences of the commercialization of genetically engineered (GE) crops have been the subject of intense debate, particularly when the GE crops are perennial and capable of outcrossing to compatible relatives (Aldhous 2003; Colwell et al. 1985; Eastham and Sweet 2002; Giles 2003; Marvier

---

A.P. Kausch and J. Hague
University of Rhode Island, 530 Liberty Lane, West Kingston, RI 02892, USA
e-mail: akausch@etal.uri.edu; joel.hague@gmail.com

M. Oliver
USDA-ARS Plant Genetics Unit, University of Missouri, Columbia. 204 Curtis Hall, Columbia, MO 65211, USA
e-mail: moliver@lbk.ars.usda.gov

L.S. Watrud
US Environmental Protection Agency, US EPA NHEERL WED, 200 SW 35th Street, Corvallis, OR 97333, USA
e-mail: Watrud.Lidia@epamail.epa.gov

C. Mallory-Smith
Oregon State University Department of Crop and Soil Science 109 Crop Science Building Oregon State University Corvallis, Corvallis, OR 79331-3002, USA
e-mail: carol.mallory-smith@oregonstate.edu

V. Meier
USDA APHIS, Biotechnology Regulatory Services, 4700 River Road Unit 147 5B45, Riverdale, MD 20737, USA
e-mail: virgil.d.meier@aphis.usda.gov

C.N. Stewart Jr
Racheff Chair of Excellence in Plant Molecular Genetics, University of Tennessee, 2431 Joe Johnson Dr. Knoxville, Knoxville, TN 37996-4561, USA
e-mail: nealstewart@utk.edu

and Acker 2005; Rogers and Parkes 1995; Tsuchiya et al. 1995). The ecological impact issues for engineered perennial crops are the following: whether (1) the techniques themselves or resulting phenotypic traits could lead to adverse ecological impacts; (2) escaped GE crop plants can persist in the environment via feral populations or hybridization with non-transgenic populations of the same or related species, depending on the source and nature of the GE trait(s) in the crop; (3) long-term environmental effects will result from commercialization of the GE crop (Eastham and Sweet 2002; Tiedje et al. 1989; Wrubel et al. 1992; Ellstrand and Hoffman 1990); (4) GE crops are grown sympatrically with wild relatives (e.g., centers of origin) or cross-compatible species (or genera); (5) GE crops have biotypes or related taxa that are already aggressive weeds; (6) GE crops can also be weeds themselves; and (7) GE crops can outcross with some degree of self-incompatibility. Most of the thousands of small-scale field tests of transgenic plants have not been designed to investigate the environmental consequences of gene flow associated with widespread commercialization (Dale et al. 2002; Eastham and Sweet 2002; Wrubel et al. 1992). However, more recent studies (Belanger et al. 2003; Christoffer 2003; Mallory-Smith and Zapiola 2008; Reichman et al. 2006; Watrud et al. 2004; Zapiola et al. 2008) demonstrate that commercialization of GE perennial grasses could lead to transgene flow via outcrossing with wild relatives and establishment of feral populations via seed escape, and may therefore present significant ecological and economic risks.

Numerous risk assessments have been conducted on transgenic plants of annual and/or self-pollinating crops (Belanger et al. 2003; Dale et al. 2002; Eastham and Sweet 2002; Ellstrand and Hoffman 1990; Ellstrand et al. 1999; Rogers and Parkes 1995). For instance, Eastham and Sweet (2002) reviewed the significance of, and the parameters affecting, gene flow in six major crop species including oilseed rape (*Brassica napus*), sugar beet (*Beta vulgaris*), potato (*Solanum tuberosum*), maize (*Zea mays*), wheat (*Triticum aestivum*), and barley (*Hordeum vulgare*). Each crop was reviewed with attention to the following points: (1) reproductive biology and crop use; (2) type of genetic modification; (3) pollen dispersal potential; (4) gene flow: crop-to-crop, including hybridization capacity and possible consequences of gene flow; (5) definition and status as a weedy species; and (6) gene flow: crop-to-wild-relative, including compatibility and distribution, hybridization, and gene flow. Using these parameters, their report focused on the significance of pollen-mediated gene flow in annual crops and provides relative risk assessments of gene flow from crop to crop and from crop to wild relatives.

However, it is also now recognized that seed-mediated gene flow is a major concern (Mallory-Smith and Zapiola 2008). Contamination of seed with non-deregulated crops, such as the StarLink and *Bt10* incidents in US maize seed, have caused serious commercial and economic impacts and negatively affected public perception and trust (Bucchini and Goldman 2002; Macilwain 2005). Grain from transgenic US corn is exported to many countries as living seeds, and this can create legal and social problems if transgenic seeds are planted in nations where they are not approved for environmental release, as occurred in Mexico. Transgenes were detected in open-pollinated landraces of corn in Oaxaca, Mexico, in 2000

(Quist and Chapela 2001) and again in 2001 and 2004 (Piñeyro-Nelson et al. 2009), although the number of locations with confirmed reports is quite low (Ortiz-García et al. 2005; Piñeyro-Nelson et al. 2009; Snow 2009). Thus, as Marvier and Van Acker (2005) note, the escape of transgenic seed via human error is quite likely, even in crops with large seeds such as maize.

It is clear that transgene flow depends on several variables: the specific crop, its location, the presence of outcrossing wild relatives/sexually compatible crops, and the fitness effect(s) of the GE trait (Daniell 2002). It is also clear that the mechanisms responsible for gene flow among crops and their related and wild relatives are: (1) dispersal of viable pollen; (2) dissemination in seed; or, in some cases (3) vegetative dispersal, e.g., by means of stolons in some perennial grasses. The major vectors for dispersal are considered to be largely wind, water, animals, and human activities.

Gene flow research is especially important in species with a high propensity for outcrossing or gene introgression. Recent studies highlight the potential for gene flow from the commercialization and large-scale seed production of perennial transgenic grasses (Mallory-Smith and Zapiola 2008; Reichman et al. 2006; Watrud et al. 2004; Wipff and Fricker 2001; Zapiola et al. 2008). Perennial grasses are grown throughout the world; furthermore, the vegetative and reproductive biology of many plants targeted for bioenergy production such as the perennial grasses switchgrass (*Panicum virgatum*) and big bluestem (*Andropogon gerardii* Vitman), and trees such as poplar (*Poplus spp.*), willow (Salix spp.), and *Paulownia* makes some gene flow to wild species or the environment inevitable. We need to come to an understanding regarding the limitations of the technologies used to mitigate gene flow and what constitutes an acceptable level of escape. Towards these ends, a review of the science of gene flow in GE perennial grasses is presented here.

10.2 Gene Flow in Glufosinate-Resistant Grasses

Field studies have been conducted to assess pollen-mediated gene-flow using open-pollinated transgenic glufosinate-resistant grasses (Bae et al. 2008; Belanger et al. 2003; Wipff and Fricker 2001) and have clearly demonstrated gene flow to non-transgenic grasses. The first of these studies (Wipff and Fricker 2001) was conducted in the Willamette Valley in Oregon, using 286 creeping bentgrass plants that were transformed with the *bar* gene, which confers resistance to glufosinate ammonium herbicides (i.e., Basta^{Tm}, Finale^{Tm}, Liberty^{Tm}). This field study was conducted under the authority, guidelines, and provisions provided by United States Department of Agriculture/Animal and Plant Health Inspection Service (USDA/APHIS) Biotechnology Regulatory Services. The objectives of the study were to (1) gather initial data on pollen movement; (2) test the effectiveness of cereal rye (*Secale cereal* L.) borders as a pollen barrier, which were used successfully in the isolation of tall fescue and perennial ryegrass nurseries; (3) study interspecific gene flow into four introduced species of bentgrass; and (4) verify the fertility of the transgenic bentgrass plants. The results of that study demonstrated that pollen from the transgenic nursery traveled at least 300 m. The most distant recipient plan, on the SW transect, had 15 seedlings survive glufosinate applications. Polymerase chain reaction (PCR) and Southern blot analyses have confirmed the presence of the *bar* gene in these individuals.

Belanger et al. (2003) measured the frequency of interspecific hybridization by pollen-mediated gene flow between transgenic creeping bentgrass (*Agrostis stolonifera* L.) and four related species: velvet bentgrass (*A. canina* L.), dryland bentgrass (*A. castellana* Boiss & Reuter), redtop (*A. gigantea* Roth), and, colonial bentgrass (*A. capillaris* L.). The transgenic creeping bentgrass plants used in this study expressed the *bar* gene. They examined two identical transgenic plots, spatially separated by 140 m, each consisting of a hexagonal array including 90 sample points for pollen reception and a central point for pollen dispersal. The center pollen dispersal array consisted of five transgenic plants and the distance between each sample point was 3 m. At each sample point, five pollen recipients were placed using one plant each of the four related *Agrostis* species and one non-GE *A. stolonifera* plant to provide an indication of where in the plot the transgenic pollen was available. Interspecific hybridization occured between transgenic creeping bentgrass and both dryland bentgrass and colonial bentgrass (at frequencies of 0.04 and 0.002%, respectively), but no hybrids were recovered between GE creeping bentgrass and velvet bentgrass or redtop. The intraspecific hybridization resulting from pollination with nontransgenic creeping bentgrass was significantly higher (0.63%) The size of these plots and the number of transgenic plants involved did not approach real world commercial plots, which disperse far greater loads of transgenic pollen that would have the capability of traveling much greater distances. However, this design presents an excellent model to examine pollen-mediated gene flow, hybridization frequencies and seed scatter using male-sterile plants as recipients at the center.

The results from the studies above established that: (1) intraspecific gene flow in creeping bentgrass is possible for much longer distances than traditionally calculated; (2) transgenes can flow considerable distances to other species of *Agrostis* (i.e., interspecific gene flow) probably via pollen; (3) the transgenic hybrid bentgrass plants are fertile and stable; and (4) neither cereal rye or spatial separation provide an effective pollen barrier for confinement. These studies also strongly implicate pollen-mediated gene flow from male-fertile open-pollinated plants as a major obstacle to transgene confinement. However, neither the *Agrostis* or *Zoysia* studies that utilized the glufosinate resistance marker (the bar gene), addressed the possibility of transgene escape via seed shatter and dispersal by wind or other abiotic or biotic means.

The potential for intra- and inter-specific hybridization via pollen-mediated gene flow from transgenic *Zoysia* grass (*Zoysia japonica* Steud.) carrying the *bar* gene to wild type (WT) *Zoysia* grass and 14 weed species was investigated from 2003 to 2005 in Nam Jeju County, Korea (Bae et al. 2008). A number of experimental plot designs were deployed to detect gene flow; in addition, 121 sites up to 3 km outside the perimeter of the 936 m^2 GE test field were screened for potential hybrids and seed escape. The authors reported significant intraspecific hybridization within distances of <3 m (6% hybrid seeds, SE = ±4%), but found that hybridization frequency effectively dropped to zero at distances greater than 3 m. There were no reported cases of interspecific hybridization with co-habitant weed species and no evidence was found for gene flow via pollen or seed from the experimental field, at least at the 121 external sites tested. The authors noted that a number of factors played a role in the above results, including: (1) *Z. japonica* is an inherently recalcitrant cross pollinator; (2) *Zoysia* seeds exhibit a very low germination rate (4%) after winter dormancy under natural conditions; and (3), the GE pollen source was relatively small. Thus, they conclude that while long distance gene flow is of lesser concern in GE *Zoysia* than in a highly outcrossing species such as creeping bentgrass further gene flow studies using larger plots of GE *Zoysia* grass are justified.

10.3 Gene Flow in Glyphosate-Resistant Creeping Bentgrass

In late 2002, under USDA-APHIS regulated status, 162 ha of a Round Up® Ready bentgrass variety (*Agrostis stolonifera* L.) were planted by The Scotts Company (http://www.scotts.com/) under permit within a 4,553 ha control area in central Oregon. An additional 2.4 ha field planted in 2003 flowered and produced seed in 2004 (Zapiola et al. 2008). This turfgrass variety contained the EPSPS (5-enolpyruvylshikimate-3-phosphate synthase) gene from *Agrobacterium tumefaciens* strain CP4 and is the first example of a transgenic perennial grass crop to attempt passage through the regulatory process. An APHIS preliminary risk assessment concluded that the genetically engineered line (ASR368) used in the study was not significantly different from its parental line or null comparators except for its tolerance to glyphosate and is not sexually compatible with any threatened or endangered species or any species on the Federal Noxious Weed List (http://plants.usda.gov/java/noxiousDriver#federal).

Glyphosate-resistant creeping bentgrass (GlyRCB) was chosen as a commercial target for use on golf courses because of its good stand persistence even with repeated close mowing, and the herbicide-tolerance trait was expected to enable better weed control. The type of end use management for GlyRCB should ideally minimize gene flow via pollen and make seed development unlikely. The 162 ha planting was comprised of eight spatially isolated fields of varying sizes, presenting a unique large-scale testing opportunity to monitor gene flow in a genetically engineered perennial grass prior to its release as a commercial product.

This experimental cultivation raised concerns among many grass seed producers and breeders in the Willamette Valley of western Oregon, which is the site of 70% of US grass seed production. Creeping bentgrass is largely self-incompatible, highly outcrossing, and wind pollinated. It can hybridize with compatible species and reproduce by vegetative stolons that can persist and propagate outside of cultivation. The issue of creeping bentgrass seed size comes up at least three times, each with a slightly different presentation of the numbers. Seems redundant

(AOSA 2002), and the mean creeping bentgrass seed yield in Oregon is 600 kg ha^{-1}, roughly 8.1×10^9 seeds (USDA-NASS 2006). Another concern was the potential production of RoundUp® resistant weeds because control of escaped plants is difficult as alternatives to RoundUp® may be less effective, more expensive, or not for that use. It was also unclear who would finance the registration of alternative herbicides if contamination occurred. Furthermore, contamination of other harvested crops with transgenic seed is a serious marketing issue, especially in domestic and international markets that are not open to GE crops (Zapiola et al. 2008).

In response to these concerns, a 4,553-ha control district was established by the Oregon Department of Agriculture in Jefferson County, OR. This control district was intentionally located >150 km east of Oregon's Willamette Valley grass seed production area with some of the following requirements: (1) non-GE *Agrostis* ssp. could not be grown, planted, or handled within the control district; (2) field borders, ditch banks, and roadsides within 50 m of the GlyRCB production fields were to be kept free of *Agrostis* ssp.; (3) GlyRCB fields were located more than 400 m away from any creeping bentgrass field outside the control district (Zapiola et al. 2008). Additional safeguards implemented to prevent seed movement included transport of seed in sealed containers to and from fields, cleaning of equipment prior to leaving the field, use of dedicated combines for the GE crop, burning of straw remaining in the field to destroy any seed left behind, and cleaning and packaging of seed produced in the control district within the same area. Thus, several precautions were to be taken to help prevent seed scatter from the Round Up® Ready production fields.

10.3.1 Gene Flow via Pollen in Glyphosate-Resistant Bentgrass

More than 30 species of *Agrostis* occur in North America, and approximately one dozen species are found in Oregon (http://plants.usda.gov). Creeping bentgrass, redtop, colonial bentgrass, dryland bentgrass, velvet bentgrass, and brown bentgrass (*A. vinealis* Schreber) form a hybridizing complex of inter-pollinating, cross-compatible species. Although naturally occurring F_1 hybrids of *Agrostis* may exhibit reduced fertility or even sterility, pollen can remain viable for 2 h, and under optimal conditions, fertile hybrids can be formed; backcrossing can restore fertility in full (Belanger et al. 2003; Fei and Nelson 2003; Pfender et al. 2007). Thus, gene flow from GlyRCB production fields to populations of *Agrostis* spp., and the establishment of feral glyphosate-resistant populations has long been a distinct possibility.

Two groups, one led by Carol Mallory-Smith at Oregon State University, and the other by Lidia Watrud at the US Environmental Protection Agency (EPA) monitored gene flow from the production fields. Outside the control district, 69 resident *Agrostis* as well as 178 "sentinel" creeping bentgrass plants were used in 2003 by the US EPA to monitor gene flow via pollen from the eight GE test fields

(Reichman et al. 2006; Watrud et al. 2004). Based on testing of seedlings in the greenhouse, these studies detected pollen-mediated gene flow at much longer, i.e., landscape level distances, measured in kilometers, rather than much shorter distances (typically measured in meters), as reported earlier. While the highest relative frequencies of gene flow were observed within 2 km of the control area perimeter, *CP4 EPSPS*-positive seedlings were recovered from resident creeping bentgrass and redtop, and sentinel creeping bentgrass at maximal distances of 8, 14 and 21 km, respectively. In 138 sentinel creeping bentgrass plants tested, 75 plants yielded positive seedling progeny (54%) and the overall incidence of *CP4 EPSPS* positive seedlings was 2.0% (625 positive / 32,000 total seedlings tested). Of the 30 resident (i.e., wild) creeping bentgrass plants, 16 also yielded positive seedling progeny (53%), and 157 positive seedling progeny of 565,000 tested (0.03%) were obtained. Resident redtop also produced glyphosate-resistant progeny, with 13 positive of 39 tested (33%); the overall incidence of positive seedlings (159 positive/397,000 seedlings tested) was 0.04%. Molecular confirmation of the presence of the *CP4 EPSPS* gene in all positive seedling progeny was obtained via PCR amplification; the 1,050 bp PCR product sequence matched the *CP4 EPSPS* sequence of a glyphosate-resistant variety of soybean (GenBank accession AF464188.1).

Based on the original 2003 data, additional searches conducted in 2004–2005 were focused on nonagronomic mesic habitats within a 4.8 km band outside the control area (Reichman et al. 2006; Watrud et al. 2004). These surveys located 55 *Agrostis* ssp. populations on publicly accessible lands, and 34 sites were newly located since the 2003 survey. Nine *CP4 EPSPS* positive plants were identified out of 20,400 tissue samples screened via Traitchek kits, eight of which were found within the new population sites. The presence of the transgene was confirmed in each plant via PCR amplification and sequencing of the PCR product, which again matched that GenBank accessions AF464188.1—*Glycine max CP4 EPSPS* (glyphosate-resistant soybean variety). To establish the parentage of each plant, sophisticated species-level molecular phylogenetic trees were constructed via sequencing of a nuclear encoded ribosomal DNA [internal transcribed spacer (ITS)] and maternally inherited chloroplast-encoded DNA (*matK*). The distribution and phylogenetic information suggested that six of the *CP4 EPSPS* positive plants resulted from pollen-mediated gene flow from the production fields to wild creeping bentgrass plants, while three arose from dispersed seeds (Reichman et al. 2006). The hybridization data with both sentinel plants of creeping bentgrass and resident *Agrostis* spp plants indicate that GE glyphosate resistance in creeping bentgrass can be transmitted to compatible wild relatives at landscape level over multi-kilometer distances. It was estimated that exposure to the *CP4 EPSPS* gene occurred over a total area of over 300 km^2 as a result of the initial year of flowering of the eight GE fields in 2003 (Reichman et al. 2006; Watrud et al. 2004). A significant caveat is that the nine wild transgenic plants above were located in publicly accessible areas limited to roughly 10% of the total estimated *Agrostis* habitat; thus, the surveys may have underestimated the establishment of wild transgenics in the study region (Reichman et al. 2006).

Surveys conducted within the control area from 2003 to 2006 found gene flow within the control area perimeter, as well as gene flow via seed to the northeast from a documented wind event (Mallory-Smith and Zapiola 2008; Zapiola et al. 2008). Glyphosate-resistant plants were identified in situ via TraitChek RUR^{TM} strips (http://www.sdix.com/). Approximately 80 km of irrigation canals, roadsides, ditch and pond banks, and pipelines in the area roughly 300 m around the production fields were surveyed in 2003. While not all the survey sites were necessarily revisited, the surveyed area was increased and extended up to a 5 km radius outside the control perimeter in 2005 and 2006. Of the 57 plants located and tested in 2003, none were herbicide resistant; however, 0.376% of the seeds collected gave rise to glyphosate-resistant seedlings in the greenhouse and had therefore received the *CP4 EPSPS* gene via pollen. In 2006, 3 years after the original GlyRCB fields were taken out of production, 62% of 585 creeping bentgrass plants tested were glyphosate resistant. Strikingly, 0.012% of 49,351 seedlings grown from seed of glyphosate-sensitive plants collected in 2006 were glyphosate resistant, thereby demonstrating that pollen-mediated transgene flow was still occurring despite intensive mitigation efforts by The Scotts Company (http://www.scotts.com/) to totally remove glyphosate-resistant plants from the area (Zapiola et al. 2007). Interestingly, two modeling studies, one based on predictions of creeping bentgrass pollen dispersal based on wind data at the time that the GE creeping bentgrass fields were growing in central Oregon in 2003 (Van de Water et al. 2007), and the other, based on counts of non-GE creeping bentgrass pollen collected near flowering fields with air-samplers in western Oregon (Pfender et al. 2007), each came up with very similar multi-kilometer distances that closely matched the maximal 21 km distance reported for live GE creeping bentgrass pollen that was based on production of F_1 seedlings tested in a greenhouse setting (Watrud et al. 2004).

10.4 Gene Flow via Seed Scatter

Seed scatter is defined as the loss of seed at any time from the beginning of production through final end use. Among perennial grasses, the possible risk of gene flow through seed scatter is high because of seed size, the potential for survival in the seed bank, and for some species subsequent vegetative reproduction. The seeds of most turf and forage grass species are much smaller than those of annual crops and therefore are very difficult to contain during production, collection, and distribution for sale. For instance, creeping bentgrass seeds are approximately 2 mm × 0.5 mm and may weigh as little as 80 µg each (Reichman et al. 2006). Also, seed viability is much longer than that of pollen. Unlike pollen, there is no "time window" for seed movement—seed movement can happen at many times (e.g., at planting, during or after harvest) and seedbanks can renew gene flow in subsequent years. Furthermore, seed does not require a sexually compatible relative to contribute to gene flow; thus, there is no need for outcrossing to compatible wild relatives.

Because of its small size, perennial grass seed can move easily via natural dispersal vectors, production practices, and in end use, e.g., in golf courses, landscapes, pastures, and forage production. Seed can be dispersed via natural dispersal vectors such as wind, water, and animals—factors over which humans have little to no control. Furthermore, perennial grass seed production involves the movement of seeding, application, and harvesting equipment, as well as seed cleaning, field irrigation and seed distribution via long-distance trucking. Thus, equipment is frequently moved in and out of the field during seed production, increasing the probability of seed escape. Ultimately, the purpose of large-scale seed production is the distribution of the product to customers who are separated by long distances. While seed scatter may be reduced in any one of these steps it cannot be entirely prevented. Unfortunately, there is a paucity of studies that address gene flow through seed scatter. This is probably because most studies to date have been concerned with annual species that generally do not survive outside of cultivation and have little or no seed dormancy. The few exceptions are gene flow studies in canola and sugar beets, neither of which are perennials. Therefore, gene flow via seed scatter presents a serious challenge to gene confinement efforts.

10.4.1 Gene Flow via Seed Escape in Glyphosate-Resistant Bentgrass

In August 2003, after swathing but before threshing, a documented strong northwesterly wind event in the production area moved creeping bentgrass seed and panicles from swathed windrows of the northernmost GlyRCB production field (Zapiola et al. 2008). Mitigation procedures were undertaken, including herbicide treatment and hand rogueing of the field, which substantially reduced the level of GlyRCB volunteers.

Additional surveys were conducted in 2004, 2005, and 2006, both within the control area and to a 5 km radius outside its perimeter. By 2004, glyphosate-resistant plants were found distributed throughout the control area along canals and irrigation ditches, often in places where they were not located in 2003 (Zapiola et al. 2008). The distances of distribution varied from adjacent to a creeping bentgrass production field to 1.9 km from the original closest production field. A total of 300 plants were tested via Traitchek RUR^{TM} strips, 49% of which were identified visually as creeping bentgrass, and 93% of these were *CP4 EPSPS* positive. In 2005, a total of 1,290 plants were tested, with 75% identified as creeping bentgrass, 19.3% redtop, 0.5% rabbitfoot grass, and the remaining 5.2% represented by *Agrostis* ssp. and potential hybrids. Of the total plants tested, 40.5% (522/1290) were glyphosate-resistant, of the creeping bentgrass plants tested, 54% (521/968) were glyphosate-resistant, and the most distant resistant plant was 4.6 km from the nearest original GlyRCB production field. By 2006, 62% of creeping bentgrass plants tested were glyphosate-resistant and the most distant GlyRCB plant was also found 4.6 km away from the nearest original GlyRCB field (Zapiola et al. 2008).

10.5 Future Impacts of Gene Flow from Glyphosate-Resistant Creeping Bentgrass

Although gene flow via pollen dispersal and seed escape occurred during seed production in 2003 and 2004, its impact in future years is still undetermined. The results of this field trial are of public and commercial interest and have significant potential regulatory and policy implications. To date, studies have measured only environmental exposure to GlyRCB, not the long-term effects of gene flow. Numerous unresolved concerns remain; creeping bentgrass seed can remain viable in seed banks for as long as 4 years (C.M.-S., unpublished data), thus its possible contribution to volunteering is uncertain; the potential for contamination of neighboring farms during GlyRCB production could create marketing issues; and the potential for establishment of hybrids and introgression of the glyphosate-tolerance trait into wild populations is uncertain. Further, contamination of irrigation ditches and drainages with herbicide-tolerant grass could make control more difficult and expensive, because glyphosate is one of a few herbicides labeled for use along waterways. It is also not known whether seed can remain viable and move through the irrigation canal system, or how much seed a volunteer plant can produce with no outside pollen sources once the creeping bentgrass fields were removed from production.

On the other hand, persistence of the glyphosate resistant trait in populations of compatible wild relatives without the selective pressure of herbicide is an open question, as the glyphosate resistant trait has not been shown to have a fitness cost (Fei and Nelson 2004). Even with safeguards in place, gene flow via seed and pollen was not contained during the 2002 and 2003 plantings as a natural dispersal mechanism (wind) coupled with hygienic production practices still led to measurable gene flow. GlyRCB plants were found in other crop fields and non-production areas that required increased control measures. Thus, while GlyRCB release into the environment will probably have little environmental impact on wild species per se, it could significantly increase the weed control costs for management of various agronomic and non-agronomic environments. Therefore, the continued development of GlyRCB requires an effective mitigation plan in place that incorporates control measures for all possible sites—crop fields, canals, ditches, non-crop fields, and non-crop areas.

10.6 Conclusions

How have our experience, data, and knowledge about gene flow with regards to regulating food, feed and fiber crops prepared us for the world of dedicated GE biofuels and biomass crops? Furthermore, how much will the creeping bentgrass story impact the future commercialization of other perennial GE grasses such as switchgrass (*Panicum virgatum*)? There are several perspectives pertinent to the future commercialization of a bioenergy feedstock such as switchgrass. These include: (1) the impacts of regulatory requirements on small scale and prospective corporate developers of GE perennial grasses for bioenergy; (2) the large potential land area for commercial production of a dedicated energy crop such as switchgrass; (3) development of effective biocontainment biotechnologies; and, (4) perceived economic, agronomic, and ecological benefits of engineered perennial grasses for use in bioenergy production.

The deregulation of transgenic plants worldwide has become increasingly more conservative and stringent in recent years, typically focusing on modes of gene transfer (i.e., transgenics), rather than phenotype. The new rules proposed by the USDA-APHIS-BRS (Biotechnology Regulatory Services) (under public comment until June 2009; the agency has said nothing further since 2009 nor issued new rules) are consistent with this trend, and would likely increase the amount of paperwork required permits allowing release into the environment, and thus require more overseeing, even for relatively environmentally benign traits and crops. Therefore, the costs of deregulating a GE plant will likely increase in the future. At the same time, large international agricultural companies are not the primary investors of research funds into the biotechnology of dedicated bioenergy crops. Rather, bioenergy investors are relatively inexperienced companies regarding deregulation—i.e., more like Scotts and less like Monsanto. So, we should expect regulators to take a very long and careful look at perennial bioenergy grasses. On the other hand, two of us (A.P.K. and C.N.S.) have pending BRS permits for releasing transgenic switchgrass into the environment, which will be the first such occurrences. There seem to be few special stipulations with regards to growth requirements, but these will both be very small trials (20 plants).

Second, the scope of potential area under commercialization of switchgrass is huge compared with a golf course grass such as creeping bentgrass. In addition, switchgrass grows over 2 m tall, which is much larger than creeping bentgrass. The potential pollen and seed production of switchgrass relative to bentgrass could translate to high levels of potential gene flow via wind and other vectors. Also, switchgrass is native across much of North America and wild populations would likely be proximate to transgenic populations. So, if unmitigated transgene movement from bentgrass into wild and non-transgenic crop varieties was undesirable, switchgrass would likely be appreciably more challenging. Pollen-mediated gene flow studies in transgenic switchgrass will provide valuable data concerning the need for gene confinement in genetically modified varieties with biofuels-specific traits.

This brings us to the third issue: the necessity for biocontainment in switchgrass, especially to limit gene flow via pollen (Stewart 2007). Fortunately, tools, such as gene deletor technologies based on site-specific recombination (Luo et al. 2007), male sterility and transplastomics (Daniell 2002) exist, and novel tools are under development (H.S. Moon, J.M. Abercrombie, A.P.K., and C.N.S.Jr., unpublished). Unfortunately, none of these seem to be ready for commercialization or have even been tested in perennial feedstock grasses such as switchgrass.

All these issues lead us to exercise caution, albeit optimistic caution, with regards to future commercialization of transgenic switchgrass or other perennial grasses such as *Miscanthus* (Stewart 2007). Biocontainment strategies should be allowed to co-mature and co-develop with traits of interest, such as domestication traits and cell wall traits for decreased recalcitrance for digestion. While cellulosic bioenergy is certainly a compelling new industry, it must play by the well-established regulatory rules. We have learned enough to know that a mature and regulated bioenergy industry will not occur quickly if it is to be sustainable.

Disclaimer: Mention of trade names does not imply endorsement of the commercial products that are mentioned nor do the views expressed herein necessarily reflect the views of USDA or USEPA.

References

Aldhous P (2003) Time to choose. Nature 425:655
AOSA (2002) Rules for testing seeds. Association of Official Seed Analysts, Stillwater, OK
Bae TW, Vanjildorj E, Song SY, Nishiguchi S, Yang SS, Song IJ, Chandrasekhar T, Kang TW, Kim JI, Koh YJ, Park SY, Lee J, Lee Y-E, Ryu KH, Riu KZ, Song P-S, Lee HY (2008) Environmental risk assessment of genetically engineered herbicide-tolerant *Zoysia japonica*. J Environ Qual 37:207–218
Belanger FC, Meagher TR, Day PR, Plumley K, Meyer WA (2003) Interspecific hybridization between *Agrostis stolonifera* and related *Agrostis* species under field conditions. Crop Sci 43: 240–246
Bucchini L, Goldman LR (2002) Starlink corn: a risk analysis. Environ Health Perspect 110: 5–13
Christoffer PM (2003) Transgenic glyphosate resistant creeping bentgrass: studies in pollen-mediated transgene flow. Masters Thesis, Washington State University
Colwell RK, Norse EA, Pimentel D, Sharples FE, Simberloff D (1985) Genetic engineering in agriculture. Science 229:111–112
Dale PJ, Clarke B, Fontes EMG (2002) Potential for the environmental impact of transgenic crops. Nat Biotechnol 20:567–574
Daniell H (2002) Molecular strategies for gene containment in transgenic crops. Nat Biotechnol 20:581–586
Eastham K, Sweet J (2002) Genetically modified organisms (GEOs): the significance of gene flow through pollen transfer. European Environment Agency, Copenhagen
Ellstrand NC, Hoffman CA (1990) Hybridization as an avenue of escape for engineered genes—strategies for risk reduction. Bioscience 40:438–442
Ellstrand NC, Prentice HC, Hancock JF (1999) Gene flow and introgression from domesticated plants into their wild relatives. Annu Rev Ecol Syst 40:434–437
Fei S, Nelson E (2003) Estimation of pollen viability, shedding pattern, and longevity of creeping bentgrass on artificial media. Crop Sci 43:2177–2181
Fei S, Nelson E (2004) Greenhouse evaluation of fitness-related reproductive traits in Roundup®-tolerant transgenic creeping bentgrass (*Agrostis Stolonifera* L.). In Vitro Cell Dev Biol Plant 40:266–273
Giles J (2003) Damned if they do, damned if they don't. Nature 425:656–657
Luo K, Duan H, Zhao D, Zheng X, Deng W, Chen Y, Stewart CN, McAvoy R, Jiang X, Wu Y, He A, Pei Y, Li Y (2007) 'GE-gene-deletor': fused *loxP*-FRT recognition sequences dramatically improve the efficiency of FLP or CRE recombinase on transgene excision from pollen and seed of tobacco plants. Plant Biotechnol J 5:263–374
Macilwain C (2005) US launches probe into sales of unapproved transgenic corn. Nature 434:423
Mallory-Smith C, Zapiola ML (2008) Gene flow from glyphosate-resistant crops. Pest Manag Sci 64:428–440

Marvier M, Acker RCV (2005) Can crop transgenes be kept on a leash? Front Ecol Environ 3:99–106

Ortiz-García S, Ezcurra E, Schoel B, Acevedo F, Soberón J, Snow AA (2005) Absence of detectable transgenes in local landraces of maize in Oaxaca, Mexico (2003–2004). Proc Natl Acad Sci USA 102:12338–12343

Pfender W, Graw R, Bradley W, Carney M, Maxwell L (2007) Emission rates, survival, and modeled dispersal of viable pollen of creeping bentgrass. Crop Sci 47:2529–2539

Piñeyro-Nelson A, Van Heerwaarden J, Perales HR, Serratos-Herández A, Rangel A, Hufford MB, Gepts P, Garay-Arroyo A, Rivera-Bustamante R, Álvarez-Buylla ER (2009) Transgenes in Mexican maize: molecular evidence and methodological considerations for GEO detection in landrace populations. Mol Ecol 18:750–761

Quist D, Chapela IH (2001) Transgenic DNA introgressed into traditional maize landraces in Oaxaca, Mexico. Nature 414:541–543

Reichman JR, Watrud LS, Lee EH, Burdick CA, Bollman MA, Storm MJ, King GA, Mallory-Smith C (2006) Establishment of transgenic herbicide-resistant creeping bentgrass (*Agrostis stolonifera* L.) in nonagronomic habitats. Mol Ecol 15:4243–4255

Rogers HJ, Parkes HC (1995) Transgenic plants and the environment. J Exp Bot 46:467–488

Snow AA (2009) Unwanted transgenes re-discovered in oaxacan maize. Mol Ecol 18:569–571

Stewart CN (2007) Biofuels and biocontainment. Nat Biotechnol 25:283–284

Tiedje JM, Colwell RK, Grossman YL, Hodson RE, Lenski RE, Mack RN, Regal PJ (1989) The release of transgenic plants into agriculture. Ecology 70:298–315

Tsuchiya T, Toriyama K, Yoshikawa M, Ejiri S-i, Hinata K (1995) Tapetum-specific expression of the gene for an endo-β-1,3-glucanase causes male sterility in transgenic tobacco. Plant Cell Physiol 36:487–494

USDA-NASS (2006) Oregon agriculture and fisheries statistics. US Department of Agriculture National Statistics Service & Oregon Department of Agriculture

Van de Water PK, Watrud LS, Lee EH, Burdick C, King GA (2007) Long-distance GE pollen movement of creeping bentgrass using modeled wind trajectory analysis. Ecol Appl 17:1244–1256

Watrud LS, Lee EH, Fairbrother A, Burdick C, Reichman JR, Bollman M, Storm M, King G, van De Water PK (2004) Evidence for landscape-level, pollen-mediated gene flow from genetically modified creeping bentgrass with *CP4 EPSPS* as a marker. Proc Natl Acad Sci USA 101:14533–14538

Wipff JK, Fricker C (2001) Gene flow from transgenic creeping bentgrass (*Agrostis stolonifera* L.) in the Willamette Valley, Oregon. Int Turfgrass Soc Res J 9:224–242

Wrubel RP, Krimsky S, Wetzler RE (1992) Field testing transgenic plants. BioScience 42:280–289

Zapiola ML, Mallory-Smith CA, Thompson JH, Rue LJ, Campbell CK, Butler MD (2007) Gene escape from glyphosate-resistant creeping bentgrass fields: past, present, and future. Proceedings of the Western Society of Weed Science, Abstract 82

Zapiola ML, Campbell CK, Butler MD, Mallory-Smith CA (2008) Escape and establishment of transgenic glyphosate-resistant creeping bentgrass *Agrostis stolonifera* in Oregon, USA: a 4-year study. J Appl Ecol 45:486–494

Chapter 11
Genetic Modification in Dedicated Bioenergy Crops and Strategies for Gene Confinement

Albert P. Kausch, Joel Hague, Melvin Oliver, Yi Li, Henry Daniell, Peter Mascia, and C. Neal Stewart Jr

11.1 Introduction

The utilization of dedicated crops as a source of bioenergy from renewable biomass resources is a goal with great relevance to current ecological, economic, and national security issues on a global scale. In the US, the Energy Policy Act of 2005 (EPAct 2005) issued a mandate for the use of up to 7.5 billion gallons of renewable fuel in gasoline by 2012. These amounts will likely increase in the future as a shift occurs toward renewable energy sources and away from foreign oil supplies (Robertson et al. 2008). Current strategies for liquid fuel production utilize fermentation of plant-derived starches and sugars to ethanol, mostly from grain and other food crops. One concern is whether sufficient amounts of these feedstock materials can be supplied without impacting the cost of agricultural land, competing with food production, and harming the environment. For a variety of reasons, production of fuel from dedicated non-food crops as cellulosic sources, such as

---

A.P. Kausch and J. Hague
University of Rhode Island, 530 Liberty Lane, West Kingston, RI 02892, USA
e-mail: akausch@etal.uri.edu; joel.hague@gmail.com

M. Oliver
USDA-ARS Plant Genetics Unit, University of Missouri, Columbia. 204 Curtis Hall, Columbia, MO 65211, USA

Y. Li
University of Connecticut, Storrs, CT 06269, USA

H. Daniell
Central Florida State University, Biomolecular Science Bldg. #20, Rm #336, 4000 Central Florida Blvd., Orlando, FL 32816, USA

P. Mascia
Ceres, 1535 Rancho Conejo Boulevard, Thousand Oaks, CA 91320, USA

C.N. Stewart Jr
Racheff Chair of Excellence in Plant Molecular Genetics, University of Tennessee, 2431 Joe Johnson Dr. Knoxville, TN 37996-4561, USA

switchgrass, Energy Cane, sorghum, Miscanthus, willow, and poplar, is widely understood as a necessary development (Sticklen 2008).

The genetic improvement of food crop species using biotechnology is well-established and, together with conventional breeding efforts, can be used to confer valuable traits. Trait enhancement and new varietal development will be useful toward the improvement of dedicated bioenergy crops. In addition, biofuels-specific traits, such as production of cellulases and other hydrolytic enzymes and biopolymers, increased cellulose, and decreased lignin can be engineered to increase fuel production per acre (Sticklen 2008). Efforts toward genetic engineering of cellulosic feedstock crops used for bioenergy have barely begun and offer significant potential improvements; however, these modifications present significant public and regulatory concerns. Commercial-scale production of some transgenic plants could lead to undesirable environmental and agricultural consequences (Altieri 2000; Dale 1993; Robertson et al. 2008; Snow and Moran Palma 1997) including transgene escape to wild and non-transgenic relatives. Thus, to realize the full potential of agricultural biotechnology for dedicated energy crops enhancement, the ecological, economic, as well as commercial impacts of gene flow must be addressed.

Currently, strategies using plant genetic engineering for biofuel production are being developed with the goal of renewable and affordable cellulosic ethanol production. Most of the plants considered as top choices for cellulosic biomass are perennial and/or have wild relatives in the areas where they will be produced commercially. Bioconfinement of engineered genes and plants used for cellulosic biofuels will likely be a prerequisite for deregulation and commercial production of these plants (Stewart 2007). Current information strongly indicates the potential for gene flow in open pollinated genetically modified (GM) bentgrass (Belanger et al. 2003; Mallory-Smith and Zapiola 2008; Reichman et al. 2006; Watrud et al. 2004; Wipff and Fricker 2001; Zapiola et al. 2007, 2008) and the need for robust gene confinement strategies (Dunwell and Ford 2005). In this chapter, we review currently viable strategies for the control of transgene flow in perennial grasses that may be useful in the engineering and commercial release of perennial dedicated biofuels crops.

11.2 Methods for Gene Confinement in Genetically Engineered Plants

11.2.1 Physical, Spatial, Mechanical and Temporal Control

One convenient method that has been proposed for gene confinement of genetically modified perennial plants would utilize agronomic practices, including physical, spatial, mechanical or temporal control. Physical containment has been proposed for specific containment requirements, such as production of plant-based biopharmaceuticals in greenhouses, underground facilities, and growth rooms, and is suitable for some crops (tomatoes, lettuce) and for research purposes, but has serious large-scale limitations for most biofuels crops (Dunwell and Ford 2005). Spatial, mechanical or temporal control strategies have been considered for genetically modified perennial plants that could be grown in areas that are outside their normal range, or in areas where there are no wild relatives. In many ways this is similar to the current large-scale control of gene flow in maize. Genetically modified perennial grasses could be routinely mowed such that they never produce fertile flowers. In addition, GM grasses could be grown in areas where their flowering time does not match that of local species. All of these mechanisms rely on human management and thus eventually will be flawed. The consequences of gene flow that have relied on management practices have already been observed in the release of open-pollinated GM creeping bentgrass in Oregon (Reichman et al. 2006; Watrud et al. 2004; Zapiola et al. 2007, 2008).

11.2.2 Pollen Sterility

Pollen-mediated transfer is widely believed to be the major contributor to gene flow in flowering plants. Interfering with the development of male reproductive structures through genetic engineering (GE) has been widely used as an effective strategy for production of male sterility in plants. These methods are distinctly different from cytoplasmic male sterility (CMS) and shown to be extremely effective and stable. The tapetum is the innermost layer of the anther wall that surrounds the pollen sac and is essential for the successful development of pollen. It has been shown that the tapetum produces a number of highly expressed messenger RNAs. Genes expressed exclusively in the anther are most likely to include those that control male fertility. Indeed, a variety of anther- and tapetum-specific genes have been identified that are involved in normal pollen development in many plant species, including maize (Hanson et al. 1989), rice (Zou et al. 1994), tomato (Twell et al. 1989), *Brassica campestris* (Theerakulpisut et al. 1991), and *Arabidopsis* (Xu et al. 1995). Selective ablation of tapetal cells by cell-specific expression of nuclear genes encoding cytotoxic molecules (Goetz et al. 2001; Jagannath et al. 2001; Mariani et al. 1990; Moffatt and Somerville 1988; Tsuchiya et al. 1995) or an antisense gene essential for pollen development (Goetz et al. 2001; Luo et al. 2000; Xu et al. 1995) blocks pollen development, giving rise to stable male sterility.

To induce male sterility in turfgrass, the 1.2-kb rice *rts* gene regulatory fragment, TAP (Lee et al. 1996) was fused with two different genes. One was the antisense of the rice *rts* gene that is expressed predominantly in the anther's tapetum during meiosis. Another was a natural ribonuclease gene from *Bacillus amyloliquefaciens* called *barnase*, which ablates cells by destruction of RNA (Hartley 1988). Both of these approaches have been shown to be effective in other plant species (De Block et al. 1997; Higginson et al. 2003; Luo et al. 2000; Mariani et al. 1990; Yui et al. 2003). Separately, both chimeric gene constructs were linked in a tandem construct to the *bar* gene driven by either a rice *ubi* promoter or the CaMV35S promoter for selection by resistance to the herbicide phosphinothricin. These two constructs—pTAP:*barnase*-Ubi:*bar* and pTAP:a*rts*-35S:*bar*—were introduced separately into bentgrass (*Agrostis stolonifera* L.), cv Penn-A-4 using *Agrobacterium tumefaciens*-mediated transformation. Transgenic plants were screened from a population of independent transformation events recovered by phosphinothricin (PPT) selection. A total of 319 primary transgenic callus lines (123 from pTAP:*barnase*-Ubi:*bar* transformation and 196 from pTAP:a*rts*-35S:*bar* transformation) were recovered and regenerated into plants. Under greenhouse conditions, the insertion and expression of the two gene constructs did not affect the vegetative phenotype. The transgenic plants were vigorous and morphologically indistinguishable from untransformed control plants. PCR assays and Southern blot analysis on genomic DNA from independent transgenic plants were carried out to assess the stability of integration of the transgenes in the host genomes. The *bar* gene was present in all the transformants, and the *barnase* or antisense *rts* gene was also detected in the respective transgenic plants. All the transgenic events had less than three copies of the inserted transgene, and a majority of them (60–65%) contained only a single copy of foreign gene integration with no apparent rearrangements.

To check the sterility/fertility status of pollen from various transgenic plants expressing *barnase* or antisense *rts*, vernalized transgenic and non-transgenic control plants were grown in the greenhouse and flowered at 25°C in artificial light under a 16/8 h (day/light) photoperiod. The pollen was taken 1 day before anthesis for viability analysis using iodine-potassium iodide (IKI) staining. More than 90% of the plants (20/23) containing *barnase* and around 50% of the plants (40/79) containing the antisense *rts* gene were completely male-sterile, without viable pollen, which are normally stained darkly by IKI as observed in the wild-type control plants and *hygromycin*-resistant control transgenic plants that do not contain the *barnase* or the antisense *rts* gene, indicating that cell-specific expression of the *barnase* or the antisense *rts* gene in transgenic plants blocks pollen development, giving rise to male sterility. Light microscopy of cross-sections through flowers at anthesis showed that tapetum development had been interrupted resulting in aborted pollen maturation. Interestingly, the single gene knockout phenotype achieved through the antisense approach appears developmentally different from barnase ablation, but both have resulted in 100% stable male sterility. Therefore, when linked to genes of agronomic interest, nuclear male sterility resulting in the lack of viable pollen grains provides an important tool to study effective mechanisms for interrupting gene flow.

11.2.3 Cytoplasmic Male Sterility, Chloroplast Transformation and Maternal Inheritance

A major concern in GE perennial grass development is the possibility of the GE trait escaping into other crops or wild/weedy relatives. The use of inherent systems, such as Cytoplasmic Male Sterilty (CMS), and/or GM approaches, such as Chloroplast transformation, may offer attractive solutions for controlling gene flow between dedicated energy crops and their wild relatives. CMS is caused by mutations in the genomes of either the chloroplast or the mitochondria and is thus inherited only maternally in many plant species. In many crop plants, nuclear genes that restore fertility (Rf) have been applied for creating hybrids. Consequently, the development of CMS systems for dedicated energy crops would be useful for gene confinement as well as providing valuable breeding tools for these crops. However, the current status for breeding these crops does not yet include these tools. An attractive option would be to genetically engineer a CMS-associated mitochondrial gene for stable nuclear expression that would cause pollen disruption (He et al. 1996).

Another GM method of gene confinement that attempts to address this concern involves introducing the transgene into chloroplasts, which are maternally inherited in most crops (Daniell 2002). In plants exhibiting *Lycopersicon*-type maternal inheritance, chloroplasts are shunted to the vegetative cell during the first pollen mitotic division in pollen formation; none are found in the generative cell from which the sperm cells arise. The paternal chloroplasts shunted to the vegetative cell are generally destroyed when the pollen tube (derived from the vegetative cell) penetrates the synergid cell prior to fertilization. Direct GE of the chloroplast genome is an advantageous approach to gene confinement since it would provide the ability for multi-gene constructs with high levels of expression without the possibility of gene silencing or position effects (Daniell 2002). However, paternal inheritance of chloroplasts has been observed in tobacco, albeit at a very low rate (Ruf et al. 2007). Additionally, there exists the possibility of transgene flow from the chloroplast to the nucleus (Stegemann et al. 2003), although it can be reasonably argued that transgenes designed to function in chloroplasts will not function if transferred to the nucleus. Thus, while not offering absolute transgene containment, confining transgenes within chloroplasts will greatly limit the passage of transgenes via pollen and therefore to other crops or relatives during outcrossing.

Chloroplast transformation not only promotes gene confinement but also confers unique molecular and expression characteristics not found in nuclear transformation. Transgenes are incorporated in a site-specific manner into "spacer DNA" within the chloroplast genome by homologous recombination using particle bombardment, thereby not disrupting native genes. The major challenge is to get the transgene into every chloroplast (homoplasmy) in each cell. However, only three rounds of selection on regeneration media are typically required to reach homoplasmy in tobacco. Southern blots and PCR are used to measure if any wildtype copies are present, and homoplasmic lines can be identified and increased. Since chloroplasts are prokaryotic compartments, they lack the silencing machinery found within the cytoplasm of eukaryotic cells. Each plant cell contains 50–100 chloroplasts and each chloroplast contains ~100 copies of its genome, so it is possible to introduce 20,000 copies of the transgene per cell as spacer DNA is present in duplicate within the chloroplast genome. This allows for very high gene expression with no silencing. For example, in overexpression of the Bt *cry 2Aa2* operon via chloroplast transformation of tobacco, nearly one half of the protein

(47%) found in leaves was foreign protein with no silencing or health effects on the plant (De Cosa et al. 2001). Other additional advantages include no position or pleiotropic effects. Thus, chloroplast transformation imparts significant advantages over nuclear transformation in addition to gene confinement. However, to date, most crops, and especially dedicated energy crops (perennial grasses, sugar cane, sorghum, maize, etc.) cannot be plastid transformed.

11.2.4 Seed-Based Gene Confinement

Seed-based gene confinement generally involves the use of genetic switch mechanisms in what have become known as genetic use restriction technologies (GURTs). This nomenclature unfortunately emphasizes only the financial or patent enforcement interests of those companies that are involved in the development of GURTs and does not reflect any of the positive aspects of their development; in particular their utility in transgene confinement. There are two major classes of GURTs, V-GURTs (varietal-level GURTs) and T-GURTS (trait-specific GURTs), which relate to the event that is triggered by the genetic switch portion of the individual technologies. When triggered, V-GURT systems prevent the propagation of the crop and its associated genetic technology without the purchase of new seed. V-GURTs allow for normal growth and full development of the desired seed; however, the progeny seed, if planted, will not germinate. Gene containment is achieved by the inability of the plants that contain the activated V-GURT mechanism to produce viable progeny either through the pollen or via seed. T-GURT systems regulate trait expression, making the value-added trait (transgene) available only if the farmer triggers the genetic switch mechanism. Plant function is normal, but when a particular engineered trait is needed in a farmer's field, a specific triggering chemical purchased from the technology provider is applied to activate transgenes expressing a desired characteristic (e.g., insect resistance). The technology would presumably be paid for and activated only when needed. Gene containment is achieved by the inability of the plants to express the transgenic trait in the absence of the activating chemical, which is presumably not freely available in the environment.

11.2.5 Perceived Risks Associated with GURTs

Since the issue of the original GeneSafe patent describing an obvious V-GURT mechanism involving the production of non-germinable seeds as a means of gene confinement, many controversies have emerged, often fueled by the ascribing of such emotion-packed monikers as "Terminator" by those opposed to the use of such mechanisms. However, almost all of these concerns present issues that are either manageable or impart a negligible risk to society, the environment, or the customer.

One of the major issues raised in objection to the use of V-GURTs is the possible impact on seed viability in compatible non-transgenic or T-GURT crops in neighboring fields as a result of the spread of pollen from a V-GURT crop. V-GURTS are at the present time designed for use in crops that preferentially self rather than outcross, e.g., cotton, soybean and wheat. In such cases, the negative effects on neighboring fields would be very restricted and would not be detectable above the background of normal germination rates for field grown crops. V-GURTs targeted for crops that readily outcross would have to contain design elements for the removal of transgenes during microsporogenesis so as to prevent transgene escape via pollen dispersal. A similar concern has been posed in regards to the possibility that pollen from V-GURT plants may prevent germination of seeds in neighboring wild species and thus reduce their long-term viability in the native habitat. Obviously, preventing the germination of hybrid seed developed from pollen outflow from a crop to a wild species is a desired outcome in the desire to contain transgenes in the environment, but it would be problematic if, in doing so, the long-term viability of a wild species could be affected. In realistic terms, this is a highly unlikely scenario because such an outcome would require that the wild species was completely compatible with the crop containing the V-GURT, and that non-V-GURT pollen was absent from the environment. Most crops do not have relatives that are sexually compatible in agricultural areas, and hybridization is very rare. In cases where there is a measure of compatibility and a problem could arise, then a change in the design of the V-GURT may be warranted (see below).

V-GURTs have also been criticized for their supposed potential for socio-economic impacts on agriculture in developing countries. The non-germinability of GeneSafe seeds and the resultant need to purchase new seed for the planting of a new crop has been suggested to be an unfair economic burden on small farmers, especially those engaged in subsistence farming. Although it is true that farmers would be required to purchase new seed every year, one has to bear in mind that, in themselves, GeneSafe and other V-GURT technologies have no value and would be in a crop only in conjunction with a valuable or advantageous transgenic trait. The farmer would not be limited to a V-GURT variety but would gain the economic value of the transgenic trait should he or she so choose. In doing so, the farmer would presumably turn a subsistence level operation into a profitable and perhaps productive concern. The initial outlay for the transgenic variety maybe a barrier to acceptance but the remedy for this problem is based on a commercial or political tenet. Another concern is that large multinational companies could monopolize seed supplies by the use of V-GURT technologies. V-GURT technologies have value only in conjunction with transgenic technologies and, as non-transgenic seed will still be freely available through public concerns, it is difficult to see how seed supplies could be monopolized. Nevertheless, GeneSafe technologies are V-GURTs that are owned jointly by the United States Department of Agriculture (USDA) Agricultural Service and a private company (Delta and Pine Land Company; http://www.deltaandpine.com), and it is the involvement of the USDA that prevents the monopolization of the technology. GeneSafe and other V-GURTs do not, in themselves, provide a competitive economic advantage. On an environmental level, concerns have been raised that the method used to prevent the germination of activated V-GURT seeds may harm other organisms. As of yet only gene products that are not toxic to animals and occur naturally in plants and microbes that are normally consumed in animal diets have been used to disrupt seed metabolism. Similarly, the chemical seed treatment used to activate the V-GURT during stand establishment would have to be, by necessity, environmentally friendly or neutral. The use of tetracycline described in the GeneSafe prototype was never targeted for commercial use in the field.

Transgenic seedless fruits (although not a complete gene containment technology) described by Tomes et al. (1998), and the GeneSafe technologies of Oliver et al. (1998, 1999a, b), are all V-GURTS and all are designed to prevent gene out-flow from GE plants. GeneSafe technology, formally the Technology Protection System (TPS), was the first gene containment V-GURT to be patented and provides a complete one-generation strategy for gene containment.

In a series of three patents, Oliver et al. (1998, 1999a, b) described two primary GeneSafe mechanisms, utilizing a single strategy, to prevent gene flow from crops where seeds are the primary production target, whether it be for food, fiber, oil, or a value-added product. The basic strategy outlined in these patents is to control the activation of a germination disruption gene sequence such that its expression prevents the establishment of a second generation of a crop that bears a value-added or production-benefit transgene. The gene activation is timed such that the transgene is available in an uncontained environment such as a farmer's field, and only after a crop is produced is the activated germination disruption gene expressed and effective. The mechanism is also designed such that pollen emanating from a plant that contains the activated germination disruption gene carries it to the ovule that it fertilizes to generate a non-germinable seed. Although this is desired for total gene containment, as mentioned below, this could be problematic in an open pollination scenario and so the GeneSafe mechanisms described here were designed for crops that reproduce under restricted or mainly closed pollination.

The genetic mechanisms designed to accomplish these goals utilize three basic elements: (1) a promoter that responds to a specific exogenous stimulus; (2) a site-specific recombinase to remove a physical block; and (3) a seed-specific promoter that is active only late in seed development. These elements were used to generate two genetic systems (basic systems from which refinements can be added), one based on a repressible promoter mechanism that is relieved by exposure to an activator and the other, simpler, system based on a chemically inducible promoter. These two mechanisms were designed originally for use in GM cotton as a technology protection system.

The original mechanism was designed as the prototypical system, and because at the time of its development there were few available chemically inducible promoters, is the one that has received most attention. The mechanism consists of two constructs or modules. The first, the LEA module, consists of a late embryogenesis abundant (LEA) protein gene promoter separated from a coding sequence for a protein synthesis inhibitor protein, either Saporin or Barnase, by a "blocking sequence", which in this case contains the gene that produces the *tet* repressor protein, flanked by LOX sites. LOX sites are recognition sequences specific for the site-specific recombinase CRE from bacteriophage PI, which is the subject of the second construct, the CRE module (Bayley et al. 1992; Boffey and Veevers 1977; Dale and Ow, 1990, 1991). The CRE gene is controlled by a 35S cauliflower mosaic virus (CaMV) promoter modified to contain three *tet* operator sites that direct binding of the *tet* repressor protein. Binding of the *tet* repressor protein to these operator sites results in the inactivation of the CRE gene.

The requirement for the precise timing of the activation of the protein synthesis inhibitor gene (germination disruptor) after seed formation and maturation necessitates the use of a LEA promoter, in particular one taken from the family of LEA genes that expresses very late in embryogenesis. In all probability such precise timing will dictate that GeneSafe technologies will be species specific. Although it is possible that LEA promoters retain their precise timing of expression when placed in a heterologous genetic environment, it is more likely that they would not and so for practical reasons one would prefer to design a GeneSafe strategy with a time-specific LEA promoter from the target crop. The original GeneSafe technologies were designed for cotton, although an attempt was made to assemble a working prototype in tobacco using cotton LEA promoters.

To establish the full repressible GeneSafe system, plants homozygous for each module have to be crossed to form a dual hemizygous plant that contains both modules. The cross has to be performed with the CRE plant as the pollen donor in order to ensure that introduced CRE gene is exposed to the *tet* repressor protein and inactivated in the fertilized egg cell. In the dual hemizygous plant, the complete GeneSafe system is inactive; the LEA promoter cannot drive the expression of the protein synthesis inhibitor during the last stages of seed maturation because of the physical presence of the blocking sequence, and the CRE gene cannot be expressed to generate the site-specific recombinase because of the binding of the *tet* repressor protein to the embedded operator sites in the 35S promoter. This allows these plants to be propagated in order to make both modules homozygous so that commercial seed stocks can be established. Transgenes can be added to either the plants that are homozygous for both modules, or they can be linked to the LEA module during the initial transformation to ensure they segregate with the germination disruption phenotype.

To activate the GeneSafe system, tetracycline, the chemical activator, is added to imbibing seeds. The tetracycline has to be able to penetrate to the cells in the L2 layer of the developing shoot apical meristem in order to activate the germline progenitor cells. The tetracycline releases the binding of the *tet* repressor protein, thus enabling transcription from the modified 35S promoter to produce the site-specific recombinase CRE. The resultant CRE enzyme locates its specific recognition sites, LOX sites (left and right), and physically removes the DNA between them. The LOX sites have been modified such that once excision occurs it is irreversible (Albert et al. 1995). This removal of the blocking sequence containing the *tet* repressor protein gene results in the permanent formation of the developmentally programmed germination-disruption (protein synthesis inhibitor) gene driven by the LEA promoter. The germination-disruption gene encodes an enzyme that, when expressed in the cytoplasm of a cell, prevents protein synthesis and thus growth. The enzymes targeted for use in the prototype of the GeneSafe system were saporin, an enzyme that cleaves a specific sequence in ribosomal RNA, which in turn inactivates the ribosome, and a translation attenuated (an added AUG codon upstream of the native start codon) barnase, a ribonuclease derived from the bacterium *Bacillus amyloliquefaciens* that digests all cellular RNAs, thus preventing protein synthesis. The germination-disruption gene is not active following exposure to tetracycline as it is under the control of the LEA promoter. As the germination-disruption proteins are synthesized only after storage proteins and oils are fully deposited, the quality of the seeds produced by the plant is unaffected even though their ability to germinate has been compromised. Since this system is activated in all germ line cells of the plant the pollen will also carry the constructed germination-disruption gene. Flowers fertilized by the pollen from an activated plant will therefore produce seed that also cannot germinate. This, in effect, makes an activated GeneSafe plant an evolutionary dead end (both seed and pollen are effectively non-viable) and incapable of spreading transgenes into the environment.

At the present time, the repressible GeneSafe technology is in place in both cotton and tobacco to varying degrees, tobacco being the most advanced. Dual hemizygous tobacco plants, containing both the LEA and CRE modules, have been utilized in tetracycline activation tests and are presently within a selfing scheme designed to generate plants homozygous for both modules. In cotton, homozygous parental lines for each module have been generated (Oliver et al. 1999a, b). Analysis of tobacco plants that arise from tetracycline-treated dual hemizygous seeds confirm that CRE activation has occurred, both by PCR analysis demonstrating the precise removal of the blocking sequence, and by northern analysis revealing a loss of *tet* repressor transcripts. Germination tests of the seed derived from selfing of these activated dual hemizygous plants did not generate the expected 3:1 ratio of non-germinable to germinable seed (assuming successful activation of CRE in all germline cells of the parental lines), in fact in only a few cases were germination percentages reduced. However, PCR analysis of the seeds used in the germination tests revealed that all were either heterozygous for the excision phenotype or homozygous for the intact module; no seeds homozygous for the excision event have been detected (360 seeds tested so far). The implication is that seeds that contain two copies of the excision event do not develop to maturity in the tobacco pods of the plants derived from *tet*-treated seeds. This would further imply that the timing of expression of the protein synthesis inhibitor driven by the cotton LEA promoter in tobacco does not mimic that seen in cotton, i.e., it occurs prior to the maturation phase of seed development, and that the level of expression of the protein synthesis inhibitors is insufficient to affect viability when only one copy of the gene is present. The analysis of these phenomena is ongoing.

The repressible GeneSafe mechanism presents some challenges within a seed production setting, the most difficult being the need to make both the LEA and CRE modules homozygous prior to transgene insertion. This can be mitigated somewhat by linking the desired transgene to the LEA module in the initial construct, but this lengthens the process to reach the desired seed production level. The solution to these difficulties became evident with the isolation and characterization of tightly controllable chemically inducible plant active promoters (Zuo and Chua 2000; Zuo et al. 2000). By replacing the *tet* repressor system elements with a chemically inducible promoter to drive the expression of the CRE gene, the GeneSafe technology can be simplified and reduced to a single construct. As a single construct it is simple to generate homozygous plants for seed production, and the more recent chemically inducible promoters are more efficient and offer tighter control than the *tet* repressor system. The inducible GeneSafe technology is being assembled in cotton at this time.

11.2.6 Gene Deletor System

The development of a highly efficient deletion mechanism that relies on site-specific recombination for removal of transgenes has been explored. Luo et al. (2007) developed a method for directing removal of transgenic cassettes from pollen and/or seed in tobacco by designing several gene cassettes using components from both FLP/*FRT* and CRE/*loxP* recombination systems. When *loxP-FRT* fusion sequences (86 bp) were used as recognition sites, simultaneous expression of both FLP and CRE reduced the average excision efficiency, but Luo et al. (2007) report that expression of either *FLP* or *Cre* alone increased the average excision efficiency, with many transgenic events being 100% efficient based on analysis of more than 25,000 T1 progeny examined per event. The deletion of all functional transgenes from pollen and seed was confirmed using three different techniques: histochemical assay for β-glucuronidase (GUS) activity, Southern blot hybridization and PCR. These studies were conducted under greenhouse conditions and have not yet been field tested. A similar system may be used to produce 'non-transgenic' pollen and/or seed from transgenic plants and to provide a bioconfinement tool for transgenic crops and perennials, with special applicability towards vegetatively propagated plants. Pollen- and seed-specific promoters could be used to control recombinase expression, whereby all functional GM genes would be deleted from these organs. If a conditionally inducible gene promoter, such as a chemically inducible or high-temperature inducible elements or conditions such as the use of inteins, were used to control recombinase expression, all functional GM genes would be deleted throughout the plant on application of the inducer.

11.2.7 Total Sterility

The introduction of novel genes by conventional or by genetic engineering is not restricted to those plants that provide food and fiber. Because of the economic and environmental importance of forage species and turf grasses, these species have been targeted for genetic improvement by GE. Improvements such as herbicide resistance, drought resistance, disease resistance, and pest resistance have all been suggested as targets for transgenic strategies. The difficulty with such species, in particular perennial grasses, is the greater potential for transgene escape given their ability to spread pollen over large distances and the large number of close relatives of the targeted commercial varieties and species used at this time. The threat of spreading herbicide resistance into weedy grass species is a real possibility and one that could have significant effects on agriculture and the environment.

Recently, H. Luo, A.P.K., J. Chandlee and M.O. (unpublished) proposed a mechanism to eliminate all possibility for gene transfer in species that are grown primarily for their green biomass, in particular turf grasses. The strategy is simply to prevent flower formation in plants that are released into the field. The mechanism makes use of a site-specific recombinase (in this case the FLP/FRT system from yeast) to activate a gene designed to down-regulate a critical gene in the initiation of floral development. The gene targeted for down-regulation is *FLORICAULA/ LEAFY*, which regulates the vegetative-to-reproductive developmental transition of meristems. The mechanism operates by establishing a transgenic line homozygous for both the transgene of interest and a genetic construct containing the following linked elements: a constitutive plant promoter—an FRT site (recognition site for FLP)—a blocking sequence—an FRT site—and an RNAi or antisense construction for *FLORICAULA/LEAFY*. In the final seed production cycle homozygous plants are crossed to plants homozygous for a constitutively expressed FLP gene to produce hybrid seed. When grown, the hybrid seeds will generate plants that express FLP constitutively, resulting in the excision of the blocking sequence contained in the initial construct. This will activate the constitutive expression of the RNAi or antisense construction for *FLORICAULA/LEAFY*. This in turn will downregulate expression of the endogenous *FLORICAULA/LEAFY* genes, rendering the plant incapable of producing flowers. The vegetative growth habit of the hybrid retains its commercial application but is incapable of transferring transgenes to neighboring grasses or weedy relatives. This is in effect a hybrid total gene containment system.

11.2.8 Total Sterility and Confinement Expression Systems

Recently, research conducted by Ceres has described a new innovative total sterility confinement strategy. The Ceres Confinement & Expression System utilizes an "Activation" line as the male, which is comprised of a proprietary promoter–yeast transcription factor $(T)^{\circledR}$. The "Target" line is then the female and pollination is by self-incompatibility, hand pollination or other male sterility systems. An upstream activation sequence (UAS) uses a Ceres gene that inhibits flowering or causes sterility $(CPG)^{\circledR}$ for example, $-UAS\text{-}CPG_1$, $UAS\text{-}CPG_2,\ldots UAS\text{-}CPG_n$ for introduction of stacked traits.

Advantages of the Ceres Expression & Confinement System include: (1) targeted gene expression dependent on the activation line; (2) multiple proteins can be driven by the same promoter without silencing; (3) transcription and protein level of individual proteins can be modulated by the number of copies of the target gene and UAS elements in the gene; (4) achieves a three- to ten-fold amplification of expression relative to direct fusions; and (5) male and female sterility can be achieved in the commercial seed while allowing breeding to occur. The benefits of this type of program are that target proteins are produced only in the production field; no pollen is produced by plants expressing the target proteins; plants do not express the target proteins unless pollinated by the activation line; and pollen that leaves the production field will express only the transcription factor. Total sterility must be selectable and highly efficient for release as a commercial product.

11.3 Regulatory Issues for Perennial Bioenergy-Dedicated Crops

Currently, The USDA APHIS regulates release of GE plants on a case-by-case basis. The process of deregulation includes lengthy reviews and data collection spanning different environments and several years, with consideration of several factors including: the biology, geography and ecology of the plant: the trait gene(s) of interest; the possibility of gene flow to wild and non-transgenic relatives; the possibility of weediness or invasiveness; and unintended consequences to other organisms. It is important to assess independently the individual species of dedicated energy crops and their novel traits or characteristics that might enhance the vigor or invasiveness of wild or weedy relatives or have other detrimental effects. While some traits may pose relatively benign risk (i.e., herbicide tolerance) others may promote unintended consequences and invasiveness (i.e., drought and pest tolerance). Many of the dedicated energy crops that are currently considered to play a major role in the developing biofuels industry are perennial and have wild relatives in areas where they will be produced and grown. To date, there is no clearly defined limit to gene flow into the environment, which begs questions concerning acceptable (if any) levels of transgene escape in these plants; zero escape is a very stringent requirement. Considering the cost of deregulation and the subsequently imposed market restrictions, some regulatory requirements may be reconsidered or modified without compromising safety (Bradford et al. 2005). These might include: deregulation of the transgenic process itself, the creation of regulatory classes in proportion to potential risk, exemption of selected transgenes and classes of transgenic modifications, and elimination of the event-specific basis of transgenic regulation.

11.4 Conclusions

Biotechnology approaches to genetic improvement of biofuels crops will undoubtedly play a large role in the development of a successful cellulosic energy industry. Certainly the development of regionally selected germplasm, marker-assisted breeding and genomics will facilitate the selection of biofuels traits. In addition, the importance of transgenic traits will further accelerate progress towards the generation of dedicated energy varieties that will allow cost-effective low-input sustainable road fuels with lower greenhouse gas emissions. However, while numerous laboratories are currently exploring expression of transgenic plants for improved biofuels, the requirements for deregulation and commercialization of these crops remains uncertain. Robust gene confinement strategies must be in place as a part of biofuels trait modification. However, even with the best technologies in place it is unlikely that any of these will achieve a zero tolerance expectation. Therefore it seems reasonable to consider now, based on existing work in transgenic grasses, environmentally acceptable levels of mitigation.

We have discussed the available strategies for GE confinement that are currently under development. There are obvious limitations to most of these strategies, most notably, physical, spatial mechanical and temporal containment, but also some of the more sophisticated transgenic approaches that have not yet been developed for most dedicated energy crops and will need to be field tested. The use of genetic modification specifically for controlled transgenic containment is at an early stage of development and there are a range of possible approaches. Pollen sterility has been accomplished in a number of transgenic plant species but may be considered to be limited in its application for controlling gene flow because of the possibility of gene flow via seed scatter. It may be argued, however, that male sterility is sufficient for mitigating gene flow, as wild type crosses would produce progeny that would then also be male sterile, but this needs to be rigorously tested in the field. Also, very little is known about the frequency of reversion of these mechanisms (i.e., ribonucleases) to fertile phenotypes. CMS systems would provide a similar level of confinement and may also provide a valuable breeding tool. Maternal inheritance through plastid transformation is relatively well developed for some dicot plants; however, it may not offer complete containment, and has not been conferred widely on monocot crops. The GeneSafe technology and other seed-based GURTS offer conditional lethality that can be induced chemically to prevent flowering or seed development but requires complete biological induction and has human management drawbacks. However, these methods provide solutions that will allow production of seeds that will contain the trait of interest and prevent the escape of non-functioning transgenes. Currently these approaches are considered to be the best and only strategies that could be deployed to prevent seed-based gene flow. The possibility of creating a hybrid system whereby a two gene system is constructed such that, when crossed, the progeny will produce seed that will never again germinate and result in total sterility may offer the most promise for perennial dedicated energy crops. Also, it may be possible to include failsafe and backup mechanisms, including transgene mitigation strategies into a platform variety that can then receive stacked genes for crop improvement.

The potential benefits of GM of dedicated energy crops are obvious from the examples of food crops already in production. Moving forward, landscape-scale field testing and monitoring of genetic containment systems for perennial dedicated energy crops must be accomplished to determine their efficacy. This should include guidelines established by regulatory agencies concerning acceptable levels of gene flow.

References

Albert H, Dale EC, Lee E, Ow D (1995) Site-specific integration of DNA into wild-type and mutant lox sites placed in the plant genome. Plant J 7:649–659

Altieri M (2000) The ecological impacts of transgenic crops on agroecosystem health. Ecosyst Health 6:13–23

Bayley CC, Morgan M, Dale EC, Ow D (1992) Exchange of gene activity in transgenic plants catalyzed by the Cre-lox site-specific recombination system. Plant Mol Biol 18:353–361

Belanger FC, Meagher TR, Day PR, Plumley K, Meyer WA (2003) Interspecific hybridization between *Agrostis stolonifera* and related *Agrostis* species under field conditions. Crop Sci 43:240–246

Boffey TB, Veevers A (1977) Balanced designs for two-component competition experiments. Euphytica 26:481–484

Bradford KJ, Van Deynze A, Gutterson N, Parrott W, Strauss SH (2005) Regulating transgenic crops sensibly: lessons from plant breeding, biotechnology and genomics. Nat Biotechnol 23:439–444

Dale EC, Ow DW (1990) Intra- and intermolecular site-specific recombination in plant cells mediated by bacteriophage P1 recombinase. Gene 91:79–85

Dale EC, Ow DW (1991) Gene transfer with subsequent removal of the selection gene from the host genome. Proc Natl Acad Sci USA 88:10558–10562

Dale PJ (1993) The release of transgenic plants into agriculture. J Agric Sci 120:1–5

Daniell H (2002) Molecular strategies for gene containment in transgenic crops. Nat Biotechnol 20:581–586

De Block M, Debrouwer D, Moens T (1997) The development of nuclear male sterility system in wheat: expression of the barnase gene under the control of tapetum specific promoters. Theor Appl Genet 95:125–131

De Cosa B, Moar W, Lee SB, Miller M, Daniell H (2001) Overexpression of the Bt cry2Aa2 operon in chloroplasts leads to formation of insecticidal crystals. Nat Biotechnol 19:71–74

Dunwell J, Ford CS (2005) Technologies for biological containment of GM and non-GM crops. DEFRA Contract CPEC 47, http://www.gmo-safety.eu/pdf/biosafenet/Defra_2005.pdf

Goetz M, Godt DE, Guivarc'h A, Kahmann U, Chriqui D, Roitsch T (2001) Induction of male sterility in plants by metabolic engineering of the carbohydrate supply. Proc Natl Acad Sci USA 98:6522–6527

Hanson DD, Hamilton DA, Travis JL, Bashe DM, Mascarenhas JP (1989) Characterization of a pollen-specific cDNA clone from *Zea mays* and its expression. Plant Cell 1:173–179

Hartley RW (1988) Barnase and Barstar: expression of its cloned inhibitor permits expression of a cloned ribonuclease. J Mol Biol 202:913–915

He S, Abad AR, Gelvin SB, Mackenzie SA (1996) A cytoplasmic male sterility-associate mitochondrial protein causes pollen disruption in transgenic tobacco. Proc Natl Acad Sci USA 93:11763–11768

Higginson T, Li SF, Parish RW (2003) AtMYB103 regulates tapetum and trichome development in *Arabidopsis thaliana*. Plant J 35:177–192

Jagannath A, Bandyopadhyay P, Arumugam N, Gupta V, Kumar P, Pental D (2001) The use of spacer DNA fragment insulates the tissue-specific expression of a cytotoxic gene (barnase) and allows high-frequency generation of transgenic male sterile lines in *Brassica juncea* L. Mol Breed 8:11–23

Lee J-Y, Aldemita RR, Hodges TK (1996) Isolation of a tapetum-specific gene and promoter from rice. Int Rice Res Newsl 21:2–3

Luo H, Lyznik LA, Gidoni D, Hodges TK (2000) FLP-mediated recombination for use in hybrid plant production. Plant J 23:423–430

Luo K, Duan H, Zhao D, Zheng X, Deng W, Chen Y, Stewart CN, McAvoy R, Jiang X, Wu Y, He A, Pei Y, Li Y (2007) "GM-gene-deletor": fused *loxP*-FRT recognition sequences dramatically improve the efficiency of FLP or CRE recombinase on transgene excision from pollen and seed of tobacco plants. Plant Biotechnol J 5:263–374

Mallory-Smith C, Zapiola ML (2008) Gene flow from glyphosate-resistant crops. Pest Manag Sci 64:428–440

Mariani C, Beuckeleer M, Truettner J, Leemans J, Goldberg R (1990) Induction of male sterility in plants by a chimaeric ribonuclease gene. Nature 374:737–738

Moffatt B, Somerville C (1988) Positive selection for male-sterile mutants of *Arabidopsis* lacking adenine phosphoribosyl transferase activity. Plant Physiol 86:1150–1154

Oliver MJ, Quisenberry JE, Trolinder N, Glover L, Keim DL (1998) Control of plant gene expression. US Patent 5723765. United States Patent and Trademarks Office, http://www.patentstorm.us/patents/5723765.html

Oliver MJ, Quisenberry JE, Trolinder N, Glover L, Keim DL (1999a) Control of plant gene expression. US Patent 5977441. United States Patent and Trademarks Office, http://www.patentstorm.us/patents/5977441.html

Oliver MJ, Quisenberry JE, Trolinder N, Glover L, Keim DL (1999b) Control of plant gene expression. US Patent 5925808. United States Patent and Trademarks Office, http://www.patentstorm.us/patents/5925808.html

Reichman JR, Watrud LS, Lee EH, Burdick CA, Bollman MA, Storm MJ, King GA, Mallory-Smith C (2006) Establishment of transgenic herbicide-resistant creeping bentgrass (*Agrostis stolonifera* L.) in nonagronomic habitats. Mol Ecol 15:4243–4255

Robertson GP, Dale VH, Doering OC, Hamburg SP, Melillo JM, Wander MM, Parton WJ, Adler PR, Barney JN, Cruse RM, Duke CS, Fearnside PM, Follett RF, Gibbs HK, Goldemberg J, Mladenoff DJ, Ojima D, Palmer MW, Sharpley A, Wallace L (2008) Sustainable biofuels redux. Science 322:49–50

Ruf S, Karcher D, Bock R (2007) Determining the transgene containment level provided by chloroplast transformation. Proc Natl Acad Sci USA 104:6998–7002

Snow AA, Moran Palma P (1997) Commercialization of transgenic plants: potential ecological risks. BioScience 47:86–96

Stegemann S, Hartmann S, Ruf S, Bock R (2003) High-frequency gene transfer from the chloroplast genome to the nucleus. Proc Natl Acad Sci USA 100:8828–8833

Stewart CN (2007) Biofuels and biocontainment. Nat Biotechnol 25:283–284

Sticklen MB (2008) Plant genetic engineering for biofuel production: towards affordable cellulosic ethanol. Nat Rev Genet 9:433–443

Theerakulpisut P, Xu H, Singh MB, Pettitt JM, Knox RB (1991) Isolation and developmental expression of Bcp1, an anther-specific cDNA clone in *Brassica campestris*. Plant Cell 3:1073–1084

Tomes DT, Huang B, Miller PD (1998) Genetic constructs and methods for producing fruits with very little or diminished seed. US Patent 5773697, http://www.freepatentsonline.com/5773697.html

Tsuchiya T, Toriyama K, Yoshikawa M, Ejiri S, Hinata K (1995) Tapetum-specific expression of the gene for an endo-beta-1,3-glucanase causes male sterility in transgenic tobacco. Plant Cell Physiol 36:487–494

Twell D, Wing R, Yamaguchi J, McCormick S (1989) Isolation and expression of an anther-specific gene from tomato. Mol Gen Genet 217:240–245

Watrud LS, Lee EH, Fairbrother A, Burdick C, Reichman JR, Bollman M, Storm M, King G, van De Waters, PK (2004) Evidence for landscape-level, pollen-mediated gene flow from genetically modified creeping bentgrass with CP4 EPSPS as a marker. Proc Natl Acad Sci USA 101:14533–14538

Wipff JK, Fricker C (2001) Gene flow from transgenic creeping bentgrass (*Agrostis stolonifera* L.) in the Willamette Valley, Oregon. Int Turfgrass Soc Res J 9:224–242

Xu H, Knox RB, Taylor PE, Singh MB (1995) Bcp1, a gene required for male fertility in Arabidopsis. Proc Natl Acad Sci USA 92:2106–2110

Yui R, Iketani S, Mikami T, Kubo T (2003) Antisense inhibition of mitochondrial pyruvate dehydrogenase E1α subunit in anther tapetum causes male sterility. Plant J 34:57–66

Zapiola ML, Campbell CK, Butler MD, Mallory-Smith CA (2008) Escape and establishment of transgenic glyphosate-resistant creeping bentgrass *Agrostis stolonifera* in Oregon, USA: a 4-year study. J Appl Ecol 45:486–494

Zapiola ML, Mallory-Smith CA, Thompson JH, Rue LJ, Campbell CK, Butler MD (2007) Gene escape from glyphosate-resistant creeping bentgrass fields: past, present, and future. Proceedings of the 60th Meeting of the Western Society of Weed Science, 13–15 March 2007, Portland, OR, Abstract 82

Zuo J, Chua NH (2000) Chemical-inducible systems for regulated expression of plant genes. Curr Opin Biotechnol 11:146–151

Zuo J, Niu QW, Chua NH (2000) An estrogen receptor-based transactivator XVE mediates highly inducible gene expression in transgenic plants. Plant J 24:265–273

Zou JT, Zhan XY, Wu HM, Wang H, Cheung AY (1994) Characterization of a rice pollen-specific gene and its expression. Am J Bot 81:552–561

Part D
Models for Uses of Biomass Feedstocks

Chapter 12
Integrated Biorefineries—A Bottom-Up Approach to Biomass Fractionation

Birgit Kamm

12.1 Introduction

One hundred and fifty years after the beginning of coal-based chemistry and 50 years after the beginning of petroleum-based chemistry, industrial chemistry is now entering a new era. An essential part of the sustainable future will be based on appropriate and innovative uses of our biologically based feedstocks. It will be particularly necessary to have a substantial conversion industry in addition to research and development investigating the efficiency of producing raw materials and product lines, as well as sustainability.

Whereas the most notable successes in research and development in the field of biorefinery system research have been in Europe and Germany (Kamm et al. 1998, 2000a; Narodoslawsky 1999), the first significant industrial developments were promoted in the United States of America by the President (US President 1999) and Congress (US Congress 2000). In the US, it is expected that by 2020 at least 25% (compared to 1995) of organic carbon-based industrial feedstock chemicals and 10% of liquid fuels will be obtained from a biobased product industry (BRDI 2006). This would mean that more than 90% of the consumption of organic chemicals and up to 50% of liquid fuel requirements in the US would be supplied by biobased products (National Research Council 2000). The US Biomass Technical Advisory Committee (BTAC)—in which leading representatives of industrial companies such as Dow Chemical, E.I. du Pont de Nemours, Cargill, Dow LLC, and Genecor International Inc., as well as corn growers' associations and the Natural Resources Defence Council are involved, and which acts as an advisor to

---

Dedicated to Michael Kamm, Founder of biorefinery.de GmbH

B. Kamm
Research Institute Bioactive Polymer Systems e.V. and Brandenburg University of Technology Cottbus, Kantstrasse 55, 14513, Teltow, Germany
e-mail: kamm@biopos.de

the US government—has made a detailed step-by-step plan of the targets for 2030 with regard to bioenergy, biofuels and bioproducts (BTAC 2002a, b, 2007).

Research and development are necessary to:

(1) increase the scientific understanding of biomass resources and improve the tailoring of those resources;
(2) improve sustainable systems to develop, harvest, and process biomass resources;
(3) improve efficiency and performance in conversion and distribution processes and technologies for a multitude of product developments from biobased products; and
(4) create the regulatory and market environment necessary for the increased development and use of biobased products.

BTAC has established specific research and development objectives for feedstock production research. Target crops should include oil- and cellulose-producing crops that can provide optimal energy content and usable plant components. Currently, however, there is a lack of understanding of plant biochemistry as well as inadequate genomic and metabolic information on many potential crops. In particular, research to produce enhanced enzymes and chemical catalysts could advance biotechnology capabilities.

In Europe there are existing regulations regarding the substitution of non-renewable resources by biomass in the field of using biofuels for transportation (European Parliament and Council, 2003) as well as the 'Renewable energy law' (Gesetz für den Vorrang erneuerbarer Energien, 2000). According to the EC Directive "On the promotion of the use of biofuels", the following products are considered as 'biofuels':

(a) 'bioethanol', (b) 'biodiesel', (c) 'biogas', (d) 'biomethanol', (e) 'biodimethylether', (f) 'bio-ETBE (ethyl-tertiär-butylether' based on bioethanol, (g) 'bio-MTBE (methyl-tertiär-butylether)' based on biomethanol, (h) 'synthetic biofuels', (i) 'biohydrogen', (j) pure vegetable oil.

Member States of the EU have been asked to define national guidelines for the minimum usage quantities of biofuels and other renewable fuels (with a reference value of 2% by 2005 and 5.75% by 2010, calculated on the basis of the energy content of all petrol and diesel fuels for transport purposes). Currently there are no guidelines for biobased products in the EU or in Germany. However, after passing directives for bioenergy and biofuels, such activities are on the political agenda. Recently the German Government has announced the biomass action plan for substantial use of renewable resources (Bundesministerium für Ernährung, Landwirtschaft und Verbraucherschutz 2009), and the German Chemical Societies have published the position paper 'Raw material change', including non-food biomass as raw material for the chemical industry (Gesellschaft Deutscher Chemiker et al. 2010). The European Technology Platform for Sustainable Chemistry has created the EU Lead Market initiative (Wittmeyer 2009). The directive for biofuels already includes ethanol, methanol, dimethylether, hydrogen and biomass pyrolysis, which are fundamental product lines of the future biobased chemical industry. A recent paper looking at future developments, published by the Industrial Biotechnology section of the European Technology platform for Sustainable Chemistry, foresaw up to 30% of raw materials for the chemical industry coming from renewable sources by 2025 [European Technology Platform for Sustainable Chemistry (ETPSC) 2005]. The ETPSC has created the EU Lead Market initiative (Wittmeyer 2009).

The European Commission and the US Department of Energy have come to an agreement for cooperation in this field (US DOE 2005). Based on the European biomass action plan of 2006 (Biomass Action Plan 2005), both strategic EU-projects, (1) BIOPOL, European Biorefineries: Concepts, Status and Policy Implications (EU-Projekt BIOPOL 2007) and (2) Biorefinery Euroview: Current situation and potential of the biorefinery concept in the EU: strategic framework and guidelines for its development (EU-Projekt Biorefinery-Euroview 2007), began preparation for the 7th EU framework.

In order to minimise food–feed–fuel conflicts and in order to use biomass most efficiently, it is necessary to develop strategies and ideas for how to use biomass fractions, in particular green biomass and agricultural residues such as straw, more efficiently. Such an overall utilisation approach is described in Sect. 12.2 below. In future developments, food- and feed-processing residues should therefore also become part of biorefinery strategies, since either specific waste fractions may be too small for a cost-efficient specific valorisation (capitalise on nature's resources) treatment in situ or the diverse technologies necessary are not available. Fibre-containing food-processing residues may then be pre-treated and processed with other cellulosic material from other sources in order to produce ethanol or other platform chemicals. Food-processing residues have, however, a particular feature one has to be aware of. Due to their high water content and endogenous enzymatic activity, food processing residues have a comparatively low biological stability and are prone to uncontrolled degradation and spoilage including rapid autoxidation. To avoid extra costs for transportation and conservation, the use of food-processing residues should also become part of a regional biomass utilisation network (Mahro and Timm 2007).

12.2 Biorefinery Technologies and Biorefinery Systems

12.2.1 Background

Biobased products are prepared for economically viable use by a suitable combination of different methods and processes (physical, chemical, biological and thermal). To this end, base biorefinery technologies need to be developed. For this reason, it is inevitable that there must be profound interdisciplinary cooperation among the individual disciplines involved in research and development. Therefore, it is appropriate to use the term 'biorefinery design', which implies that well-founded scientific and technological principles are combined with technologies, products and product lines inside biorefineries that are close to practice. The basic conversions of each biorefinery can be summarised as follows:

In the first step, the precursor-containing biomass is separated by physical methods. The main products (M_1–M_n) and by-products (B_1–B_n) will subsequently be subjected to further processing by microbiological or chemical methods. The subsequent products (F_1–F_n) obtained from the main products and by-products can be further converted or used in a conventional refinery.

Four complex biorefinery systems are currently under testing at the research and development stage:

(1) Lignocellulosic feedstock biorefinery using naturally dry raw materials such as cellulose-containing biomass and wastes.
(2) Whole crop biorefinery using raw material such as cereals or maize (whole plants).
(3) Green biorefineries using naturally wet biomasses such as green grass, alfalfa, clover, or immature cereal (Kamm and Kamm 2004a, b).
(4) The two platforms biorefinery concept, which includes the sugar platform and the syngas platform (Werpy and Petersen 2004).

12.2.2 Lignocellulosic Feedstock Biorefinery

Among the potential large-scale industrial biorefineries, the lignocellulosic feedstock (LCF) biorefinery will most probably achieve the greatest success. First, there is optimum availability of raw materials (straw, reed, grass, wood, paper waste, etc.), and second, the conversion products are well-placed on the traditional petrochemical as well as on the future biobased product market. An important factor in the utilisation of biomass as a chemical raw material is its cost. Currently, the cost for corn stover or straw is US $50/metric ton, and for corn US $80/metric ton (Tiffany 2007).

Lignocellulose materials consist of three primary chemical fractions or precursors: (1) hemicellulose/polyoses—a sugar-polymer of predominantly pentoses; (2) cellulose—a glucose-polymer; and (3) lignin—a polymer of phenols (Fig. 12.1). The lignocellulosic biorefinery system has a distinct ability to create genealogical trees. The main advantages of this method are that the natural structures and structure elements are preserved, the raw materials are cheap, and many product varieties are possible (Fig. 12.2). Nevertheless, there is still a requirement for development and optimisation of these technologies, e.g. in the field of separating cellulose, hemicellulose and lignin, as well as in the use of lignin in the chemical industry.

$$\text{Lignocellulose} + H_2O \rightarrow \text{Lignin} + \text{Cellulose} + \text{Hemicellulose}$$
$$\text{Hemicellulose} + H_2O \rightarrow \text{Xylose}$$
$$\text{Xylose } (C_5H_{10}O_5) + \text{Acid Catalyst} \rightarrow \text{Furfural } (C_5H_4O_2) + 3H_2O$$
$$\text{Cellulose}(C_6H_{10}O_5) + H_2O \rightarrow \text{Glucose } (C_6H_{12}O_6)$$

Fig. 12.1 A possible general equation of conversion at the lignocellulosic feedstock (LCF)-biorefinery

12 Integrated Biorefineries—A Bottom-Up Approach to Biomass Fractionation

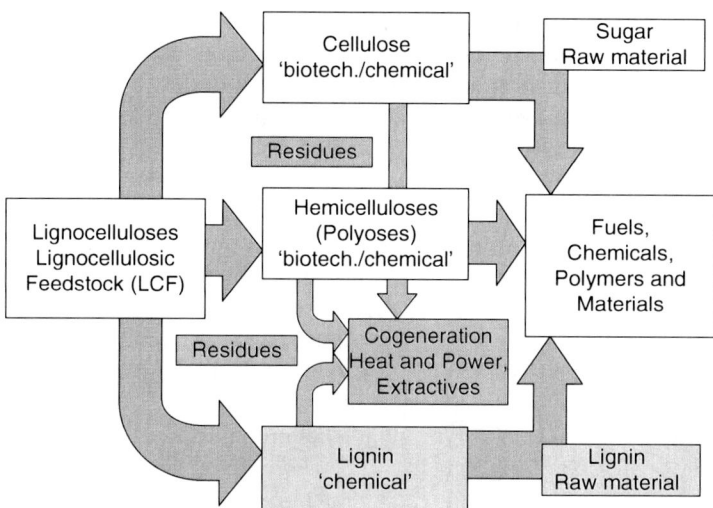

Fig. 12.2 Lignocellulosic feedstock biorefinery (Kromus et al. 2006)

Furfural and hydroxymethylfurfural, in particular, are interesting products. Furfural is the starting material for the production of Nylon 6,6 and Nylon 6. The original process for the production of Nylon 6,6 was based on furfural. The last of these production plants in the US was closed in 1961 for economic reasons (the artificially low price of petroleum). Nevertheless, the market for Nylon 6 is still very large.

However, some aspects of the LCF system, such as the utilisation of lignin as a fuel, adhesive or binder, remain unsatisfactory because the lignin scaffold contains considerable amounts of mono-aromatic hydrocarbons which, if isolated in an economically efficient way, could add significant value to the primary process. It should be noted that there are no obvious natural enzymes to split the naturally formed lignin into basic monomers as easily as polymeric carbohydrates or proteins, which are also naturally formed (Ringpfeil 2001).

An attractive accompanying process to the biomass-nylon process is the previously mentioned hydrolysis of cellulose to glucose and the production of ethanol. Certain yeasts produce a disproportionate amount of the glucose molecule during their generation of ethanol to glucose (ethanol to glucose), which effectively shifts the entire reduction ability into the ethanol and makes the latter obtainable at 90% yield (w/w; with regard to the formula turnover).

Based on recent technologies, a plant was designed for the production of the main products furfural and ethanol from LC-feedstock in West Central Missouri. Optimal profitability can be reached with a daily consumption of about 4,360 t feedstock. Annually, the plant produces 47.5 million gallons ethanol and 323,000 t furfural (Van Dyne 1999).

Ethanol may be used as a fuel additive. Ethanol is also a connecting product for a petrochemical refinery, and can be converted into ethylene by chemical methods.

As is well-known from the use of petrochemically produced ethylene, nowadays ethanol is the raw material for a whole series of large-scale technical chemical syntheses for the production of important commodities, such as polyethylene, or polyvinylacetate. Other petrochemically produced substances, such as hydrogen, methane, propanol, acetone, butanol, butandiol, itaconic acid and succinic acid, can similarly be manufactured by substantial microbial conversion of glucose (Zeikus et al. 1999; Vorlop et al. 2006; Werpy et al. 2006). DuPont has entered into a 6-year alliance with Diversa in a biorefinery to produce sugar from husks, straw and stovers, and to develop processes to co-produce bioethanol and value-added chemicals such as 1,3-propandiol. Through metabolic engineering, the microorganism *Escherichia coli* K12 produces 1,3-propandiol (PDO) in a simple glucose fermentation process developed by DuPont and Genencor. In a pilot plant operated by Tate and Lyle, the PDO yield reaches 135 g l^{-1} at a rate of 4 g $l^{-1} h^{-1}$ (Du Punt 2004). PDO is used for the production of polytrimethylene-terephthalate (PTT), a new polymer used in the production of high-quality fibres with the brand name Sorona (DuPont 2004). Production is predicted to reach 500 kt/year in 2010.

12.2.3 Whole Crop Biorefinery

Raw materials for whole crop biorefineries are cereals such as rye, wheat, triticale and maize (Fig. 12.3). The first step is their mechanical separation into grain and straw, where the portion of grain is approximately 1 and the portion of straw is 1.1–1.3 (straw is a mixture of chaff, stems, nodes, ears and leaves). The straw represents an LCF and may be processed further in an LCF biorefinery system.

Initial separation into cellulose, hemicellulose and lignin is possible, with their further conversion within separate product lines, as described above for LCF-biorefineries. Furthermore, straw is a raw material for the production of syngas via pyrolysis technologies. Syngas is the base material for the synthesis of fuels and methanol (Figs. 12.3, 12.4).

The corn may either be converted into starch or used directly after grinding into meal. Further processing can take one of four routes: (1) breaking up, (2) plasticisation, (3) chemical modification, or (4) biotechnological conversion via glucose. The meal can be treated and finished by extrusion into binder, adhesives or filler. Starch can be finished via plasticisation (co- and mix-polymerisation, compounding with other polymers), chemical modification (etherification into carboxy-methyl starch; esterification and re-esterification into fatty acid esters via acetic starch; splitting reductive amination into ethylene diamine, and hydrogenative splitting into sorbitol, ethylene glycol, propylene glycol and glycerine (Moris and Ahmed 1992; Bozell 2004; Webb et al. 2004). In addition, starch can be converted by a biotechnological method into poly-3-hydroxybutyric acid in combination with the production of sugar and ethanol (Nonato et al. 2001; Rossel et al. 2006). Biopol, the copolymer poly-3-hydroxybutyrate/3-hydroxyvalerate, developed by ICI is produced

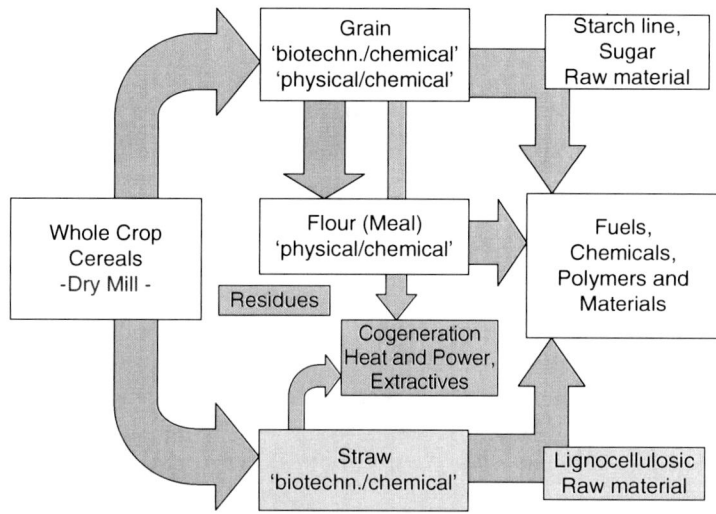

Fig. 12.3 Whole crop biorefinery—based on dry milling (Kamm et al. 2006)

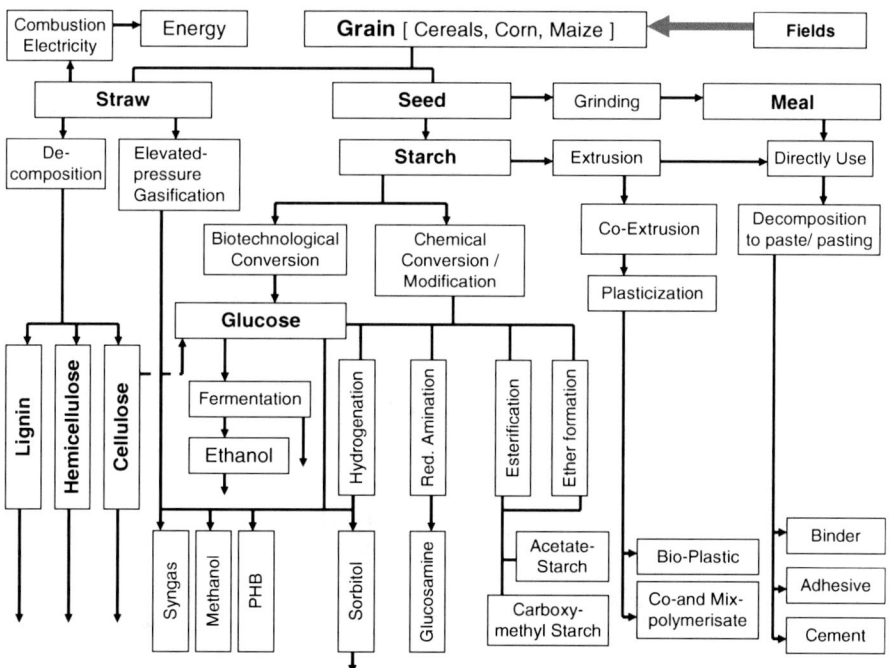

Fig. 12.4 Products from the whole crop biorefinery (Kamm and Kamm 2004a, b)

from wheat carbohydrates by fermentation using *Alcaligenes eutropius* (Fiechter 1990).

An alternative to the traditional dry fractionation of mature cereals into sole grains and straw has been developed by Kockums Construction Ltd (Sweden), now called Scandinavian Farming Ltd. In this whole crop harvest system, whole immature cereal plants are harvested and all the harvested biomass is conserved or dried for long-term storage. When convenient, it can be processed and fractionated into kernels, straw chips of internodes, and straw meal, including leaves, ears, chaff, and nodes (see also Sect. 12.2.4).

Fractions are suitable as raw materials for the starch polymer industry, the feed industry, the cellulose industry and particle-board producers, as gluten for the chemical industry and as a solid fuel. This kind of dry fractionation of the whole crop to optimise the utilisation of all botanical components of the biomass has been described in Rexen (1986) and Coombs and Hall (1997). An example of such a biorefinery and its profitability is described in Audsley and Sells (1997).

The whole crop wet mill-based biorefinery expands the product lines into grain processing. The grain is swelled and the grain germs are pressed, generating highly valuable oils.

The advantages of the whole crop biorefinery based on wet milling are that the natural structures and structure elements like starch, cellulose, oil, and amino acids (proteins) are retained to a great extent, and well-known base technologies and processing lines can still be used. The disadvantages are the high raw material costs and costly source technologies required for industrial utilisation. On the other hand, many of the products generate high prices, e.g. in pharmacy and cosmetics (Figs. 12.5, 12.6).

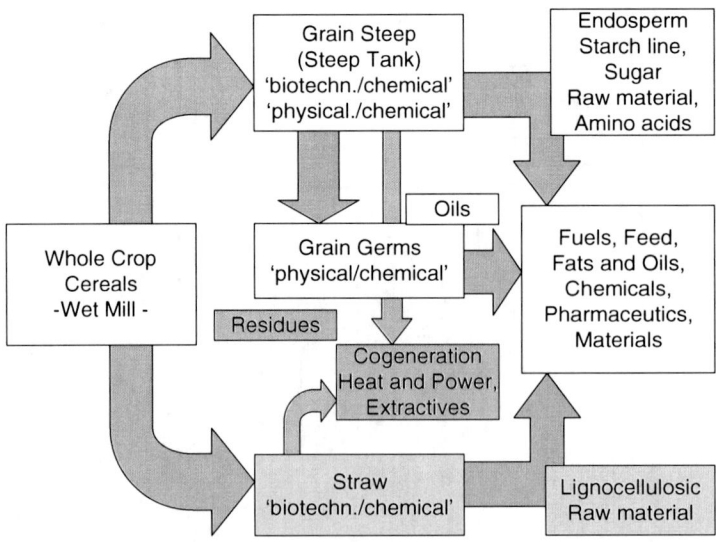

Fig. 12.5 Whole crop biorefinery, wet-milling (Kromus et al. 2006)

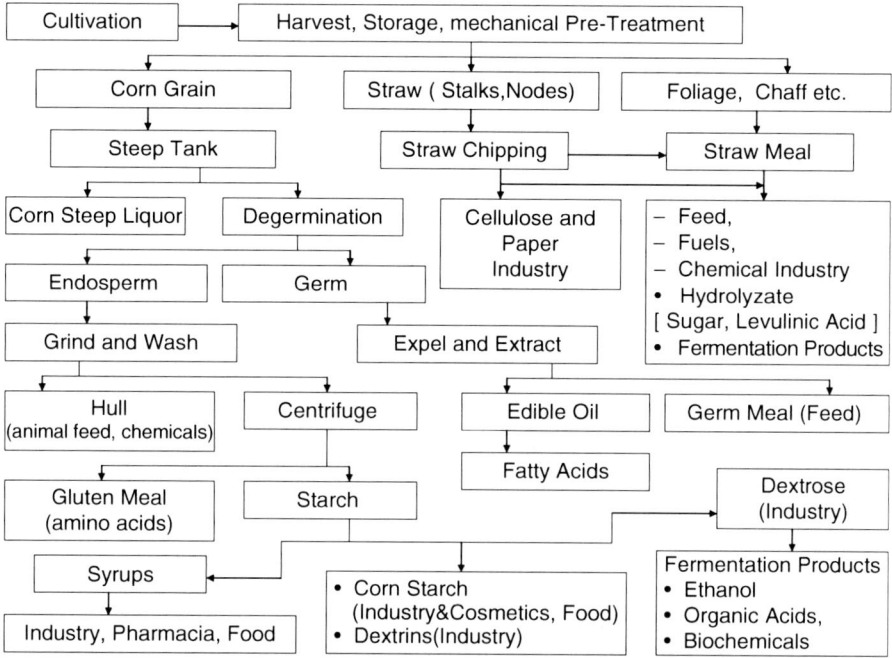

Fig. 12.6 Products from the whole crop wet mill based biorefinery

The wet milling of corn yields corn oil, corn fibre, and corn starch. The starch products obtained from the US corn wet-milling industry are fuel alcohol (31%), high-fructose corn syrup (36%), starch (16%), and dextrose (17%). Corn wet milling also generates other products (e.g. gluten meal, gluten feed, oil; Hacking 1986). An overview of the product range is shown in Fig. 12.6.

12.2.4 Green Biorefinery

Often, it is the economics of bioprocesses that are the main problem because the price of bulk products is affected greatly by raw material costs (Willke and Vorlop 2004). The advantages of green biorefineries are a high biomass profit per hectare and a good coupling with agricultural production, combined with low prices for raw materials. On the one hand, simple base technologies can be used, with good biotechnical and chemical potential for further conversions (Fig. 12.7). On the other hand, either fast primary processing or the use of preservation methods like silage or drying is necessary for both the raw materials and the primary products. However, each preservation method changes the content of the materials.

Green biorefineries are also multi-product systems and operate with regard to their refinery cuts, fractions and products in accordance with the physiology of the

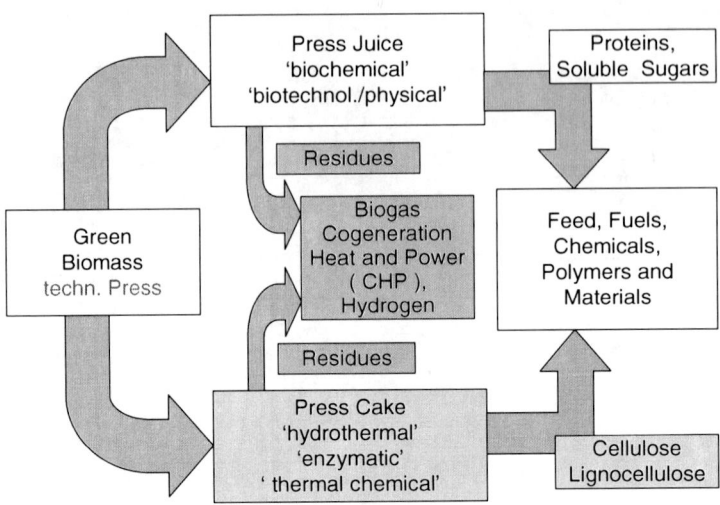

Fig. 12.7 A green biorefinery system (Kromus et al. 2006)

corresponding plant material; in other words, maintaining and utilising the diversity of syntheses achieved by nature. Green biomass consists of, for example, grass from the cultivation of permanent grassland, closed fields, nature preserves or green crops such as lucerne (alfalfa), clover, and immature cereals from extensive land cultivation. Today, green crops are used primarily as forage and a source of leafy vegetables. In a process called wet-fractionation of green biomass, green crop fractionation can be used for the simultaneous manufacture of both food and non-food items (Carlsson 1994). Thus, green crops represent a natural chemical factory and food plant.

Scientists in several countries in Europe and elsewhere have developed green crop fractionation (Pirie 1971, 1987; Carlsson 1998); indeed, green crop fractionation is now studied in about 80 countries (Carlsson 1994). Several hundred temperate and tropical plant species have been investigated for green crop fractionation (Carlsson 1983, 1998; Telek and Graham 1983). However, more than 300,000 higher plant species remain to be investigated (for reviews, see Kamm et al. 1998; Pirie 1971, 1987; Wilkins 1977; Tasaki 1985; Fantozzi 1989; Singh 1996).

By fractionation of green plants, green biorefineries can process from a few tonnes of green crops per hour (farm-scale process) to more than 100 t/hour (industrial-scale commercial process). Wet fractionation technology is used as the first step (primary refinery) to carefully isolate the contained substances in their natural form. Thus, the green crop goods (or humid organic waste goods) are separated into a fibre-rich press cake (PC) and a nutrient-rich green juice (GJ).

Besides cellulose and starch, PC contains valuable dyes and pigments, crude drugs and other organics. The GJ contains proteins, free amino acids, organic acids, dyes, enzymes, hormones, other organic substances, and minerals. In particular, the application of biotechnology methods is ideally suited for conversions because the

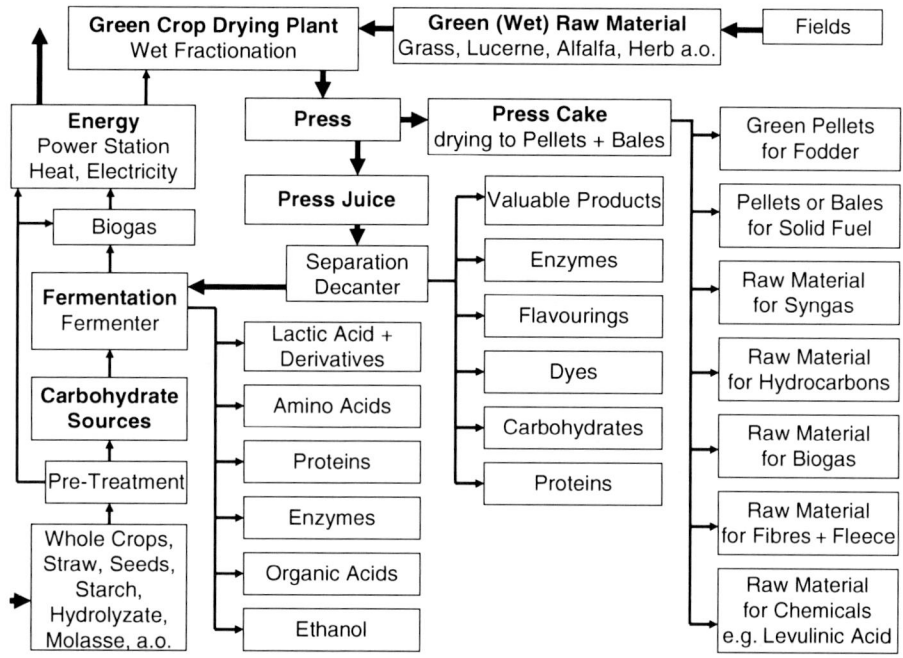

Fig. 12.8 Products from a green biorefinery system, combined with a green crop drying plant (Kamm and Kamm 2004a, b)

plant water can simultaneously be used for further treatments. When water is added, the lignin-cellulose composite bonds are not as strong as they are in dry lignocellulose feedstock materials. Starting from GJ, the main focus is directed to producing products such as lactic acid and corresponding derivatives, amino acids, ethanol, and proteins. The PC can be used for the production of green feed pellets and as a raw material for the production of chemicals such as levulinic acid, as well as for conversion to syngas and hydrocarbons (synthetic biofuels). The residues left when substantial conversions are processed are suitable for the production of biogas combined with the generation of heat and electricity (Fig. 12.8). Reviews of green biorefinery concepts, contents and goals have been published (Kamm et al. 1998, 2000a, b; Narodoslawsky 1999; Kromus et al. 2006).

12.2.5 The Two Platforms Biorefinery Concept

The "two platform concept" means firstly that biomass consists on average of 75% carbohydrates, which can be standardised over an intermediate sugar platform as a basis for further conversions, and secondly that the biomass is converted thermochemically into synthesis gas and further products.

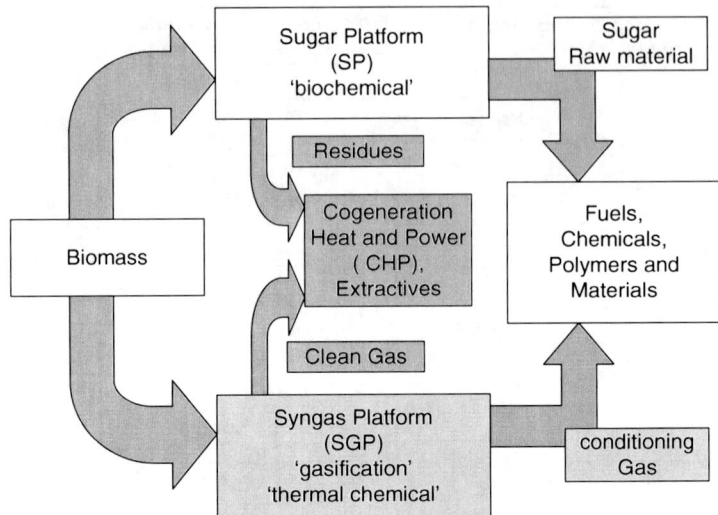

Fig. 12.9 Sugar platform and Syngas platform (NREL 2005)

- The "sugar platform" is based on biochemical conversion processes and focusses on the fermentation of sugars extracted from biomass feedstocks.
- The "syngas platform" is based on thermochemical conversion processes and focusses on the gasification of biomass feedstocks and by-products from conversion processes (White and Wolf 1988; Werpy and Petersen 2004; Pirie 1971). In addition to gasification, other thermal and thermochemical biomass conversion methods have also been described: hydrothermolysis, pyrolysis, thermolysis, and burning. The application used depends on the water content of the biomass (Okkerse and van Bekkum 1999).

Gasification and all the thermochemical methods concentrate on utilisation of the precursor carbohydrates as well as their inherent carbon and hydrogen content. The proteins, lignin, oils and lipids, amino acids and general ingredients, as well as the N- and S-compounds occurring in all biomass, are not taken into account in this case (Fig. 12.9).

12.3 Platform Chemicals

12.3.1 Background

A team from the Pacific Northwest National Laboratory (PNNL) and the National Renewable Energy Laboratory (NREL) submitted a list of 12 potential biobased chemicals (Werpy and Petersen 2004). The key areas of the investigation were

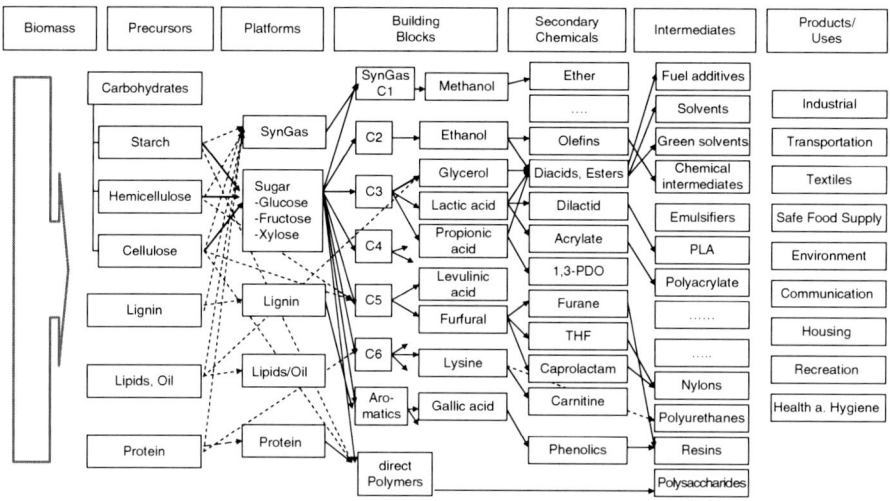

Fig. 12.10 Model of a biobased product flowchart for biomass feedstock (from Werpy and Petersen 2004)

biomass precursors, platforms, building blocks, secondary chemicals, intermediates, products and uses (Fig. 12.10).

The final selection of 12 building blocks began with a list of more than 300 candidates. A shorter list of 30 potential candidates was selected using an iterative review process based on the petrochemical model of building blocks, chemical data, known market data, properties, performance of the potential candidates and the prior industry experience of the team at PNNL and NREL. This list of 30 was ultimately reduced to 12 by examining the potential markets for the building blocks and their derivatives, and the technical complexity of the synthesis pathways.

The selected building-block chemicals can be produced from sugar via biological and chemical conversions. The building blocks can subsequently be converted to a number of high-value biobased chemicals or materials. Building-block chemicals, as considered for this analysis, are molecules with multiple functional groups that possess the potential to be transformed into new families of useful molecules. The 12 sugar-based building blocks (Fig. 12.10) are: 1,4-diacids (succinic, fumaric and malic); 2,5-furan dicarboxylic acid; 3-hydroxy propionic acid; aspartic acid; glucaric acid; glutamic acid; itaconic acid; levulinic acid; 3-hydroxybutyrolactone; glycerol; sorbitol; and xylitol/arabinitol (Werpy and Petersen 2004).

A second-tier group of building blocks was also identified as viable candidates. This group included gluconic acid; lactic acid; malonic acid; propionic acid; the triacids, citric and aconitic acids; xylonic acid; acetoin; furfural; levuglucosan; lysine; serine; and threonine. Recommendations for moving forward include examining top-value products from biomass components such as aromatics, polysaccharides, and oils; evaluating technical challenges related to chemical and biological conversions in more detail; and increasing the number of potential pathways to these candidates. No further products obtained from syngas were selected. For the purposes of this study, hydrogen and methanol are the best short-term prospects for biobased commodity chemical production because obtaining simple alcohols, aldehydes, mixed alcohols and Fischer-Tropsch liquids from biomass is not economically viable and requires additional development (Werpy and Petersen 2004).

12.3.2 The Role of Biotechnology in Production of Platform Chemicals

The application of biotechnological methods will be of great importance, and will involve the development of biorefineries for the production of base chemicals, intermediate chemicals and polymers (EuropaBio 2003; BIO 2004). The integration of biotechnological methods must be managed intelligently with respect to the physical and chemical conversions of the biomass. Therefore biotechnology cannot continue to be restricted to glucose from sugar plants and starch from starch-producing plants (Fig. 12.11).

One of the main goals is the economical processing of biomass containing lignocellulose and the provision of glucose in the family tree system. Glucose is a key chemical for microbial processes. The preparation of a large number of family-tree-capable base chemicals is described in the following sections. Among the variety of possible product family trees that can be developed from glucose-accessible microbial and chemical sequence products are the C-1 chemicals methane, carbon dioxide, and methanol; C-2 chemicals ethanol, acetic acid, acetaldehyde, and ethylene; C-3 chemicals lactic acid, propandiol, propylene, propylene oxide, acetone, acrylic acid; C-4 chemicals diethylether, acetic acid anhydride, malic acid,

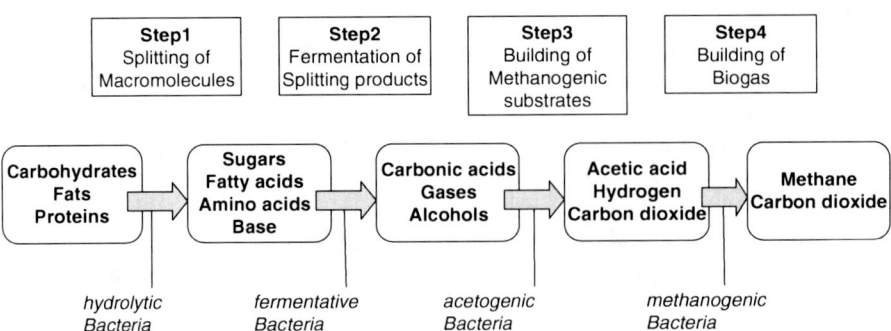

Fig. 12.11 Simplified presentation of a microbial biomass-breakdown regime (Kamm and Kamm 2004a)

12 Integrated Biorefineries—A Bottom-Up Approach to Biomass Fractionation

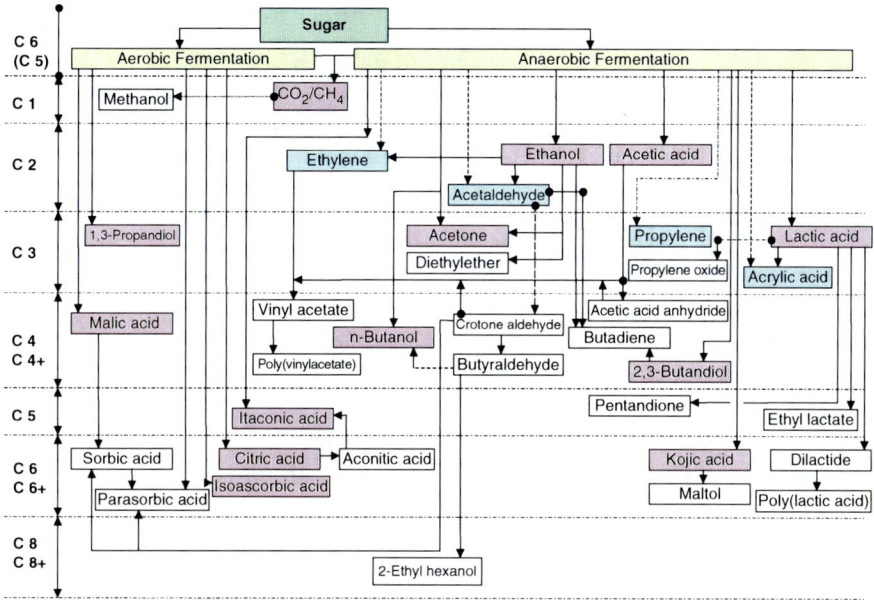

Fig. 12.12 Biotechnological sugar-based product family tree

vinyl acetate, n-butanol, crotone aldehyde, butadiene, and 2,3-butandiol; C-5 chemicals itaconic acid, 2,3-pentane dione, and ethyl lactate; C-6 chemicals sorbic acid, parasorbic acid, citric acid, aconitic acid, isoascorbinic acid, kojic acid, maltol, and dilactide; and the C-8-chemical 2-ethyl hexanol (Fig. 12.12).

Currently, guidelines are being developed for the fermentation section of a biorefinery. An answer needs to be found to the question of how to produce an efficient technological design for the production of bulk chemicals. The basic technological operations for the manufacture of lactic acid and ethanol are very similar. The selection of biotechnology-based products from biorefineries should be done in a way that they can be produced from the substrates glucose or pentoses. Furthermore, the fermentation products should be extracellular. Fermentors should have a batch, feed batch or continuous stirred-tank reactor (CSTR) design. Preliminary product recovery may require steps like filtration, distillation or extraction. Final product recovery and purification steps may possibly be product-unique. In addition, biochemical and chemical processing steps should be efficiently connected.

Unresolved questions for the fermentation facility include: (1) whether or not the entire fermentation facility can/should be able to change from one product to another; (2) can multiple products be run in parallel, with shared use of common unit operations; (3) how should scheduling of unit operations be managed; and (4) how can in-plant inventories be minimised, while accommodating any changeovers required between different products for the same piece of equipment (Van Dyne et al. 1999).

12.3.3 Green Biomass Fractionation and Energy Aspects

Today, green crops are used primarily as forage and as a source of leafy vegetables. In a process called wet-fractionation of green biomass, green crop fractionation can be used for simultaneous manufacture of both food and non-food items (Carlsson 1994).

The power and heat energy requirements of a forage fractionation of a protein concentrate production system are within practical limits for large farms and dehydrating plants (Bruhn et al. 1978). Mechanical squeezing of the fresh crop results in energy savings of 1.577 MJ/t crop input, equal to 52% of the total energy input (compared to energetic drying of green biomass; Ricci et al. 1989). Three simplified systems of wet green crop fractionation, which are characterised by the direct use of GJ or deproteinised juice as feeding supplements for pigs or liquid fertiliser, have been described (Favati et al. 1989). Wet green crop fractionation involves an energy saving of 538 MJ/t fresh crop, equal to 17.7% of the total energy input of crop drying (Ricci et al. 1989). Compared with conventional fractionation technology, membrane filtration results in an energy saving of 370 MJ/t crop input, which corresponds to 14.8% of the total energy input (Favati et al. 1989).

Via fractionation of green plants, green biorefineries are able to process amounts in the range of a few tons of green crops per hour (farm scale process) to more than 100 t/h (industrial-scale commercial process). Careful wet fractionation technology is used as a first step (primary refinery) to isolate the ingredients in their natural form. Thus, the green crops (or wet organic wastes) are separated into a fibre-rich press cake and a nutrient-rich GJ.

Beside cellulose and starch, the PC contains valuable dyes and pigments, crude drugs and other organics. The GJ contains proteins, free amino acids, organic acids, dyes, enzymes, hormones, further organic substances, and minerals. The application of biotechnological methods is particularly appropriate for conversion processes since the plant water can be used simultaneously for further treatments. In addition, the pulping of lignin-cellulose composites is easier compared to LCF materials. Starting from GJ, the main focus is directed to products such as lactic acid and corresponding derivatives, amino acids, ethanol, and proteins.

The PC can be used for production of green feed pellets; as raw material for production of chemicals, such as levulinic acid; and for conversion to syngas and hydrocarbons (synthetic biofuels). The residues of substantial conversion are applicable to the production of biogas combined with generation of heat and electricity. Special attention is given to the mass and energy flows of the biorefining of green biomass.

12.3.4 Mass and Energy Flows for Green Biorefining

Green biorefining is described as an example of a type of agricultural factory in greenland-rich areas. Key figures are determined for mass and energy flow, feedstock and

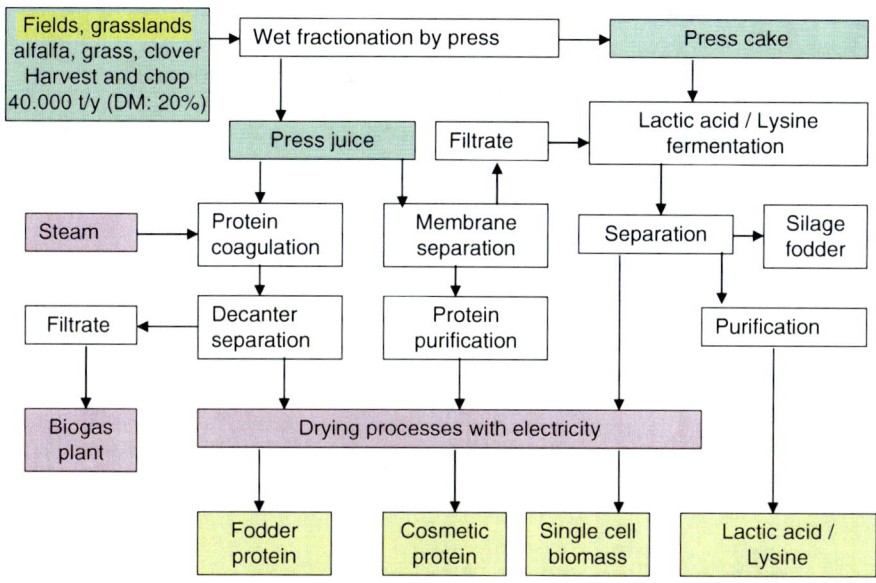

Fig. 12.13 Selected and simplified processes of a green biorefinery (Kamm et al. 2009)

product quantities. Product quantities vary depending on the market and the demand for quality products. Mass flows (Scenario 1, Scenario 2) can be constructed from our own experimental results combined with market demand in the feed, cosmetic, and biotechnology industries. The technical and energy considerations of the fractionation processes of a green biorefinery, and production of the platform chemicals lactic acid and lysine are shown in Fig. 12.13.

Using a mechanical press, about 20,000 t press juice [dry matter (DM): 5%] can be manufactured from 40,000 t biomass. Firstly, the juice is the raw material for further products; and secondly, the green cut biomass contains much less moisture. Through fractionation of GJ proteins by different separation and drying processes, high quality fodder proteins and proteins for the cosmetic industry can be produced (Bruhn et al. 1978; Bohlmann 2002; Reismann 1988). The fodder proteins would be a complete substitute for soy proteins. They even have a nutritional physiological advantage due to their particular amino acid patterns (Schwenke 1998). Utilisation of the easily fermentable sugar in the biomass and the available water offers an excellent biotechnological-chemical potential and makes possible the use of basic technologies such as the production of lactic acid or lysine.

In the next step (fermentation), the carbohydrates of the juice and one part of the PC can be used (after hydrolysis) for the production of lactic acid (scenario 1; Kamm et al. 2006) or lysine (scenario 2; Thomsen et al. 2004). Thus, single cell biomass, which can be applied after appropriate drying as a fodder-protein, is produced. The fermentation base in lactic acid fermentation is sodium hydroxide. By means of ultrafiltration (Patel et al. 2006), reverse osmosis (Patel et al. 2006),

Table 12.1 Combined production of lactic acid, lysine, cosmetic-protein, single cell biomass, fodder, and biogas with energetic input (Kamm et al. 2009). *DM* Dry matter

Green biorefinery	Scenario 1:	Lactic acid	
Input:	DM	Quantity	Unit
Green biomass (lucerne, clover, grass)	DM: 20%	40,000	t
Steam		2,268	GJ
Electricity		1.3 million	kWh
Output:			
Silage fodder	DM: 40%	13,000	t
Fodder protein 80%	DM: 90%	400	t
Cosmetic protein 90%	DM: 90%	29.6	t
Lactic acid 90%	DM: 90%	660	t
Residue to biogas plant TS: 2%		17,690	t
Single cell biomass (as fodder protein 60%)	DM: 90%	33	t
Green biorefinery	Scenario 2:	Lysine	
Input:		Quantity	Unit
Cut green biomass (lucerne, clover, grass)	DM: 20%	40,000	t
Steam		2,268	GJ
Electricity		0.492 million	kWh
Output:			
Silage fodder	DM: 40%	13,000	t
Fodder protein 80%	DM: 90%	400	t
Cosmetic protein 90%	DM: 90%	29.6	t
Lysine-HCl, 50%	DM: 90%	620	t
Residue to biogas plant DM: 2%		17,770	t
Single cell biomass (as fodder protein 60%)	DM: 90%	31	t

bipolar electrodialysis (Kim and Moon 2001; Datta and Tsai 1997) and distillation (Lavis 1996), lactic acid (90%) is recovered from sodium lactate fermentation broth. Lysine hydrochloride is the product of lysine fermentation (Reismann 1988). After separation of the single cell biomass by ultrafiltration (Patel et al. 2006) and a membrane separation of water followed by a drying process, lysine hydrochloride (50%) is recovered (Thomsen et al. 2004). The broth that is left after separation from lactic acid or lysine, respectively, and single cell-biomass can be supplied to a biogas plant. Input and output data including required energy were estimated for the production of lysine-hydrochloride, lactic acid, proteins for fodder and cosmetics and the utilisation of the residue (PC) as silage fodder from 40,000 t green cut biomass (Table 12.1).

By drying, the PC could be manufactured into fodder-pellets. However, this drying is energetically very expensive. From an energy point of view it is far better to suggest that the PC be used as silage-feed. From an ecological and economical viewpoint, at this stage it has to be concluded that coupling of green biorefineries with green crop drying industry is necessary.

12.3.5 Assessment of Green Crop Fractionation Processes

Green biorefineries use different kinds of energy (steam and electricity) for the treatment of PC and press juice (intermediate products) to produce valuable end products. It is also possible to use the PC together with the press juice as a source of carbohydrate for the fermentation. For the separate processes mass balances were set up and thus the consumption of energy can be calculated by means of power consumption of the facilities (plants and machinery).

A linear programming model used to optimise the profitability and determine an optimised planning process for biorefineries is described in Annetts and Audsley (2003). The raw materials are wheat (straw and grain) and rape, and therefore this would be a model for a whole crop biorefinery and hardly applicable to a green biorefinery. At a capacity of 40,000 tons per year [t/annum (a)] fresh biomass (lucerne, wild-mixed-grass), and a operation time of 200 working days per year, an average of 200 t are converted per day. Under these conditions, the screw extrusion press used has an energy consumption of 135,000 kWh per year (kWh/a). It generates 100 t/day PC with DM $\sim$35% and 100 t/day press juice with DM $\sim$5%.

Around 10 t of the 100 t press juice are fed to membrane-separation for a cosmetic-protein-extraction. For separation of feed protein, 90 t press juice are put into a steam-coagulation. The required heat quantity as steam is 2,268 GJ/a. The freshly pressed juice is preheated up to 45°C in a heat exchanger within a counter current process. Via steaming, a temperature rise of the freshly pressed juice of up to 30°C is reached. Steam coagulation occurs at a temperature of 75°C. The following calculations are carried out according to Bruhn et al. (1978). For the separation of feeding proteins the following energy input is required: 1,500 kWh/a for skimming; 15,000 kWh/a for dehydration to $\sim$50% DM; and 32,000 kWh/a for drying up to DM $=$ 90%. Separation of cosmetic-proteins via ultra filtration needs an energy input of 9,700 kWh/a. For subsequent solvent extraction, a further energy input of about 507 kWh/a is generated via stirring (Petrides et al. 1989).

For the separation via centrifugation 101 kWh/a (Bohlmann 2002) are required and 2,360 kWh/a for the subsequent spray-drying to DM $=$ 90% (Bartholomew and Reismann 1979). If the press-juice contains 2% proteins, 400 t feed proteins as protein concentrate and 29.6 t cosmetic proteins can be produced per year. Correspondingly increased quantities can be produced if the press juice contains a higher proportion of proteins. After protein-separation, 100 t fermentation-broth ($\sim$96.6 m^3 at a density of about 1,035 kg/l) are available per working day. The energy input required for stirring during fermentation amounts to 150,000 kWh/a (Thomsen et al. 2004).

For lactic-acid fermentation, NaOH is added as a base, resulting in sodium lactate. The purification of lactic acid occurs with the following steps and corresponding energy yields: ultra filtration 97,000 kWh/a, reverse osmosis 171,000 kWh/a (Patel et al. 2006) and bipolar electro-dialysis 660,000 kWh/a (Kim and Moon 2001). Bipolar electro-dialysis is particularly energy-intensive. Subsequently, the lactic acid solution (45%) is concentrated up to a 90% lactic acid via vacuum distillation. The energy consumption for this single-stage distillation will amount 26,400 kWh/a (Lavis 1996). The energy consumption for 660 t of 90% lactic-acid amounts 1,104 MWh/a using this procedure.

If lysine fermentation is chosen instead of lactic acid, ultra filtration and reverse osmosis are required for purification with the following corresponding energy-yields: ultra filtration (97,000 kWh/a), and reverse osmosis (171,000 kWh/a; Patel et al. 2006). Afterwards the lysine-hydrochloride is dried to DM = 90% with an energy requirement of 49,000 kWh/a (Annetts and Audley 2003). The energy consumption of 620 t lysine-hydrochloride using this method results in 296,000 kWh/a.

In a biorefining plant processing 40,000 t green biomass for the combined production of 660 t lactic acid, 29.6 t cosmetic-protein, 33 t single cell biomass, 400 t fodder-protein, 13,000 t silage fodder and 17,690 t liquid residues for biogas production, the following energy input is required: 2,268 GJ heat, and 1.3 million kWh electricity. The combined production of 620 t lysine, 29.6 t cosmetic-protein, 31 t single cell biomass, 400 t fodder-protein, 13,000 t silage fodder and 17,700 t liquid residues to produce biogas requires the following energy input: 2,268 GJ heat, and 0.492 million MWh electricity.

These results clearly demonstrate the quantity of products a green biorefinery can provide with the help of biotechnology, and the corresponding required energy input. The economic benefits of biorefining green biomass are the high yields of biomass per hectare and year, and synergetic effects via combination with established production processes in the agriculture and feed industries. Therefore, in the mid-term, it is reasonable to combine the economic potential of green agriculture and green-crop-drying-plants.

These data concerning quantity, quality and required process energy form the basis of further economic considerations in connection with calculation of break-even points when planning and establishing a green biorefinery. In future, energy inputs will be reduced further due to optimisation of the corresponding biorefinery technology. The combination of biotechnological and chemical conversion processes will be a very important aspect in decreasing process energy input. Thus, the biotechnological production of aminium lactates, such as piperazinium dilactates as starting material for high purity lactic acid and polylactic acid could be a new approach (Kamm et al. 2006).

12.4 Green Biorefinery: Economic and Ecologic Aspects

Plant biomass is the only foreseeable sustainable source of organic fuels, chemicals and materials. A variety of forms of biomass, notably many LCFs, are potentially available on a large scale and are cost-competitive with low-cost petroleum, whether considered on a mass or energy basis, in terms of price defined on a purchase or net basis for both current and projected mature technologies, or on a transfer basis for mature technology (Fowler et al. 2003). Green plant biomass in combination with LCF represents the dominant source of feedstocks for biotechnological processes for the production of chemicals and materials (Kamm and Kamm 1999; Thomsen et al. 2004; Werpy and Petersen 2004; Kamm and Kamm 2007; Tullo 2005). The development of integrated technologies for the conversion of biomass is essential for the economic and ecological production of products. The biomass industry, or bio-industry, at present produces basic chemicals like ethanol (15 million t/a); amino acids (1.5 million t/a), of which L-lysine amounts to 500,000 million t/a (Elements 2005); and lactic acid (200,000 million t/a). The target of a biorefinery is to establish a combination of a biomass-feedstock mix with a process and product mix (Werpy and Petersen 2004; Kamm and Kamm 2007). A life cycle assessment (LCA) is available for the production of polylactic acid (capacity 140,000 t/a; Vink et al. 2003). For total assessment of the utilisation of biomass, one has to consider that cultivation of the plant has to fulfil certain economic and ecological criteria. Agriculture both creates pressure on the environment and plays an important role in maintaining many cultural landscapes and semi-natural habitats (IEEP 2004). Green crops, in particular, provide especially high yields.

Additionally, grassland can be cultivated in a sustainable way (Kromus 2004; Kromus et al. 2004). European grassland experiments have shown that species-rich grassland cultivation provides not only ecological but also economic advantages. With greater plant diversity, grassland is more productive and the soil is protected against nitrate leaching. Of the 71 species examined thus far, 29 had a significant influence on productivity. *Trifolium pratense*has an especially important function regarding productivity. On sites where this species occurs, more than 50% of the total biomass has been produced by this species. Legumes like clover and herbs also play an important role, as do fast growing grasses (Hector et al. 1999). An initial assessment of the concept of a green biorefinery has been carried out by Schidler and colleagues (Schidler 2003; Schidler et al. 2003) for the Austrian system approach. Furthermore, an Austrian-wide concept for the use of biomass and cultivable land for renewable resources has yet to be developed in Austria, which also holds true for Europe (Narodoslawsky and Kromus 2004). The size of such plants depends on the rural structures of the different regions. Concepts with more decentralised units would have a size of about 35,000 t/a (Halasz et al. 2003) and central plants could have sizes of about 300,000–600,000 t/a (Van Dyne et al. 1999).

References

Annetts JE, Audsley E (2003) Modelling the value of a rural biorefinery. Part I: the model description. Part II: analysis and implications. Agric Syst 76:39–76

Audsley E, Sells JE (1997) Determining the profitability of a whole crop biorefinery. In: Campbell GM, Webb C, McKee SL (eds) Cereals—novel uses and processes. Plenum, New York, pp 191–294

Bartholomew WH, Reismann HB (1979) Economics of fermentation processes. In: Peppler HJ, Perlman D (eds) Microbial technology, 2nd edn, vol 2. Academic, New York

BIO (2004) New biotech tools for a cleaner environment—industrial biotechnology for pollution prevention. Resource Conservation and Cost Reduction Biotechnology Industry Organisation; http://www.bio.org/ind/pubs/cleaner2004/cleanerReport.pdf

Biomasse Action plan (2005) http://www.euractiv.com/en/energy/biomass–action–plan/article–155362

Bohlmann G (2002) Several reports on White Biotechnology processes. Stanford Research International, Menlo Park, CA

Bozell JJ (2004) Alternative feedstocks for bioprocessing. In: Goodman RM (ed) Encyclopedia of plant and crop science. Dekker, New York, doi: 10.1081/E-EPCS-120010437

BRDI (2006) Vision for bioenergy and biobased products in the United States. Biomass Research and Development Initiative. http://www1.eere.energy.gov/biomass/pdfs/final_2006_vision.pdf

Bruhn HD, Straub RJ, Koegel RG (1978) A systems approach to the production of plant juice protein concentrate. In: Proceedings of the International Grain and Forage Harvesting Conference, American Society of Agricultural Engineers, St. Joseph, MI

BTAC (2002a) Roadmap for biomass technologies in the United States. Biomass Technical Advisory Committee, Washington DC. http://www.bioproducts–bioenergy.gov/pdfs/FinalBiomassRoadmap.pdf

BTAC (2002b) Vision for bioenergy and biobased products in the United States, Biomass Technical Advisory Committee, Washington DC, www.bioproducts–bioenergy.gov/pdfs/ Bio Vision_03_Web.pdf

BTAC (2007) Roadmap for bioenergy and biobased products in the United States, October 2006. Biomass R&D Technical Advisory Board, http://www1.eere.energy.gov/biomass/pdfs/obp_roadmapv2_web.pdf

Bundesministerium für Ernährung, Landwirtschaft und Verbraucherschutz (2009) Aktionsplan der Bundesregierung zur stofflichen Nutzung nachwachsender Rohstoffe. BT–Drucksache 16/14061 vom 03.09.2009

Carlsson R (1983) Leaf protein concentrate from plant sources in temperate climates. In: Telek L, Graham HD (eds) Leaf protein concentrates. AVI, Westport, pp 52–80

Carlsson R (1994) Sustainable primary production. Green crop fractionation: effects of species, growth conditions, and physiological development. In: Pessarakli M (ed) Handbook of plant and crop physiology. Dekker, NY, pp 941–963

Carlsson R (1998) Status quo of the utilization of green biomass. In: Soyez S, Kamm B, Kamm M (eds) The green biorefinery. Proceedings of 1st International Green Biorefinery Conference, Neuruppin, Germany, 1997, GÖT, Berlin, ISBN 3–929672–06–5

Coombs J, Hall K (1997) The potential of cereals as industrial raw materials: legal technical, commercial considerations In: Campbell GM, Webb C, McKee SL (eds) Cereals—novel uses and processes. Plenum, New York, pp 1–12

Datta R, Tsai S-P (1997) Lactic acid production and potential uses: a technology and economics assessment. In: Saha BC, Woodward J (eds) Fuels and chemicals from biomass. American Chemical Society, Washington DC 1997, p 224

DuPont (2004) US patent 5 686 276, http://www.dupont.com/sorona/home.html

Elements Degussa Science Newsletter (2005) 7:35

ETPSC (2005) European Technology Platform for Sustainable Chemistry, Industrial Biotechnology Section, http://www.suschem.org

EU-Projekt BIOPOL (2007) Specific Support Action, Priority Scientific Support to Policies, http://www.biorefinery.nl/biopol

EU-Projekt Biorefinery-Euroview (2007) Specific Support Action, Priority Scientific Support to Policies, http://www.biorefinery–euroview.eu

EuropaBio (2003) White biotechnology—gateway to a more sustainable future. EuropaBio, Lyon

European Parliament and Council (2003) Directive 2003/30/EC on the promotion of the use of biofuels or other renewable fuels for transport; Official Journal of the European Union L123/42, 17.05.2003, Brussels

Fantozzi P (1989) (ed) Proceedings of the 3rd International Leaf Protein Research Conference, Pisa–Perugia–Viterbo, Italy

Favati F et al (1989) Energy evaluation of a wet green crop fractionation process utilizing reverse osmosis. Third International Conference on Leaf Protein Research. 1–7 October 1989, Pisa, Perugia, Viterbo, Italy

Fiechter A (1990) Plastics from bacteria and for bacteria: poly(β–hydroxyalkanoates) as natural, biocompatible, and biodegradable polyesters, Springer, New York, pp 77–93

Fowler PA, McLauchlin AR, Hall LM (2003) The potential industrial uses of forage grasses including miscanthus. BioComposites Centre, University of Wales, Bangor, Gwynedd 2003, http://www.nnfcc.co.uk/library/reports/download.cfm?id=60

Gesellschaft Deutscher Chemiker (2010) Dechema, DGMK, VCI, Positionspapier Rohstoffbasis im Wandel, Frankfurt, Januar 2010, http://www.vci.de/default~cmd~shd~docnr~126682~lastDokNr~-1.htm

Gesetz für den Vorrang erneuerbarer Energien (2000) Erneuerbare Energiegesetz, EEG/EnWGuaÄndG., 29 March 2000, BGBI, 305

Hacking AJ (1986) The American wet milling industry. In: Economic aspects of biotechnology. Cambridge University Press, New York, pp 214–221

Halasz L, Povoden G, Narodoslawsky M (2003) Process synthesis for renewable resources. Presented at PRES 03, Hamilton, Canada

Hector A, Schmid B, Beierkuhnlein C, Caldeira MC, Diemer M, Dimitrakopoulos PG, Finn JA, Freitas H, Giller PS, Good J, Harris R, Högberg P, Huss-Danell K, Joshi J, Jumpponen A, Körner C, Leadley PW, Loreau M, Minns A, Mulder CPH, O'Donovan G, Otway SJ, Pereira JS, Prinz A, Read DJ, Scherer-Lorenzen M, Schulze E-D, Siamantziouras A-SD, Spehn EM, Terry AC, Troumbis AY, Woodward FI, Yachi S, Lawton JH (1999) Plant diversity and productivity experiments in European grasslands. Science 286:1123–1127

IEEP (2004) Contribution to the background study agriculture. The Institute for European Environmental Policy, Brussels

Kamm B, Kamm M (1999) The green biorefinery—principles, technologies and products. In: Proceedings of 2nd International Symposium Green Biorefinery, 13–14 October 1999, SUSTAIN, Verein zur Koordination von Forschung über Nachhaltigkeit (Hrsg.), Feldbach, Austria, pp 46–69

Kamm B, Kamm M (2004a) Principles of biorefineries. Appl Microbiol Biotechnol 64:137–145

Kamm B, Kamm M (2004b) Biorefinery systems, Chem Biochem Eng Q 18(1):1–6

Kamm B, Kamm M (2007) Biorefineries—multi product processes. In: Ulber R, Sell D (eds) White Biotechnology (Advances in biochemical engineering/biotechnology, vol 105). Springer, Heidelberg, pp 175–204

Kamm B, Kamm M, Soyez K (1998) (eds) Die Grüne Biraffinerie/The Green Biorefinery. Technologiekonzept. Proceedings of the 1st International Symposium Green Biorefinery/Grüne Biraffinerie, October 1997, Neuruppin, Berlin

Kamm B, Kamm M, Richter K, Linke B, Starke I, Narodoslawsky M, Schwenke KD, Kromus S, Filler G, Kuhnt M, Lange B, Lubahn U, Segert A, Zierke S (2000a) Grüne BioRaffinerie Brandenburg-Beiträge zur Produkt- und Technologieentwicklung sowie Bewertung. Brandenburgische Umwelt Ber 8:260–269

Kamm B, Kamm M, Richter K, Reimann W, Siebert A (2000b) Formation of aminium lactates in lactic acid fermentation, fermentative production of 1,4-piperazinium-(L.L)-dilactate and its use as starting material for the synthesis of dilactide (part 2). Acta Biotechnol 20:289–304

Kamm B, Venus J, Kamm M (2006) Principles of biorefineries—the role of biotechnology, the example lactic acid fermentation. In: Hearns EC (ed) Trends in biotechnology research. Nova Science, New York, pp 199–223

Kamm B, Schönicke P, Kamm M (2009) Biorefining of green biomass—technical and energetic considerations. Clean 37(1):27–30

Kim YH, Moon S-H (2001) Lactic acid recovery from fermentation broth using one-stage electro dialysis. J Chem Technol Biotechnol 76:169–178

Kromus S (2004) Die Grüne Bioraffinerie Österreich—Entwicklung eines integrierten Systems zur Nutzung von Grünlandbiomasse. Dissertation, TU Graz

Kromus S, Wachter B, Koschuh W, Mandl M, Krotschek C, Narodoslawsky M (2004) The green biorefinery Austria—development of an integrated system for green biomass utilization. Chem Biochem Eng Q 18(1):13–19

Kromus S, Kamm B, Kamm M, Fowler P, Narodoslawsky M (2006) In: Kamm B, Kamm M, Gruber P (eds) Biorefineries—industrial processes and products, vol 1. Wiley-VCH, Weinheim, pp 253–294

Lavis G (1996) Evaporation. In: Schweitzer PA (ed) Handbook of separation techniques for chemical engineers, 3rd ed. McGraw–Hill, New York

Mahro B, Timm M (2007) Potential of biowaste from the food industry as a biomass resource. Eng Life Sci 7(5):457–468

Morris DJ, Ahmed I (1992) The carbohydrate economy, making chemicals and industrial materials from plant matter. Institute of Local Self Reliance, Washington DC

Narodoslawsky M (1999) (ed) Green biorefinery. In: Proceedings of 2nd International Symposium Green Biorefinery, 13–14 October 1999, Feldbach, Austria. Proceedings, SUSTAIN, Verein zur Koordination von Forschung über Nachhaltigkeit, Graz TU, Austria

Narodoslawsky M, Kromus S (2004) Development of decentral green biorefinery in Austria. In: Kamm B, Hempel M, Kamm M (eds) Biorefinica 2004, Proceedings and Papers, 27–28 October, Osnabrück, biopos, Teltow, p 24

NREL (2005) National Renewable Energy Laboratory, http://www.nrel.gov/biomass/biorefinery.htm

National Research Council (2000) Biobased industrial products: priorities for research and commercialization, National Academic Press, Washington DC

Nonato RV, Mantellato PE, Rossel CEV (2001) Integrated production of biodegradable plastic, sugar and ethanol. Appl Microbiol Biotechnol 57:1–5

Okkerse C, van Bekkum H (1999) From fossil to green. Green Chem 4:107–114

Patel M, Crank M, Dornburg V, Hermann B, Roes L, Hüsing B, Overbeek L, Terragni F, Recchia E (2006) Medium and long-term opportunities and risks of the biotechnological production of bulk chemicals from renewable resources. The BREW Projekt, prepared under the European Commission's GROWTH Programme, Utrecht, pp 120–122

Petrides DP, Cooney CL, Evans LB (1989) An introduction to biochemical process design. In: Shuler ML (ed) Chemical engineering problems in biotechnology. American Institute of Chemical Engineers, New York

Pirie NW (1971) Leaf protein—its agronomy, preparation, quality, and use. Blackwell, Oxford

Pirie NW (1987) Leaf protein and its by-products in human and animal nutrition. Cambridge University Press, UK

Reismann HB (1988) Economic analysis of fermentation. CRC, Boca Raton

Rexen F (1986) New industrial application possibilities for straw. Documentation of Svebio Phytochemistry Group (in Danish) [Fytokemi i Norden, Stockholm, Sweden, 1986–03–06] 12

Ricci A et al (1989) Energy evaluation of a conventional wet green crop fractionation process. In: Proceedings of 3rd International Conference on Leaf Protein Research: Pisa, Perugia, Viterbo (Italy) 1–7 October 1989

Ringpfeil M (2001) Biobased industrial products and biorefinery systems—Industrielle Zukunft des 21. Jahrhunderts? http://www.biopract.de

Rossel CEV, Mantellato PE, Agnelli AM, Nascimento J (2006) Sugar-based biorefinery—technology for an integrated production of poly(3–hydroxybutyrate), sugar and ethanol. In: Kamm B, Kamm M, Gruber P (eds) Biorefineries—industrial processes and products, vol 1. Wiley-VCH, Weinheim, pp 209–226

Schidler S (2003) Technikfolgenabschätzung der Grünen Bioraffinerie, Teil I: Endbericht, Institut für Techikfolgen-Abschätzung, Österreichische Akademie der Wissenschaften

Schidler S, Adensam H, Hofmann R, Kromus S, Will M (2003) Technikfolgenabschätzung der Grünen Bioraffinerie, Teil II: Materialsammlung, Institut für Technikfolgen-Abschätzung, Österreichische Akademie der Wissenschaften

Schwenke K-D (1998) Das funktionelle Potential von Pflanzenproteinen. In: Kamm B, Kamm M, Soyez K (eds) Die Grüne Bioraffinerie; Beiträge zur ökologischen Technologie, vol 5. Gesellschaft für ökologische Technologie und Systemanalyse, Berlin, pp 185–195

Singh N (1996) (ed) Green vegetation fractionation technology. Science, Lebanon, NH

Tasaki I (1985) (ed) Recent advances in leaf protein research. In: Proceedings of the 2nd International Leaf Protein Research Conference. Nagoya, Japan

Telek L, Graham HD (eds) (1983) Leaf protein concentrates. AVI, Westport, CN

Thomsen MH, Bech D, Kiel P (2004) Manufacturing of stabilised brown juice for L-lysine production—from university lab scale over pilot scale to industrial production. Chem Biochem Eng Q 18(1):37–46

Tiffany DG (2007) Economic comparison of ethanol production from corn stover and grain. AURI Energy Users Conference, 13 March 2007, Redwood Falls, MN,

Tullo A (2005) Renewable materials, two pacts may help spur biomass plastics. Chem Eng News, 28 March 2005, http://www.CEN–ONLINE.org

US DOE (2005) 1st International Biorefinery Workshop, July 20 and 21, US Department of Energy, Washington D.C.; http://www.biorefineryworkshop.com

US Congress (2000) Biomass research and development, Act of 2000, June

US President (1999) Developing and promoting biobased products and bioenergy. Executive Order 13101/13134, William J. Clinton, The White House, 12 August 1999, http://www.newuse.org/EG/EG–20/20BioText.html

Van Dyne DL (1999) Estimating the economic feasibility of converting ligno-cellulosic feedstocks to ethanol and higher value chemicals under the refinery concept: a phase II study, OR22072-58. University of Missouri

Van Dyne DL, Blasé MG, Clements LD (1999) A strategy for returning agriculture and rural America to long-term full employment using biomass refineries. In: Janeck J (ed) Perspectives on new crops and new uses. ASHS, Alexandria, VA, pp 114–123

Vink ETH, Rabago KR, Glassner DA, Gruber PR (2003) Applications of life cycle assessment to NatureWorks™ polylactide (PLA) production. Polym Degrad Stability 80:403–419

Vorlop KD, Willke Th, Prüße U (2006) Biocatalytic and catalytic routes for the production of bulk and fine chemicals from renewable resources, In: Kamm B, Kamm M, Gruber P (eds) Biorefineries—industrial processes and products, vol 2. Wiley-VCH, Weinheim, pp 385–406

Webb C, Koutinas AA, Wang R (2004) Developing a sustainable bioprocessing strategy based on a generic feedstock. Adv Biochem Eng Biotechnol 87:195–268

Werpy T, Petersen G (eds) (2004) Top value chemicals under the refinery concept: a phase II study. US Department of Energy, Office of Scientific and Technical Information. No.: DOE/GO-102004-1992, http://www.osti.gov/bridge

Werpy T, Frye J, Holladay J (2006) Succinic acid—a model building block for chemical production from renewable resources. In: Kamm B, Kamm M, Gruber P (eds) Biorefineries—industrial processes and products, vol 2. Wiley, Weinheim, pp 367–379

White DH, Wolf D (1988) In: Bridgewater AV, Kuester JL (eds) Research in thermochemical biomass conversion. Elsevier, New York

Wilkins RJ (1977) (ed) Green crop fractionation. British Grassland Society, Hurley, Maidenhead, UK

Willke Th, Vorlop KD (2004) Industrial bioconversion of renewable resources as an alternative to conventional chemistry. Appl Microbiol Biotechnol 66(2):131–142

Wittmeyer D (2009) EU lead market initiative in the frame of European technology platform for sustainable chemistry. Deutscher Bioraffineriekongress, 8 July 2009, Industrieclub Potsdam, http://www.biorefinica.de

Zeikus JG, Jain MK, Elankovan P (1999) Biotechnology of succinic acid production and markets for derived industrial products. Appl Microbiol Biotechnol 51:545–552

Chapter 13
Heat and Power Production from Stover for Corn Ethanol Plants

Shahab Sokhansanj, Sudhagar Mani, Cannayen Igathinathane, and Sam Tagore

13.1 Introduction

In 2006, ethanol plants in the US used corn grain as the feedstock for producing nearly 18 billion L (4.8 billion gal) ethanol (American Coalition for Ethanol 2006). Most of the ethanol produced was blended with gasoline. The ethanol represented roughly 1.5% of the annual petroleum consumption of 1,150 billion L (7.3 billion barrels; CSLF 2006). National plans call for increasing ethanol production to levels that would offset at least 30% of the annual transportation fuel or roughly 345 billion L (90 billion gal) within the next 20–30 years. It is forecast that the current and projected increase in corn grain yield may support starch-based ethanol production of up to 56 billion L (15 billion gal) ethanol. Lignocellulosic biomass feedstock can further support ethanol production beyond the 56 billion L from corn grain. It is therefore important to develop secure sources of biomass and supply infrastructure to support the projected growth of bioenergy.[1]

An ethanol plant requires 9.67 MJ process heat and 0.288 kWh electricity to produce 1 L ethanol. Most of the existing ethanol plants in the US use natural gas as the source of process heat. Electricity comes from the grid. Prices for natural gas and electricity have climbed in recent years, making ethanol production economics

---

[1] Statements in this paper represent the authors' research and personal view points. The statements and data are not official

S. Sokhansanj
Environmental Sciences Division, Oak Ridge National Laboratory, P.O. Box 2008, Oak Ridge, TN 37831, USA
e-mail: sokhansanjs@ornl.gov
Department of Chemical and Biological Engineering, University of British Columbia, 2360 East Mall, Vancouver, BC V6T 1Z3, Canada

S. Mani
Driftmier Engineering Center, University of Georgia, Athens, GA 30602, USA

C. Igathinathane
Department of Chemical and Biological Engineering, University of British Columbia, 2360 East Mall, Vancouver, BC V6T 1Z3, Canada

S. Tagore
Office of Biomass Program, U.S. Department of Energy, Washington, DC 20585, USA

less attractive. Some plants use, or are planning to use, coal (Clayton 2006), which has created a negative impact on public acceptance of ethanol as a "green or clean fuel". The technology of producing process heat and electricity from direct combustion of biomass is now well developed. Biomass can provide heat and power to existing and future starch-based biofuel (ethanol) plants. We envision the following benefits from using biomass as the fuel source.

- Reduction of costs associated with natural gas prices and price volatility
- Generation of new revenue for producing electricity from excess heat
- Generation of new countryside industries in processing and handling of biomass and improving the economics of farming
- Creation of demands on feedstock supply that will encourage the development of biomass harvest, storage, and handling infrastructure—a pre-requisite for large-scale deployment of cellulosic ethanol production
- Improvement in the net energy input:output ratio of ethanol plants, and minimization or elimination of the use of fossil energy in the ethanol production process
- Decreased dependence on imported natural gas, while mitigating greenhouse gases and reducing global warming

Table 13.1 lists the estimated future number of ethanol plants in the US. The list includes an estimate of the required grain (corn) as feedstock and the amount of process heat and electrical energy required to keep the plants running. An estimate of the required grain and stover and the amount of co-product, distillers dried grain

Table 13.1 Overall ethanol production statistics in the US for current, under construction, and planned corn-starch-based ethanol plants. *DDG* Distillers dried grain

Description[a]	Per typical plant	No. of current plants (2006)	Under construction	In planning stage	Total
Number of plants	1	97	40	150	287
Billion gallons of ethanol	0.045	4.4	1.8	6.8	12.9
Liters of ethanol (GL)[b]	0.169	16.4	6.8	25.3	48.4
Mass of grain as feedstock (Tg)[c]	0.411	39.9	16.5	61.7	118.1
Mass of DDG co product produced (Tg)[d]	0.129	12.5	5.2	19.3	37.0
Process heat input (PJ)[e]	1.632	158.3	65.3	244.8	468.3
Electricity energy input (PJ)[f]	0.175	17.0	7.0	26.2	50.2
Gross energy requirement (PJ)[g]	2.759	267.6	110.4	413.8	791.8
Natural gas (Gm^3)[h]	0.073	7.1	2.9	11.0	20.8
Biomass (Tg)[i]	0.167	16.2	6.7	25.1	48.0

[a] Units prefix $k=10^3$, $M=10^6$, $G=10^9$, $T=10^{12}$, $P=10^{15}$
[b] Conversion factor 3.75 L in a US gallon
[c] A conversion efficiency of 0.41 L ethanol per kg corn. Tg is equal to 1 million Mg
[d] 0.313 kg DDG per kg corn processed
[e] 9.67 MJ of process heat per liter ethanol
[f] 0.288 kWh of electricity per liter ethanol
[g] Conversion efficiency of combustion to heat 80% and conversion efficiency to electricity 30%
[h] Conversion factor of 38 MJ m^{-3} for natural gas
[i] Conversion factor of 16.5 GJ Mg^{-1} for stover

(DDG) is also presented in Table 13.1. The last two rows are an estimate of net energy input to the plants for process heat and electricity. Assuming that 287 plants become operational in the next 5–7 years, production of ethanol from corn will amount to around 48 billion L. Roughly 48 million Mg (metric ton = 1,000,000 g) biomass will be required to heat and power these plants provided they all use biomass as a sole fuel.

Several researchers (Savola and Fogelholm 2006; Morey et al. 2005; Wang et al. 1999; Shapouri et al. 1999, 2002) have reported energy analyses of corn-to-ethanol processes based on entire life cycle analyses of corn grain to ethanol. Although, according to most analysts, the corn-to-ethanol net energy balance is positive, the ethanol conversion process has a relatively high heat and power requirement. The energy input:output ratio in plants that use external source has been calculated at 1.1–1.3. This ratio increases when energy conservation measures such as heat recycling is implemented in the plant and may reach 2–3 when biomass is used as a source of heat and power.

Morey et al. (2005) proposed the use of DDG or corn stover for heat and electricity generation for the ethanol plants and found that there was a significant annual energy cost savings for the 170 million L ethanol plant. Use of DDG for power and heat generation may not be the best option due to its high nutritional value as a choice source of protein and fiber for animal feed (>US $80/t). In order to compare and make biomass competitive with existing heating fuels, a techno-economic analysis of the use of biomass to supply heat and power systems should be conducted. Any biomass-based heat and power production system relies mainly on a continuous supply of cheap biomass delivered to the plant. The main objectives of this chapter are to provide an estimate of the cost of supplying corn stover to existing dry mill ethanol plants to produce heat and/or power, and to calculate the cost of on-site fuel preparation for delivering it to the burner. This chapter also examines the economics of producing heat and/or power using stover.

13.2 Economics of Stover Supply to the Ethanol Plant

Almost all of the present ethanol plants in the US use corn grain as source of feedstock. The concept is to integrate the harvest and supply of corn stover with the well-developed grain harvest and delivery system as pictured in Fig. 13.1. The grain is combined and stored on the farm or sent directly to a central depot (country elevator). The grain is transported to the ethanol plant either directly from the farm store or from the depot. Similar to the way grain is combined and stored in a bin, stover is baled and stacked on the farm or transported to a larger storage area (depot). The supply of biomass to the ethanol plant will be in one of three forms: (1) baled, (2) ground, and (3) pelletized. We assume a 170 million L (45 million gal) ethanol plant. The annual heat and power requirement of the plant is estimated to be 5.6 MW power and 52.3 MW process heat. The plant requires as much as 150 million Mg stover annually.

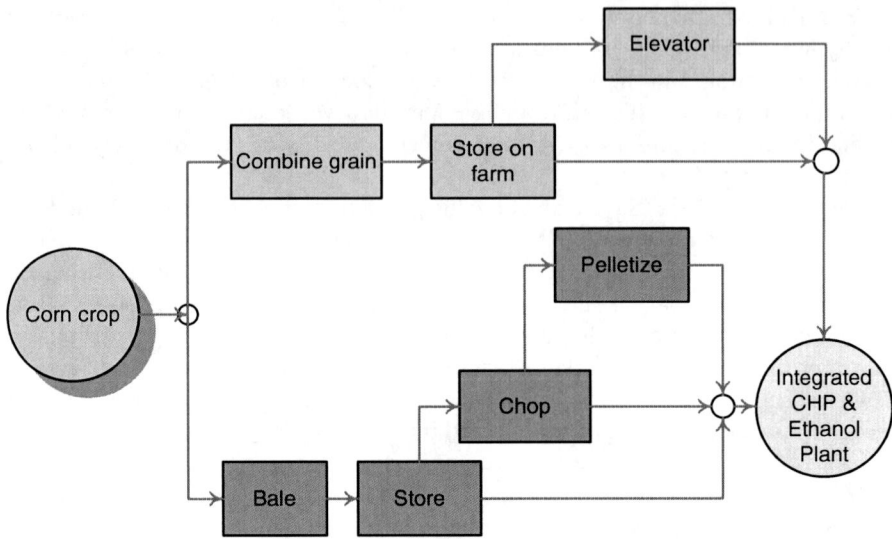

Fig. 13.1 The co-flow paths of grain and stover to an ethanol plant. *CHP* Combined heat and power

13.2.1 Stover Collection

We define collection as operations for picking up the biomass, packaging, and transporting to a nearby site for temporary storage. The most conventional method for collecting biomass is baling (Cundiff and Marsh 1996). Bales can be either round or square. Limited experience with using round bales for biomass applications indicates that round bales are not suitable for large scale biomass handling (Cundiff 1995, 1996). Because of their round shape, round bales tend to deform under static loads in a stack. Bales that are not perfectly round cannot be loaded onto trucks to form a transportable load over open roads. Therefore, square baling operation was considered in this study. The integrated biomass supply analysis and logistics (IBSAL) model (Sokhansanj et al. 2006; Sokhansanj and Mani 2006) was used to estimate the resource requirement and cost of biomass collection. We used a net stover yield of 5.7 Mg ha^{-1}. This represents roughly 60% of the above-ground biomass associated with a corn yield of 180 bu/ac—currently a typical stover yield in the US Midwest (Sokhansanj et al. 2002; Sokhansanj and Turhollow 2002). The assumption is that the mass of above-ground biomass is equal to the mass of grain yield at 15% moisture content. This calculation allows 40% of the biomass to remain in the field for its nutritive value and soil conservation. The collection sequence consists of shredding and baling it into large rectangular bales 1.2 m × 1.2 m × 2.4 m. The bales are collected using a truck-mounted bale collector (Stinger Model 6500; http://www.stingerltd.com/). The Stinger transports the collected biomass to the side of the farm and stacks it in rows four bales high (4.8 m high).

13.2.2 Preprocessing

Preprocessing of biomass may consist only of grinding. Loose-cut biomass has a low bulk density ranging from 50 to 120 kg/m^3 depending on the particle size. Biomass ground to a size of less than 6 mm could have a bulk density of 160 kg/m^3 in the truck box (Wright et al. 2006). This density may be suitable for short hauls. For longer hauls and long-term storage, denser biomass in the form of pellets or cubes may be desired. The bulk density of pellets can be as high as 700 kg m^{-3} and cubes as high as 500 kg m^{-3} (Sokhansanj and Turhollow 2004).

To make pellets, the biomass is ground and extruded using a pellet mill. The pellets, which have a bulk density of 600–700 kg/m^3, are cooled and stored. Corn stover can be easily compacted into pellets for easy transport and storage (Mani 2005; Mani et al. 2006a). Economics and energy input to the pelleting operation was obtained from Mani (Mani et al. 2006b). To process into chopped form, a mobile grinder/shredder is used to chop the biomass bales. The machine is similar to agricultural tub grinders and can be transported to the stacks. A bale is placed on the tub grinder; the ground biomass is transferred to a waiting truck box using an attached belt conveyor. The cost of the biomass chopping process was estimated using the IBSAL simulation model.

13.2.3 Stover Transport

Stover in the form of bales, chops and pellets are usually transported by trucks. Bales are transported using flatbed trucks. The ground (chopped) and pelletized biomass is transported using truck boxes. In our analysis, the biomass is loaded onto trucks and transported to the biofuel plant daily throughout the year. For bale transport, bales are stacked on a flatbed trailer to a maximum height of 4 m (above ground). Roughly 36 rectangular bales (1.2 m x 1.2 m x 2.4 m) are placed on a 14.6 m long flat bed. Larger trailers with more axles are available when allowed by road transport regulations. For bale loading and unloading, loaders equipped with bale grabbers are used. The bale grabbers remove one or two bales from the stack at a time and place them on the deck of the trailer. A full bale load weighs roughly 18 dry Mg assuming 0.5 dry Mg per bale.

The truck box is 2.4 m wide x 2.4 m high x 12.6 m long, with a capacity of roughly 70 m^3. Tests with grinders have shown that grinding biomass to a size less of than 5 mm can result in a bulk density of about 160 kg m^{-3} (Wright et al. 2006). This will yield a load of roughly 11 dry Mg per truck box. The pellets can be loaded onto truck boxes and transported to the ethanol plant. We assume a bulk density of 560 kg m^{-3} for pelletized or cubed stover. A 70 m^3 truck box can hold roughly 40 dry Mg. This is an efficient way of moving biomass assuming that carrying heavy loads is permitted on local roads.

13.2.4 On-Site Storage and Fuel Preparation

For storage, a covered building with a flat floor for bales and chops is considered. For pellets we specify steel bins with a flat floor. The square area for three forms of biomass is 2,780 m^2 for bales, 3,870 m^2 for ground stover and 348 m^2 for pelletized biomass. The difference in area is due to bulk density and the area allowed for movement of handling equipment. The storage building costs were estimated at US $107 m^{-2}, $161 m^{-2} and $215 m^{-2}, for storing bales, ground (chops) and pellets, respectively. The storage building for chopped biomass was more than that for bales because retaining walls are needed to hold the pile of biomass. The most expensive storage is steel bins with filling and emptying auxiliaries and aeration equipment.

Solid fuels require on-site preparation to be used in the boiler. The combustion system for this project will accommodate pulverized biomass/solid fuels. Biomass corn stover delivered to the ethanol plant will be in one of three forms: square bales, ground or pelletized. The existing grain-handling system can easily accommodate pelletized biomass. Ground biomass may also be stored in grain silos if the particles are small and dry enough not to bridge. Ground biomass and square bales can be stored in covered sheds or in the open. Each of the biomass formats (Fig. 13.2) requires different pre-treatment before it can be fed to the burner. Square bales require de-stringing, de-baling (primary shredding), and fine grinding (hammer milling), and are then conveyed to the boiler by a pneumatic feeding system. To avoid dusting during fine grinding, the ground biomass is collected in a cyclone system. A similar type of fuel preparation system was used in the Chariton Valley power generation system (CVRC&D 2002, 2004). Dry chops and pelletized biomass require only a single stage of grinding. Ground biomass may not require re-grinding.

13.3 Costs

13.3.1 Cost of Biomass Collection

Table 13.2 summarizes IBSAL's output for collection operations. Size, speed, and numbers of equipment units make it possible to complete each operation on a

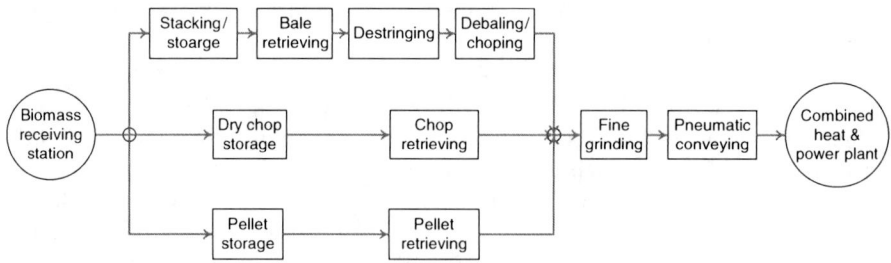

Fig. 13.2 On-site stover fuel preparation for use in the boiler

13 Heat and Power Production from Stover for Corn Ethanol Plants

Table 13.2 Completion day, cost, energy input, emissions and dry matter for a conventional baling system for corn stover at a yield of 5.7 Mg ha^{-1}

Operation[a]	Date completed[b]	Mass (Mg)	Cost (US $ Mg^{-1})	Energy input (MJ Mg^{-1})	Carbon emissions (kg C Mg^{-1})
Payment to producer			10.00		
Shredding	10 December	141,550	4.11	47.4	1.0
Baling[c]	10 December	140,359	12.36	136.9	2.9
Stacking	16 December	139,819	7.89	123.5	2.7
Profit (15% of the collection costs)			3.65		
Overall			38.01	313.7	6.7

[a]Equipment (number of units): combines 48; grain trucks 16; shredders 22; balers 36; stackers 6
[b]Start of harvest is 15 September in Iowa
[c]Large rectangular baler / square bale

specific date. Collection operation requires 22 shredders, 36 balers, and 6 bale stackers to complete the collection operations by 16 December in Iowa (in some years the harvest season is shorter due to snow and rain). The output includes completion dates for each operation. The net tonnage collected was 139,819 Mg at 38.01 US $ Mg^{-1}. The cost includes $10 payment to the producer and 15% of total collection costs as profit, assuming the work is done by a custom operator. The unit operation costs per metric ton are $4.11 for shredding, $12.36 for baling and $7.89 for stacking. Table 13.2 lists the amount of energy input for the power equipment. The input energy required to collect corn stover is 313.7 MJ Mg^{-1}. To put this energy use into perspective, the amount of energy in 1 Mg dry stover is roughly 17 GJ. The amount of energy used to power equipment is roughly 2% of the energy content of biomass. Table 13.2 also lists the amount of carbon emission as a result of using diesel fuel.

13.3.2 Preprocessing Costs

The costs for grinding involve a loader with an operator ($1.42 dry Mg^{-1}), and an industrial grinder $5.72 dry Mg^{-1}, giving a total of $7.14 dry Mg^{-1}. To account for profit (return on investment), we increase the cost by 15% to $8.21 dry Mg^{-1}. Table 13.3 shows the breakdown of cost and energy demand for the biomass pelleting process. The pelleting plant has a production capacity of 6 Mg h^{-1}, with annual production of 45,000 Mg (Mani et al. 2006a). Roughly three mills are needed to process the entire biomass for the ethanol plant. The plant operates 24 h for 310 days annually (annual utilization period 85%). The ground biomass is used as a feedstock, which eliminates the drying operation. The total cost of pelleting is about $23.61 Mg^{-1} pellets with an energy requirement of 1.69 GJ Mg^{-1}. About 10% of the biomass energy is used to produce pellets. The costs of pelleting biomass can be further reduced if larger plant capacity (12–15 Mg h^{-1}) is considered (Mani et al. 2006a).

Table 13.3 Cost and energy analysis of biomass pelleting process

Processes	Cost ($ Mg^{-1})	Energy (GJ Mg^{-1})
Hammer mill	0.95	0.59
Pellet mill	3.31	0.31
Pellet cooler	0.34	0.46
Screens	0.16	0.07
Miscellaneous equipment	2.77	0.08
Labor	12.74	0.18
Land and building use	0.26	
Sub-total	20.53	1.69
Total (plus 15% profit)	23.61	

Table 13.4 The cost of transporting biomass in three forms. Travel distance is assumed 70 km and with a speed of 50 km/h; transport load dimensions 2.43 m x 3.10 m x 12.2 m; bales are 1.2 m x 1.2 m x 2.4 m; bale density 176 kg m^{-3}, chop density 128 kg m^{-3}, pellet density 650 kg m^{-3}, transport container fill fraction 0.7

	Mg load^{-1}	No. of truckloads	Loads h^{-1}	No. of trucks	Total $ h^{-1}	Cost $ Mg^{-1}
Bales	18	9,722	1.16	3.24	249.54	9.98
Chop	11	15,909	1.89	5.30	408.33	16.33
Pellet	40	4,375	0.52	1.46	112.29	4.49

13.3.3 Transport Costs

The transport cost of biomass is one of the major cost components in the entire supply system. These costs vary with the format of the biomass and transport distance. The footnote in Table 13.4 specifies the assumed load dimensions for a transporter at 2.43 m x 3.1 m x 12.2 m [8 feet(') x 10' x 40']. The dimensions of bales are 1.2 m x 1.2 m x 2.4 m (4' x 4' x 8 '). The bale bulk density is assumed at 176 kg m^{-3}. The bulk density of chopped biomass is taken as 128 kg m^{-3} and that of pellets at 650 kg m^{-3}. The container is not filled with pellets because the weight of a full container would surpass the allowable weight on most public roadways.

Table 13.4 lists Mg/load, total number of truck loads required to transport 140,000 dry Mg year^{-1}, and the number of required loads per hour. The number of truckloads required for a given average speed of 50 km h^{-1} and 140 km distance traveled (70 km each way) is calculated. Total cost ($ h^{-1}) is the product of the required number of trucks and the cost of one truck load at ($77 h^{-1}). Total cost ($ h^{-1}) divided by 25 Mg h^{-1} gives the cost in dollars per metric ton, which yields $9.98 for transporting bales, $16.33 for transporting ground biomass, and $4.49 for transporting pellets.

For these calculations, we did not consider any specific geometry for the supply basin with respect to the location of the biofuel plant. For example costs per metric ton increase with two-thirds of the increase in radius of a circular supply basin (Gallagher et al. 2003).

13.3.4 On-Site Fuel Storage and Preparation

Table 13.5 itemizes the costs of receiving, storing, reclaiming and final fine grinding of biomass for immediate delivery to the boiler room. Bales are received at the ethanol plant and are stacked under a roof. Chopped (shredded) biomass is received in silage-type truck boxes. The load is dumped on a concrete floor and piled using a front end loader. The existing grain pit and grain elevator legs at the ethanol plant can serve to unload and distribute pellets to the pellet bin. We assume that 1,500 Mg storage space is adequate for 3 days of operation at 500 Mg day^{-1}.

The cost of reclaiming bales is higher than those for chops and pellets because each bale has to be placed individually onto a conveyor belt. The strings are cut and removed automatically as the bales advance towards a primary shredder (debaler). These processes are not necessary for the shredded or pelletized feedstock. The shredded feedstock or pellets are ground to powder size in a hammer mill. Table 13.5 also estimates the cost of pneumatic conveying of ground biomass to the burner at US $2.21 per dry Mg. An overhead of 30% is added to the costs for contingency, and any extra labor requirement.

13.3.5 Total Cost of Biomass Fuel Delivered to the Burner

Table 13.6 summarizes the total delivered cost of different biomass formats to the burner. At $84.31 Mg^{-1}, shredded or ground biomass has the highest delivery cost, mostly because of the high cost of its handling, transportation, and storage. Next, at $79.91 Mg^{-1}, is the delivered cost of pellets. The square bale format has the lowest delivered cost of $72.63 Mg^{-1}. Biomass can be supplied to produce heat and power to the ethanol plant at a price lower than the current market price of DDG ($100 Mg^{-1}; Morey et al. 2005). A cost difference of $10–20 Mg^{-1} feedstock would make a large difference in the annual energy saving for the ethanol plant. Overall, for these scenarios, and assuming a distance of 70 km transport, chops are the most expensive form of biomass delivered to the burner, followed by pellets and square bales.

13.4 Heat and Power Production

Combined heat and power (CHP) or cogeneration systems produce both heat and power simultaneously from the same primary source of fuel. Many kinds of CHP systems are available on the market for both heat and power production. Detailed reviews of the different available CHP technologies are reported in EDUCOGEN (2001) and ONSITE SYCOM (1999). Overall thermal efficiency of the CHP system varies from 70% to 80%, compared to a power plant efficiency of about 35%. Wahlund (2003) reported a detailed overview of 14 cogeneration plants in Sweden. The electrical capacity varies from 2 MW$_e$ to 39 MW$_e$ with an electrical

Table 13.5 Estimated cost of receiving, storing, and preparation of biomass for boiler use. Throughput at 500 t/day, reserve for 3 days

Operations	Baled stover		Chopped stover		Pelletized stover	
	Explanation	Cost ($ Mg^{-1})	Explanation	Cost ($ Mg^{-1})	Explanation	Cost ($ Mg^{-1})
Receiving	Unload and stack bales	2.22	Front end loader and piler	1.10	Dump in pit, elevate[a]	0.00
Storing	Enclosed, free standing, stacked bales	4.77	Enclosed reinforced bearing walls for piled stover	6.62	Steel bin with overhead distribution	1.27
Reclaiming biomass	Stacker, belt conveyor, destringer, debaler	4.34	Front-end loader, conveyor	1.40	Bin unloader	1.40
Fine grinding	Hammer mill	5.41	Hammer mill	5.41	Hammer mill	5.41
Delivery	Pneumatic, airlock, blower	2.21	Pneumatic, airlock, blower	2.21	Pneumatic, airlock, blower	2.21
Total (+30% overhead)		24.64		21.76		13.38

[a]Use existing grain pit and elevator

Table 13.6 Total cost for different forms of biomass delivered to the biomass burner

	Bale ($ Mg^{-1})	Chop ($ Mg^{-1})	Pellet ($ Mg^{-1})
Collection	38.01	38.01	38.01
Pre-process	0.00	8.21	23.61
Transport	9.98	16.33	4.49
Total delivered cost to the plant	48.35	62.91	66.47
On site fuel preparation	24.64	21.76	13.80
Total cost	72.63	84.31	79.91
Total cost ($ GJ^{-1})[a]	4.40	5.12	4.84

[a]Total delivered cost in $ GJ^{-1} calculated by assuming heating value of biomass as 16.5 GJ Mg^{-1}

efficiency of 20–30%, and the systems' configuration includes a steam turbine with different combustion boilers (centrifugal fluidized, boiling fluidized bed, grate-fired type). The US Environmental Protection Agency (EPA) promotes the use of CHP production systems for corn ethanol producers in the US, mainly because of increased fuel efficiency, additional cost savings from onsite production of heat and power and alternative fuels, and reduced greenhouse gas emissions (Cundiff 1995). Steam-turbine-based CHP systems have many advantages over other technologies, especially for corn ethanol plants. Fuel flexibility, plant availability for use throughout the year and the high heat:power ratio (3–10 kW$_{th}$/kW$_{el}$) of the steam-turbine-based CHP system make it a very attractive option for corn ethanol plants (EPA-CHP 2002).

13.4.1 Process Heat Generation

We carried out a technical and economic evaluation of biomass-fired heat generation for the corn ethanol plant to supply 52.3 MW$_{th}$ process heat (Fig. 13.3); the results presented in Table 13.7 are based on the fuel cost and heat value data in Table 13.8. In Table 13.7, biomass- and coal-based heat generation systems were compared and analyzed relative to a natural-gas-based heat generation system. The operating conditions for the system are given in Table 13.9. Corn stover was delivered to the heat generation plant in three different forms: bale, chop or pellet. The cost of corn stover delivered to the burner varies depending on the form of stover delivered to the plant. An increase in the corn stover costs results in high annual plant operating costs. The economic data from the stover-fired heating system were compared with both coal-fired and the existing natural-gas-fired heating system. Total investment costs for the biomass-fired CHP plant were relatively high compared to the existing natural-gas-fired system. The annual fuel requirement for the biomass-fired heating plant was 121,274 Mg, whereas the coal-fired heating plant requires 64,428 Mg. Capital investment costs for the biomass- and coal-fired plants were similar. Although the capital costs of the coal-fired boiler were less than for the biomass-fired boiler, coal boilers require an additional pollution control device to remove sulfur dioxide from the flue gas. The capital investment costs of a natural-gas-fired plant were almost 50% lower than solid fuel

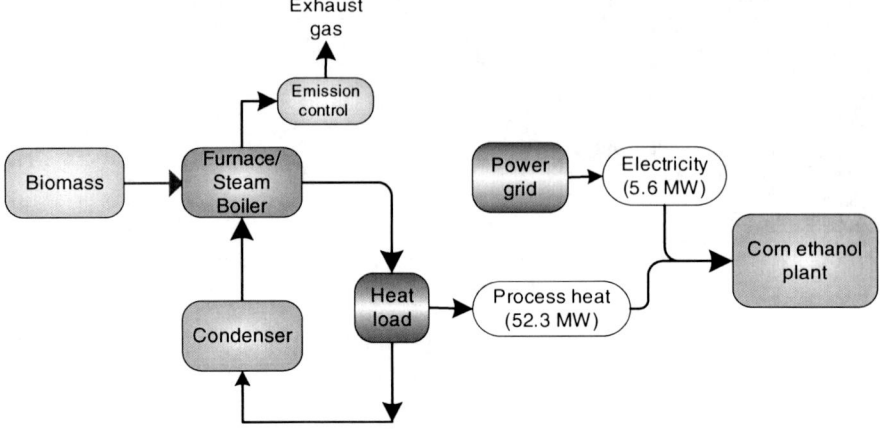

Fig. 13.3 Block diagram of a biomass-fired thermal heat plant

Table 13.7 Technical and economic data for the process heat generation system. *O&M* Operating and maintenance

Parameter	Corn stover			Coal	Natural gas[c]
	Bale	Chop	Pellet		
Annual fuel consumption (Mg)	121,274	121,274	121,274	73,631	56,611
Total investment cost ($M)	18.89	18.89	18.89	18.89	8.62
Annual O&M cost ($M)[a]	3.24	3.24	3.24	3.24	2.02
Annual fuel cost ($M)	8.81	10.22	9.69	3.98	15.52
Annual ash disposal cost ($M)	0.14	0.14	0.14	0.16	–
Total annual cost ($M)	12.19	13.50	13.07	7.37	17.54
Simple payback (years)[b]	4.0	5.7	4.9	2.0	–
Total annual savings ($M)	3.59	1.94	2.56	9.18	–
Benefit-cost ratio	7.57	4.55	5.69	17.55	–

[a]Annual O&M cost includes internal electricity consumption costs, labor charge, debt payment and plant maintenance costs
[b]Simple payback period represents the length of time required for heating plant to recoup its own investment costs
[c]Natural-gas-based process heat generation system is the base case for comparing and evaluating cost of biomass- and coal-based heat generation systems

Table 13.8 Cost of different fuels used in this study. *HHV* Higher heating value

	Corn stover			Coal	Natural gas
	Bale	Chop	Pellet		
Total fuel cost up to the burner ($/t)	72.63	84.31	79.91	54	424
HHV of fuel (GJ/t)[a]	17.0	17.0	17.0	28	53
Total energy cost up to the burner ($/GJ)	4.27	4.96	4.70	1.93	8

[a]HHV of fuel corrected for corn stover at 10% moisture content and for coal at 2% moisture content

Table 13.9 Selected operating conditions and efficiencies of different combined heat and power (CHP) units

Parameters	Selected conditions	Sources
Power boiler efficiency (%)	80	EPA-CHP (2002); van den Broek et al. (1995)
Steam turbine inlet pressure (MPa)	10	Wahlund (2003)
Steam turbine inlet temperature (°C)	400	Wahlund (2003)
Steam turbine outlet pressure (MPa)	0.5	Assumed
Steam turbine outlet temperature (°C)	152	Assumed
Turbine isentropic efficiency (%)	78	Bhatt (2001)
Generator efficiency (%)	97	Bhatt (2001)

boilers due to there being no requirement for additional pollution control devices such as bag filters or ESPs. On the other hand, annual fuel consumption costs for the natural-gas-fired system were about 40–50% higher than solid fuel burners, which resulted in high annual savings of $2–$9 million and a relatively low payback period of 2–6 years. A coal-fired heating plant has relatively low annual operating costs compared to biomass-fired systems, due mainly to the low energy cost of the fuel.

13.4.2 Combined Heat and Power Generation

The CHP generation system proposed for the corn ethanol plant produces 9.5 MW$_e$ power and 52.3 MW$_{th}$ process heat based on the operating conditions given in Table 13.9. Figure 13.4 shows a typical CHP system using biomass showing heat and power. The CHP system has a heat:power ratio of 5.5. A corn ethanol plant requires only 5.6 MW$_e$ power, and the excess power is exported to the electrical grid. A power boiler steam production rate of 74 Mg h^{-1} at 10 MPa is required to meet the total process heat demand. Process heat is extracted in the form of steam from a back pressure port of the steam turbine under the same conditions used for the process heat generation system (0.5 MPa pressure and 152°C). The fuel requirement for the power boiler was calculated from knowledge of the thermal efficiency of the boiler. Five different fuel combinations were used to compare economic and emission rates in the CHP system. Table 13.10 shows the stover and coal consumption rate of the CHP system for five scenarios. Scenario 1 represents the stover-fired CHP system, whereas scenario 5 represents the coal-fired CHP system. The remaining scenarios represent co-firing CHP systems. All five scenarios produce the same amount of power and process heat with an overall CHP efficiency of 83.3%. CHP electrical efficiency was about 12.7%. Many of the biomass-fired CHP plants with 1–10 MW$_e$ power capacity have a CHP electrical efficiency ranging from 8% to 24% depending on the turbine operating conditions (Cundiff and Marsh 1996).

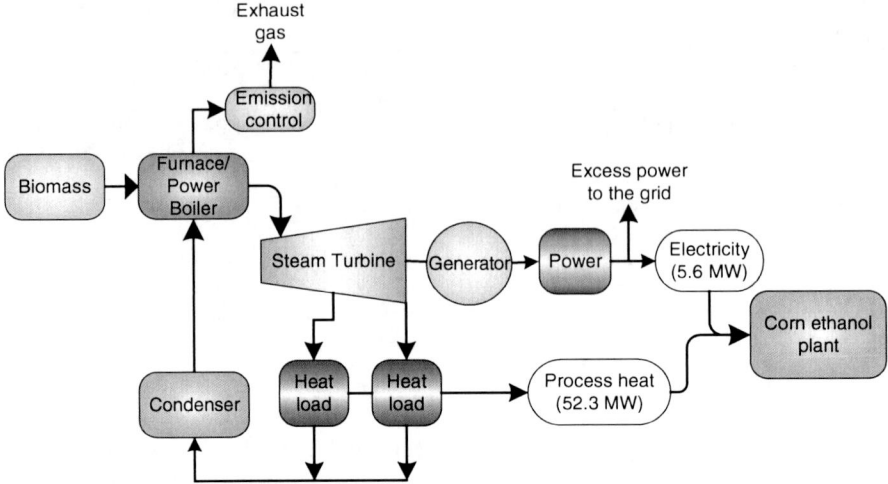

Fig. 13.4 Block diagram of a biomass fired CHP plant

Table 13.10 Combinations of fuel options for the CHP system

Scenario	Percent biomass used	Percent coal used
1	100	0
2	75	25
3	50	50
4	25	75
5	0	100

Economic performance data for the CHP system with the five different fuel combinations is given in Table 13.11. Capital investment costs for the CHP system were about $38.15 million. Annual operating and maintenance costs include internal power consumption costs, labor costs and annual maintenance costs for the CHP system. Total annual operating and maintenance costs vary from US $15.71 to $10.24 million depending on the different fuel combinations. Total annual operating costs for the stover-fired CHP system were about 33% higher than for the coal-fired CHP system. The major cost difference was due to the low energy cost of the coal fuel. Annual saving costs of the CHP system were calculated compared to existing heat and electricity input conditions (process heat from the natural-gas-fired heating system and electricity from the power grid) of the corn ethanol plant. Annual savings in CHP system costs varied from US $4 to $10 million depending on the fuel options. Simple payback for the stover-fired CHP system was about 6 years, which can be reduced to almost 3.3 years if coal is used as a fuel. Co-firing of coal with stover in the CHP system shows relatively lower payback periods than those seen with the biomass-fired CHP system. The benefit: cost ratio for the coal-fired CHP system was also the highest of all five fuel combinations.

13 Heat and Power Production from Stover for Corn Ethanol Plants

Table 13.11 Economic performance data for the CHP system using the scenarios from Table 13.10

	Scenario 1	Scenario 2	Scenario 3	Scenario 4	Scenario 5
Total investment cost ($M)	38.15	38.15	38.15	38.15	38.15
Annual O&M cost ($M)[a]	5.56	5.56	5.56	5.56	5.56
Annual fuel cost ($M)	9.98	8.61	7.25	5.88	4.51
Annual ash disposal cost ($M)	0.16	0.17	0.17	0.17	0.18
Total annual cost ($M)	15.71	14.34	12.98	11.61	10.25
Total annual savings ($M)	3.61	5.20	6.79	8.39	9.98
Simple payback (years)[b]	6.0	4.9	4.2	3.6	3.2
Benefit-cost ratio	4.22	5.64	7.06	8.49	9.91

[a]Annual O&M cost includes internal electricity consumption costs, labor charge, debt payment and plant maintenance costs
[b]Simple payback period represents the length of time required for the heating plant to recoup its own investment costs

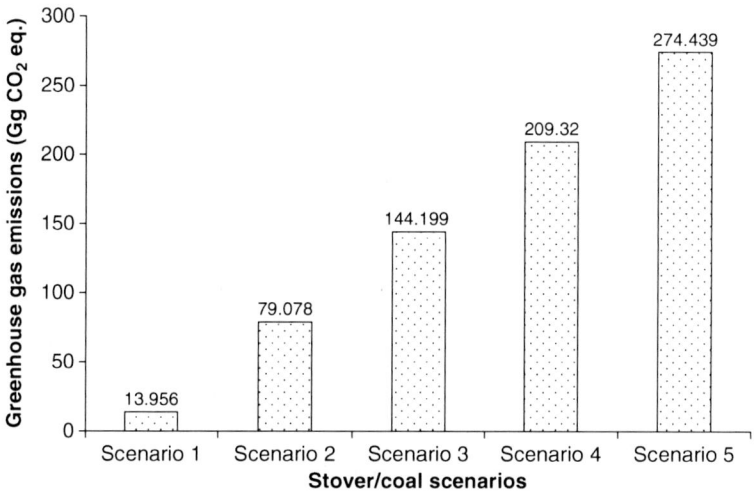

Fig. 13.5 Greenhouse gas emissions of different stover and coal combinations (scenarios from Table 13.10) for the CHP plant

Table 13.11 shows that coal-fired CHP systems are more economically attractive than stover-fired CHP systems, but the environmental impact of the coal-fired CHP system was the highest among the five scenarios. Greenhouse gas emissions from the coal-fired CHP system were 275 Gg CO_2 eq., which is about 20 times higher than that of the stover-fired CHP system (Fig. 13.5). Coal emissions can be reduced further if stover is co-fired with coal in the CHP system. An increase in percentage of stover in the CHP system reduces greenhouse gas emissions considerably. If all corn ethanol plants, including those under construction and in the planning stage (total 287 plants), used stover as a fuel, it would create an annual stover demand of 40 million Mg, which is about 50% of the total stover available in the US (Perlack et al. 2006) If stover is used instead of coal, it would reduce the greenhouse gas emissions by 75 Tg CO_2 eq., which is equal to more than 1% of the total greenhouse gas emissions (7,074 Tg CO_2 eq.) of the US (USEPA 2006).

13.5 Concluding Remarks

A systematic cost analysis of corn stover collection, pre-processing and transport estimated US $48, $63 and $66.5 Mg^{-1} for baled, chopped and pelleted biomass, respectively. An additional $25 Mg^{-1} is required to prepare baled biomass to make it suitable for combustion. Pelleted biomass has the least fuel preparation cost, $14 Mg^{-1}, due to its low on-site storage cost. The collection cost of biomass ($38 Mg^{-1}) was the largest cost factor in the entire supply system, which could be reduced further if a whole crop harvesting system is developed and used. Research work has been initiated already to estimate stover collection costs using a whole crop harvesting system. The transport cost of chopped biomass was highest due to low bulk density. Compaction of chopped biomass in the truck to increase bulk density may reduce transport costs further by increasing the payload per truck. Although the transport cost of pelleted biomass is lower, pellet production costs remain a technical barrier. Granulation of biomass to increase bulk density may reduce the cost of pre-processing. Biomass granulation will also reduce the on-site fuel preparation cost as granules do not require fine grinding and an expensive pneumatic handling system. Apart from high cost of compacting biomass into pellets, pellets can be stored easily for long periods, similar to grains, and are an attractive form of fuel for domestic and central heating applications. Further reduction in biomass delivered cost is possible if more efficient collection and transport systems are developed. We envision that creating demand for biomass will build an infrastructure for biomass fuels, which may further reduce the cost of biomass. Successful use of biomass in Europe is due mainly to a well established infrastructure for collecting, transporting and storing biomass fuels. For example, in Holland, Denmark and Sweden, the use of waterways to transport bulk biomass has considerably reduced the cost of biomass transport.

In summary, this chapter has presented a techno-economic analysis of corn-stover-based process heating and CHP generation systems compared with coal- and natural-gas-based systems. Stover-fired heating systems were relatively expensive compared to coal-fired heating systems, but very attractive compared to natural-gas-fired heating system. Coal-fired heating systems had the highest environmental and human toxicity impacts compared to natural gas and stover. The economic and environmental impacts of CHP systems for corn ethanol plants were examined using stover, coal and a combination of the two as a primary fuel. The proposed CHP system produced 9.5 MW_e power and 52.3 MW_{th} process heat with an overall CHP efficiency of 83.3%. Excess electricity is sold to the power grid. Stover-fired CHP systems can produce an annual saving of $3.6 million and a payback period of 6 years compared to existing power and process heat input systems. The coal-fired CHP system showed the highest annual saving cost, with the lowest payback period of 3.3 years. Although the coal-fired CHP system was economically attractive, greenhouse gas emissions from coal combustion were almost 20 times higher than those of stover. Corn ethanol produced using coal energy may be considered as an environmentally friendly fuel. Co-firing of stover with coal may be the most attractive option to maximize both the economical and environmental benefits of the CHP system.

Acknowledgments The authors acknowledge Oak Ridge National Laboratory (ORNL) and the Office of Biomass Program, US Department of Energy (DOE), for providing funding to conduct this research.

References

American Coalition for Ethanol (2006) US Ethanol Production. Available via DIALOG. http://www.ethanol.org/production.html. Cited 15 June 2006
Bhatt MS (2001) Mapping of general combined heat and power systems. Energy Convers Manage 42:115–124
Clayton M (2006) Carbon cloud over a green fuel. Available via DIALOG. http://www.csmonitor.com/2006/0323/p01s01-sten.html. Cited 15 June 2006
CSLF (2006) An energy summary of the United States of America. Carbon Sequestration Leadership Forum. Available via DIALOG. http://www.cslforum.org/usa.htm. Cited 15 June 2006
Cundiff JS (1995) Delayed harvest of switchgrass. 1994–1995 Annual Report. Biological Systems Engineering Department. Virginia Tech. Blacksburg, VA
Cundiff JS (1996) Simulation of five large round bale harvesting systems for biomass. Bioresour Technol 56:77–82
Cundiff JS, Marsh LS (1996) Harvest and storage costs for bales of switchgrass in the southern United States. Bioresour Technol 56:95–101
CVRC&D (2002) Chariton Valley biomass project—design package. Prepared for USDOE under Contract #DE–FC36–96 G010148. Chariton Valley RC&D Centerville, IA
CVRC&D (2004) Chariton Valley biomass project—Department of Energy project update. Available via DIALOG. http://biomass.ecria.com/technical~engineering.html. Cited 15 March 2006
EDUCOGEN (2001) A guide to cogeneration. Contract No. XV11/4. 1031/P/99-159. The European Association for the Promotion of Cogeneration, Brussels, Belgium. Available via DIALOG http://www.cogen.org. Cited 15 June 2006
EPA-CHP (2002) Catalogue of CHP technologies. Environmental Protection Agency, Washington DC
Gallagher P, Dikerman M, Fritz J (2003) Biomass crop residues: cost and supply estimates. Report no. 819. USDA Office of Chief Economist, Office of Energy Policy and New Uses, Washington, DC
Mani S (2005) A systems analysis of biomass densification process. PhD Dissertation, Department of Chemical and Biological Engineering, University of British Columbia, Vancouver, BC:
Mani S, Sokhansanj S, Bi X, Turhollow AF (2006a) Economics of producing fuel pellets from biomass. Appl Eng Agric 22(3):421–426
Mani S, Tabil LG, Sokhansanj S (2006b) Effects of compressive force, particle size and moisture content on mechanical properties of biomass pellets from grasses. Biomass Bioenergy 30:648–654
Morey RV, Tiffany DG, Hatfield DL (2005) Biomass for electricity and process heat at ethanol plants. ASABE paper # 056131. ASABE, St. Joseph, MI

ONSITE SYCOM (1999) Review of CHP technologies. Grant no. 98R020974. US Department of Energy, Washington DC

Perlack RD, Wright LL, Turhollow AF, Graham RL, Stokes B, Erbach DC (2006) Biomass as feedstock for a bioenergy and bioproducts industry: the technical feasibility of a billion ton annual supply. Report # ORNL/TM-2005/66. Oak Ridge National Laboratory (ORNL), Oak Ridge, TN, 2005. Available via DIALOG. http://www.osti.gov/bridge. Cited 15 June 2006

Savola T, Fogelholm CJ (2006) Increased power to heat ratio of small scale CHP system using biomass fuels and natural gas. Energy Convers Manage 47:3105–3118

Shapouri H, Duffield JA, Wang M (1999) The energy balance of corn ethanol revisited. Trans ASAE 46(4):959–968

Shapouri H, Duffield JA, Wang M (2002) The energy balance of corn ethanol: an update, Agricultural Economic Report No. 813, Washington, DC

Sokhansanj S, Mani S (2006) Modeling of biomass supply logistics. In: Bridgewater AV, Bobcock BDG (eds) Science in thermal and chemical biomass conversion, vol 1. CPL, Newbury Perks, UK, pp 387–403

Sokhansanj S, Turhollow AF (2002) Baseline cost for corn stover collection. Appl Eng Agric 18 (5):38–43

Sokhansanj S, Turhollow AF (2004) Biomass densification—cubing operations and costs. Appl Eng Agric 20(4):495–499

Sokhansanj S, Turhollow AF, Cushman J, Cundiff J (2002) Engineering aspects of collecting corn stover for bioenergy. Biomass Bioenergy 23:347–355

Sokhansanj S, Kumar A, Turhollow AF (2006) Development and implementation of integrated biomass supply analysis and logistics (IBSAL) model. Biomass Bioenergy 30:838–847

US EPA (2006) Inventory of US greenhouse gas emissions and sinks: 1990–2004. USEPA report # 430-R-06-002. United States Environmental Protection Agency, Washington, DC

Van den Broek R, Faaij A, van Wijk A (1995) Biomass combustion: power generation technologies. Background report 4.1. Department of Science, Technology and Society, Utrecht University, The Netherlands

Wahlund B (2003) Regional bioenergy utilization in energy systems and impacts on CO_2 emission. PhD Dissertation, Royal Institute of Technology, Sweden

Wang M, Saricks C, Santini D (1999) Effects of fuel ethanol use on fuel-cycle energy and greenhouse gas emissions. Argonne National Laboratory, ANL/ESD-38; 1999. Available via DIALOG. http://www.ipd.anl.gov, Cited 12 December 2004

Wright CT, Pyrofogle PA, Stevens NA, Hess JA, Radtke CW (2006) Value of distributed preprocessing of biomass feedstock to a bioenergy industry. ASABE Paper No. 066151. ASABE, St. Joseph, MI

Part E
Agricultural Fit of Biomass Crops and Lifecycle Analysis and Criteria

Chapter 14
The Problem is the Solution: the Role of Biofuels in the Transition to a Regenerative Agriculture

Daniel G. De La Torre Ugarte and Chad C. Hellwinckel

14.1 Introduction

During the food price crisis of 2008, world commodity markets reached their highest nominal levels in 30 years, food prices skyrocketed, and shortages emerged in many regions of the world. After many years of having to deal with the negative consequences of chronically low prices, poorer nations suddenly had to deal with the opposite. The high prices and outright shortages led to food riots affecting more than 40 countries. Although not wholly caused by increased biofuel production, biofuels were seen as the major contributing cause.

With the global economic downturn starting in late 2008, both energy and food commodity prices have receded from their high-water mark, yet forecasts of declining conventional oil production suggest it is only a matter of time before oil prices rise once again, and renewed emphasis is put upon the search for sustainable energy production to fill the drop in conventional energy sources (IEA 2008).

Among the list of potential renewable fuels, biofuels have attracted great interest for many reasons. Ethanol is a liquid fuel easily integrated into the existing transportation infrastructure, its technological feasibility has been proven, and, until the recent price bubble, it was relatively cheap to produce. Yet looking one layer deeper, we must evaluate the sustainability of the agriculture upon which biofuels depends. And it is, in fact, the environmental footprint of agriculture as it is currently practiced that precludes biofuels from being the 'sustainable' replacement for depleting fossil fuel resources. If the non-sustainability of agriculture's environmental and socio-economic impacts are not addressed in the near future, nothing that depend upon agriculture will be sustainable, and this includes biofuel production.

D.G. De La Torre Ugarte and C.C. Hellwinckel
Department of Agricultural and Resource Economics, The University of Tennessee, 2621 Morgan Circle, 310 Morgan Hall, Knoxville, TN 37996-4519, USA
e-mail: danieltu@utk.edu

We believe that the proper course of action should not begin by asking industrial agriculture to provide a sustainable source of energy, but rather to first ask whether biofuels can help make agriculture sustainable.

'Regenerative agriculture' refers to systems of agriculture that mimic the dynamics found in nature that allow natural systems to maintain their own fertility, build soil, resist pests and diseases and be highly productive. Regenerative agriculture uses the natural dynamics of the ecosystem to construct agricultural systems that yield for human consumption. The Rodale Institute first used this term over 30 years ago to refer to systems that continually recreate the resources that they use. Other terms—such as natural farming, permaculture, agro-ecology, integrated agriculture, perennial polyculture, wholistic management, forest gardening, natural systems agriculture and sustainable agriculture—refer to similar principles. We use the term 'regenerative' because its direct meaning is more precise than the others, and avoid the term 'sustainable' agriculture because it now refers to a large array of methods and techniques that are quite removed from mimicking natural processes (such as monoculture irrigated organic lettuce in the desert, genetically engineered cotton, or farming at a 'tolerable' erosion rate). Regenerative agriculture is precise, because its methods regenerate the soil, the fertility, and the energy they use in semi-closed nutrient cycles, and by capturing, harvesting and reusing resources such as sunlight, rain, and nutrients that fall within the system's (farm's) boundary. Additionally, these systems act to regenerate communities and economies that have been bypassed by the large cycles of the global economy made possible by the availability of cheap energy.

If a biofuels policy is implemented within a larger agricultural policy matrix, then biofuels can be part of the solution rather than the problem. The demand for biofuels can be the catalyst that creates the right conditions for a transition that will allow a truly regenerative agriculture to take hold and grow. Yet if agriculture, in its current form, is used simply as part of the matrix of a larger energy production policy, then biofuels will continue to be seen as a major cause of an expanding food and environmental crisis. In this chapter, we discuss the role of biofuels in the recent price crisis, the short- and long-term drivers of agriculture, and how biofuels policy can help in transitioning modern agriculture to regenerative practices. If a biofuels policy is implemented as one component of a set of agricultural policies used simultaneously to address the long-term problems of agriculture, then the 'problem' of biofuels can become part of the solution to a sustainable food supply.

14.1.1 The Recent Price Bubble

Biofuel production has formed a small part of the liquid fuels supply for over 25 years, but it was the rapid oil price increase beginning in 2003 that really propelled their expansion and attracted the attention of policymakers. The Renewable Fuels Standard Program increased mandated quantities of US biofuels to 7.5 billion gallons by 2012, and the Energy Independence and Security Act of

2007 mandated the expansion of US biofuels to 36 billion gallons by 2022 along with the introduction of cellulosic ethanol. Although not wholly caused by increased biofuel demand, simultaneous increases in commodity prices and the resulting world food crisis have led many to question the future expansion of biofuels.

While the recent expansion of biofuels was an immediate cause of the recent food price bubble, other factors, both short- and long-term, contributed to setting up the situation for a rapid crisis.

One major short-term cause was the low stock-to-use ratios that resulted from recent US policy choices. In 2003—the year before the boom in ethanol began—ending stocks of grains fell below 25% for the first time since the early 1980s. The Food and Agriculture Organization (FAO) considers an ending stock of around 17% to be the lowest safest threshold—this is equivalent to about 2 months food at current demand levels (a 25% year end stock means there is enough grain for 3 months, or one-quarter of the year's demand). The trend is shown in Fig. 14.1. The decline was due largely to the result of the US government's decision in 1996 to eliminate the US inventory management policy that had prevailed since the 1930s.

The new policy was to leave it up to the market to define the appropriate level of inventories. The US is a big enough producer and exporter of a number of grains and oilseeds for its policies to have a significant impact on world inventory levels and world prices. The EU also abandoned stockholding as a tool of its agricultural policy. At the same time, the World Bank and International Monetary Fund (IMF) conditions attached to structural adjustment programs pushed developing countries to abandon local and regional reserves of grain as expensive and unnecessary.

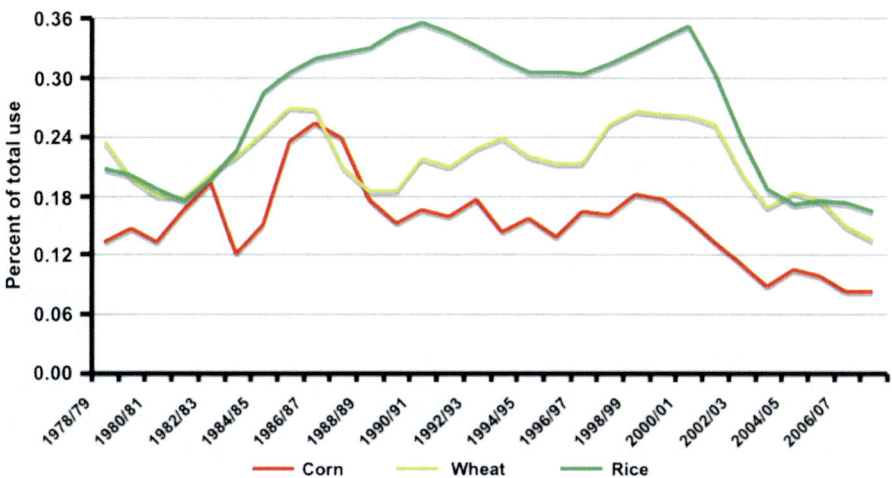

Fig. 14.1 Stock-to-use ratio trends

Advocates of this policy change insisted that globalization had reduced the need for local inventories because there would always be a supply somewhere in the world. In practice, agriculture started to operate the "just-in-time" inventory policy common in the manufacturing sector. This shift in thinking was reflected in the fact that while the stock-to-use ratio decreased significantly, the resulting increase in prices was not significant, nor was it in line with previous price responses to drops in stocks. Even when the stock-to-use ratios dropped to their lowest level in 25 years, prices stayed low (for a time). When demand for biofuels started to climb sharply in 2004, significant new pressure was put on the already low levels of stocks. The result was the significant price increases that started to affect world food prices from 2005.

Added to this, persistent drought and other weather-related problems in some of the major commodity production regions of the world, especially Australia, led to shortages of supply just when demand was taking off.

The crisis also showed how current policies do not include correctives to price spikes—in fact, quite the opposite. The particularly dramatic increases in rice prices provided a good illustration of how governments reacted to rising prices initially: panic buying, hoarding by some traders, and sudden shifts in trade policies all contributed to the problem. For instance, India banned all rice exports except of basmati varieties (export duties on which were raised sharply, and raised prices sharply for a number of poor neighboring importers, including Bangladesh, whose own harvest had been curtailed severely by a devastating cyclone. A number of importing countries lowered import tariffs on cereals, but to little avail.

The spike in agricultural prices has been further exacerbated by the simultaneous dramatic increase in oil prices. Some researchers would argue that this is not a coincidence, as higher oil prices are also contributing to higher biofuels use.

14.1.2 Long-Term Factors

While the acceleration in the production and use of biofuels at a time of historic low stock levels triggered the sudden jump in agricultural prices, factors that exacerbated the magnitude of the price increases were already in place. Although it may seem contradictory, longer-term policies of producing food as cheaply as possible has played a key role in the rapid price rises. Extremely low prices throughout the past 30 years has led to the concentration of agriculture in a handful of geographic areas of the world and, therefore, the dependence of the rest. With the encouragement of international organizations like the World Bank and the IMF, countries reduced protective policies, bought cheaply available grains and reduced investment in their own agriculture. Now they find themselves at the mercy of agricultural policies outside their control.

Prices are a critical factor affecting the long-term investment and performance of agriculture. The evolution of the nominal and real agricultural commodity price

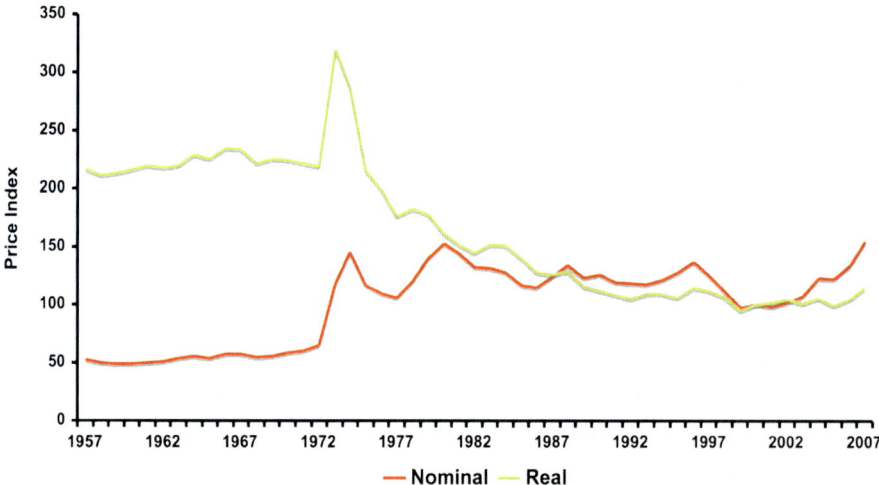

Fig. 14.2 Trend of agricultural real and nominal price indices

indexes is described in Fig. 14.2. The index of nominal prices refers to actual season average prices, while the index of real prices indicates that actual season average prices have been corrected for inflation. Agricultural commodity prices have remained flat at a fraction of the real prices that prevailed before the price increases of the early 1970s. Low real prices for agricultural commodities, illustrated in Fig. 14.2, have been possible because of the continued expansion in the agricultural productive capacity of a small number of countries that have the major share of the planet's agricultural resource base—Argentina, Australia, Brazil, Canada, the European Union, the United States, together with China and India. The expansion in productive capacity in these countries, driven by public investment in agriculture and agricultural research, has outpaced growth in demand for the last 30 years. This in turn has kept agricultural prices at historically low levels.

Agriculture's productive capacity is determined largely by four elements: natural endowment of resources; public and private investment in infrastructure, research and technology; and public policy towards agricultural producers (De La Torre Ugarte 2007; De La Torre Ugarte and Dellachiesa 2007). Since natural resources—land, climate, topography, water—are largely fixed, the level of public investment in infrastructure, research and development, and support to farmers are all indicators of how the productive capacity of the sector is likely to evolve.

Decades of low prices have discouraged investment, both public and private, in the agriculture of developing countries. Low prices have also limited the ability of the sector to generate adequate income and economic activity for the 2.5 billion people worldwide that depend on agriculture to survive. Depressed global prices have undermined production and markets at every level, local through global. For example, cheap rice exports from Japan and Thailand into West Africa have depressed not just local rice production, but also production of more traditional staple foods, such as millet. Governments in pursuit of cheap food for largely urban markets have encouraged this trend, to the detriment of local food production and rural development.

14.2 The Socio-Economic Impacts of Industrial Agriculture

Although the last 30 years of industrial agricultural policy has produced food calories at record low prices, it has done little to alleviate poverty. The global socio-economic impacts of industrial agriculture are probably far too complex to be addressed fully here, but perhaps two of the biggest open challenges are its performance in ensuring a growing degree of food security, and its role in poverty reduction.

The 2008 World Development Report (World Bank 2007) provides a starting point to describe the performance of the agricultural sector in terms of poverty reduction and food security. The report indicates that 3 billion people out of the 5.5 billion people that populate the developing world live in rural areas. Of the 3 billion, 2.5 billion people are in households whose livelihoods depend on agriculture; and of these 1.5 billion are in smallholder households. It is these rural smallholder families that industrial agricultural policies are leaving behind. According to the latest figures on world rural poverty, the 2.1 billion people that survive on less than US $2 a day live in rural areas, and 880 million of those live on less than $1 a day. In summary, three of every four poor people in developing countries live in rural areas (World Bank 2007).

The World Bank Report also states that, as economies develop and the population becomes more urban, rural poverty still persists. Studies evaluating policies to reduce rural poverty show that 81% of the world's reduction in rural poverty during the years 1993 though 2002 was due to improved conditions in the rural areas themselves, whereas migration accounted for only 19% of the reduction (World Bank 2007). Furthermore, in the poorest regions investments in agriculture yielded the most significant marginal benefit in terms of lifting people out of poverty.

Turning to the subject of undernourishment, the 2006 State of Food Insecurity in the World Report (FAO 2006) indicated that in the period 2001–2003 there were 854 million undernourished people in the world; of those, 820 million were in developing countries. Ten years earlier this same number in developing countries was 823 million, thus the reduction over a decade was extremely marginal. In 1990, 65% of undernourished people were under the $1 a day poverty line, while that share increased to almost 70% in 2002.

The above statements summarize the status of poverty and food security prevailing at the time of the biofuels expansion and price crisis. At the same time they reinforce the role of agriculture in reducing poverty and improving food security. Cheap, industrially produced grains have done nothing to improve the lives, health or nutrition of the poorest one-third of humanity living in rural areas. Their lives will not be improved by policies aimed at continually lowering the relative price of the commodities. To reach these people, more innovative thinking is necessary.

But even if industrial agriculture could promise to lift the poor out of poverty, it would still not be a prudent path to walk. Agriculture is soon approaching some limitations that will not allow expansion of the industrial model. To understand these approaching limitations, and to see the appropriate role of biofuels in addressing them, we must look more closely at the sustainability problem of agriculture.

14.3 The Environmental Footprint of Industrial Agriculture

Over the past century, innovations in mechanization, fertilizers, chemicals, and seeds have created a system where food could be produced cheaply and abundantly in a few ideal geographic locations. Yet the model that has been so successful at increasing production and feeding a growing population is not sustainable due to (1) the rapid loss of our soils, (2) agriculture's dependence on fossil fuels, and (3) agriculture's emissions of greenhouse gases (GHGs).

14.3.1 Soil Loss

Every year, 75 billion metric tons of soil are eroded from the Earth's agricultural lands, and 30 million acres are abandoned due to over-exhaustion of the soil (Myers 1993; Faeth and Crosson 1994). This is equivalent to losing an area the size of Ohio every year. Erosion has been a problem that has followed cultivation for 10,000 years. Its slow effects are evident in the lands surrounding fallen civilizations such as in the Tigris Euphrates valley, Israel, Greece or the hills of Italy. Over time, agriculture has led to the loss of nearly one-third of global arable land, with much of this loss taking place within the past 40 years (Montgomery 2007a). Green revolution methods of mechanization and the use of fertilizers have sped the rate of erosion in many regions, and have led to the abandonment of traditional practices, such as integrated crop-animal systems or polyculture plantings, that had slowed erosion and enabled traditional systems to function for centuries (IFPRI 2002; King 1911).

Soils are a depletable resource that form over thousands of years. It is estimated that it takes an average of 500 years for one inch (2.54 cm) soil to form (Troeh 2003). Modern agriculture is depleting soils at a rate 1–2 magnitudes faster than that with which they are formed (Montgomery 2007b). Once soil is eroded, it cannot be recreated easily or quickly. Our use of soils can be thought of as spending the accumulated capital of millennia—not unlike our use of fossil fuels. Whereas, in the past, if one culture exhausted its soils and declined, civilization could always reemerge in newly settled fertile areas, today, with 3.7 billion acres under cultivation, there are few remaining virgin soils. If we continue this historic trend and deplete our soils, we will be faced with an increasingly hungry world, even without the added burden of biofuels production.

14.3.2 Fossil Energy Dependence

Agriculture, like all other industries over the past century, has taken great advantage of the extraction and refining of plentiful, energy-dense fossil fuels. Today, industrial agriculture has evolved into a net energy user for the first time in its 10,000-year history, where, instead of being a means of converting free solar energy into metabolizable energy, it is now a means of transforming finite fossil energy into metabolizable energy. The system has allowed for the cheap production of plentiful food to feed a growing population but, as the total annual quantity of oil physically capable of being extracted from the Earth begins to decline over the next several years, agriculture may find itself dependent upon a scarce and expensive resource.

Some energy experts believe that global oil production has indeed peaked in August of 2005 at 84 million barrels per day and will decline annually from now into the future regardless of how high the price of oil climbs (Koppelaar 2008; Simmons 2007). Although demand for, and therefore the price, of oil has fallen during the current recession, depleting oil resources will likely reemerge as a problem when the economy begins to grow again.

To meet the needs of a growing population, modern US food system uses 10.25 quads fossil energy inputs, which represents 10% of US annual fossil fuel consumption. Industrialization of agriculture has, for the first time in history, led to the situation where agriculture actually uses more energy than it creates, with 7.3 units of energy going to grow, process, transport, store and cook 1 unit of metabolizable energy (Heller and Keoleian 2000). This energy deficit of agriculture is an historic anomaly. Up until the past 50 years, agriculture has always yielded more energy than it uses (Green 1978). Indeed, this has been the whole point of agriculture. By producing more energy than the farmer needs, others were freed from food production, and civilizations were built upon the small positive gains in energy from agriculture. It has been estimated that traditional Egyptian agriculture had an energy return on energy invested (EROEI) ratio of 1.8. The EROEI of US agriculture in 1920 has been estimated at 3.1, but by the 1970s had fallen to 0.7 (Gifford 1976). Add to this the energy required to move, process, package, deliver and cook food in the modern food economy, and we arrive at an EROEI of 0.14 (Heller and Keoleian 2000), indicating that agriculture has lost its traditional role as an energy production system and become simply another user of fossil fuels.

Whereas historically, the foundation of civilization rested upon the consistent and ever falling solar radiation, now it rests upon the annual extraction of finite fossil fuels. One solution to the situation is to find another energy subsidy for our energy intense agriculture, such as wind or solar. Yet, when comparing the EROEI ratios of alternative fuels, it becomes apparent what an incredibly good deal we have had with oil (Fig. 14.3). We are still running our economies off the large oil discoveries of the 1950s and 1960s with EROEI ratios of 50+ (Hall et al. 1986). Alternative fuels will likely have an increasing role in meeting the energy needs of the larger economy, but to believe agriculture can continue to function under the current energy balance is folly. It is imperative that agriculture return to a more balanced energy ratio over the next century.

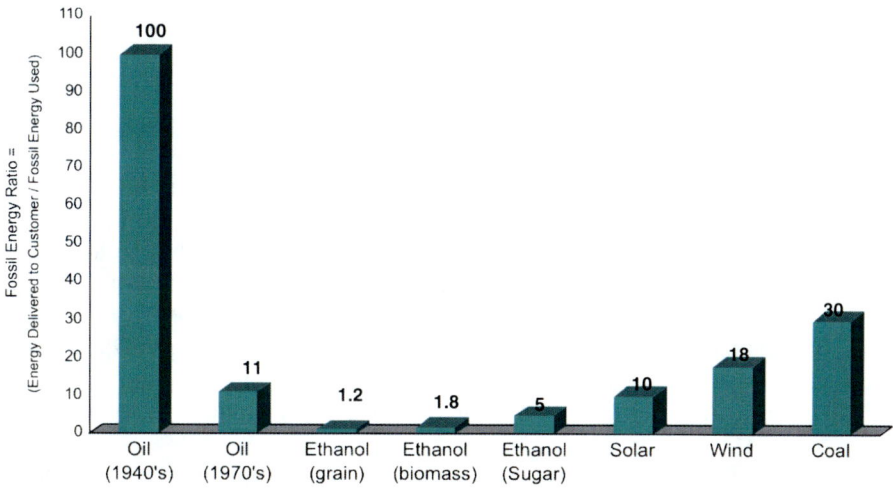

Fig. 14.3 Energy returned on energy invested (Bender 2002; Wang et al. 1999)

14.3.3 Greenhouse Gas Emissions

Data from the fourth assessment report of the Intergovernmental Panel on Climate Change (IPCC 2007) suggest that, while anthropogenic GHG emissions from forestry and agriculture have not grown as fast as emissions from the energy supply and transportation sector, they account for 30.9% of total emissions—second only to the energy sector. Most of the emissions from agriculture are a result of nitrogen used as fertilizer, and enteric methane from ruminants' digestion. While most current and proposed climate change legislation exempts emissions from agriculture, in the future agriculture could lose that exemption. For example, in Australia, where very aggressive GHG reductions are beginning, agriculture is currently exempt, but only until 2015 when the subject will be revisited.

Unfortunately, agriculture is not exempt from the ramifications of climate change. The IPPC also reports that agriculture is feeling the effects of more severe and frequent extreme weather events such as droughts and floods. Furthermore, models predict that, as climate change continues, "subsistence sectors at low latitudes" will experience the most severe variability in weather events (IPPC 2007). By 2100, the annual costs to agriculture of climate change could reach US $1 trillion for the US alone and halt yield increases completely (Ackerman and Stanton 2008).

Agriculture could also play a significant part in abating climate change through the soil's ability to sequester carbon in organic matter. Through appropriate land use changes, agriculture has the potential to sequester 207 million metric tons per year in US soils alone, or about 10% of our total annual emissions (Lal et al. 1998) but, as the climate warms, the soil's ability to sequester carbon may decrease (Davidson and Janssens 2006). Future agricultural policies must steer agriculture towards reducing climate change gases while simultaneously building resilience to the effects of climate change.

14.4 Future Agricultural Policy: What is Needed?

As global production of fossil fuels peaks and begins to decline in the coming years, it becomes evident that agriculture, as currently practiced, is not sustainable. Within the next 50–100 years, the practices of agriculture must transform so that they (1) stop erosion and begin to regenerate soil; (2) reverse the energy ratio and once again become a net source of energy; and (3) meet human food needs. In other words, agriculture must regenerate fertility, capture solar energy, and produce in abundance

In setting biofuels policy, meeting these future objectives of agricultural transition must be the primary guide. We must ask, how can biofuels help facilitate the transition of agriculture to a sustainable system that will adequately feed people, build soil, and meet its own energy needs? Simply viewing agriculture as a potential source for meeting the greater economy's fuel demand will not guarantee the necessary transition of agriculture, and could even exacerbate the destruction of our soils, increase input consumption of agriculture, and lead to food shortages. While humans may be able to get along without liquid fuels for transportation, we cannot survive without adequate food production. If appropriate, biofuels could be a vital part of long-term agricultural policy, but agriculture should not simply become a part of energy policy.

So, how can biofuels policy help the transition of agriculture to sustainability over the next 50–100 years? Most obviously, the production of biofuels could help to correct the negative energy deficit of industrial agriculture by producing some of its own energy needs. But the energy use of agriculture is so large that any significant contribution of biofuels to filling this demand would be infeasible under current agriculture models. The relevant question in setting biofuels policy is not what is the potential contribution of biofuels to reduce the dependence of fossil fuels, but rather what is the optimal level of biofuels production that would encourage the transition of agriculture to a system that enhances food security, reduces poverty, and improves the environmental footprint of the sector?

14.4.1 How do Agricultural Prices Impact Food Security, and Environmental Performance?

To understand the role of biofuels, we must first understand how commodity prices affect food security and environmental performance. Although agricultural prices are not the only, and perhaps not the most important, determinant of food security

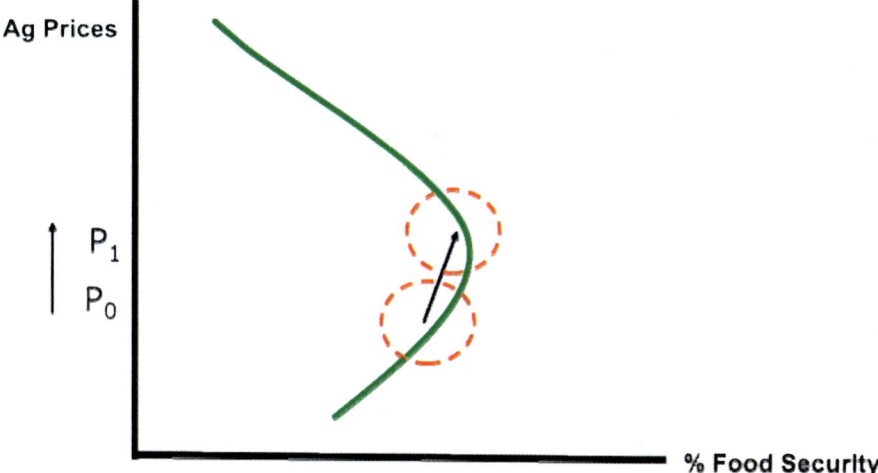

Fig. 14.4 Food security and agricultural prices. Movement indicates an increase in food security as agricultural prices increase

and/or environmental performance, they are, without doubt, the shortest link between increases in demand—such as biofuels use—and food security and environmental performance.

Let us start by looking at a simple relationship between agricultural prices and food security. Consider the significance of the agricultural sector in the economy of many developing countries, and at the same time the role of food in sustaining healthy and productive human lives; this implies that agricultural prices are linked to two effects. (1) As agricultural prices increase, incomes in rural areas are also likely to increase, and consequently the degree of food security would improve as 80% of the food-insecure population is in rural areas. (2) As agricultural prices increase, the cost of food would also increase, so the degree of food security is likely to decrease, especially in urban areas. In Fig. 14.4 this relationship is described by the solid curve, and it implies an increasing level of food security as prices increase—everything else remaining constant—from very low levels to the inflection point. This behavior is based on the fact that the livelihood of 2.5 billion households is tied to agricultural economic activity. Consequently, as prices increase, their incomes increase and farmers have an incentive to produce more for the market. The degree of food security increases as the rural population produce food not only for themselves but also for urban areas, and the increase in rural income increases investments to access to food markets for rural households.

The inflection point marks the price level at which the gains in rural areas do not compensate for the losses in food security among the urban poor as agricultural prices continue to increase. Beyond this point, further increases in price would result in losses in the degree of food security, as the high prices restrict access to food for the poor in urban areas. Additionally, the rural poor may be hurt as higher prices give incentives for land consolidation for the expansion of industrially produced food. Investors seeking short-term profits would apply the typical technology-intensive, labor-minimizing methods that produce plenty in the short term at the expense of the long-term productivity of the land. This situation would force small farmers off the land and into urban areas, exacerbating the burden on already tight food supplies. The severity of the inflection point and the price level at which it occurs will be determined by the overall vulnerability of the population to higher food prices (one would expect that higher income countries would show an inflection point at a higher price level, and also countries with a larger income disparity may show a more dramatic inflection slope). The location of the inflection point would depend on the importance of the agricultural sector in a given country, and the distribution of the food insecure population. The larger the agricultural sector, the higher the inflection point; conversely, the less significant the agricultural sector, the lower the inflection point will be. This means that, at the extremes, a country where agriculture is the only productive sector, the curve will be only positively sloped, while in a country with no agricultural resources, the curve would be negatively sloped.

However, the shape of this relationship is not the only significant factor: the other key element is the position of the curve. The specific "location" of this solid curve is determined by the structural characteristics of the agricultural and food production, distribution, and consumption system. That means that one can hypothesize curves to the left and right of the solid curve; the ones at the left would indicate that, at the same level of prices, the degree of food security is lower, while the ones at the right would indicate higher levels of food security at the same level of prices. Obviously, it is better to be located further to the right.

Moving to the relationship of agricultural prices and environmental performance, or in this case cost, one could consider the solid line presented in Fig. 14.5. The two-tailed solid line indicates that, for the tail above the inflexion point, as agricultural prices increase the environmental costs or the environmental footprint of the agricultural system increase. This attempts to represent an agricultural system that relies on the intensive use of fossil-fuel-based inputs and monoculture. But this figure also indicates that, as agricultural prices increase, there would be significant pressure to expand agricultural activity—based on the characteristics described above—into environmentally sensitive areas like tropical forests, wetlands, conservation reserves, or steep slopes, which would result in a higher environmental footprint. This can be illustrated by the expansion of agricultural activities into the tropical forests of Southeast Asia and Brazil. Not that agricultural prices are the major cause or contributor to deforestation but, everything else being held constant, an increase in agricultural prices would add pressure to expand agricultural activities into tropical forests.

On the other hand, for the tail below the inflexion point, as prices decrease, environmental costs also increase. In this situation, prices are low enough that farmers forgo investments in soil improvements and, in effect, 'mine' the soil to get much needed short-term cash. The dramatic deforestation in Haiti illustrates this extreme situation. Again one can hypothesize a series of curves to the right and left of the solid line, with similar implications to the ones derived above. But one could

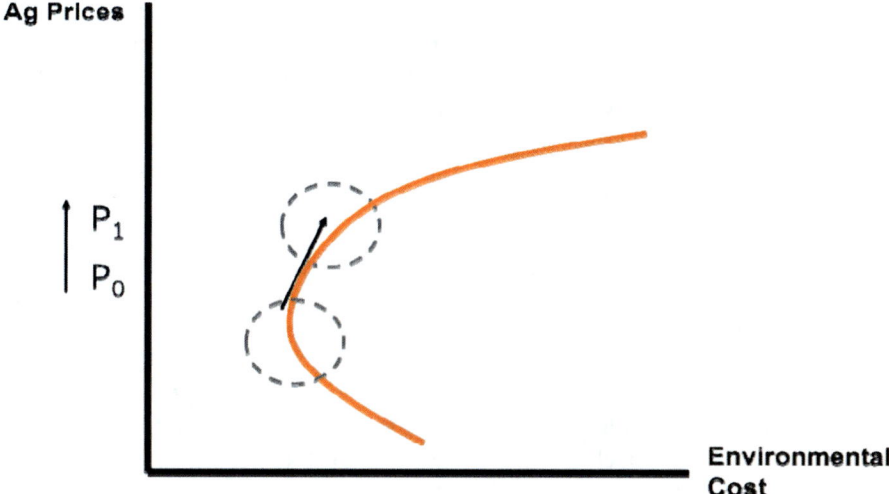

Fig. 14.5 Environmental cost and agricultural prices. Movement indicates a greater environmental cost as agricultural prices rise

also hypothesize curves of different shape, even reversing the initial shape of the solid line, all of them representing different relationships between agricultural activity and practices with the environment. For instance, if the predominant system were a regenerative system, it is possible that, as prices increase, the availability of additional economic resources allow a higher investment in practices with higher private cost, which would result in the regeneration of the environment in terms of soil productivity, carbon sequestration, water use and quality, among other factors.

14.4.2 The Role of Biofuels

When addressing the impacts of the expansion of biofuels on food prices and poverty, most studies assume the replication of the current agricultural system, along with its consequent negative impacts on food security, poverty reduction, and environmental performance. This assumption leads to the common conclusion that biofuels will exacerbate the failures of the current agricultural and food production, distribution and consumption systems.

There is no question that the increased utilization of biofuels implies an increase in the demand on agricultural (and forest) resources, and consequently would result in higher agricultural prices. Under the prevalent agricultural system, higher agricultural prices will trigger higher investments in agriculture, and (for the medium-term) increase production and reduce prices. But this increase in production will be the result of investing in an agricultural system based on the increased use of fossil fuel inputs and consumption of precious soil resources. It is not enough to say that the expansion of biofuels could trigger a massive investment in the agricultural sector. What matters is the type of investment and whether the new investment is expanding the current model of agriculture or transforming it. To meet the long-term goals of food security and environmental performance in a manner that does not ignore the long-term costs, it is necessary to guide the new investment.

Biofuel demand should be used to raise long-term prices of all agricultural products so that necessary investments can be made in agricultural systems that (1) stop erosion and begin to regenerate soil; (2) reverse the energy ratio so that agriculture once again becomes a net source of energy; and (3) adequately meet human food needs. Biofuels demand can accomplish this by increasing the total demand for agricultural land, which, in turn leads to higher prices across the commodity spectrum.

Opponents of higher commodity prices argue that although net sellers of commodities may benefit, net buyers of food will be hurt. A recent study by Aksoy and Isik-Kikmelik (2008) concludes that, even though there are more net buyers than sellers, most net buyers are only marginal net buyers. Additionally, net buyers have a higher income than net sellers. Therefore, small increases in food prices will have a smaller marginal negative impact on net buyers and a larger marginal positive impact on net sellers.

The welfare of net buyers may also be improved. Higher prices, combined with the introduction of new methods, could improve the diversity and nutritional quality of food calories consumed by net buyers. Additionally, the higher prices could present the opportunity to invest in urban agriculture, where a significant portion of 'net buyers' food can be produced in small plots within the city. Havana invested heavily in urban agriculture after the collapse of the Soviet Union and the rapid decline in imported fossil fuels. Today, 50% of the fruit and vegetables consumed by Havana's residents are grown within the city limits (Quinn 2006).

If prices are sustained at this higher level, then investments can be made in land improvements and extension that could work toward meeting the three policy imperatives. The investments would then act to increase productivity, decrease input use, and improve the regenerative capacity of the soil.

14.4.3 Transformative Investments in a New Agriculture

The demand for biofuels can stimulate the right conditions for a new agricultural policy to be implemented, namely, higher sustained prices. Next, investments must be made to transform agriculture from its current non-sustainability into practices that regenerate fertility, capture solar energy, and produce in abundance. For this to occur, transformative investments must be made in (1) creating regenerative practices appropriate for each ecosystem, (2) extension education to prove the value of regenerative practices to farmers, and (3) infrastructure to help farmers capture more of the value of their goods.

14.4.3.1 Establishing Regenerative Practices

To address the three policy criteria and improve food security, investments must be made in practices that transform the productive capability of agricultural lands. There is increasing evidence indicating that there are practices that are not input intensive yet are more productive than industrial agriculture (Altieri 2008; Pretty 2005). Such practices are used by small landholders who are capable of managing more intensive and complex systems that rely on the integration of crop–animal–human functions, the use of perennial species, and the growing of multiple crops in the same field (Gitau et al. 2009). Many of these practices are, in fact, traditional cultural land-use practices, but others are newly forged systems. They fall under the general categories of agro-ecology, integrated agriculture, permaculture, perennial polyculture, wholistic management, and natural systems agriculture.

One of the most promising and most easily scalable to improve the health and productivity of large amounts of land is the use of intensive grazing, which consists of dividing a pasture into several small fields and closely managing the time livestock are allowed to graze each field (Savory 1998; Dagget 2000). Evidence indicates that by finely managing when herbivores are placed upon a field to graze, the total primary productivity of the landscape can increase dramatically (Brundage and Petersen 1952; Salatin 1996). Grassland productivity can be augmented through Keyline plowing, which is a simple method of widely spaced deep chisel plowing with the contour of hillsides. This acts to shed water away from the eroding valleys and out to the water-poor ridges while increasing ground absorption (Yeomans 1964). Intensive grazing with Keyline tilling could increase grass productivity, allow more animals to feed on less land, build soil carbon and reduce erosion. Furthermore, increased grassland productivity and the regenerative capabilities of the grasses would lift a considerable burden off feedlot production, which, in turn, would reduce the demand for input-intense row crops. Intensive grazing and Keyline design are practices that have already been developed and are spreading on their own. Newly immigrated dairy farmers from New Zealand, who brought intensive grazing techniques to the US are expanding in Georgia, while traditional confined operations are going out of business (Schupska 2007). It may not take much of a push for these practices to become widespread.

Another proven practice is the traditional highland Vietnamese production system (VAC) that integrates aquaculture, garden, livestock, and forest agriculture within small plots; VAC could serve as a template for other tropical regions (FAO 2001). The VAC system illustrates a key principle of regenerative practices, which is the use of the waste stream of one component to feed another component. Food scraps are dumped into the pond to feed the fish, pond growth is removed and fed to pigs, and pig manure is used to fertilize the garden and fruit trees. In this manner, regenerative systems conserve energy and maintain their own fertility. The VAC system also makes full use of vertical space by planting vegetables and fruit bushes below fruit trees. Riparian zones, such as these managed ponds, are known to be the most productive ecosystems on Earth, yielding more net primary productivity per unit of area than any other.

Other regenerative systems already in use include the Zai methods used in the Sahel of Africa (FAO 2008), the no-till rice–legume–rye system developed by Masanobu Fukuoka in Japan (Fukuoka 1978), and the edible forest system indigenous to the Kerala region of India. Efforts are also underway to develop new regenerative systems. The Land Institute, located in central Kansas, has been working for over 25 years to perennialize grain agriculture (Jackson 1980). By breeding in rhizome roots that over-winter, the soil would not need to be tilled and planted annually. Midwestern agriculture would resemble the native prairie ecosystem in form and function, but produce grains for human consumption. Planting many species in the same field would allow the land to provide its own fertility while resisting pests and diseases (Soule and Piper 1992).

Successful regenerative systems will differ depending upon local ecosystem capabilities and constraints. By studying the fundamental elements of already existing systems, new practices unique to individual ecosystems can be developed fairly rapidly. Some of these elements or principles, such as the integration of components, stacking of functions, the use of redundancy in the system, and the use of natural systems as a model for our constructed systems, are drawn out in the permaculture literature (Mollison 1990; Holmgren 2002). Research investments should be made in bringing along locally adapted regenerative systems.

14.4.3.2 Extension Education

Although there is great potential in the widespread application of existing successful regenerative systems, they are not being adopted on a large scale. It takes time, initial resources, and knowledge to transform land into regenerative systems. One primary reason for the non-adoption of regenerative systems is the long-term trend in low commodity prices, which gives little slack to farmers for making long-term investments to their land; biofuels policy could help this by increasing the long-term returns. However, without massive efforts in the realm of extension education, we may not see widespread applications.

Development of regenerative systems must go hand-in-hand with extension education. Demonstration farms should be established within traveling distance of every farmer that demonstrate and test locally adapted regenerative practices. By seeing the new practices in action, farmers will more likely adopt them, leading to further adoption by neighbors. The old adage 'seeing is believing' holds very true for the world's farmers.

14.4.3.3 Infrastructure Investments

Agricultural infrastructure investment has typically meant more roads, ports and large storage facilities. But, in determining future infrastructure needs, we must make investments in line with future energy decline and the needs of successful agricultural systems in such an environment. Investments in electric railways and waterway transportation may be more in line with declining traditional liquid transportation fuels. In addition, paving or graveling smaller roadways will increase access to populous markets that may be inaccessible during certain times of the year.

These large-scale national investments will be important, but possibly the most important infrastructure needs are at the local and even farm level. Regional farmers' cooperatives should be created with locally shared storage and transportation infrastructure to enable farmers to capture more of the market share. Investments in small-scale appropriate technologies, like simple bicycles, can have significant effects upon poorer farmer's profitability (Kwibuka 2008). Micro-processing technologies, such as canning equipment or oil presses, could enable farmers to process their harvest into higher value and more monetarily dense commodities close to home. At the farm level, the transition to a locally adapted regenerative agriculture may entail the upfront construction of ponds and swales, or the planting of orchards. Although higher prices will facilitate farmers to make these investments themselves, access to capital through loans is a key component of creating regenerative systems, and should be viewed as a long-term investment in agricultural infrastructure.

14.4.4 Structural Shift

Higher sustained commodity prices combined with simultaneous investments in regenerative agriculture will have the effect of moving food security from point A to point C on Fig. 14.6. The solid curve represents industrial agriculture and the dashed curve represents regenerative agriculture. Note that without transformative

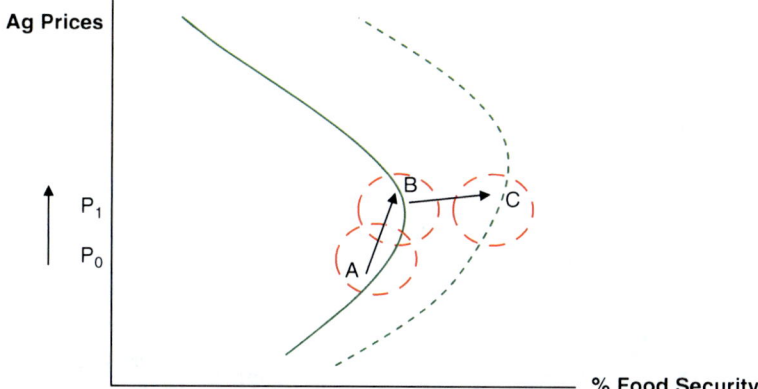

Fig. 14.6 Food security curve and its outward movement as a result of investments in regenerative agriculture and distribution infrastructure (for explanation see text)

investments in regenerative agriculture, higher prices will not increase food security significantly. But the investments will act to move the food security curve outward and increasing food security at all price levels relative to industrial agriculture. At point C, farmers will have used the higher prices to invest in practices that improve their soils, provide a significant portion of their own fertility, and are high yielding and diverse. Government investments that lead to the propagation of regenerative systems and infrastructure to support them have also enabled farmers to capture a larger share of the market price. Many urban residents will also see an increase in food security as their food supply is more local and diverse, and not as susceptible to short-term agriculture or energy policies of other countries.

Environmental costs would decrease as a result of higher sustained prices and simultaneous investments in regenerative agriculture. This is illustrated as a move from point A on to point B on Fig. 14.7, with the solid line representing industrial agriculture and the dashed lines representing potential regenerative systems. Although we are certain that regenerative agriculture would be to the left, the shape could vary. Due to regenerative agriculture's diversity and lessoned dependence on inputs, its environmental performance will likely be more resilient to price changes. This is the case at point B, where, once the structural shift has occurred, there is little change in environmental performance across prices. However, we can imagine a system of agriculture where the tails of the curve could actually curve backwards to where higher prices could mean more investments in soil regeneration, and very low prices could mean a withdrawal from production and a return of lands to their wild states. Musing aside, the likely flattening of the curve, as with point B, is a necessary and attainable goal of transformative investments.

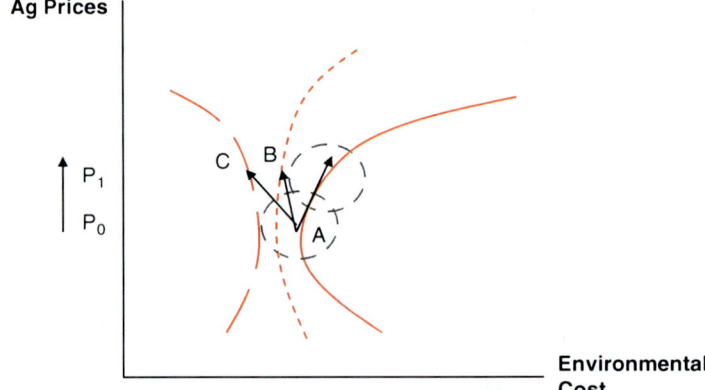

Fig. 14.7 Environmental costs curves. Movements depict potential outcomes of the transition to regenerative systems

14.5 Final Remarks

Our intent was to illustrate the non-sustainability of agriculture as it is currently practiced and the absolute imperative of coordinating international agricultural and energy policies to transition agriculture to regenerative practices. We believe biofuels could play a key role in this transition by increasing long-term prices of all commodities and therefore creating the opportunity for farmers to invest in regenerative practices. Such a policy must be implemented simultaneously with national and international investments in research, extension and infrastructure of regenerative agriculture.

References

Ackerman F, Stanton EA (2008) What we'll pay if global warming continues unchecked. Natural Resources Defense Council. May

Aksoy A, Isik-Dikmelik A (2008) Are low food prices pro-poor? net food buyers and sellers in low-income countries. World Bank—Policy Research Working Paper, vol 4642

Altieri M (2008) Small farms as a planetary ecological asset: five key reasons why we should support the revitalization of small farms in the global south. Food First. Institute for Food and Development Policy, Oakland, CA

Bender MH (2002) Energy in agriculture: lessons from the Sunshine Farm project. In: Proceedings of the Third Biennial International Workshop, Advances in Energy Studies: Reconsidering the Importance of Energy, 24–28 September 2002, Porto Venere, Italy

Brundage AL, Petersen WE (1952) A comparison between daily rotational grazing and continuous grazing. J Dairy Sci 35:623–630

Dagget D (2000) Beyond the rangeland conflict: toward a West that works. University of Nevada Press

Davidson EA, Janssens IA (2006) Temperature sensitivity of soil carbon decomposition and feedbacks to climate change. Nature 440:165–173

De La Torre Ugarte DE (2007) The contributions and challenges of supply management in a new institutional agricultural trade framework. In: EcoFair trade dialogue. Heinrich Boll Foundation. Discussion Paper 6, March

De La Torre Ugarte DE, Dellachiesa A (2007) Advancing the agricultural trade agenda: beyond subsidies. Georgetown Int Environ Law Rev 19:775–796

Faeth P, Crosson P (1994) Building the case for sustainable agriculture. Environment 36(1):16–20

FAO (2001) Integrated agriculture–aquaculture: a primer. Series title: FAO Fisheries Technical Paper, T407. FAO, Rome

FAO (2006) The state of food insecurity in the World 2006: eradicating world hunger—taking stock ten years after the World Food Summit. FAO, Rome

FAO (2008) Climate, climate change and agropastoral practices in the Sahel region. Natural Resources Management and Environment Department Publication, September. FAO, Rome

Fukuoka M (1978) One straw revolution. Rodale, Emmaus, PA

Gifford RM (1976) An overview of fuel used for crops and national agricultural systems. Search 7:412–417

Gitau T, Gitau MW, Waltner-Toews D (2009) Integrated assessment of health and sustainability of agroecosystems. CRC/Taylor & Francis, Boca Raton

Green M (1978) Eating oil: energy use in food production. Westview, Boulder

Hall CAS, Cleveland CJ, Kaufmann R (1986) Energy and resource quality: the ecology of the economic process. Wiley, New York
Heller M, Keoleian G (2000) Life-cycle based sustainability indicators for assessment of the US food system. Center for Sustainable Systems. University of Michigan, Ann Arbor, MI
Holmgren D (2002) Permaculture: principles and pathways beyond sustainability. Holmgren Design Services, Hepburn, Australia
IEA (2008) World Energy Outlook 2008. International Energy Agency, December
IFPRI (2002) Green Revolution: curse or blessing? Issue Brief 11. International Food Policy Research Institute, Washington, DC
IPCC (2007) Climate Change 2007: Impacts, adaptation, and vulnerability. Contribution of Working Group II to the Third Assessment Report of the Intergovernmental Panel on Climate Change. Cambridge University Press, Cambridge, UK
Jackson W (1980) New roots for agriculture. University of Nebraska Press
Kwibuka E (2008) Coffee farmers get bikes on credit. The New Times, Rwanda, 15 March, http://projectrwanda.org/news/coffee-farmers-get-bikes-on-credit. Cited 30 March 2009
King FH (1911) Farmers of forty centuries: permanent agriculture in China, Korea, and Japan. Organic Gardening Press, Emmaus, PA
Koppelaar R (2008) Monthly report. OilWatch Monthly. http://www.peakoil.nl/wp-content/uploads/2008/12/2008_december_oilwatch_monthly.pdf. Cited 25 March 2009
Lal R, Kimble JM, Follett RF, Cole CV (1998) The potential of US cropland to sequester carbon and mitigate the greenhouse effect. Ann Arbor Press, Chelsea
Mollison B (1990) Permaculture: a designers' manual. Island Covelo, CA
Montgomery DR (2007a) Soil erosion and agricultural sustainability. Proc Natl Acad Sci USA 104:13268–13272
Montgomery DR (2007b) Dirt: the erosion of civilizations. University of California Press
Myers N (1993) Gaia: an atlas of planet management. Anchor/Doubleday, Garden City, NY
Pretty J (2005) The Earthscan reader in sustainable agriculture. Earthscan, UK
Quinn M (2006) The power of community: how Cuba Survived peak oil. Permaculture Activist Issue #59, Spring
Salatin J (1996) Salad bar beef. Polyface, Swoope, VA
Savory A (1998) Holistic management: a new framework for decision making. Island, Covelo
Schupska S (2007) New Zealand partnership may boost milk in Georgia. Georgia Face, University of Georgia College of Agriculture. 1 Nov http://georgiafaces.caes.uga.edu/storypage.cfm?storyid=3264. Cited 20 March 2009
Simmons MR (2007) Another nail in the coffin of the case against peak oil. White paper, http://www.simmonsco-intl.com. Cited 25 March 2009
Soule JD, Piper JK (1992) Farming in nature's image: an ecological approach to agriculture. Island, Covelo
Troeh FR, Hobbs JA, Donahue RL (2003) Soil and water conservation for productivity and environmental protection, 4th edn. Prentice Hall, Englewood Cliffs, NJ
Wang MC, Saricks D, Santini D (1999) Effects of fuel ethanol use on fuel-cycle energy and greenhouse gas emissions. US Department of Energy, Argonne National Laboratory, Center for Transportation Research, Argonne, IL
World Bank (2007) 2008 World Development Report, Agriculture Development. Washington, DC
Yeomans PA (1964) Water for every farm: Yeomans keyline plan. CreateSpace, https://www.createspace.com

Chapter 15
Life-Cycle Analysis of Biofuels

Michael Wang

15.1 Introduction

Biofuels that are produced from different bio-based feedstocks via different production technology platforms are promoted for their benefits of reducing fossil energy use and greenhouse gas (GHG) emissions, among other environmental benefits relative to petroleum fuels. Such benefits have long been assessed on a life-cycle basis to fully take into account energy consumption and emissions from all stages of biofuel life cycles.

Life-cycle analysis (LCA) of biofuels has been applied widely to examine biofuel energy and environmental effects since 2000, and the methodologies used have advanced in the past 20 years. In the early years of examining biofuels, the so-called energy balance (energy contained in biofuels minus the fossil energy consumed to make them through the whole life cycle) was estimated for biofuels, especially for corn-based ethanol (Chambers et al. 1979; Pimentel and Patzek 2005). LCA models have been developed since the early 1990s, and detailed LCAs have been conducted to examine energy and emission effects of biofuels, especially corn-based and cellulosic ethanol in comparison to petroleum fuels (Delucchi 1991; Wang 1996; Wang et al. 1997). Most recently, the debate on biofuel energy and environmental effects has focused on expansion of the system boundary of LCAs by including indirect effects at a global scale and by including other environmental sustainability issues such as water consumption, biodiversity, and soil erosion, among many other issues (Searchinger et al. 2008; Kim et al. 2009; Wu et al. 2009).

This chapter presents the stages, results, and key issues of biofuel LCAs.

M. Wang
Center for Transportation Research, Argonne National Laboratory, 9700 South Cass Avenue, Argonne, IL 60439, USA
e-mail: mqwang@anl.gov

15.2 Potential Biofuel Production Pathways

At present, the two major biofuels that are produced worldwide are ethanol and biodiesel. Ethanol is used in spark-ignition engines (or gasoline engines) from low-level blends such as E10 (10% ethanol and 90% gasoline by volume) to high-level blends such as E85 in the US and E100 (pure ethanol) in Brazil. While low-level ethanol blends can be used in gasoline vehicles without vehicle modifications, the use of high-level blends requires modifying gasoline vehicles to make so-called flexible-fuel vehicles (FFVs), as in the US and Brazil now, or to make dedicated ethanol vehicles, as in Brazil in the 1970s to late 1990s.

While pure biodiesel (B100) can be used in compression-ignition engines (diesel engines), biodiesel is currently used in the form of blends with petroleum diesel up to 20% by volume (B20). In most cases, and with adequate biodiesel fuel quality, diesel vehicles can use biodiesel blends with petroleum diesel up to B20 without vehicle modifications.

Ethanol is currently produced from fermentation of sugars in corn, sugarcane, cassava, wheat, sugar beets, and other crops. In the US, where the largest amount of ethanol production occurs, ethanol has been produced from corn since 1980 when the US started its fuel ethanol program. In Brazil, sugarcane ethanol has been produced for almost 100 years. Recently, China and Southeast Asia began to produce fuel ethanol from cassava (China also produces a significant amount of corn ethanol). In Europe, ethanol is produced from corn, wheat, and sugar beet.

Biodiesel is produced from vegetable oils and animal fats via the transesterification process. In the US, biodiesel is produced from soybeans, while in Europe it is produced primarily from rapeseed. In Southeast Asia (particularly in Malaysia), biodiesel is produced from palm oil.

Other biomass feedstocks are being explored to produce ethanol and biodiesel. In the US, cellulosic biomass such as crop residues, forest residues, and energy crops are being considered for producing cellulosic ethanol. Jatropha, a sub-tropical and tropical perennial bush, is being considered for biodiesel production.

Besides the fermentation and transesterification processes, there are many other technology paths available for producing biofuels. For example, cellulosic biomass can be gasified to produce synthetic gas (syngas). Syngas can be then used to produce Fischer-Tropsch (FT) diesel via the FT synthesis process or ethanol via fermentation of syngas. Renewable hydrocarbon fuels such as gasoline and diesel could be produced from vegetable oils and animal fats via hydrogenation. Butanol, a fuel with a higher volumetric energy content than ethanol, could be produced from sugars via fermentation processes. Recently, interest has heightened in producing hydrocarbon fuels from algae. Table 15.1 summarizes existing and potential biofuel production pathways.

15 Life-Cycle Analysis of Biofuels

Table 15.1 Existing and potential production pathways of motor biofuels from biomass

Biomass feedstock	Motor fuel	Notes
Existing biofuel production pathways		
Corn	Ethanol	In North America, Europe, and China
Sugarcane	Ethanol	In Brazil, Southeast Asia, and India
Wheat	Ethanol	In Europe and Canada
Sugar beets	Ethanol	In Europe
Cassava	Ethanol	In China and Southeast Asia
Soybeans	Biodiesel	Via transesterification; in the US
Rapeseeds	Biodiesel	Via transesterification; in Europe
Palm oil	Biodiesel	Via transesterification; in Malaysia and Europe
Potential biofuel production pathways		
Sweet potato	Ethanol	Pilot plants in China
Sweet sorghum	Ethanol	Pilot plants in China
Crop residues	Ethanol	Corn stover, wheat straw, rice straw, etc.
Forest residues	Ethanol	
Dedicated energy crops	Ethanol	Switchgrass, miscanthus, fast growing trees, mixed prairie grasses, etc.
Municipal solid waste	Ethanol	
Jatropha	Biodiesel	Sub-tropical and tropical perennial bush; via transesterification; dedicated plantations in India and China
Animal fats	Biodiesel	Via transesterification
Waste cooking oil	Biodiesel	Via transesterification
Corn	Butanol	Via fermentation
Sugar beets	Butanol	Via fermentation
Cellulosic biomass	Butanol	Via pretreatment and fermentation
Soybeans	Renewable diesel	Via hydrogenation
Rapeseeds	Renewable diesel	Via hydrogenation
Palm oil	Renewable diesel	Via hydrogenation
Jatropha	Renewable diesel	Via hydrogenation
Animal fats	Renewable diesel	Via hydrogenation
Waste cooking oil	Renewable diesel	Via hydrogenation
Cellulosic biomass	Ethanol	Via gasification and fermentation
Cellulosic biomass	FT diesel	Via gasification and synthesis
Cellulosic biomass	Oils	Via pyrolysis
Algae	Oils	

15.3 Biofuel Life Cycle Analysis Boundary

For biofuel LCAs, the analytic system boundary needs to be defined so as to include key activities or stages of biofuel life cycles. Figure 15.1 shows the LCA boundary defined for the corn ethanol pathway. For other biofuel cycles, the boundary is usually defined similarly.

In the corn ethanol case, the life cycle includes fertilizer manufacture, corn farming, ethanol production, and ethanol use in vehicles. All transportation activities involved in moving goods from one location to another (such as corn movement from farms to ethanol plants) are included. Co-product distillers' grains and soluables (DGS) and their emission effects are also included. Most recently,

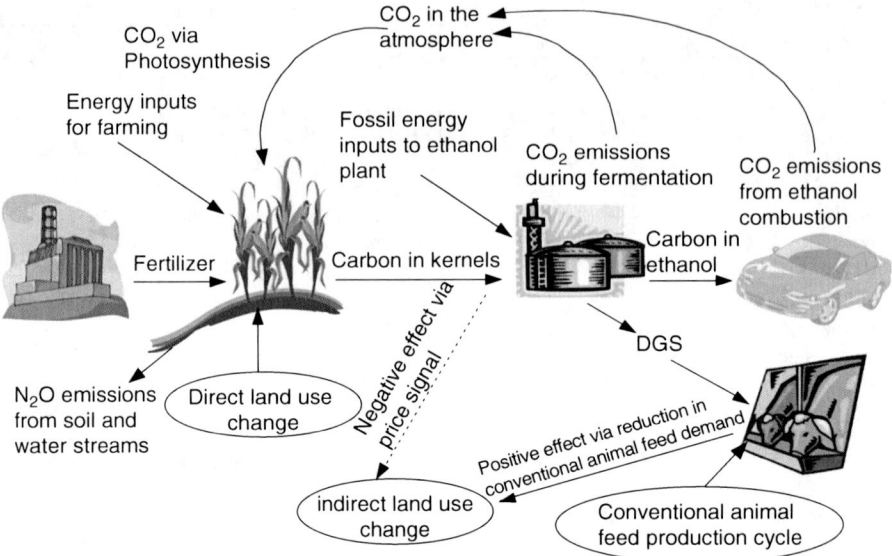

Fig. 15.1 Life-cycle analysis (LCA) system boundary of corn-to-ethanol life cycle as applied to calculating greenhouse gas (GHG) emissions

potential direct and indirect land use changes by large-scale corn ethanol production have begun to be included as well.

While every operation stage of the corn-to-ethanol cycle is included in an LCA, infrastructure-related activities such as building farming equipment and ethanol plants are usually not included. Some maintained that the contribution of constructing farming equipment to corn ethanol LCA results could be significant (Pimentel and Patzek 2005). The effects of farming equipment construction (as well as ethanol plant construction) on LCA results are determined by a few key factors such as energy use and emissions of building farming equipment, the lifetime of the equipment (affecting amortization of one-time energy use and emissions from building the equipment over its lifetime), and the acreage of land that the farming equipment serves. A detailed analysis by Wu et al. (2006) examined these key issues with updated data. They concluded that the contribution of farming equipment construction to corn ethanol life-cycle GHG emissions is only 1%, well within the uncertainty range of corn ethanol LCA results.

An LCA of biofuels is usually comparative to an LCA of baseline fuels such as petroleum gasoline and diesel, so the life-cycle system boundary needs to be defined for gasoline and diesel. To make the comparison between biofuels and petroleum fuels valid, the system boundary between them needs to be defined as consistently as possible. Figure 15.2 shows the LCA system boundary usually defined for petroleum gasoline and diesel.

As the figure shows, the life cycle of petroleum fuels begins with petroleum recovery in oil fields and ends with gasoline and diesel combustion in motor

Fig. 15.2 LCA system boundary for petroleum gasoline and diesel

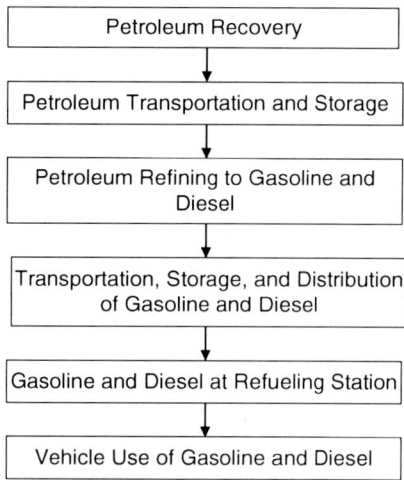

vehicles. Besides production-related activities, all transportation-related activities to move goods from one location to another (such as crude oil from oil fields to petroleum refineries) are included. Again, infrastructure-related activities such as construction of drilling rigs and petroleum refineries are not included in the LCA of petroleum gasoline and diesel. Oil exploration, which occurs well before oil recovery, is also usually not included in petroleum fuel LCAs.

15.4 Life-Cycle Analysis Models for Biofuels

Extensive LCAs of transportation fuels began in 1980s when the US and some other countries were promoting battery-powered electric vehicles and other alternative-fuel vehicles for their air pollution and energy benefits. LCAs for transportation fuels are often called fuel-cycle analyses or well-to-wheels (WTW) analyses. In the early 2000s, LCAs were extended to examine the energy and environmental effects of hydrogen fuel cell vehicles. Most recently, LCAs have been widely applied to examine energy and emission effects of biofuels. As a result of these research efforts, LCA models applicable to transportation fuels have been developed. The following is a summary of a few key LCA models for transportation fuels.

15.4.1 The GREET Model at Argonne National Laboratory

Argonne National Laboratory has been examining energy and environmental benefits of alternative transportation fuels and advanced vehicle technologies for the US Department of Energy (DOE) since the middle of the 1980s. In 1995, with DOE support, Argonne began to develop the Greenhouse gases, Regulated Emissions, and Energy use in Transportation (GREET) model. The first version of the GREET model was released in 1996 (Wang 1996). Since then, Argonne has continued to update, upgrade, and expand the GREET model.

The GREET model is free to download and use and is available from Argonne's GREET website (http://www.transportation.anl.gov/modeling_simulation/GREET/index.html). The current GREET version includes more than 100 production pathways for transportation fuels, many of which are biofuel pathways. It also includes major vehicle propulsion technologies such as internal combustion engine-powered vehicles, hybrid electric vehicles, battery-powered electric vehicles, and fuel cell vehicles.

The GREET model generates the following output items for a given vehicle technology and fuel combination:

- Energy use separately for total energy, fossil energy, petroleum energy, natural gas energy, and coal energy.
- Emissions of GHGs including carbon dioxide (CO_2), methane (CH_4), and nitrous oxide (N_2O); these GHGs are weighted together with their global warming potentials to produce CO_2-equivalent (CO_2e) GHG emissions.
- Emissions of six criteria pollutants including volatile organic compounds (VOCs), carbon monoxide (CO), nitrogen oxides (NO_x), sulfur oxides (SO_x), particulate matter with size smaller than 10 microns (PM_{10}), and particulate matter with size smaller than 2.5 microns ($PM_{2.5}$). These emissions are separated into total emissions and urban emissions (the latter is a subset of the former).

15.4.2 The Lifecycle Emissions Model at the University of California at Davis

The Lifecycle Emissions Model (LEM) was developed in the early 1990s at the University of California at Davis (Delucchi 1991). The original intent of the model—the first type of comprehensive LCA models for transportation fuels—was to evaluate battery-powered electric vehicles and alternative-fuel vehicles. The model has been updated and upgraded on a continuous basis. Relative to other LCA models, the LEM takes the liberty of expanding the LCA system boundary by including some market mitigation effects, among other effects. The LEM includes the following pollutants (Delucchi 2003):

- CO_2
- CH_4
- N_2O
- CO
- NO_x
- Nonmethane organic compounds (NMOCs)

- Sulfur dioxide (SO_2)
- Total particulate matter
- PM_{10}
- Chlorofluorocarbons (CFC-12)
- Hydrofluorocarbons (HFC-134a), and
- The CO_2-equivalent GHGs of all of the above pollutants

At present, the LEM is generally not available to other researchers. Efforts are underway that may result in the model becoming available for use by the public.

15.4.3 The GHGenius Model in Canada

The GHGenius model was originally developed from an LEM version in 1999. Since then, Natural Resources Canada has supported the maintenance, updates, and upgrades of the model [(S&T)^{2}Consultants 2008]. The model includes the following output items:

- Emissions of GHGs including CO_2, CH_4, N_2O, CFC-12, and HFC-134a
- Emissions of criteria pollutants including CO, NO_x, NMOCs, SO_2, and total particulate matter, and
- Energy use by total energy and fossil energy.

The GHGenius model is available for free download from its website at http://www.ghgenius.ca/. The model has been applied widely in Canada to examine biofuels and Canadian oil sands, among many other transportation fuels.

15.4.4 The E3 Database from Ludwig-Bölkow-Systemtechnik

The E3 database has been developed by the German consulting company Ludwig-Bölkow-Systemtechnik (LBST; see http://www.e3database.com/). For a given fuel pathway, the model estimates the following output items:

- Emissions of GHGs including CO_2, N_2O, CH_4, perfluoromethane (CF_4), sulfur hexafluoride (SF6), and HFC-134a;
- Emissions of criteria pollutants including NO_x, SO_2, CO, NMVOC, and dust/particulate matter;
- Energy use; and
- Costs

The precursor of the E3 database was developed primarily for evaluation of hydrogen production pathways. The model was expanded and updated to include many other fuel production pathways in the past 7 years (LBST 2008) and is used widely in Europe to examine WTW energy and emission effects of vehicle/fuel systems. The model is proprietary to LBST and is not available to the public.

Besides these LCA models designed specifically for examining transportation fuels, generic LCA models developed for evaluation of consumer products exist, which can be tailored for LCA applications to transportation fuels. Such models include the Ecobalance model by PricewaterhouseCoopers. These models do not, in general, address detailed issues specific to transportation fuels well.

15.5 Life-Cycle Energy and Greenhouse Gas Emission Results of Key Biofuel Pathways with GREET Simulations

Over the past several years, Argonne National Laboratory has expanded, updated, and applied its GREET model to examine life-cycle energy and GHG emissions of a few key biofuel production pathways. This section summarizes results of these efforts.

15.5.1 Corn and Cellulosic Ethanol

Since the beginning of the US corn ethanol program in 1980, production of US corn ethanol has risen to 34 billion L (9 billion gallons) in 2008 (Renewable Fuels Association 2009). The US 2007 Energy Independence and Security Act established a goal of 56.8 billion L (15 billion gallons) per year of corn ethanol production by 2015. The corn ethanol industry has expanded quickly to reach that goal.

Historically, corn ethanol has been produced from both dry and wet milling plants with different front-end milling technologies and with different co-products. Wet milling plants were built large in size and with the flexibility of producing multiple co-products, but required large capital investment. Dry milling plants initially were built small in size and with a single co-product (i.e., DGS), and required small capital investment. Since 2000, virtually all newly built corn ethanol plants in the US have been dry milling plants. The size of dry milling plants has approached that of wet milling plants. That is, some of the newly built dry milling plants produce more than 378 million L (100 million gallons) ethanol per year.

Corn ethanol plants require a large amount of steam for fermentation and distillation. Natural gas (NG) and coal are the two primary process fuels to generate steam. Historically, large wet milling plants were fueled with coal, and dry milling plants were fueled primarily with NG. The price of NG skyrocketed a few years ago, making owners of new dry milling plants consider coal as a plant process fuel. However, while coal is inexpensive to use, the emission controls needed in ethanol plants would add to plant capital costs, and air emission permission for coal-fired ethanol plants takes a longer time to obtain. These factors have precluded widespread use of coal in new dry milling plants. On average, more than 90% of US corn ethanol capacity is fueled with NG and the remaining 10% with coal.

Since the beginning of the US corn ethanol program, the energy use intensity of corn ethanol plants has been reduced from more than 19.5 MJ/L [70,000 British thermal units (BTU) per gallon] of ethanol (Chambers et al. 1979) to less than 8.4 MJ (30,000 BTU; Liska et al. 2009). The more than 57% reduction in energy intensity has been achieved by high ethanol yield, increased production of wet DGS in lieu of dry DGS, and better process designs, all of which were driven by the economics of ethanol plant operation. In addition, biomass-based process fuels such as wood chips and corn stover could be used in ethanol plants (Wang et al. 2007); since the carbon in these fuels is renewable carbon, GHG emissions from corn ethanol production can be reduced significantly by switching to these fuels from NG or coal.

Corn farming requires large amounts of nitrogen fertilizer and fuels, although since the 1970s, usage intensities for chemicals and fuels of US corn farming have been reduced significantly. For example, US corn productivity in terms of the amount of corn yielded per unit of fertilizer input to farms increased by 88% between 1970 and 2005 (Wang et al. 2007). This was accomplished by a continuous increase in corn yield per unit of land without a corresponding increase in fertilizer use. Due mainly to this corn yield per unit of land increase, farming energy use per unit of corn yielded was reduced by 34% between 1996 and 2001 (the two most recent years in which the US Department of Agriculture conducted farming energy expenditure surveys; Wang 2008).

Figure 15.3 shows GHG emission shares by key activities for corn ethanol. Ethanol plants are by far the largest source of GHG emissions. N_2O emissions from nitrogen fertilizer nitrification and denitrification in cornfields are the second largest source of GHG emissions. GHG emissions from nitrogen fertilizer plants, farming energy consumption, and farming chemicals such as phosphorous fertilizer, potash fertilizer, and lime are significant contributors as well.

Various cellulosic biomass feedstocks could be used for ethanol production, including crop residues such as corn stover, wheat straw, and rice straw; forest residues; dedicated energy crops such as switchgrass, miscanthus, willow trees, and hybrid poplars; and municipal solid wastes. The LCAs of cellulosic ethanol that have been completed at Argonne National Laboratory include ethanol from corn stover, forest waste, and switchgrass.

Wu et al. (2006) examined the corn stover-to-ethanol pathway. The production ratio of corn stover to corn grain is about 1:1 on a dry matter basis. Corn stover is usually left in cornfields for soil protection and as a nutrient supplement for the next growing season. Extensive research has been done to examine how much stover can be removed from cornfields without causing soil quality deterioration. Within the LCA context, the operation of collecting and transporting corn stover from fields to cellulosic ethanol plants needs to be taken into account. In addition, the nutrients removed from cornfields as corn stover is removed need to be supplemented during the next season for growing crops. These factors were considered in Argonne's LCA for the corn stover-to-ethanol pathway. Wu et al. (2006) also examined the

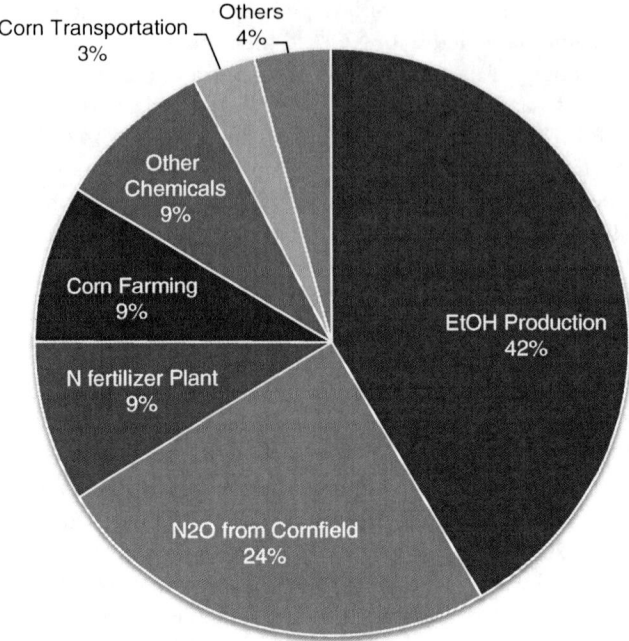

Fig. 15.3 Shares of GHG emission sources for corn ethanol [estimated from the greenhouse gases, regulated emissions, and energy use in transportation (GREET) model]

forest wastes-to-ethanol pathway. Major activities for this pathway include stumping, collecting, and transporting forest wastes from fields to ethanol plants. In fact, the amount of diesel fuel used for these activities could be significant.

Switchgrass can be farmed as a dedicated energy crop. Managed switchgrass farms may require fertilizer applications in order to maintain a good yield per unit of land, though the amount of fertilizer used is much less for switchgrass farming than for corn farming. Also, if switchgrass farming occurs in arid regions such as the US Pacific Northwest, irrigation may also be required. If switchgrass is grown on marginal land or unmanaged prairie land, it is possible that growth of switchgrass could indeed help increase soil carbon content—a benefit for additional GHG emission reductions by switchgrass-based cellulosic ethanol.

In ethanol plants, cellulosic biomass goes through a pretreatment process so that cellulose and hemi-cellulose can be broken down into simple sugars for hydrolysis and fermentation. The lignin portion of the biomass cannot be fermented. Because of its high energy content, lignin can be used as a process fuel in cellulosic ethanol plants to provide needed steam. In fact, mass balance calculations indicate that the amount of lignin available in cellulosic ethanol plants can exceed the amount of lignin needed for steam generation. Combined heat and power systems are proposed to generate both steam and electricity in cellulosic ethanol plants. Some of the generated electricity can be exported to the electric grid to displace conventional electric power generation, which brings some additional GHG emission reductions.

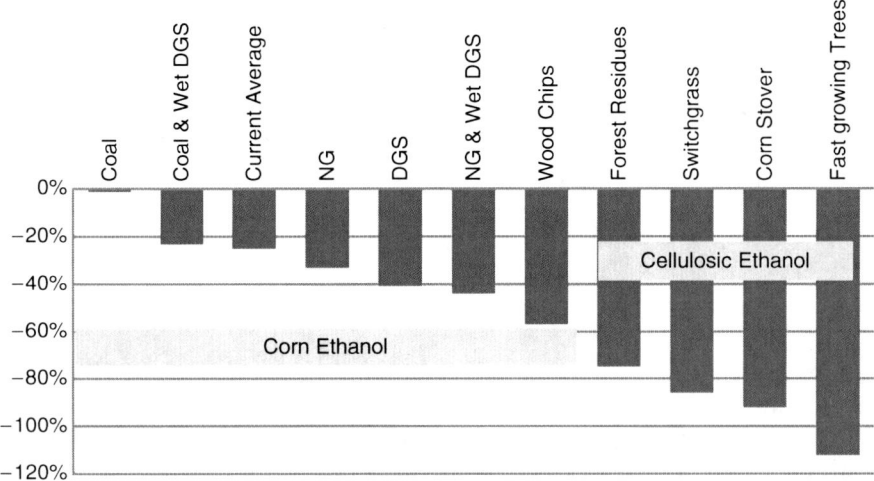

Fig. 15.4 GHG emission reductions of corn and cellulosic ethanol (relative to gasoline, on the per-energy unit basis, estimated with the GREET model)

Figure 15.4 shows the GHG emission reductions of corn and cellulosic ethanol relative to gasoline for each unit of energy used for each type of ethanol to displace a same unit of energy of gasoline. GHG emission effects of corn ethanol vary from no change to up to 57% reductions, depending on the type of process fuels and on production of wet or dry DGS. This shows that ethanol plant designs can significantly impact corn ethanol GHG emission results. On the other hand, cellulosic ethanol can reduce GHG emissions by more than 75%, depending on the type of cellulosic biomass feedstocks.

15.5.2 Sugarcane Ethanol

Wang et al. (2008) evaluated the GHG emission reduction potentials of Brazilian sugarcane ethanol for use in both Brazil and the United States. Similarly, Macedo et al. (2008) evaluated sugarcane ethanol production and use in Brazil.

Sugarcane is a tropical and sub-tropical crop. Once sugarcane sets have been planted on sugarcane farms, the sugarcane can be harvested for five to seven seasons. After that, sugarcane farms are replanted. As for any crop, fertilizers are applied to sugarcane plantations, and fuels are required for planting, harvesting, and sometimes irrigation. Traditionally, sugarcane is harvested by laborers, comprising the so-called "manual harvest." To ease manual labor's efforts, sugarcane fields are burned before harvest. After harvest, the remaining stalks are often burned to control disease and promote cane growth in the next season. Primarily because of concerns about air pollution caused by open-field burning, Brazil, especially the

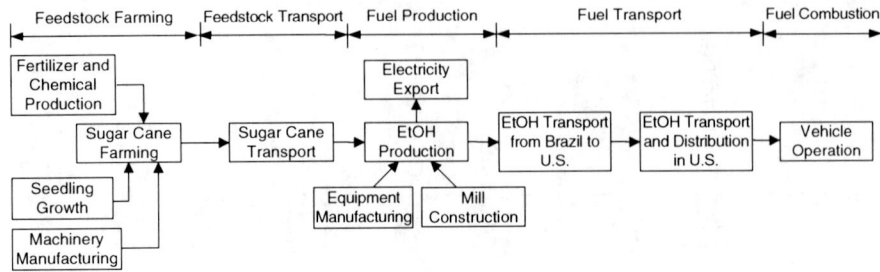

Fig. 15.5 System boundary of sugarcane ethanol life-cycle analysis with the GREET model

State of Sao Paulo, is phasing out open burning. As a result, mechanical harvests with farming machinery will replace manual harvests.

In sugarcane mills, sugarcane is washed and crushed and cane juice is extracted. The juice is then treated to produce ethanol and/or sugar. The split between the two products is based on market demand. The leftover after juice extraction is a material composed primarily of fiber that is called bagasse. Steam demand accounts for most energy use in sugarcane ethanol plants and is met through bagasse combustion. In Brazilian sugarcane ethanol plants, bagasse is combusted in biomass boilers to produce steam to meet the plant's needs and to generate electricity with steam turbines to meet plant requirement for electricity and to provide electricity for export. Figure 15.5 shows the system boundary for an LCA of sugarcane ethanol in the GREET model.

Figure 15.6 shows GHG emission shares by key activities of the sugarcane ethanol life cycle. CH_4 and N_2O emissions from open-field burning in sugarcane plantations alone are responsible for 24% of total GHG emissions for sugarcane ethanol. Overall, the five major contributors to sugarcane ethanol GHG emissions are open-field burning, N_2O emissions from sugarcane fields, fertilizer production, GHG emissions from sugarcane ethanol plants, and farming energy consumption.

Figure 15.7 presents GHG emission reductions of Brazilian sugarcane ethanol vs gasoline. Of the four sugarcane ethanol cases representing use of ethanol in Brazil vs the US, sugarcane ethanol achieves GHG emission reductions of 75–80%, which is similar to the GHG emission reductions by cellulosic ethanol.

15.5.3 Biodiesel and Renewable Diesel from Soybeans

Biodiesel is produced from seed oils or animal fats via the transesterification process. Biodiesel can be derived from various biological sources such as seed oils (e.g., soybeans, rapeseeds, sunflower seeds, palm oil, jatropha seeds, and waste cooking oil) and animal fats. In the US, most biodiesel is produced from soybean oil. In Europe (especially in Germany), biodiesel is produced primarily from rapeseeds. Biodiesel can be blended with conventional diesel fuel in any proportion and used in diesel engines without significant engine modifications (Keller et al. 2007).

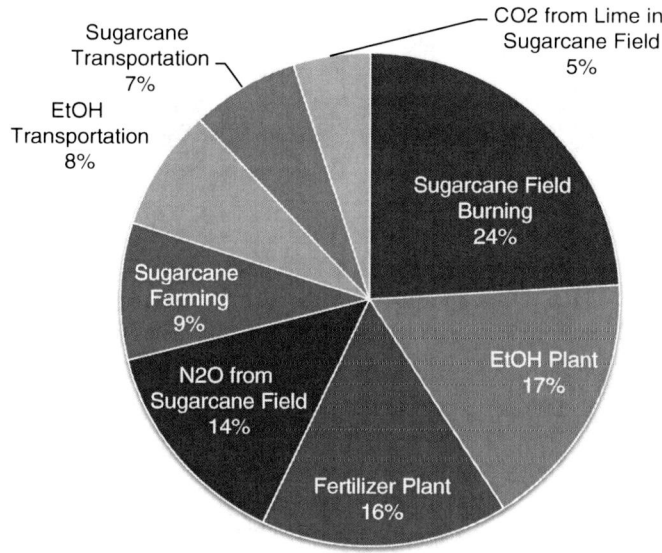

Fig. 15.6 GHG emission sources of sugarcane ethanol

In recent years, the sales volume for biodiesel in the US has increased dramatically, from about 8 million L in 2000 to 284 million L in 2005 and 2.65 billion gallons in 2008 (National Biodiesel Board 2009).

New process technologies based on hydrogenation to convert seed oils and animal fats to diesel fuels with properties similar to petroleum diesel fuels have emerged recently (Huo et al. 2008). One of the renewable diesel production pathways is the SuperCetane™ process, which is based on adapting a conventional hydrotreating process for renewable diesel production. This technology was developed by the Canadian Energy Technology Center of Natural Resources Canada. Another production pathway was developed by UOP (Universal Oil Products) based on a hydroprocessing technology.

Huo et al. (2008) conducted an LCA of biodiesel and renewable diesel that are produced with soybeans. In the US midwest, soybean farming is usually rotated with corn farming. Because the soybean plant is a legume it has the ability to fix nitrogen in the soil. Thus, soybean farming requires much less nitrogen fertilizer than corn farming does, which helps increase the energy and emission benefits of soybean-based biodiesel and renewable diesel.

Before production of biodiesel or renewable diesel, soybeans are crushed to separate soy meals and soy oil. Soy meals are a high-value animal feed. On a mass basis, 82% of soybeans ends up in soy meals and the remaining 18% in soy oil. Soy oil is then used to produce biodiesel or renewable diesel.

In biodiesel plants, glycerin, a specialty chemical, is produced together with biodiesel. On a mass basis, 82% of soy oil ends up in biodiesel and 18% in glycerin. In renewable diesel plants with the SuperCetane process, fuel gas and heavy oils, both of which are energy products, are produced with renewable diesel. For each

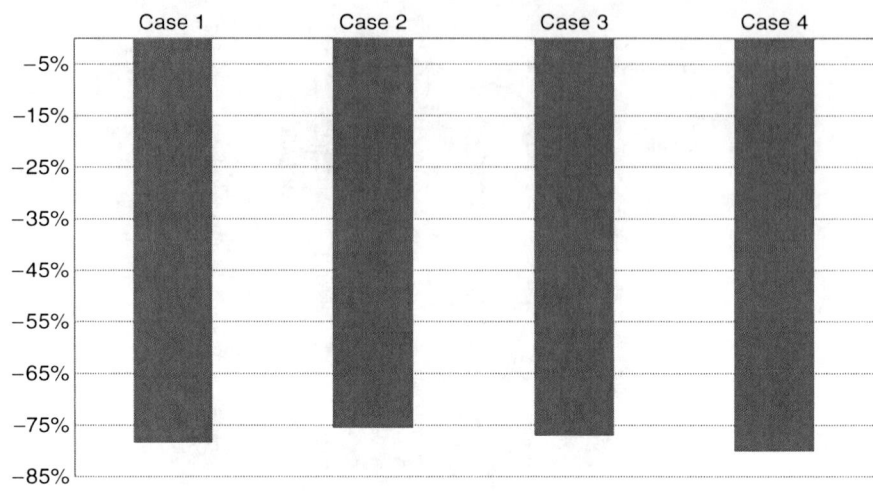

Notes:
Case 1: sugarcane ethanol is produced in Brazil and used in the United States; energy embedded in farming equipment manufacturing and sugarcane mill construction is not included.
Case 2: sugarcane ethanol is produced in Brazil and used in the United States; energy embedded in farming equipment manufacturing and sugarcane mill construction is included.
Case 3: sugarcane ethanol is produced in Brazil and used in the United States; energy embedded in farming equipment manufacturing is included.
Case 4: sugarcane ethanol is produced and used in Brazil; energy embedded in farming equipment manufacturing is included.

Fig. 15.7 GHG emission reductions of sugarcane ethanol (relative to gasoline, on a per-energy unit basis (from Wang et al. 2008)

kilogram of renewable diesel produced, 16.5 MJ fuel gas and 8.4 MJ heavy oils are produced. With the UOP process, propane is produced with renewable diesel. For each kilogram of renewable diesel produced, about 2.6 MJ propane fuel mix are produced. Table 15.2 presents inputs and outputs in biodiesel and renewable diesel plants as developed in Huo et al. (2008).

Figure 15.8 presents LCA results of GHG emission reductions by biodiesel and renewable diesel. In general, biodiesel and renewable diesel can reduce GHG emissions by more than 60% relative to petroleum diesel. A large amount of soy meals are produced with soy oil. In addition, biodiesel and renewable diesel plants produce multiple co-products in significant amounts (see Table 15.2). The methods used in LCAs to address these co-products have a significant effect on LCA results for biodiesel and renewable diesel. For this reason, Huo et al. (2008) evaluated several methods in dealing with co-products, as shown in Fig. 15.8 . The

15 Life-Cycle Analysis of Biofuels

Table 15.2 Energy use and amount of fuel product and co-products for biodiesel and renewable diesel production (per metric ton of soybeans input)

Inputs and outputs	Biodiesel	Renewable diesel	
		SuperCetane™ Process	UOP process
Outputs			
Fuel (kg)	175	116	150
Fuel (MJ)	6,582	5,070	6,582
Co-products			
Soy meal (kg)	786	786	786
Glycerin (kg)	38	None	None
Energy product (MJ)	None	2,884	384
Inputs			
Natural gas (MJ)			
For soy oil extraction	2,093	None	2,093
For fuel production	360	None	35
Electricity (MJ)			
For soy oil extraction	226	226	226
For fuel production	19	36	33
Methanol (MJ)	352	None	None
Hydrogen (MJ)	None	419	570

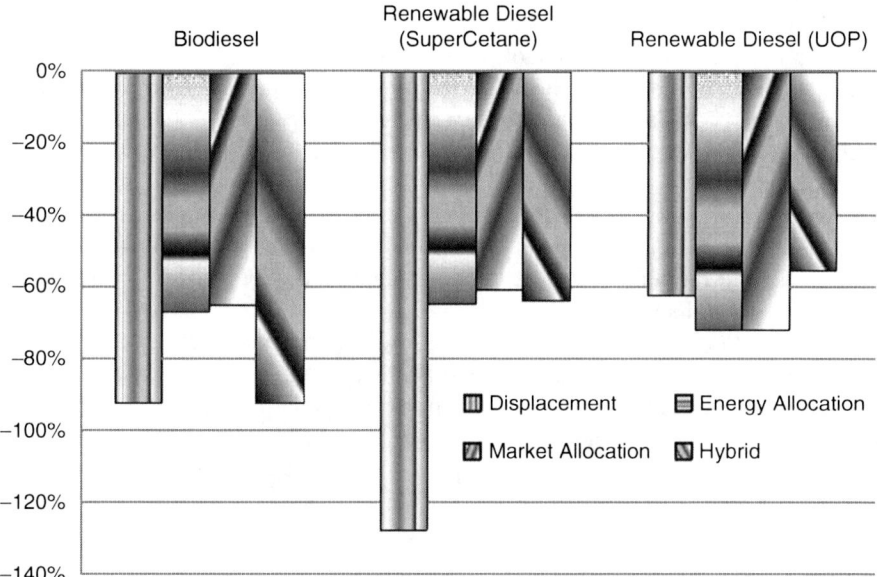

Fig. 15.8 GHG emission reductions of soybean-based biodiesel and renewable diesel (relative to petroleum diesel, on a per-energy unit basis (from Huo et al. 2008)

displacement method assumes products to be displaced by the co-products (such as soy meals and glycerin), and estimates the energy and emission credits of the displacement. The energy allocation method allocates energy and emission burdens of soybean farming and fuel production plants among different products according to their energy output shares. Similarly, the market allocation method is conducted according to the market revenue shares of different products. The hybrid method for the two renewable diesel cases uses the displacement method for soy meals and the energy allocation method for the products in renewable diesel plants.

15.5.4 Corn Butanol

Butanol (BtOH) produced from starch has gained interest in recent years as a replacement for gasoline. The energy content of butanol—27.8 MJ/L [low heating value (LHV)]—is 86% of the energy content of a liter of gasoline but 30% higher than the energy content of a liter of ethanol. Butanol has low water solubility that could minimize the co-solvency concern associated with ethanol, consequently decreasing the tendency of microbial-induced corrosion in fuel tanks and pipelines during transportation and storage. Butanol is much less evaporative than gasoline or ethanol, making it safer to use and generating fewer VOC emissions. The majority of butanol is currently used as a chemical for surface coatings and is produced from petroleum propylene through a process in which syngas is reacted with propylene.

The most dominant bio-butanol production process has been acetone-butanol-ethanol (ABE) fermentation. In the ABE fermentation process, corn is fed into a conventional corn dry mill for conversion to glucose through liquefaction and saccharification. The glucose is then fermented to ABE. After fermentation, ABE compounds are removed by means of in-situ gas stripping. ABE products are recovered through molecular sieve adsorption and a distillation process that separates acetone, butanol, and ethanol. Solids and biomass that are removed from grain processing and fermentation undergo centrifugation and proceed to drying, along with syrup from distillation, to produce DGS.

Wu et al. (2007) conducted an LCA of corn-based butanol with the ABE process. They developed an ABE process simulation model with Aspen Plus® to estimate the energy and mass balance of the ABE process. Table 15.3 presents corn-based butanol plant energy use, and Table 15.4 presents the outputs of acetone, butanol, and ethanol from corn-based butanol plants. Recent technology development of butanol fermentation process may result in elimination or much smaller production of acetone.

Table 15.3 Energy requirements in corn-based butanol plants (per liter of butanol produced). *ABE* Acetone-butanol-ethanol

Process	Natural gas (MJ)	Electricity (MJ)
Cooking	5.08	0.0
Drying	5.49	1.17
ABE fermentation and processing	21.52	0.50
Total	32.09	1.67

Table 15.4 Yields of acetone, butanol, and ethanol from corn-based butanol plants with the ABE fermentation process

Product	Yield (L/kg corn)	Yield (MJ/kg corn)	Energy output share (%)
Acetone	0.129	2.888	31.4
Butanol	0.223	6.200	67.4
Ethanol	0.006	0.117	1.2
Total	0.358	9.205	100.0

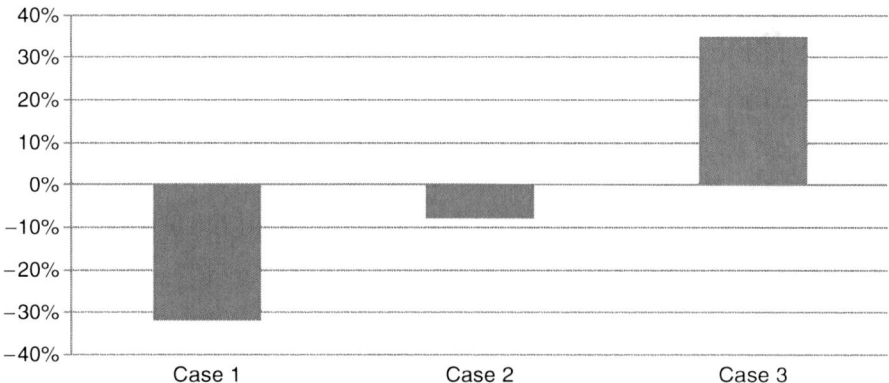

Fig. 15.9 GHG emission reductions of corn-based butanol (relative to petroleum gasoline, on the per-energy unit basis, based on Wu et al. 2007). Cases 1, 2 and 3 are explained in the text

Figure 15.9 presents LCA results of GHG emissions of corn-based butanol vs petroleum gasoline. A large amount of acetone and DGS are produced from butanol plants. Ways of dealing with these co-products significantly influence corn-based butanol results. For this reason, Wu et al. (2007) developed three cases of dealing with acetone and DGS in the butanol evaluation. With Case 1, bio-acetone and DGS are regarded as energy products, and are thus credited on the basis of the energy allocation method among butanol, acetone, ethanol, and DGS. With Case 2, bio-acetone is regarded as a chemical to displace petroleum-based acetone, and is therefore credited by product displacement, as DGS and ethanol displace animal feed and gasoline, respectively. With Case 3, bio-acetone is regarded as waste, and therefore no acetone credit is assigned.

Figure 15.9 shows that GHG reductions by corn-based butanol can vary significantly, depending how co-products acetone, DGS, and ethanol are treated in a butanol LCA. This is because the amount of acetone, besides DGS, produced in butanol plants with the ABE fermentation process examined is large (see Table 15.4 above). Acetone is a specialty chemical with a small market size. If a large amount of acetone is produced from butanol plants, the market value of acetone could be reduced significantly, with it having no value in the extreme case (Case 3). Realizing this potential problem, butanol technology developers are working on fermentation processes to reduce or eliminate the amount of acetone produced.

15.6 Key Life-Cycle Analysis Issues and Uncertainties

As illustrated in the above sections, LCA results for biofuels are influenced heavily by a few key analytical issues. This section presents three key issues that are being debated currently and that can significantly affect LCA results for biofuels.

15.6.1 Direct and Indirect Land Use Changes

Although the land use change (LUC) issue for biofuels is not new, Searchinger et al. (2008) were the first to develop quantitative results for this issue and advocated against biofuel policies based on their results. Conceptually, production of biofuels will require that biomass feedstocks are grown on land. Growth of a given feedstock on a piece of land changes the original land use pattern of that piece of land. This land use change is referred to as a direct land use change. Direct land use changes are identifiable and measurable, since such changes can be observed directly and attributed to biofuel production. On the other hand, use of agricultural commodities such as corn causes an imbalance between supply and demand of agricultural commodities, which can trigger commodity price increases. The price increase signal can have ripple effects, causing cultivation of additional land for growing agricultural commodities somewhere in the world. This land use change is referred as to an indirect land use change. As one can see, indirect land use changes are thought to be caused by increased commodity prices. Indirect land use changes may be simulated with computational general equilibrium (CGE) models to take into account all interrelationships among all economic sectors and activities via price elasticities of commodities. Searchinger et al. (2008) used the Food and Agricultural Policy Research Institute (FAPRI) model (which is a partial, not general, equilibrium model) to develop estimates of direct and indirect land use changes. The Global Trade Analysis Program (GTAP) model developed by Purdue University is a general equilibrium model that is being used by several organizations to address LUC issues.

After publication of the Searchinger study, the simulation and data problems of that study were identified, as were the problems of LUC modeling by CGE models in general. Major efforts have been made to expand and upgrade these models to better simulate LUC issues for biofuels. While models have been improved and LUC results have become more reliable, at present, the LUC results are still not definitive.

Furthermore, one could argue that even if CGE models identify statistical relationships among different events through price elasticities, these relationships may not necessarily mean causal effects from biofuel production on indirect land use changes that may occur remotely somewhere else at the same time scale. There are many social and political factors determining the seemingly related events that occurred during the same period of time.

15 Life-Cycle Analysis of Biofuels

In addition, at best, CGE models predict additional land requirements for commodity production due to the presence of biofuel production. They are generally not capable of determining how the additional land requirements are to be met in a given region. CGE models are generally designed to address macro-scale trade issues, not land supply issues at regional levels. Often, the additional land requirements estimated with CGE models are prorated with past LUC patterns in given regions (which may have nothing to do with biofuel production) to match the new demand and supply of land in those given regions.

Next, carbon emissions are determined from the LUC results from CGE models. Carbon emissions (in some cases, carbon sequestration) are determined using changes in above-ground biomass and below-ground biomass and soil carbon contents of different land cover types. Even though there are data sources available regarding these, they are not comprehensive enough to cover all major land cover types in different global regions. Often, scarce data from a small set of regions are applied to different global regions (IPCC 2006). In addition, soil carbon content data cover only a shallow depth of soil (such as the top 30 cm). Adequate, comprehensive data on above-ground biomass, below-ground biomass, and soil carbon content will take considerable effort to collect and accumulate for use.

With these fundamental methodology and data issues, results of carbon emissions from direct and indirect LUCs are tentative for now, even though regulatory agencies such as the California Air Resources Board (CARB) and US Environmental Protection Agency (EPA) have adopted or proposed carbon emission values for direct and indirect LUCs (CARB 2009; US EPA 2009). Major efforts are being made to address these issues, and these efforts will certainly be continued.

15.6.2 Co-Product Issues for Biofuel Life-Cycle Analyses

Section 15.5 presents biofuel LCA results with different ways of dealing with co-products of biofuels. As shown by the results for biodiesel, renewable diesel, and butanol, the LCA results for biofuels can vary significantly, depending on the method selected in dealing with biofuel co-products.

Wang et al. (2010) examined five methodologies for dealing with co-products in biofuel LCAs. First, with the mass-based allocation method, energy and emission burdens of a given biofuel pathway are allocated among all products according to their mass output shares. This allocation method is based on the presumption that energy use and emissions are somewhat related to the amount of mass processed. This method is used widely in LCAs of consumer products and is embedded in some generic LCA models. The method is applicable as long as all products are used for their mass values (e.g., 1 kg steel for use). However, this method becomes problematic when products have distinctly different uses. For example, in the cases of sugarcane ethanol and cellulosic ethanol production, electricity is co-produced but cannot be allocated by mass.

Second, with the energy-content-based method, the energy and emission burdens of a given fuel production pathway are allocated among products according to their energy output shares. The energy outputs of all products are calculated using the amount of products and their energy content (usually heating content of the products). This method is applicable where most, if not all, of the products are used for their energy content. The method becomes problematic when products have distinctly different uses. For example, starch-based ethanol plants produce ethanol and animal feeds. Even though animal feeds have energy content, they are used because of their significant nutritional values, not their heating values, which are on par with conventional animal feeds (such as corn and soybean meal).

Third, the market-value-based method allocates energy and emission burdens based on economic revenue shares of individual products. The economic revenue of a given product is calculated from the product yield of a given pathway and the price of the product. Economists generally advocate use of this method. In fact, some LCA applications of general equilibrium models adopt this method. This method assumes that activities and decisions are driven by economics, and thus burdens should be disbursed according to economic benefits. One unique advantage of this method is that it normalizes all products to a common basis—their economic values. However, in practice this method is subject to great fluctuations in product prices.

Fourth, the process-purpose-based method estimates energy use and emissions of individual processes in a fuel production facility. The energy use and emissions of a given process are allocated to a given product, if the purpose of that process is solely for the production of the given product. An example is the dryer in a corn ethanol plant. The dryer is installed to dry DGS. Thus, energy use and emissions from the dryer operation are allocated to DGS. However, in many cases, individual processes in a facility may produce multiple products, causing the need to allocate energy and emissions of a given process among all products from the process. Furthermore, this method requires energy and emission data at the process level, not at the facility level, which may not be available to researchers for many biofuel facilities. Even if the process-purpose-based method is applied to a given facility, the activities upstream of the facility still need to be allocated. For example, this method can be used to allocate energy use and emissions of corn ethanol plants between ethanol and DGS. But the allocation of energy use and emissions of corn farming between ethanol and DGS still needs to be decided. This decision, in turn, might be based on the mass-, energy-content-, or market-based method.

Fifth, with the displacement method (also called the "system boundary expansion method"), the products that are to be displaced by non-fuel co-products are determined first. Energy and emission burdens of producing the otherwise displaced products are then estimated. The estimated energy and emission burdens are credits that are subtracted from the total energy and emission burdens of the biofuel production cycle. While the displacement method is generally advocated for LCAs, it poses some major challenges to implement. The method requires conducting LCAs for the conventional products that will be displaced, which could be time- and resource-intensive. Another major problem with the displacement method is that when non-fuel products are a large share of the total output, the method generates distorted LCA results for fuels (see Wang et al. 2010).

It is far from being settled whether a given method can be uniformly and blindly recommended for LCA studies. Consistency of co-product method choices for evaluation of different biofuel production pathways may not serve the purpose of providing reliable LCA results well. Transparency of LCA methods is important in LCA studies, and sensitive cases with multiple co-product methods may be warranted in LCA studies where co-products can significantly impact study outcomes.

15.6.3 Other Environmental Sustainability Issues

Environmental sustainability issues of biofuel production and utilization now are an important topic. Such issues include fresh water consumption for both feedstock growth and biofuel production, soil erosion effects of growing certain feedstocks, biodiversity implications of feedstock growth, and air pollution and its health effects of producing and using biofuels. So far, these issues have not been addressed on the life-cycle basis, rather than on the basis of including key stages of the biofuel life cycle. In addition, some of the issues (such as biodiversity) are difficult to address quantitatively. Eventually, these issues should be addressed along the whole life cycle of biofuels. That is, all stages of the life cycle should be considered. More importantly, these issues should be addressed on a comparative basis, so that biofuels can be compared with baseline petroleum gasoline and diesel for relative environmental sustainability implications.

Wu et al. (2009) recently estimated the consumptive water requirements of ethanol and gasoline production. For biofuel production, the key determinants are feedstock and the amount of irrigation water needed to generate reasonable yields. For gasoline production, the key determinants are the characteristics of individual oil reservoirs, the recovery technology used, and the degree of produced water recycling. On average, corn ethanol production tends to consume more water than cellulosic ethanol production does on a life-cycle basis. Net water use for cellulosic ethanol production is comparable to that of gasoline. Biofuels production exhibits significant regional differences in water use. Consumptive water use for corn ethanol production varies significantly in the major US corn-growing regions. Producing a liter of corn ethanol can consume as little as 10 or as much as 324 L water, depending on the amount of irrigation water used for corn growing. On average, more than half of US corn ethanol is produced at a water use rate of 10 L water per liter of ethanol. Switchgrass-based cellulosic ethanol production, when grown in its native habitat in the US, can consume from 1.9 to 9.8 L water per liter of cellulosic ethanol, depending on process technology. In comparison, net water use to produce a liter of gasoline varies from less than 3 L to nearly 7 L.

Hill et al. (2009) examined the economic values of GHG emissions and health effects of air pollution from biofuel production relative to petroleum gasoline in the US. They included petroleum gasoline, corn ethanol, and cellulosic ethanol from corn stover, switchgrass, prairie biomass, and miscanthus. They estimated monetary values of GHG emissions and health effects of fine PM emissions (both primary and secondary formation) from these fuel production pathways. Their emission estimates were generated for the life cycle of each of the fuel pathways. They concluded that, while corn ethanol and petroleum gasoline may have similar GHG and health costs, cellulosic ethanol has much lower costs from these effects.

15.7 Conclusions

Biofuels are being promoted for their energy and GHG reduction benefits. In general, they can be produced regionally and locally to provide fuels for motor vehicle use, thus reducing reliance on imported petroleum for many countries. Since the carbon in biofuels is taken from the air during biomass growth, biofuels can potentially reduce GHG emissions. It is well recognized that the life cycle of biofuel production and utilization is associated with fossil energy use and GHG emissions. LCA of biofuels has become an integral part of thorough evaluation of the energy and environmental effects of biofuels.

Over the past 20 years, LCAs of transportation fuels in general and biofuels in particular have advanced considerably. While LCA results of biofuels have generally shown energy and GHG benefits of biofuels relative to petroleum fuels, the magnitudes of the benefits are determined by the types of feedstocks and production technologies. The practice of biofuel LCAs has helped identify the opportunities and challenges of biofuels for becoming a solution to energy insecurity and GHG emission problems. For example, biofuel LCAs have shown clearly that all biofuels are not created equal. Depending on feedstocks and fuel plant production technologies, energy and GHG emission effects among different types of biofuels can vary dramatically. It is critical to select appropriate feedstocks and to use improved technologies for biofuels to truly achieve energy and emissions reduction benefits. Biofuel policies are being developed on the basis of their LCA energy and emission performances, so that development of the biofuel industry will move in an energetically and environmentally sustainable direction.

In addition, LCA results are influenced heavily by decisions in regarding the system boundary of a given analysis and method of dealing with co-products of biofuels, among many other factors. LCA practitioners need to make LCAs transparent and reliable, to prevent the use of misguided information for biofuel development.

Acknowledgments The author is with Argonne National Laboratory. Argonne National Laboratory's work is supported by the US Department of Energy, Assistant Secretary for Energy Efficiency and Renewable Energy, under contract DE-AC02-06CH11357.

References

CARB (2009) Proposed regulation for implementing low carbon fuel standards, vol 1. Staff Report: Initial Statement of Reasons, 5 March. California Air Resources Board, Sacramento, CA

Chambers RS, Herendeen RA, Joyce JJ, Penner PS (1979) Gasohol: does it or doesn't it produce positive net energy? Science 206:789–795

Delucchi MA (1991) Emissions of greenhouse gases from use of transportation fuels and electricity, vol 1. ANL/ESD/TM-22. Center for Transportation Research, Argonne National Laboratory, Argonne, IL

Delucchi MA (2003) A lifecycle emissions model (LEM): lifecycle emissions from transportation fuels, motor vehicles, transportation modes, electricity use, heating and cooking fuels, and materials—documentation of methods and data. UCD-ITS-RR-03-17. UC Davis, Davis, CA

Hill J, Polasky S, Nelson E, Tilman D, Huo H, Ludwig L, Nuemann J, Zheng H, Bonta D (2009) Climate change and health costs of air emissions from biofuels and gasoline. Proc Natl Acad Sci USA 106:2077–2082

Huo H, Wang M, Bloyd C, Putsche V (2008) Life-cycle assessment of energy and greenhouse gas effects of soybean-derived biodiesel and renewable fuels. ANL/ESD/08-2. Center for Transportation Research, Argonne National Laboratory, Argonne, IL

IPCC (2006) 2006 IPCC guidelines for national greenhouse gas inventory: vol 4—agriculture, forestry, and other land use. Intergovernmental Panel on Climate Change, Hayama, Japan.

Keller G, Mintz M, Saricks C, Wang M, Ng H (2007) Acceptance of biodiesel as a clean-burning fuel: a draft report in response to Section 1823 of the Energy Policy Act of 2005. Prepared for Office of Energy Efficiency and Renewable Energy, the US Department of Energy, Washington, DC, by Center for Transportation Research, Argonne National Laboratory, Argonne, IL

Kim H, Kim S, Dale B (2009) Biofuels, land use change, and greenhouse gas emissions: some unexplored variables. Environ Sci Technol 43:961–967

LBST (2008) E3 database: an introduction into the life-cycle analysis calculation tool. Ottobrunn, Germany

Liska AJ, Yang HS, Bremer VR, Klopfenstein TJ, Walters DT, Erickson GE, Cassman KG (2009) Improvements in life cycle energy efficiency and greenhouse gas emissions of corn-ethanol. J Ind Ecol 13:58–74

Macedo IC, Sebra JEA, Silva JEAR (2008) Greenhouse gases emissions in the production and use of ethanol from sugarcane in Brazil: the 2005/2006 averages and a prediction for 2020. Biomass Bioenergy 32:582–595

National Biodiesel Board (2009) Estimated US biodiesel production. http://www.biodiesel.org/pdf_files/fuelfactsheets/Production_Graph_Slide.pdfCited 3 March 2009

Pimentel D, Patzek TW (2005) Ethanol production using corn, switchgrass, and wood; biodiesel production using soybean and sunflower. Nat Resour Res 14:65–76

Renewable Fuels Association (2009) 2009 Ethanol industry outlook: growing innovation. Washington, DC

Searchinger T, Heimlich R, Houghton RA, Dong F, Elobeid A, Fabiosa J, Tokgoz S, Hayes D, Yu TH (2008) Use of US croplands for biofuels increases greenhouse gases through emissions from land use change. Science 319:1235–1238

(S&T)^{2}Consultants (2008) 2008 GHGenius update. Natural Resources, Canada

US EPA (2009) Regulations of fuels and fuels additives: changes to renewable fuel standard program; proposed rule. Federal Register, vol 74, no 99; 24904–25143, May 26

Wang MQ (1996) Development and use of the GREET model to estimate fuel-cycle energy use and emissions of various transportation technologies and fuels. ANL/ESD-31. Center for Transportation Research, Argonne National Laboratory, Argonne, IL

Wang M (2008) Life-cycle analysis of biofuels: issues and results. American Chemical Society Congressional Briefing, Washington, DC

Wang MQ, Saricks C, Wu M (1997) Fuel-cycle fossil energy use and greenhouse gas emissions of fuel ethanol produced from US midwest corn. Prepared for Illinois Department of Commerce and Community Affairs by Center for Transportation Research, Argonne National Laboratory, Argonne, IL

Wang M, Wu M, Hong H (2007) Life-cycle energy and greenhouse gas emission impacts of different corn ethanol plant types. Environ Res Lett 2:024001

Wang M, Wu M, Huo H, Liu J (2008) Well-to-wheels energy use and greenhouse gas emissions of Brazilian sugarcane ethanol production simulated by using the GREET model. Int Sugar J 110:527–545

Wang M, Huo H, Arora S (2010) Methodologies of dealing with co-products of biofuels in life-cycle analysis and consequent results within the US context. Energy Policy (in press)

Wu M, Wang M, Huo H (2006) Fuel-cycle assessment of selected bioethanol production pathways in the United States. ANL/ESD/06-7. Center for Transportation Research, Argonne National Laboratory, Argonne, IL

Wu M, Wang M, Liu J, Huo H (2007) Life-cycle assessment of corn-based butanol as a potential transportation fuel. ANL/ESD/07-10. Center for Transportation Research, Argonne National Laboratory, Argonne, IL

Wu M, Mintz M, Wang M, Arora S (2009) Consumptive water use in the production of ethanol and petroleum gasoline. ANL/ESD/09-01. Center for Transportation Research, Argonne National Laboratory, Argonne, IL

Chapter 16
Criteria for a Sustainable Bioenergy Infrastructure and Lifecycle

Jürgen Scheffran

16.1 Introduction

Around 10% of global primary energy use is based on bioenergy sources (Ladanei and Vinterbäck 2009), of which biofuels for transportation account for 2.2%, with strong growth over the last decade. In the US, ethanol production increased from 1.63 billion gallons in 2000 to 10.75 billion gallons in 2009, an increase of more than 6-fold in 10 years (RFA 2010). It is projected that by the year 2022, 36 billion gallons of ethanol will be produced, 15 billion gallons from grain and the remaining 21 billion gallons are expected to come from second generation processes that convert cellulose and hemi-celluloses into ethanol (US EPA 2007). Much of this unprecedented rise in demand for biofuels is the result of public support rather than market forces (Scheffran 2010a). A number of countries have set ambitious mandates for the substitution of fossil fuels by biofuels in the transportation sector, attracting public and private investment to stimulate biofuel production and use. The total sustainable technical potential of bioenergy is estimated to be around one-quarter of current global energy use.

The major driving forces behind the biofuels boom are concerns about energy security and global warming that have triggered the search for low-carbon energy alternatives to fossil fuels. In addition, home-grown energy sources offer development opportunities for structurally weak rural areas (Rosillo-Calle and Walter

---

J. Scheffran
Institute for Geography, KlimaCampus, University of Hamburg, ZMAW, Bundesstraße 53, 20146 Hamburg, Germany
e-mail: juergen.scheffran@zmaw.de

2006), attracting support from agricultural communities who will benefit from new income and job opportunities. Advanced bioenergy development also finds interest in developing countries hoping to reduce poverty through modernization of traditional and often harmful bioenergy use.

On the other hand, rapid growth in bioenergy use has raised concerns about its impacts on the environment, land use, water resources and food security (Scheffran 2009). In 2008, a debate on the sustainability and economic viability of ethanol and other biofuels emerged. Studies and media reports questioned the energy and carbon balance of the current generation of corn-based ethanol, and highlighted the adverse impacts of the biofuel boom on land use competition, water availability, food prices and biodiversity. Some have questioned not only bioenergy, but other renewable energy sources as well because of their large ecological footprint (Ausubel 2007), others take the pathway and the specific environmental impact into consideration (Jacobson 2008; Gallagher 2008; WBGU 2009). The fading political support for biofuels became obvious when the European Union adjusted its mandates for biofuels and considered conditions and criteria for the sustainable use and certification of biofuels. Similarly, the US Energy Independence and Security Act of 2007 requested that cellulosic biofuels must offer a greenhouse gas (GHG) emission reduction rate of at least 60% relative to conventional gasoline, taking into account direct and indirect emissions during the lifecycle.

To succeed in the long run, bioenergy systems have to demonstrate their environmental sustainability and economic viability, compared with fossil fuels and alternative energy sources. To address some of the concerns, it is important to improve the energy and carbon balance of existing ethanol pathways and accelerate the transition towards more sustainable energy uses based on cellulosic biofuels and integrated biorefineries. Integrated assessment approaches and lifecycle analysis offer tools to improve scientific understanding of the value chain and provide sound input into policymaking on bioenergy futures. At the same time, sustainability principles, criteria and standards are being developed to evaluate the various aspects of the bioenergy lifecycle, including feedstock production, harvesting and transportation, processing, distribution and use. This is a complex task because bioenergy covers a wide range of issues, including energy, agriculture and climate change; transportation and foreign trade; and environment, development and security policy (Scheffran 2010b).

This chapter will discuss principles and criteria for establishing and evaluating a sustainable bioenergy lifecycle, covering the components of the lifecycle as well as specific aspects. Evaluation criteria include the technical feasibility and economic viability of the bioenergy infrastructure, environmental dimensions of sustainability and socio-economic criteria that affect societal acceptability. Based on various ideas expressed in the literature, initiatives for bioenergy certification will be discussed as an instrument to distinguish sustainable products from non-sustainable products on the market.

16.2 Optimizing Bioenergy Lifecycle and Infrastructure

16.2.1 Bioenergy Supply Chain and Lifecycle

Developing biomass-based resources into a viable and sustainable alternative to petrochemical sources for chemicals and energy requires an integrated infrastructure to realize the bioenergy supply chain from sunlight to bioproducts (Fig. 16.1). Modeling tools can help to analyse and optimize the bioenergy infrastructure, and find the best routes in the transportation network, taking into account regional feedstock production patterns, the design and location of biorefineries and the location of ethanol demand. Multicriteria analysis ranks alternative infrastructure designs of feedstock logistics and bioprocessing. By utilizing decision tools, integrated models incorporate the best mix of feedstock, farm, biorefinery site, size, and technology.

Based on land allocation for feedstock production, transportation patterns and refinery site selection, options can be analyzed that are consistent with regional feedstock production patterns and the location of demand for ethanol. Alternative transportation modes and trans-shipment terminals for feedstocks from agro-zones producing them, and the optimal locations for ethanol plants and bio-refineries need to be assessed in conjunction, depending on distances from production sources of

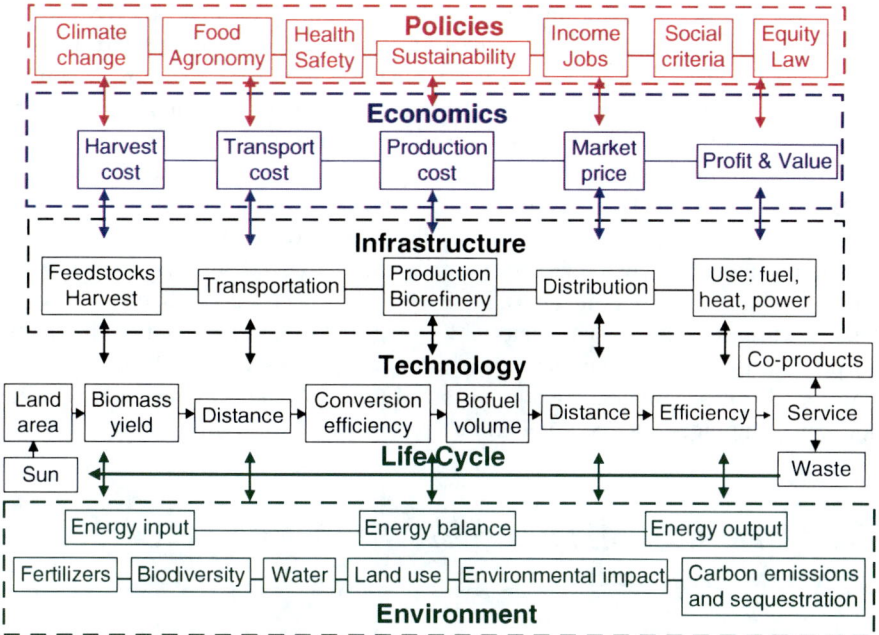

Fig. 16.1 Framework, variables and criteria of the bioenergy lifecycle

the feedstock and the road/railroad network. A comprehensive assessment compares alternative pathways for various scenarios of ethanol from corn grain, corn stover, and other cellulosic feedstocks such as wood and perennial grasses. Analyzing the trade-off between economies of scale of plant size and increasing transportation costs, the overall cost of production and transportation of feedstock to biorefineries and transportation of biofuels to consumers is minimized, taking into account the conversion of feedstock materials into fuel and higher value chemicals.

Expanding cellulosic ethanol from the current near-zero levels to take a leading role in the next decade poses enormous challenges to the infrastructure needed. According to GAO (2007) there are significant barriers to producing biofuels at a lower cost than petroleum fuels. Considerable investments are required to make biofuel production cost competitive with petroleum-based transportation fuels and overcome the technical and economic barriers at all stages of the production and supply chain (CRS 2007). In addition, the environmental impact of biofuels must be addressed, considering material, water, land and energy inputs as well as emissions, and waste streams along the entire life-cycle.

Along the learning curve of the biofuel industry there is significant potential to optimize the production path of corn ethanol in terms of improved efficiency of resource use (energy, water, land), reduction of environmental impacts (in particular GHG emissions) and the generation of co-products that improve the overall economic and environmental balance in the bioenergy lifecycle (Fig. 16.1). The new generation of biofuels based on cellulosic materials, as well as integrated biorefineries that take any organic material, such as lignin, as an input to produce co-products and electricity are expected to have a better cost effectiveness, energy ratio, water use and GHG balance. While progress in these areas is expected in the coming decade, the energy and carbon efficiency of the current generation of biofuels may be improved by short-term intermediate solutions during a transition period.

A sustainability analysis addresses the environmental consequences of biofuels, including material, water, land, and energy inputs as well as emissions and waste streams along the entire lifecycle. Among the benefits of bioenergy is the increased natural resource efficiency for energy from waste that would otherwise be burned or left to rot, and the creation of useful by-products such as fertilizers. The cost-effective use of organic waste material from agriculture, municipalities and industry plays a significant role in the transition towards more extensive bioenergy uses. Environmental benefits can also be expected through reduced emissions of sulfur and nitrogen oxides and carbon dioxide.

16.2.2 The Integrated Biorefinery—From Feedstocks to Bioproducts

An integrated biorefinery would take a large variety of organic feedstocks as input and apply multiple conversion processes to produce various bioproducts, including fuels, power, and chemicals from biomass (see Chap. 12 by Kamm, this volume). Due to the wide range of options it is a complex task to design a biorefinery that optimally converts biomass supply into bioproducts in a way that is adaptive to market demands and specific regional conditions. Next generation biorefineries would be able to utilize plant cell wall matter derived from waste material such as corn stover and dedicated energy crops such as switchgrass and Miscanthus, which need to be harvested and prepared to feed the available conversion processes in the biorefinery in an optimal way. Productivity and optimization potential for each of these materials is characteristic for the given local conditions. This implies that future biorefineries need to adapt to their environment and its variability to find the best possible mix in terms of benefit-cost efficiency, generation of co-products, minimal resource use (energy, water, land), and reduction of environmental impacts.

Current biorefineries are based largely on a single input (corn, sugarcane or plant oil) and a single output (ethanol or biodiesel), often also generating co-products (distillers grains, bagasse). Ethanol plants use two types of fermentation processes (dry and wet mills). Driven by governmental subsidies, the capacity of ethanol plants in the US Midwest has been increasing in recent years, from 15 million to 50 million gallons per year for dry grind ethanol plants to the 100-million-gallon plants that are the current standard. Some smaller-scale plants have been expanded by increasing efficiency or taking advantage of economies of scale (Brown et al. 2007).

Ethanol becomes more competitive if distillers dried grains and solubles (DDGS) and higher valued chemicals are produced as co-products that can provide incremental increases in cost effectiveness and energy balance of ethanol production. Currently, DDGS are used as animal feed, while most of the bagasse from sugarcane production in Brazil is burned for power generation (see Chap. 4).

Moving beyond corn ethanol, several competing conversion technologies would enable the use of lignocellulosic biomass as biorefinery feedstock, including gasification, pyrolysis, liquefaction, hydrolysis, fermentation, and anaerobic digestion. Finding the best mix can lead to potential cost reductions for the future biorefinery but most of these processes are still too expensive to be commercially attractive and have to compete with the cost of ethanol plants. According to Brown (2007), the capital cost of building a dry grind ethanol plant has increased in recent years, "ranging from US $1.50 to $2.00 per gallon, so the capital required to build a 100-million-gallon dry grind ethanol plant can be in the US $200 million range with rail access included." Techno-economic analyses of dry grind ethanol in the US and Europe estimate production costs to be between US $0.80 and $1.36 per gallon ethanol depending on various assumptions, especially feedstock costs (Wallace et al. 2005; Wooley et al. 1999). Compared to a total project investment of US $76 million for a 50 million gallon capacity per year corn ethanol plant, the National Renewable Energy Laboratory estimated the total project investment for a cellulosic ethanol plant with similar capacity at about US $250 million dollars (NREL 2007). The cost of producing a gallon of cellulosic ethanol is estimated to be about twice the cost of corn-based ethanol.

In a more detailed assessment, Wright and Brown (2007) compare the costs of the current generation of ethanol plants with those of first-generation lignocellulose biorefineries, using enzymatic hydrolysis as the biochemical platform and three types of thermochemical platforms (biomass gasification and upgrade to hydrogen, methanol, or Fischer-Tropsch liquids). Including economies of scale, capital costs were scaled to a common plant size of 150 million gallons per year gasoline equivalent, using a simple power law for both grain ethanol and cellulosic ethanol platforms. Capital costs of advanced biochemical and thermochemical biorefineries are four to five times as much as comparable grain ethanol plants.

To overcome the bottlenecks, research into cellulosic bioprocessing has been growing extensively in recent years, in particular to find low-cost enzymes to break down cellulose into sugar to make ethanol. Whether and when these initiatives can overcome the cost barrier to make cellulosic ethanol cost efficient, remains open.

Ethanol is currently the most important renewable liquid biofuel in the US, which produces about half of the world's ethanol, compared to about 38% in Brazil and 4.3% in the European Union (RFA 2010). As the worldwide demand for fuels and chemicals surges and petroleum deposits are becoming depleted, producers of fuel ethanol are increasingly looking beyond corn, potatoes, and other starchy crops as substrates for fuel ethanol production. Especially promising is cellulosic ethanol, which can capitalize on microbial engineering and biotechnology to reduce costs. Derived from low-cost and plentiful feedstocks, it can achieve high yields, and has high octane and other desirable fuel properties.

16.2.3 Biomass Transportation Infrastructure

Feedstocks are collected over large areas and transported to biorefineries. From there, bioproducts are distributed to the consumers. Transportation of feedstocks and products are important cost factors in a regional biofuel assessment. Field harvested corn and cellulosic biomass has a low energy density compared to solid fossil fuel sources such as coal. Since biorefineries are more likely to be profitable in areas closer to production and demand centers, finding the right balance between the location in rural areas and the demand in urban areas is a task for optimization.

Rural areas growing biomass require a good infrastructure, including roads, and access to electricity, natural gas, and water, which can differ significantly between regions. Transportation is often a limiting factor, possibly leading to supply disruptions and sometimes congestion. Significant growth in the distribution system faces a number of impediments and concerns about the system's ability to effectively transport the biomass and biofuels needed if production significantly increases.

The costs and financial returns to farmers and refineries vary with the travel distance of biomass (Chap. 13). Given the transportation demands for large-scale distribution of biomass and biofuels, it is important to find mechanisms to avoid transportation from becoming a bottleneck and limiting cost factor. An efficient distribution network combines cost-effective and sustainable transport of biomass, biofuels and biogas by road, rail, sea and pipeline with properly located storage and conversion plants. Decision-making on the siting and sizing of plants and refineries for power, fuel, chemical and food production has to address tradeoffs between large-scale facilities that take advantage of economies of scale and decentralized network nodes closer to producers and consumers.

From the biorefinery, ethanol moves to petroleum terminals where it is blended with gasoline. An inadequate blending and distribution infrastructure could become a serious constraint to the marketing of ethanol as the industry expands. To address these constraints, capital investment "in tanks, rail siding and cars, barges, piping, blending, metering, terminals, etc., is needed to sustain the expected national diffusion of demand" (Brown et al. 2007). Each of the transportation modes has its own characteristics, making it appropriate in a particular environment (Kang et al. 2010).

Given the limits of pipelines, ethanol is distributed to blending terminals and further to gasoline retailers primarily by rail, and less often by truck and barge, which is similar for biodiesel. This distribution system is more complicated and more costly than for petroleum fuels. While nationwide petroleum transportation from refineries to fueling stations is estimated to cost about 3–5 cents per gallon, the overall cost of transporting ethanol is estimated at between 13 and 18 cents per gallon, depending on distance traveled and mode of transportation (GAO 2007, p 23). The increase in ethanol production has contributed to regional supply shortages. For example, the number of ethanol carloads has tripled between 2001 and 2006, and this number is expected to further increase (Brat and Machalaba 2007). The freight rail system has limited capacity and may not be able to meet the growing demand. Replacing, maintaining, and upgrading the existing aging rail infrastructure is extremely costly (GAO 2007).

A major limit is also given by the "blending wall", i.e., the maximum fraction of ethanol blending of 10–15% that can be mixed with gasoline without diminishing car fuel economy. An alternative could be flex fuel vehicles (FFVs) that can use any mix efficiently, including E85. In early 2007, about 1% of fueling stations in the US, primarily in the Midwest, offered E85 or high blends of biodiesel. Significant growth in the number of E85 stations and FFVs beyond the production regions requires considerable investment.

Since transportation and distribution are important limiting factors in the biofuel supply chain, efforts to find the best transportation network and location of biorefineries can make the whole process more efficient. To support optimal decision-making on infrastructure design, advanced modeling tools help to simulate the supply chain to a biorefinery and beyond. Location models of ethanol plants and biorefineries incorporate and integrate factors such as land use, transportation and optimal plant size. The problem is to locate the processing facilities so as to minimize the total transportation cost and the plant cost adjusted by the returns from by-product sales, such as DDGS and higher value co-products.

Building on research on biofuels and land use in Illinois (Khanna et al. 2008; Scheffran and Bendor 2009), Kang et al. (2010) have developed a model to identify the optimal locations for ethanol plants and biorefineries. The approach minimizes the total system costs for transportation and processing of biomass, transportation of ethanol and DDGS from refineries to the blending terminals and demand destinations, capital investment in refineries, and by-product credits. A multi-year trans-shipment and facility location model determines the optimal size and time to build each plant in the system, the amount of raw material processed by individual plants, and the distribution of bioenergy crops, ethanol and co-products. Based on the distances (and hence transportation costs) from production sources of the feedstock and the road/railroad network, certain locations were found to be more suitable for corn and corn-stover-based ethanol plants, while others are more suitable for producing ethanol by using perennial grasses (Miscanthus).

16.3 Biogenic Wastes and Residues

Waste biomass is attractive because it can displace fossil fuel with material that typically would otherwise decompose, with no additional land use and GHG emissions required for its production (Scheffran et al. 2004). Using biogenic wastes and residues for biofuel production offers the greatest benefits with the smallest impacts and costs and would therefore have the highest priority. Using their energy content avoids methane emissions from slurry or landfills. Energy sources derived from biogenic wastes and residues are more relevant to climate change mitigation than farmed energy crops and therefore have higher priority in bioenergy production. The challenge is to improve the availability and accessibility of primary resources. Conversion technologies are available, as is the know-how to operate them, including biogas and biodiesel facilities, biomass-fired cogeneration systems or pellet heating systems.

Different from biomass specifically cultivated for energy purposes, residues and wastes are available as a by-product of other activities. These occur in different sectors (agriculture and forestry, manufacturing, municipal enterprises and private households) and at different stages of the value chain. Primary residues are generated in the field during production of food crops and forest products (e.g., straw, corn stalks and leaves, or wood thinnings from commercial forestry). Secondary residues become available in industries during the processing of food products or biomass materials. Examples are nut shells, sugarcane bagasse and sawdust. Tertiary residues remain after a biomass-derived commodity has been used, e.g., the organic fraction of municipal solid waste (MSW) or waste wood (Worldwatch 2007, p 48).

The physical and chemical characteristics of biomass wastes and residues vary widely, leading to different suitability for different conversion technologies (Faaij et al. 1997). Because of their higher lignin content, woody crops are considered more attractive for gasification and conversion to synthetic diesel fuel. Agricultural residues with a higher sugar content and less lignin (e.g., sugarcane leaves) tend to be bulkier and typically have greater amounts of ash than woody crops, making them more difficult to gasify and better suited for enzymatic conversion to ethanol. Certain sources such as sewage sludge, manure from dairy and swine farms and residues from food processing are very wet, with moisture contents over 60–70%, making them more suitable for producing biogas. It is expected that lower-cost residue and waste sources of cellulosic biomass, such as wood, tall grasses, forestry and crop residues, will provide an initial amount of "next-generation" feedstocks.

Large amounts of biomass from wastes and residues could potentially be used as feedstock for biofuel production. Much of the plant matter produced by common crops is left on the fields after harvest and decomposes into carbon dioxide. Harvested residues are generated at agricultural processing facilities and forestry mills. These include sugarcane bagasse, rice hulls, nut shells, sawdust and black liquor (at paper mills). In urban areas, cellulosic residues include portions of municipal solid waste, grass clippings, and wood from tree trimmings and land clearing activities (Worldwatch 2007).

Perlack et al. (2005) have estimated that over 1.3 billion dry tons per year of biomass from forestland and agricultural land alone are potentially available in the US, a large fraction of which is from wastes and residues. This amount is sufficient to meet more than one-third of the current demand for transportation fuels while still meeting food, feed, and export demands. This biomass resource potential can be produced with relatively modest changes in land use, or agricultural and forestry practices.

The technical potential of biogenic wastes and residues worldwide is estimated to be 40 Exajoule (EJ = 10^{18} Joule) to 170 EJ energy/year (Worldwatch 2007) or 80 EJ/year (WBGU 2009) which is roughly comparable to the 100 EJ used to meet global transportation needs today. However, only a few global assessments of waste potential exist, and these are subject to considerable uncertainty. How much of this technical potential can be utilized in a sustainable and cost-effective way is still the subject of research. This includes ecological research into the significance of residues for biodiversity, soil conservation and climate change mitigation for forests and agricultural land located in various climate zones and used in various ways. It also covers research into the infrastructure required, and into economically and technically appropriate ways of integrating this potential into global energy systems. The extent of the exploitable potential depends on the economic importance of biomass-intensive sectors and the infrastructure of the waste management sector (e.g., capacities for reprocessing, recycling, conversion for energy production, or landfilling) in different countries. In some countries, such as Germany, the biogenic waste potential is, to a large degree, already tapped.

There are limits for the removal of residues like straw that would impair soil fertility (humus content), depending on the site, crop rotation and input of other organic fertilizers. It has been recommended that, for sustainability, 67–80% of straw should be left on arable land (Knappe et al. 2007), and 40% of the exploitable potential of timber in forests for sustainability reasons (EAA 2007). Unseparated municipal waste, which consists largely of organic components, is often burnt for waste disposal purposes rather than for energy production. In some places, disposal is limited to landfilling waste without any separation at all—a practice that is banned in Europe and in other parts of the world is increasingly regulated to create incentives for the controlled capture and use of landfill gas. Greater promotion of biogenic wastes and residues for use as advanced energy sources could develop the unutilized potential more efficiently. Potential competition for waste resources can be reduced by cascade use where subsequent stages are optimally integrated, steered by waste management laws. To find the most efficient use, more research is needed to determine how much of this technical potential can be utilized in a sustainable and cost-effective way.

To reduce the impact on land resources available for the production of food crops, a further increase in biofuel production will require the use of agricultural materials not directly tied to food. Lignocellulosic biomass such as corn stover and corn fiber, perennial grasses. woody plants, mixture of prairie grass, agricultural residues, and municipal waste have been proposed to offer environmental and economic benefits. Compared to current biofuel sources, these biomass feedstocks require fewer agricultural inputs than annual crops and can be grown on agricultural marginal lands. After crop harvesting, the residues usually represent relatively large amounts of cellulosic material that could be returned to the soil for its future enrichment in carbon and nutrients or could be made available for further conversion to biofuels. Similarly, animal wastes are high in cellulose content and can also be converted to liquid biofuels. Such agricultural by-products can play an important role in triggering the transition to sustainable biofuels (Blaschek et al. 2010).

The use of by-products or co-products is widely recommended because it improves the cost effectiveness and energy balance for biofuel production. It has been estimated that DDGS production in the US could increase rapidly, from 15 million tons in 2006/2007 to more than 35 million tons in 2009/2010, and eventually to about 43 million tons by 2013/2014 (Babcock et al. 2008). This expansion could go together with further improvements in prefractionation, corn oil and corn fiber use, changing the current average 17.5 pounds of DDGS gained per bushel of corn in dry mills. While DDGS dried to 10% moisture can be stored and shipped across larger distances and into export markets, wet distillers grains with solubles (WDGS) has a wet moisture content of 65–70%, which reduces the amount of thermal heat used (Dooley et al. 2008). It cannot be stored for more than 5–7 days and is usually shipped to users within a 50 mile radius. In addition to starch, distillers grains contain fiber composed of cellulose, xylan and arabinan, which could be further hydrolyzed and converted into liquid fuels or other bioproducts. New approaches such as the Quick Germ-Quick-Fiber process (Singh et al. 2004) increase the efficiency and profitability of ethanol plants and reduce energy requirements and GHG emissions (Kim et al. 2008). A study by the Midwest Consortium for Biobased Products (Ladisch and Dale 2008) on the utilization of DDGS shows that it is possible to add value to distillers' grains by further processing them into additional fermentable sugars and ethanol, while leaving a solid that is reduced in weight and rich in protein. Corn coarse fiber is another co-product that can be utilized to generate heat and electricity when the fiber is burned in a steam generator/turbine. Burning of biomass from other waste material or crops would further lower the carbon footprint of ethanol plants by replacing coal and natural gas (Chap. 14).

It is often argued that once by-products become scarce and valuable, producing them will become more profitable, leading to increased production and thus higher risks for sustainability. One approach is to define by-products as having an economic value of less than 10% of the value of the crop as a whole as it leaves the farm, or of the total value of product leaving the factory (Dehue et al. 2008), exempting them from sustainability reporting. If the value of the by-product rises above the 10% threshold, it would no longer be counted as a by-product and therefore reporting on sustainability criteria would become a requirement.

16.4 Energy Balance and Efficiency

Bioenergy has been highlighted as being carbon neutral because the carbohydrates used to manufacture these fuels originate from atmospheric carbon fixed by photosynthesis. However, if the fossil energy input during the biofuel lifecycle is taken into consideration, this assumption becomes disputable. A key measure of the environmental impacts of biofuel is the fossil energy input into the bioenergy production path, which is measured by the ratio of the energy output from the fuel to the fossil energy input to produce and use the fuel. These net values are likely to vary substantially, depending on plant selection, growth and harvesting methods, as well as transportation and conversion processes. In various studies, the energy ratio for corn grain ethanol ranges from 0.8 to 1.45, while gasoline achieves 0.8 at best (Pimentel 2003; Kim and Dale 2004; Sheehan et al. 2004; Brinkman et al. 2005; Farrell et al. 2006; Hill et al. 2006; see also Chap. 15 by Wang, this volume). With advanced production techniques and process optimization methods, the energy balance can be considerably improved. With breakthroughs in cellulosic ethanol expected in the coming years, net biomass energy output could potentially increase by a factor of ten (Solomon et al. 2007). By more efficiently using biomass to produce energy, it may be possible to minimize land competition and reduce the associated conflicts.

Lifecycle analysis provides a framework to estimate the environmental impact of biofuels. Existing lifecycle models include the Greenhouse gases, Regulated Emissions and Energy use in Transportation (GREET) model of Argonne National Laboratory (Wang 2004), the Berkeley Energy and Resources Group Biofuel Analysis Meta-Model (EBAMM; Farrell et al. 2006), and the Biofuels Emissions And Cost Connection (BEACCON) model (Mueller and Plevin 2008).

The second generation of biofuels is assumed to be based on organic waste, residual crops, and nonfood sources of cellulosic feedstocks that are processed in integrated biorefineries. Although major progress in these areas is expected in the coming decade, the energy and carbon efficiency of the current generation of biofuels may be improved by short-term intermediate solutions during a transition period. Examples are the use of non-fossil sources (e.g., perennial grasses like Miscanthus) for energy input into ethanol production, or utilizing co-products from corn-ethanol production (DDGS) as animal feed in the US Midwest, or of bagasse from sugarcane in Brazil.

Ethanol plants today are powered largely by the fossil energy sources coal and natural gas, which contribute to an unfavorable energy and carbon balance of ethanol. Direct combustion may provide a near-term market and could help to improve the economic viability of energy feedstocks. Biomass can be co-fired with coal to produce electricity in ratios as high as 15% without loss of thermal efficiency and can be burnt in furnaces to produce heat using existing technologies.

Using electricity generated by co-firing biomass with coal and using biomass to generate heat in existing corn ethanol plants (Chap. 13) reduces the operating costs of ethanol plants and increases the savings in utility bills. Additionally, carbon savings will likely be generated under various climate policy scenarios with cap and trade programs. In the US Midwest, perennial grasses and corn stover are considered as major feedstocks of cellulosic biofuels in the coming decades but have no established markets as yet. This approach could offer a win–win solution, making existing ethanol production more sustainable by improving energy and carbon balance, and helping to trigger a market for efficient energy uses of cellulosic feedstocks that could stimulate landowners to convert some land from existing row crops to bioenergy crops.

Based on projected cost reductions for biomass-based energy systems, as well as a likely valuation of carbon in the fuel, ethanol plants may have incentives to switch to biomass-based fuel that is provided either as solid fuel for boilers or gasifiers or converted to biogas in integrated biogas energy systems using wet cake or agricultural waste such as manure from animal feedlots. Additionally, by reducing the GHG intensity of biofuels by more than 50% relative to gasoline, they may also be able to use corn ethanol to meet the mandate for advanced biofuels. Widely available technologies already exist to use biomass to heat buildings or generate steam and electricity on an industrial scale. Biomass from grass polycultures or perennial grasses such as switchgrass and Miscanthus can be used for direct combustion. Since transportation is an important cost factor, there is spatial variability in the costs of using biomass to produce heat and power.

16.5 Carbon Intensity and Conservation of Carbon Stocks

Reducing GHG emissions and preventing the destruction of carbon stocks are important sustainability criteria for the biofuel lifecycle. Although the carbon released during energy use was captured initially from the atmosphere by plants, bioenergy is not climate-neutral because additional emissions are caused throughout the lifecycle. GHG savings vary significantly across biofuels, partly because of different fossil fuel inputs, partly due to specific attributes such as sugar content (OECD 2008). Several studies have examined the lifecycle emissions of corn and cellulosic ethanol (for a survey, see Jacobson 2008). One study (Delucci 2006) also accounted for the emissions of soot, cooling aerosol particles, nitric oxide gas, carbon monoxide gas or detailed treatment of the nitrogen cycle. A few studies (Delucci 2006; Fargione et al. 2008; Searchinger et al. 2008) considered the change in carbon storage due to direct or indirect land conversion for fuel crops or price increases for soy or corn that may encourage biofuel expansion or land clearing in another country.

Relative to the fossil fuels displaced by biofuels, Hill et al. (2006) concluded that GHG emissions are reduced 12% by the production and combustion of ethanol over its whole lifecycle and 41% by biodiesel. For comparison, ethanol produced from sugarcane or second-generation processes may reduce GHG emissions by 80% or more (Chap. 4). Of the life-cycle carbon emissions generated by corn-ethanol in the United States, about two-thirds are generated during the process of conversion and one-third during the process of producing the corn. Two-thirds of the emissions during the process of conversion are generated by the use of coal and one-third by natural gas to power and heat the ethanol refinery. With current policy support, GHG emission reductions and use of fossil fuels amount to around 1% of the total, making biofuels based on current technologies a rather expensive path to energy security and mitigation of climate change (around US $ 1,000 per metric ton of CO_2-equivalent saved; OECD 2008).

Overall, the substitution of bioenergy for fossil fuels could reach 2–5 Gigatons (Gt) CO_2 equivalent (CO_2eq) per year total, under very optimistic conditions up to 4–9 Gt CO_2eq (WBGU 2009). This corresponds to an increase of a factor of roughly two to four over the mitigation efforts currently under discussion in the EU as a standard for biofuels in the transport sector. Compare this with around 50 Gt CO_2eq per year of global anthropogenic GHG emissions. WBGU (2009) considers that the use of bioenergy carriers reduces GHG emissions by at least 30 t CO_2eq per Terajoule (TJ $= 10^{12}$ Joule) raw biomass used in comparison with fossil fuels. For biofuels, this is equivalent to a GHG reduction of around 50% compared to fossil fuels. Additional GHG reduction from co-products can also be factored in.

An important indirect effect is the displacement of uncultivated areas into agricultural production, in particular the clearing of land for biofuel production, which releases GHGs from the soil and thus contributes to global warming. The effects of these indirect land-use changes (iLUC) are also referred to as leakage. Emissions created by the conversion of ecosystems that contain a high proportion of carbon (such as forests and wetlands, as well as some natural grasslands) generally can negate the climate change mitigation effects of bioenergy and may even exacerbate climate change. An indicator for the destruction of the carbon stock is the "carbon payback time", i.e., the number of years needed to grow a bioenergy crop to compensate for the loss of the carbon storage resulting from LUC. Thus, the release of a carbon stock becomes more acceptable with higher yield and global warming potential (GWP) and lower carbon intensity of the bioenergy chain, favoring the growth of perennial crops that store more carbon on average than annual crops.

Searchinger et al (2008) studied the effect of price changes on land use change with an econometric model and found that conversion from gasoline to ethanol

(E85) vehicles could increase lifecycle CO_2eq emission by over 90% when the ethanol is produced from corn, and around 50% when it is produced from switchgrass. According to Delucci (2006), ethanol from switchgrass might reduce US CO_2eq emissions by about 52.5% compared with light-duty gasoline in the US.

It has been estimated that for some biofuel pathways (such as forest clearing for soy to biodiesel production in the US) payback time can be several decades to compensate for the clearing of land. In other cases, payback time could be less than a decade—for example, grassland conversion to palm for biodiesel or sugarcane for bioethanol (Gallagher 2008). Maximum payback time has been proposed as 10 years (Dehue et al. 2008), based on calculations on emissions from land use change worked out in the IPCC Technical Guidance. WBGU (2009) argues against direct or indirect conversion of woodland, forests and wetlands into land used for energy crops. Information on GHG emissions associated with different forms of cultivation and land use is inadequate; reliable data are rare and reporting insufficient. Thus, indirect LUC are difficult to account for in a bioenergy standard using current methodologies, increasing the need for more in-depth studies of the main cultivation systems in major producing countries. One possible approach is to include the risk of a potential indirect LUC through an additional iLUC factor that puts the additional GHG emissions onto the GHG balance of the bioenergy carrier. Such a factor has been used by the Öko-Institut, for example, and has already been used in model calculations (Fritsche and Wiegmann 2008).

In the future, biomass could play a significant role in removing carbon dioxide from the atmosphere. Biological carbon sequestration can make production, deconstruction, and fermentation processes of biofuels even "carbon negative" through photosynthesis and CO_2 storage in biomass, soils, and sediments (Post and Kwon 2000). Current research focuses on studying the opportunities for enhancing terrestrial sequestration by photosynthetic organisms for CO_2 fixation, by accumulation of soil organic matter, by reducing CO_2 emissions from soils, and using degraded lands to sequester carbon. Research in molecular bioengineering searches for new means to enhance terrestrial CO_2 uptake and storage processes. Another area of research focuses on the potential of using perennial bioenergy crops such as switchgrass and Miscanthus for increasing soil organic matter, sequestering carbon and replacing fossil fuels and annual row crops such as corn and soybean (Dohleman et al. 2010; Khanna et al. 2010).

The possible contribution of black carbon sequestration to climate change mitigation has been suggested and debated (e.g. Marris 2006; Lehmann 2007). Converting biomass into biomethane by fermentation or gasification through pyrolysis allows one to separate the CO_2 from these processes and deposit it in storage sites, thus simultaneously using the biomass for energy generation and for CO_2 sequestration. The remaining carbon can be inserted as charcoal into the soil, where it decomposes very slowly and can thus be removed from the atmosphere for a long time. The charcoal serves as a soil conditioner by improving the structure and fertility of the soil, an effect that is known from the very fertile *terra preta* (dark earth) soils of the Amazon basin (Denevan and Woods 2004; Fowles 2007). Char can sequester or store carbon in the soil for hundreds or thousands of years, improving soil fertility and stimulating plant growth, which then consumes more CO_2 from the atmosphere.

Biochar is a fine-grained charcoal produced from pyrolysis of plants and organic waste that is high in organic carbon and largely resistant to decomposition. Soil amended by biochar has an enhanced nutrient retention capacity and thus can reduce fertilizer, climate and environmental impact of croplands (see http://www.biochar-international.org). According to Lehmann (2007), incorporating charcoal into the soil would result in emission reductions of 12–84% compared to using the charcoal to replace fossil fuels for energy production. The biochar process offers a possible win–win solution for energy and climate change (Laird 2008) but its global potential, as well as its cost-effectiveness and environmental impact, is as yet unclear. To determine the associated GHG balance and any further ecological impacts from this technology requires further research to measure how long the carbon stored in the soil remains removed from the atmosphere.

A comprehensive agreement on the limitation of GHG emissions and conservation of terrestrial carbon reservoirs could be combined with the gradual introduction of minimum standards and the reform of accounting procedures that could be embedded into international standards and certification for bioenergy carriers and sustainable land use.

To minimize GHG emissions in the cultivation process, possible approaches are to avoid tillage of the soil, maintain year-round ground cover, reduce the use of primary energy, increase the use of biogenic fertilizers (e.g., green manure), and reduce leaching of nutrients. Biomass production should not destroy or damage large carbon stocks. This requires evidence that biomass production does not cause LUC, with a possible carbon payback time exceeding 10 years. Certain soil types, such as peat lands, mangroves, wetlands and certain grasslands, would be exempted from biofuels production.

16.6 Soil Protection, Land Use and Food Security

Due to its low energy density, bioenergy requires large land areas, making land a critical resource for bioenergy production. Possible direct land use effects of bioenergy include the local environmental impacts upon air, water and soil quality, exacerbation of local water supply and the destruction of habitats. Since protection of soil quality is essential, bioenergy crops should not impair soil functions or soil fertility, e.g., through erosion, salinization, compaction or nutrient depletion. When removing agricultural residues for energy recovery, an adequate proportion of the residues needs to be left in the fields to maintain nutrient cycles and for humus formation. Bioenergy should comply with the provisions of national or regional laws of agricultural, forestry and fisheries sectors, for instance regarding the correct use of fertilizers, restrictions on pesticide use and avoidance of sediment input into neighbouring ecosystems (WBGU 2009). To apply principles of good agricultural practices for soil protection, specific indicators are required for control of erosion, salinization, use of fertilizers and pesticides, and for improving soil nutrient balance, soil organic matter, soil pH, soil structure and soil biodiversity (Dehue et al. 2008).

Recent decades have seen a significant increase in global food production, attributed largely to improved germplasm and increases in land productivity caused by the modernization and mechanization of farming. For instance, corn yields increased by more than 50% between 1975 and 2008, and this increase is supposed to continue in the future (DOE 2008). To feed an expected world population of 8 billion people, the Food and Agricultural Organization (FAO) estimates that by 2030 some 50% more food will be needed (FAO 2003a). Given the limits and constraints on the total area of agricultural land, about 80% of this increase will have to be achieved through more intensive agriculture without compromising the environment (WBGU 2009). Another source of demand for land are the meat-based food consumption patterns of the industrialized countries, which are spreading to emerging economies such as China.

Energy crops compete with food and animal feed for land and other production factors in agriculture (such as machinery, fertilizers, seed, feed and fuel). One of the indirect land use issues of bioenergy discussed since 2008 is the displacement of food production, potentially causing rising food commodity prices on the global market and food insecurity for the poor (Fargione et al. 2008; Searchinger et al. 2008; Gallagher 2008; Sylvester-Bradley 2008). Runge and Senauer (2007) criticized the competition from biofuels for land and food crops, which, in their view, has already contributed to increasing the price of staple foods, e.g., the sharp rise in the price of white corn and tortillas made from this corn in Mexico in 2006. Other examples are the possible expansion of soy production in South and Latin America as a consequence of increasing production of corn in the US (and decreasing production of soy), or increasing demand and prices for rape oil seeds for biodiesel production in the EU, possibly leading to expansion of palm oil production in South-East Asia. Some suggest that expansion of sugarcane production in Brazil could lead to displacement of cattle ranching and accelerated deforestation in Amazonia (Barreto 2008). In a study for the International Food Policy Research Institute (IFPRI), von Braun and Pachauri (2006) predicted that an aggressive biofuel scenario—without technological breakthroughs increasing productivity—could lead to significant price increases for some food crops.

The cultivation of energy crops "implies a close coupling of the markets for energy and food. As a result, food prices will in the future be linked to the dynamics of the energy markets" (WBGU 2009). The overall effect will depend on the type of bioenergy, the natural, agricultural and social conditions and the development of global food markets.

That food price is strongly connected to the price of petroleum was confirmed by the sharp price drop in both commodities during the 2008 economic crisis, which occurred despite the continued boom in biofuels (Tyner 2009) and seems to have had only a minor effect on food prices. In its 2008 Economic Assessment of Biofuel Support Policies, the OECD suggested that the medium-term impacts of current biofuel policies on agricultural commodity prices are important, but should not be overestimated (OECD 2008).

Since the relationship between food security and bioenergy production is complex and not sufficiently understood, further research and monitoring is needed. Higher food prices are not necessarily a threat to food security because farmers and net exporters of food benefit from higher commodity prices while the burden on consumers increases (Chap. 14). For people living in poverty and spending the majority of their income on food, high food prices pose a threat to life and livelihood. Most likely affected by rising global market prices is the group of the Low-Income-Food-Deficit Countries (LIFDCs), many of which have a significant bioenergy potential. To strengthen agricultural markets in poorer developing countries, trade barriers to agricultural goods from these countries should be removed, including tariffs and export subsidies in the industrialized countries (WBGU 2009).

Perennial crops could reduce land demands and improve environmental quality due to lower fertilizer requirements compared to established farming practices in corn and soybeans. Extensive perennial root systems and winter harvest may improve water quality, decrease soil erosion and increase soil organic matter (SOM). To power all US on-road vehicles with E85, Jacobson (2008) estimated that cellulosic ethanol from switchgrass could require between 430,000 and 3,240,000 km^2 (106–800 million acres), using a wide range of switchgrass yields of 2.23–10 t dry matter/acre. More efficient use of the land could significantly reduce the demand for land. For instance, to replace 10% of its gasoline demand with ethanol, the US would need to devote approximately 15% of its agricultural land area to ethanol-generating bioenergy crops. With possible efficiency gains (higher yields, more efficient energy conversion, lower fuel consumption in cars), land needs could be possibly reduced to one-sixth of this land area (National Commission on Energy Policy 2004). These efficiency improvements in the use of biomass for energy purposes could make a contribution to avoiding land use conflicts. Total area could be further reduced by using a high-yield perennial grass such as Miscanthus, which could double the yield compared to switchgrass and would offer several other advantages (Dohlemann et al. 2010). Miscanthus meets 12 out of 13 ideal biofuel feedstock characteristics: high yielding, perennial, C4 photosynthesis, long canopy duration, low fertilization requirements, low herbicide requirements, low pesticide requirements, sterile, non-invasive, winter standing, easily removable, high water use efficiency, use of existing farm equipment.

A comprehensive review of the indirect impacts of biofuels has been prepared by Gallagher (2008) for the UK Renewable Fuels Agency. Despite concerns that biofuels contribute to rising food prices and adversely affect the poorest, the report concludes "that there is a future for a sustainable biofuels industry but that feedstock production must avoid agricultural land that would otherwise be used for food production... The introduction of biofuels should be significantly slowed until adequate controls to address displacement effects are implemented and are demonstrated to be effective." The report asserts that there is probably sufficient land for

Table 16.1 Land use categories distinguished for sustainability reporting (Dehue et al. 2008)

Land use	Description
Cropland	Includes cropped land, including rice fields and set-aside, and agroforestry systems where the vegetation structure falls below the thresholds used for the Forest Land category
Forest land	Land spanning more than 0.5 ha with trees higher than 5 m and a canopy cover of more than 10%, or trees able to reach these thresholds in situ. It does not include land that is predominantly under agricultural (or urban) land use
Grassland (and other wooded land not classified as forest) with agricultural use	Includes rangelands and pasture land that are not considered Cropland but which have an agricultural use. It also includes systems with woody vegetation and other non-grass vegetation such as herbs and brushes that fall below the threshold values used in the Forest Land category and which have an agricultural use. It includes extensively managed rangelands as well as intensively managed (e.g., with fertilization, irrigation, species changes) continuous pasture and hay land
Grassland (and other wooded land not classified as forest) without agricultural use	This category includes grasslands without an agricultural use. It also includes systems with woody vegetation and other non-grass vegetation such as herbs and brushes that fall below the threshold values used in the Forest Land category and which do not have an agricultural use

food, feed, and biofuels, and that biofuels production must target idle and marginal land and use of wastes and residues (Gallagher 2008).

Specific incentives are required to stimulate advanced technology, and stronger, enforced global policies are needed to prevent deforestation. A carbon and sustainability reporting scheme would monitor fuel supplier performance for each of the land use categories (see the classification by Dehue et al. 2008 in Table 16.1). One approach is to define boundaries for admissible action (so called guardrails). For instance, WBGU (2008) recommends that the "amount of agricultural land available globally must at least be sufficient to enable all people to receive food with an average calorie content of 2,700 kcal per person per day." Since current global food production amounts to an average of approximately 2,800 kcal per person per day (FAO 2003b), hunger and malnutrition are primarily problems of access and distribution of food.

The food crisis and the coupling between agricultural and energy markets have spurred efforts in development cooperation in the UN system to draw up strategies and action plans for limiting the emerging risks to food security. Possible elements could include (WBGU 2009):

- Conditions for food production in regions at risk must be directly and immediately improved (e.g., through provision of seed for the next harvest).
- Attention must focus on improving conditions for food security and food production in the medium to long term (e.g., through conversion to more productive farming systems).
- Activities must be coordinated and integrated with other policy areas, such as climate change mitigation and biodiversity conservation.
- Food and feed production take precedence over energy crop cultivation.

To regulate the displacement problem of land, standards and certification systems, including monitoring processes, are still missing. Appropriate models are needed to represent the complex causal relationships. With current methodologies, iLUC are difficult to account for in a bioenergy standard. A proposed criterion limits the cultivation of energy crops to marginal land, e.g., fallow land and land with low productivity in its previous use. The impacts of indirect land-use change on biodiversity and food security, in addition to the GHG emissions, could be represented by an iLUC factor. As WBGU (2009) notes, "the problem of indirect land-use change in connection with bioenergy crops can only be completely resolved if all the countries and all types of biomass are included within a uniform standard, or if binding international agreements are concluded on national land-use planning criteria (including systems of protected areas)."

A possible international framework is the United Nations Convention to Combat Desertification (UNCCD), which can serve as a platform for programs and strategies supporting sustainable land use that help to reduce poverty in the regions affected by drought and desertification. In this context, production of energy crops is seen as an opportunity to generate income and export revenues from new energy sources that strengthen motivation and capability to combat desertification. This could help to improve the condition of degraded dryland ecosystems and improve the living conditions of people affected by drought and desertification (WBGU 2009). The National Action Programmes (NAP) to Combat Desertification and the UNCCD 10-year strategic plan adopted in 2007 offer possibilities for standard-settings for sustainable bioenergy use.

16.7 Water Needs and Water Crisis

One consequence of bioenergy expansion is greater demand for freshwater. Currently, global biofuel crops account for about 1% of total crop water requirement and 1.67% of irrigation water use (IWMI 2007). The largest cumulative crop water requirements are from corn and sugarcane. Unsustainable production of energy crops may increase the pressure on available water resources (Varghese 2007). Together with the impact of climate change and soil degradation this could further increase the pressure on water resources and aggravate water insecurity in many parts of the world (IPCC 2007; FAO 2008).

Limited attention has yet been paid to the impact of biofuels expansion on water use and criteria for the sustainable management of water resources. Lifecycle analyses has focused largely on the net energy balance or greenhouse emissions in biofuel production, while only a few studies have assessed the environmental impacts of water use in biofuel production (Environmental Defense 2007; NRC 2007). The biofuel lifecycle and the water cycle are interconnected in multiple ways. Water inputs are required in growing feedstock and for the production process in biofuel plants. In addition, water quality is affected due to water pollution (Varghese 2007).

Water requirements and impacts vary significantly for different feedstocks. Most important is the intensity of irrigation, and crop water use increases substantially in fully irrigated agriculture. Other factors that determine water use are the crop type and evapo-transpiration at different stages of crop growth in a particular agro-climatic zone. What also matters is the cultivation method and the use of fertilizers and pesticides. It makes a difference whether existing native vegetation is used as feedstock, or whether it is converted and production practices are shifted.

There is wide range of water requirements across the US. For Iowa, the unirrigated crop water requirement for producing a gallon of ethanol has been estimated to be between 1,081 and 1,121 gallons water (Al-Kaisi 2000). If corn is irrigated, as in Southwestern Nebraska, the estimated average crop water requirement increases to about 1,568 gallons water for producing a gallon of ethanol (Varghese 2007). Generally, intensive mono-cultural crops have a higher water demand, but most corn in the US is grown with natural rainfall and is not irrigated. An important question is whether cellulosic biofuel feedstocks would increase or decrease water consumption. According to Jacobson (2008), the use of switchgrass for ethanol production would most likely reduce irrigation in comparison with the use of corn. However, since agricultural productivity increases with irrigation it is likely that some growers will irrigate to increase productivity. For instance, irrigated corn produced 178 bushels per harvested acre in the US in 2003, compared to 139.7 bushels per harvested acre for a mix of irrigated and non-irrigated corn (Jacobson 2008).

A study by the International Water Management Institute (IWMI) compared average crop water requirements for major biofuel crops in selected countries (de Fraiture et al. 2008). Due to higher irrigation, maize-based ethanol in China consumes almost 1.5 times that in the United States (Varghese 2007). To address concerns about food security, China prefers non-food crops such as sweet sorghum and jatropha to produce ethanol and bio-diesel (Xinhua 2007).

Compared to water use in feedstock production, water requirements in biofuel processing appear to be minor. Much of the water consumed is from evaporation during cooling and wastewater discharge. In Minnesota, the water use efficiency in some of these plants has improved from about 5.8 gallons water per gallon of ethanol produced in 1998 to 4.2 gallons water per gallon of ethanol produced in 2005 (Keeney and Muller 2006). Thus, water requirements in modern plants is less of a concern than for growing biofuel crops. However, the water used in biofuel processing plants is drawn from a smaller area, and can have localized impacts on water quality and quantity. The siting of these plants in water-scarce regions and localities can affect the water available for other basic needs.

Less attention has been paid to the impacts of biofuel production on water quality, which depends on the practices applied in growing feedstock, and the regulations of effluents from crop production and processing, such as nitrogen, nutrients and pesticides. Corn is the most nitrogen-intensive of major field crops. Excess nitrates leach through the soil into ground water and contaminate both soil and water sources. For instance, nutrient leaching from farm land has affected the Mississippi River and its tributaries. This contributed to high rates of algae growth in the Gulf of Mexico and caused oxygen depletion in the Gulf (Varghese 2007). Sediment erosion also affects water quality.

A report on the water implications of biofuel production in the US warns that "if projected increases in the use of corn for ethanol production occur, the harm to water quality could be considerable, and water supply problems at the regional and local levels could also arise" (NRC 2007). The report by Keeney and Muller (2006) indicates that "shortage of water could be the Achilles heel of corn-based and perhaps cellulose-based ethanol." In its 2008 strategy paper, the United Nations High Level Task Force on the Global Food Crisis points out the rising competition for use of freshwater resulting from the global bioenergy boom. Among its recommendations are the development of standards on the sustainable use of freshwater, prioritizing water use for food production (UN 2008).

As indicated by these reports, an excessive and indiscriminate biofuel growth may aggravate the water crisis, possibly making access to water a primary limiting factor in the production of biofuels. In regions already under water stress, biofuel production may further decrease the freshwater availability for other development options and basic needs (Varghese 2007). Over 1.2 billion people do not have access to safe drinking water, and almost 40% of world population does not have access to water to meet their daily sanitation needs. Unless drastic changes are made, there will not be enough water to meet the world's food, feed and fiber needs in the coming five decades (IWMI 2007). Climate change will further aggravate the water crisis and potentially affect several billion people (IPCC 2007). Increasing production of thirsty crops and livestock have brought severe strains on water resources in many parts of the world, including parts of North America and the European Union. In China, 550 of its largest 600 cities already face water shortages, and compete with farmers and rural areas for water (Vidal 2006).

The growing pressure on water access and the increasing demand for bioenergy highlight the need for regulation of the sustainable use of freshwater resources. While international agreements exist for climate protection, conservation of biodiversity or soil protection in drylands, global freshwater resources have yet been largely unregulated. However, provisions of national or regional (e.g., EU) law exist concerning the protection of water resources in the agricultural, forestry and fisheries sector. The World Water Council organizes the World Water Forum, a gathering of international water experts, decision-makers, scientists and representatives of international organizations. The consequences of energy crop production for water use were on the agenda for the fifth World Water Forum in Istanbul in 2009. Various UN organizations are active in the water sector (e.g., WHO, FAO, UNDP). The UN Commission on Sustainable Development (CSD) is engaged in discussion of integrated water resources management that may contribute to policy recommendations to achieve the Millennium Development Goals in the water sector although binding regulations are unlikely due to lack of power (WBGU 2009). Several principles and criteria of sustainable water use are subject to debate: water use should be efficient, water access and quality not be significantly impaired. Groundwater resources should not be overused. Agro-chemicals should be used responsibly. If irrigation is used, this should be based on an effective integrated water resources management plan. Salinization and waterlogging should be avoided. Good agricultural practices are documented annually.

16.8 Wildlife, Biodiversity and Environmental Impact

Expansion and intensification of bioenergy production may have significant environmental impacts upon air, water and soil quality and cause ecological problems that directly or indirectly affect wildlife and biodiversity, including bird and insect habitats, as well as the spread of genes and plants into native communities. The intensive cultivation of monoculture cash crops such as sugarcane and corn is often associated with environmental impacts from increased agrichemical uses (especially nitrogen and phosphorus from fertilizers and pesticides) via leaching and surface flow from farms that causes damage to habitats and aquifers (Goolsby et al. 2000). Soil erosion, degradation of water resources or the conversion of land from natural vegetation to cropland could affect natural ecosystems and reduce available habitats. Removal of significant proportions of crop residues, such as corn stover, that are currently returned to soil not only increases atmospheric carbon but can decrease mineral nutrients and soil organic matter, and disturb the biodiversity of soil organisms. For instance, when switchgrass replaces a non-biofuel crop and the lignocellulose is removed to produce ethanol, microorganisms, which normally process the lignocellulose, cannot replenish soil nutrients, reducing biota in the soil (Jacobson 2008).

Possible direct land use effects of bioenergy expansion include the destruction of habitats, where impacts are greatest when biomass-rich ecosystems, such as tropical or other forests, are affected. Particular concerns have been raised for soy production in Amazonia and for palm oil in South-East Asia. The impacts could also occur indirectly when biofuel farming in other areas causes cattle ranchers or soy farmers to move and clear rainforest areas. Another indirect effect would occur if the use of non-native species or genetically modified organisms (GMOs) changes the regional composition of species or leads to biodiversity loss. A related factor is the impact on bird and insect habitats, and whether genes and plants spread into native communities (Keeney and Nanninga 2008). An important question is how perennials affect soil microbial diversity, including pathogens and pests.

As long as the expanding bioenergy sector is increasing the pressure on ecological systems and biodiversity, large-scale conversion of land for bioenergy production will require criteria to conserve biological diversity and ensure its sustainable use. A framework for such measures could be the Convention on Biological Diversity (CBD), which refers explicitly to conservation and sustainability objectives. Some parties are concerned about possible trade barriers imposed by industrialized countries that are in conflict with WTO free trade rules (WBGU 2009). At the Conference of the Parties (COP-9) meeting in Bonn, a EU-led resolution on biofuels was adopted, which reaffirmed the principle of sustainability and acknowledged the role of the CBD in this area. An important issue is the enlargement and effective management of the global system of protected areas that can help to limit ecosystem conversion from bioenergy utilization. Concrete biodiversity guidelines for the development of bioenergy standards are subject to debate. These could cover the following key dimensions (WBGU 2009):

Expanding land use for bioenergy production should not endanger ecologically valuable sites, such as primary forests or wetlands, species-rich grasslands or savannahs. Protected areas and elements of protected area systems and ecosystems of high biodiversity value would be exempt from conversion for bioenergy purposes. The World Conservation Union (IUCN) and the World Database on Protected Areas can serve as a basis for the necessary monitoring and possible expansion to a global satellite-based land-use register (WDPA 2008). Critical areas need to be "identified, encircled with buffer zones and, where reasonable, networked with corridors with a view to establishing habitat connectivity" (WBGU 2009). Before unused land (such as marginal, degraded or fallow land) is converted for bioenergy, its ecological value and nature conservation value should be assessed, exclusion zones must be identified prior to cultivation of energy crops, which should be embedded in the landscape to link with protected area systems, and preserve landscape diversity and agrobiodiversity. Large-scale conversion of land from current uses will require environmental impact assessments. Removal of significant proportions of crop residues, such as corn stover, which are currently returned to soil, should not destroy the biodiversity of soil organisms.

Use of biogenic residues or energy crops should be sustainable. In agricultural and forest ecosystems, the accompanying flora and fauna and genetic diversity should be safeguarded and adverse impacts on other ecosystems should be avoided. Regulations are tailored to the production system and to local conditions, with due regard to such factors as observing crop rotations, use of water and agrochemicals, and avoiding the cultivation of potentially invasive species. Perennial crops should be considered to reduce land demands and improve environmental quality due to lower fertilizer requirements compared to established farming practices in corn and soybeans. Extensive perennial root systems and winter harvest may improve water quality and nitrogen fixation, decrease soil erosion and increase soil organic matter and soil biodiversity. For the use of GMOs, the CBD and the provisions of the Cartagena Protocol on Biosafety provide a framework for action, yet to be worked out in detail. The task is to prevent such risks as the spread of modified genes into wild populations or to rule out contamination of the food and animal feed chain. Cropping systems with maximum possible diversity are preferable. Introduction of potentially invasive species should be avoided.

In practice the criteria on biodiversity will have a strong overlap with the criteria on LUC and ground carbon storage, where measuring biodiversity is more complex than measuring carbon storage. In many cases where forest conversion is involved, conversion to plantations will be excluded based on the carbon storage criteria and where more complex debate on biodiversity is not needed for assessing compliance.

16.9 Health, Safety and Social Criteria

The lifecycle of bioenergy affects human beings in various ways. On the one hand, bioenergy serves human well-being by providing economic growth, jobs and income. In rural areas and in developing countries, bioenergy can fuel development and reduce energy poverty where small-scale initiatives are often more appropriate than large projects (FAO/PISCES 2009). On the other hand, biofuels could be produced in a way that adversely affects human health and working or living conditions (Hill et al 2009). To strengthen the benefits and reduce the risks, there are increasing demands that bioenergy should be produced in a socially and economically viable manner. To compete with fossil fuels, biofuels will have to become more competitive and market-oriented to avoid some of the distortions from subsidies and tariffs. During production, unfair labor practices or working conditions that endanger health or safety should be avoided. Basic social standards for bioenergy carriers need to be defined and observed, building on labor standards of the International Labour Organization (ILO). Particular social criteria for the sustainable production of biofuel feedstocks have been defined by Social Accountability in Sustainable Agriculture (SASA) regarding labor conditions (not on land right issues). Some recommendations will be highlighted in the following, not all of which are proposed as minimum requirements (for a comprehensive list, see Dehue et al. 2008).

16.9.1 Health and Safety

Most of human history has relied on traditional uses of biomass that are often inefficient, unhealthy, and non-sustainable, such as the burning of straw, dung, and wood to satisfy basic human needs. More than one-third of the world's population still depends on these forms of bioenergy (Ezzati and Kammen 2001), and it is estimated that the pollution caused by open fires claims the life of more than 1.5 million people each year. Thus, replacing these energy uses by more advanced forms could make a major contribution to human health (ICRISAT 2007). This does not imply that biofuels and other advanced forms of bioenergy use have no health impacts. Like for any fuel, emissions during the lifecycle of biofuels (CO, NO_2, N_2O, black carbon) are potentially harmful to human health when inhaled. Jacobson (2008) indicates that corn- and cellulosic-E85 emissions from light- and heavy-duty gasoline on-road vehicles could even cause more casualties than gasoline and other renewable energy sources. To reduce air pollution risks requires environmental impact assessments and compliance with national laws and regulations relevant to air emissions and burning practices. Open burning due to land clearing or waste disposal should be abandoned. Health and safety are important requirements in the workplace, including "potable drinking water, clean latrines or toilets, a clean place to eat, adequate protective equipment and access to adequate and accessible (physically and financially) medical care" (Dehue et al. 2008). Workers would receive regular health and safety training, and be trained to avoid and manage accidents and hazards. The work would avoid activities that are hazardous or dangerous to the health and safety, in particular of young workers.

16.9.2 Decent Working Conditions

To address concerns about unacceptable working conditions in parts of the biofuels production process (notably in harvesting of crops), standards for fair working conditions are suggested. These could include the payment of a living wage, regular working hours (usually 8 h a day and 48 h a week), sufficient rest time and overtime payment. According to ILO conventions, workers are guaranteed the rights to organize and negotiate their working conditions and should not be discriminated or suffer repercussions when exercising this right. Discrimination that denies or impairs equal opportunity, conditions, or treatment based on individual characteristics and group membership or association such as race, nationality, religion, disability, gender, age, sexual orientation or political affiliation, is not permitted (Dehue et al. 2008). Forced labor should not be supported in any way. Child labor is restrained. Children can work on family farms if not interfering with their educational, moral, social and physical development.

16.9.3 Fair Feedstock Production and Land Rights

In the past, new plantations of feedstocks have violated or restrained the interests of indigenous peoples in parts of the world. Generally, it is essential to respect land tenure and ownership rights and not violate human rights. Bioenergy production should support the well-being of local communities and stakeholders, and balance crucial issues such as equity and gender. In contract farming, fair market relations and price building mechanisms should be observed as well as reliable and transparent business relations with smallholders and other local businesses established.

16.10 Sustainability Standards and Certification Schemes for Bioenergy

There is an emerging debate on principles and criteria for a sustainable bioenergy lifecycle that minimizes competition with food crops and adverse environmental impacts (UN 2007; NRC 2009; Tilman et al. 2009). If countries and the international community can agree on the appropriate set of sustainability standards these can be applied to the certification of bioenergy products, similar to standards that apply to other areas. This implies that only those feedstocks and production processes for which the producer can provide evidence for compliance with these standards would be promoted. The purpose is to channel bioenergy use into sustainable pathways, to "maximize the use of potentials while minimizing risks" (WBGU 2009). A certification process would be particularly relevant for renewable energy sources that receive policy support through mandates, subsidies and tariffs to meet widely accepted societal goals such as energy security and preventing climate change. Not surprisingly, as the number and stringency of criteria increases, it becomes harder to satisfy all of them, which restrains the spectrum of available options. On the other hand, further investments and research efforts are required to meet these standards, which may become a driving force for innovations. The challenges are highest where conflicting goals have to be addressed.

Until recently, international markets have not come up with a label for bioenergy and other biomass products. Thus, consumers did not have the possibility to choose sustainable products, and no incentives for sustainable production existed for farmers and bioenergy producers. To overcome this deficit, various initiatives were launched to develop standards and certification mechanisms to distinguish sustainable from non-sustainable products on the market. New regulatory frameworks are taking shape, at both the fuel and feedstock levels, and a few systems regarding bioenergy are operational (Smeets et al. 2008; Smith et al. 2009).

One of the oldest is the Forest Stewardship Council (established in 1994), which has certified about 7% of the world's productive forests, and which may expand "its solutions to non-timber management objectives such as climate change and biofuels". Some of the initiatives are voluntary, such as the Roundtable on Sustainable Biofuels, which includes a large number of stakeholders, including major companies and environmental groups, and has developed a first set of principles and criteria for certification of sustainable biofuels, specifically for ethanol production. Since 2004, the Roundtable on Sustainable Palm Oil established principles and criteria for sustainable palm oil production, used to certify over 1.3 million tons palm oil, of which less than 15,000 tons have been sold (WWF 2009). A similar initiative is the Round Table on Responsible Soy, and the US Sustainable Biodiesel Alliance has discussed principles and baseline practices for biodiesel sustainability. A consensus-based Sustainable Agriculture Practice Standard has been under debate in the US that went through redefinition under a ruling of the American National Standards Institute. In the international context, a "Biopact" for a North-- South Trade in biofuels (Mathews 2007), aims at establishing ecological and social standards instead of trade barriers to open a fair market access and implement sustainability standards for tropical biofuels. Finally, the Global Bioenergy Partnership (GBEP) brings together public, private and civil society stakeholders at a multilateral level in a joint commitment to promote negotiation processes and accelerate the formulation of bi- and multilateral policies on global standards. Implementation could be facilitated with political support from the G8 (WBGU 2009). At the climate-change talks in Copenhagen there was increasing recognition of the Reduced Emissions from Deforestation and Degradation (REDD) initiative, which suggests that biomass criteria could be supported worldwide.

A comprehensive set of criteria was assembled by Dehue et al. (2008) as part of the Gallagher Review. To steer the production of bioenergy products along sustainable trajectories, the German Advisory Council on Global Change (WBGU 2009) recommended minimum standards that must be met in the production process, and so-called "guardrails" that define critical limit or threshold values that should not be exceeded. These could serve as benchmarks for the use of all bioenergy products and carriers, including biomethane, biofuels, electricity from biomass and wood pellets, energy crops, wood products, vegetable oils and crop residues such as straw and forest residues used in energy generation. Bioenergy use should be promoted only if it complies with the sustainability criteria while non-sustainable use would be restricted worldwide in the long term by a comprehensive and effective global regulatory framework. Sustainability standards could be connected to existing control regimes, e.g., the United Nations Framework Convention on Climate Change (UNFCCC) and the CBD. Country-specific sustainable bioenergy strategies would strengthen capacities for action, governance and monitoring, including application-oriented research and measures to improve food security. International development cooperation in bioenergy is essential to significantly reduce poverty and build partnerships for climate-friendly energy systems. In particular, WBGU suggests that the use of bioenergy should not endanger food security or the goals of nature conservation and climate protection. Pilot projects could focus on particularly efficient, innovative energy technologies, sustainable cultivation systems and the use of wastes and residues. Since some bioenergy feedstocks (e.g., rapeseed and palm oil, soya or grain) can be used for production of energy, food and animal feed, a minimum standard would affect all related products.

A number of European countries—including Germany, the UK, the Netherlands and Switzerland—have proposed the development of sustainability standards for bioenergy carriers, especially liquid biofuels. In Switzerland and the UK, the promotion or import of biofuels legislation is conditional on compliance with appropriate sustainability criteria. The world's first carbon and sustainability reporting scheme in the UK includes acceptable levels of environmental performance and average GHG savings for the proportion of feedstocks. The UK Renewable Transport Fuel Obligation (RTFO) is a requirement on transport fuel suppliers to supply 5% of all road vehicle fuel from sustainable renewable sources by 2010. To meet the full RTFO Sustainable Biofuel Meta-Standard, a set of 'minimum requirement' criteria and indicators are considered. Other criteria are seen as 'recommendations' that are not required but are considered good practice, including for instance a number of criteria on working conditions. For the purpose of the RTFO, idle land is land that meets the Meta-Standard on carbon storage (no destruction of large carbon stocks), on biodiversity (no conversion in or near areas with high conservation values), on land rights and community relations (no violation of local people's rights). Furthermore, the land should not be used for any other significant productive function that causes significant LUC (Gallagher 2008).

Until common standards are defined, a Meta-Standard would make maximum use of existing regulations and sub-standards for sustainable agriculture and forestry.

In 2008, the European Commission (EC) introduced a set of sustainability criteria that would apply to liquid biofuels produced in the EU or imported from other countries, although these criteria do not impose a general ban on the import and use of bioenergy products that do not meet these standards. In order to establish comprehensive and more stringent sustainability criteria, the EU Council encourages development of bi- and multi-lateral agreements with producer countries, as well as voluntary standards (Council of the European Union 2008).

In February 2010, the EC confirmed that legally binding sustainability criteria for biomass used to generate heat and power are not necessary in Europe, thus ending a long debate about the utility of a supranational scheme. However, the Commission adopted a report on sustainability requirements for the use of solid biomass and biogas in electricity, heating, and cooling, making recommendations on sustainability criteria at the national level. This approach minimizes the risk of varied and possibly incompatible binding criteria, possibly imposing substantial costs and barriers to trade that limit the growth of the bio-energy sector in the EU. The decision is linked to a Renewable Energy Directive adopted in 2009, which sets up sustainability criteria for biofuels and bioliquids. The recommended criteria relate to (Mackinnon 2010):

(a) "a general prohibition on the use of biomass from land converted from forest, other high carbon stock areas and highly biodiverse areas;
(b) a common greenhouse gas calculation methodology which could be used to ensure that minimum greenhouse gas savings from biomass are at least 35% (rising to 50% in 2017 and 60% in 2018 for new installations) compared to the EU's fossil energy mix;
(c) the differentiation of national support schemes in favor of installations that achieve high energy conversion efficiencies; and
(d) monitoring of the origin of biomass."

The first certification system for sustainable biomass and bioenergies is the International Sustainability and Carbon Certification (ISCC). Established by German Law (Biokraftstoff-Nachhaltigkeitsverordnung) and supported by the German Federal Ministry of Food, Agriculture and Consumer Protection via the Agency for Renewable Resources (FNR), ISCC was recognized 18 January 2010 by the German Federal Institute for Agriculture and Food (Bundesanstalt für Landwirtschaft und Ernährung, BLE). The system is operational, though still under development. ISCC describes the rules and procedures for certification that are issued by the approved Certification Bodies with the ISCC Label (Seal).

The objectives of the ISCC are the establishment of an internationally oriented, practical and transparent system for the certification of biomass and bioenergy, allowing a differentiation of sustainable from non-sustainable products at different stages of the value chain. Regarding the key issues (reduction of GHG emissions; sustainable land use; protection of natural biospheres; social sustainability), six principles are defined, specified by their respective criteria (see Table 16.2). These are categorized as "major musts" and "minor musts", where, for a successful audit, all major musts and 80% of the minor musts have to be complied with (Table 16.2).

16 Criteria for a Sustainable Bioenergy Infrastructure and Lifecycle

Table 16.2 Principles, criteria and requirements of the International Sustainability and Carbon Certification (ISCC) system

Principles and types of criteria	Criterion	M1	M2
Principle 1: Biomass shall not be produced on land with high biodiversity value or high carbon stock and not from peat land. HCV areas shall be protected			
	Biomass is not produced on land/grassland with high biodiversity value/carbon stock	X	
	Biomass is not produced on land that was peat bog in January 2008 or thereafter	X	
	For land converted after 1 January 2008, conversion/use should not be contrary to principle 1	X	
Principle 2: Biomass shall be produced in an environmentally responsible way. This includes the protection of soil, water and air and the application of Good Agricultural Practices			
EIA, stakeholder consultation	Environmental aspects are considered in planning buildings, drainage etc.	X	
Natural water courses	Natural vegetation areas around springs and natural watercourses are maintained or re-established		X
Hunting	Hunting done according to local legislation	X	
Soil erosion	Field cultivation techniques used to reduce the possibility of soil erosion	X	
Soil organic matter	Soil organic matter is maintained/preserved	X	
	Organic matter, if used, is evenly spread throughout the production area	X	
	There is a restriction on burning as part of the cultivation process	X	
Soil structure	Techniques to improve or maintain soil structure, and to avoid soil compaction	X	
Ground water	Chemicals are stored in an appropriate manner, which reduces the risk of contaminating the environment		
Seed/rootstock quality and origin	Purchased seeds are accompanied by records of variety name, batch number, supplier, seed certification details and are seed treatment records retained		X
	Home-saved seed have available records of the identity, source, treatments applied (e.g., cleaning and seed treatments)		X
Irrigation	Producer can justify irrigation in light of accessibility of water for human consumption	X	
	Producer respects existing water rights, both formal and customary	X	
	Producer can justify the method of irrigation used in light of water conservation		X
	To protect the environment, water is abstracted from a sustainable source		X
	If ground water is used for irrigation, the level of the ground-water table is monitored		X
	Advice on abstraction has been sought from water authorities, where required by law	X	
Quantity/type of fertilizer	Recommendations for application of fertilizers (organic or inorganic) are given by competent, qualified persons	X	
	During the application of fertilizers with a considerable nitrogen content care is taken not to contaminate the surface and ground water	X	

(*continued*)

Table 16.2 (continued)

Principles and types of criteria	Criterion	M1	M2
	Fertilizers with a considerable nitrogen content are applied only onto absorptive soils	X	
Records of fertilizer application	Complete records of all fertilizer applications are available		X
Fertilizer application machinery	The fertilizer application machinery is kept in good condition and verified annually to ensure accurate fertilizer application		X
Fertilizer storage	Inorganic fertilizers are stored in a covered area, clean area and dry area, in an appropriate manner, which reduces the risk of contamination of water courses		X
	Inorganic fertilizers are stored		X
IPM	Assistance with implementation of IPM systems has been obtained through training or advice		X
	The producer can show evidence of implementation of at least one activity that falls in the category of Prevention, Observation and Monitoring, Intervention		X
PPP	Is the choice of PPP made by competent persons?	X	
	All workers handling and/or administering PPP have certificates of competence, and/or details of other such qualifications	X	
	Producers only use PPP that are registered in the country of use for the target crop where such official registration scheme exists	X	
	There is a process that prevents chemicals that are banned in the European Union from being used on crops for biomass	X	
	The producer follows the label instructions, all application equipment is calibrated	X	
	Invoices of registered PPP kept		X
	If there are local restrictions on the use of PPP they are observed	X	
	All the PPP applications have been recorded	X	
Disposal of surplus application mix	Surplus application mix or tank washings is disposed of according to national or local law		X
	Surplus application mixes or tank washings are applied onto designated fallow land, where legally allowed, and records kept		X
PPP storage	PPP are stored in accordance with local regulations and in a secure location	X	
	PPP are stored in an appropriate location and store is able to retain spillage		X
	All PPP storage shelving is made of non-absorbent material		X
	There are facilities for measuring and mixing PPP and to deal with spillage	X	
	The product inventory is documented and readily available		X
	All PPP are stored in their original package	X	
	Liquids are not stored on shelves above powders		X
Obsolete PPP	Obsolete PPP are securely maintained and identified and disposed of by authorised or approved channels	X	

(*continued*)

16 Criteria for a Sustainable Bioenergy Infrastructure and Lifecycle

Table 16.2 (continued)

Principles and types of criteria	Criterion	M1	M2
Empty PPP containers	The re-use of empty PPP containers for purposes other than containing and transporting of the identical product is avoided		X
	The disposal of empty PPP containers does occur in a manner that avoids exposure to humans and the environment		X
	Official collection and disposal systems are used when available		X
	Empty containers are rinsed either via the use of an integrated pressure rinsing device on the application equipment, or at least three times with water	X	
	The rinsate from empty containers is returned to the application equipment tank		X
	All local regulations regarding disposal or destruction of containers are observed	X	
Waste disposal	The premises have adequate provisions for waste disposal		X
	There a documented farm waste management plan to avoid or reduce wastage and pollution and avoid the use of landfill or burning, by waste recycling		X
Energy efficiency	The producer can show monitoring of energy use on the farm		X
Principle 3: Safe working conditions through training and education, use of protective clothing and proper and timely assistance in the event of accidents			
Workers health, safety and welfare	The farm has a written risk assessment for safe and healthy working conditions		X
	The farm has a written health, safety and hygiene policy and procedures including issues of the risk assessment		X
	First Aid kits are present at all permanent sites and in the vicinity of fieldwork		X
	Workers (including subcontractors) are equipped with suitable protective clothing in accordance with legal requirements and/or label instructions or as authorised by a competent authority	X	
	Protective clothing is cleaned after use and stored so as to prevent contamination of the clothing or equipment		X
	Potential hazards are clearly identified by warning signs and placed where appropriate		X
	Safety advice is available for substances hazardous to worker health, when required		X
	There are records kept for training activities and attendees		X
	All workers handling and/or administering chemicals, disinfectants, PPP, biocides or other hazardous substances and all workers operating dangerous or complex equipment as defined in the risk assessment have certificates of competence, and/or details of other such qualifications	X	
	All workers received adequate health and safety training and are they instructed according to the risk assessment		X
	Workers have access to clean food storage areas, designated dining areas, hand washing facilities and drinking water		X

(continued)

Table 16.2 (continued)

Principles and types of criteria	Criterion	M1	M2
	On-site living quarters are habitable and have the basic services and facilities		X
PPP handling	The accident procedure is evident within ten meters of the PPP/chemical storage facilities		X
	Facilities to deal with accidental operator contamination		X
	Procedures dealing with re-entry times on the farm		X

Principle 4: Biomass production shall not violate human rights, labour rights or land rights. It shall promote responsible labour conditions and workers' health, safety and welfare and shall be based on responsible community relations (Based on core ILO standards: ILO 29, 105, 138, 182, 87, 98, 100, 111)

	Criterion	M1	M2
	A self-declaration on good social practice regarding human rights has been communicated to the employees and signed by the farm management and the employees' representative		X
	Employment conditions comply with equality principles	X	
	There is no discrimination (distinction, exclusion or preference) practiced that denies or impairs equality of opportunity, conditions or treatment based on individual characteristics and group membership or association. Based on race, caste, nationality, religion, disability, gender, sexual orientation, union membership, political affiliation, age, marital status, working status (i.e. temporary, migrant, seasonal), HIV/AIDS	X	
	There is no forced labour at the farm	X	
	Workers have the freedom to join labour organizations or organize themselves to perform collective bargaining. Workers must have the right to organize and negotiate their working conditions. Workers exercising this right should not be discriminated against or suffer repercussions	X	
	The farm pays a living wage which meets at least legal or industry minimum standards	X	
	The person responsible for workers' health, safety and good social practice and the elected person(s) of trust have knowledge about and/or access to recent national labour regulations/collective bargaining agreements		X
	All impacts for surrounding communities, users and land owners taken into account and sufficiently compensated for		X
	The management does hold regular two-way communication meetings with their employees where issues affecting the business or related to worker health, safety and welfare can be discussed openly		X
	There are at least one worker or a workers' council elected freely and democratically who represent the interests of the staff to the management		X
	There is a complaint form and/or procedure available on the farm, where employees and affected communities can make a complaint		X

(*continued*)

Table 16.2 (continued)

Principles and types of criteria	Criterion	M1	M2
	All children living on the farm have access to quality primary school education	X	
	There are records that provide an accurate overview of all employees (including seasonal workers and subcontracted workers on the farm). Do they indicate full names, a job description, date of birth, date of entry, wage and the period of employment?		X
	No minors are employed on the farm	X	
	All employees are provided with fair legal contracts. Copies of working contracts can be shown for every employee indicated in the records. These have been signed by both the employee and the employer		X
	There is a time recording system that shows daily working time and overtime on a daily base for all employees		X
	The working hours and breaks of the individual worker are indicated in the time records that comply with legal regulations and/or collective bargaining agreements		X
	Pay slips document the conformity of payment with at least legal regulations and/or collective bargaining agreements		X
	Other forms of social benefits are offered by the employer to employees, their families and/or community		X
	Mediation is available in case of a social conflict		X
	Provisions are in place to compensate impact on workers and land (ecosystem quality) on exit or bankruptcy of farm operations		X
	Fair and transparent contract farming arrangements are in place		X
	The biomass production does not impair food security		X
Principle 5: Biomass production shall take place in compliance with all applicable regional and national laws and shall follow relevant international treaties			
	The producer can proof that the land is used legitimately and that traditional land rights have been secured	X	
	There is awareness of, and compliance with, all applicable regional and national laws and ratified international treaties	X	
Principle 6: Good management practices shall be implemented			
Record keeping and internal self assessment	All records requested during the external inspection are accessible and kept for a minimum period of time of 2 years, unless a longer requirement is stated in specific control points	X	
	The producer or producer group takes responsibility to undertake a minimum of one internal self-assessment or producer group internal inspection, respectively, per year against the ISCC Checklist	X	
	Effective corrective actions taken as a result of non-conformances detected during the internal self-assessment or internal producer group inspections	X	

(continued)

Table 16.2 (continued)

Principles and types of criteria	Criterion	M1	M2
Site history and site management	A recording system is established for each unit of production undertaken at those locations. Are these records kept in an ordered and up-to-date fashion?	X	
	Records are kept for the description of the areas in use	X	
Subcontractors	In case of the engagement of subcontractors they must comply fully with the ISCC standard and provide the respective documentation and information	X	

Source: http://www.iscc-system.org. *M1* Major Must, *M2* Minor Must, *HCV* high conservation value, *EIA* environmental impact assessment, *IPM* integrated pest management, *PPP* plant protection products

ISCC initially is about a centrally organised certification system where the standards are meant to be international and valid in all countries and regions that are part of the value added chain. As needed, a national or regional initiative can adapt the ISCC standards to local conditions (for further information see http://www.iscc-system.org).

The efficiency of regulations will depend on the methodology used for the assessment of criteria and standards and on the verification of compliance with the promotion criteria, which could take place directly with the feedstock producers concerned. ISCC defines a checklist for the verification of requirements on traceability and provides guidelines for the certification bodies on how to verify the requirements. The Management System shall be established on a plant level and, in case of outsourcing shall involve relevant suppliers, and corresponding service providers in the traceability trail.

A methodology for mass balance calculation and GHG emissions calculation is introduced, and a risk management procedure is defined based on a number of monitored risk indicators. The effectiveness of a standard increases the more bioenergy carriers and countries are covered. Furthermore, the establishment of protected areas and networks of protected areas, safeguarding global food security, and agreements on agricultural land and land use can further improve the effectiveness of certification.

While coordinated multilateral measures are preferable to unilateral measures in the long run, in the short term it is likely that global minimum standards will remain relatively weak and ineffective due to differing interests of the individual countries. Compared to voluntary certification schemes, which are likely to occupy only a niche in a market for bioenergy carriers, unilateral minimum standards appear to be the most practical option in the short term, provided these are compatible with the international trade rules of the WTO. To address the difficulties and different interests, a phased approach that combines unilaterally binding minimum standards with the integration of sustainability standards in bi- and multilateral agreements between major producer and buyer countries of bioenergy products could be appropriate (WBGU 2009).

16.11 Conclusion

There is an emerging debate on the principles, standards and criteria that can be applied to bioenergy evaluation, justification and certification. New regulatory frameworks are taking shape, both at the fuel and feedstock levels, and a few systems regarding bioenergy are already operational. If key players agree on a number of such measures, they are shaping the pathways of bioenergy development and use towards a more sustainable direction, promoting those feedstocks and production processes for which the producer can provide evidence for compliance with these standards. A certification process is relevant for renewable energy sources that receive policy support through mandates, subsidies and tariffs to meet societal goals such as energy security and preventing climate change. Rather than blocking promising development paths, criteria should provide incentives for investment and innovation, selecting bioenergy pathways that are technically efficient, economically viable, environmentally sustainable and acceptable for society. Most promising are use of biogenic residues and wastes materials from agriculture, forestry, industry and municipalities. Cellulosic bioenergy crops also bear a significant potential if they offer high yield at low cost and diminished environmental impact. Altogether, there is considerable potential for optimization in the bioenergy value chain that can be tapped by improved modeling and lifecycle analysis tools to achieve the maximum energy output at lowest resource input and environmental impact. Combining scientific approaches with certified sustainability standards developed in stakeholder dialogues is a promising route for future bioenergy systems.

Acknowledgements This work was supported in part by the German Science Foundation (DFG) through the Cluster of Excellence 'CliSAP' (EXC177).

References

Al-Kaisi M (2000) Crop water use or evapotranspiration. Integrated Crop Management, Iowa State University, http://www.ipm.iastate.edu/ipm/icm/2000/5-29-2000/wateruse.html
Ausubel JH (2007) Renewable and nuclear heresies. Int J Nuclear Gov Econ Ecol 1(3):229–243
Babcock BA, Hayes DJ, Lawrence JD (eds) (2008) Using distillers grains in the US and international livestock and poultry industries. Midwest Agribusiness Trade Research and Information Center, Iowa State University, Ames, IA
Barreto P (2008) Implications of the climate change debate on land tenure in the Brazilian Amazon. Speech text for the Conference on New Challenges for Land Policy and Administration, 14–15 February 2008, The World Bank, Washington DC

Blaschek H, Ezeji T, Scheffran J (eds) (2010) Biofuels from agricultural wastes and byproducts. Blackwell, Oxford (in press)

Brat I, Machalaba D (2007) Can ethanol get a ticket to ride? Wall Street Journal, 1 February, p B1

Brinkman N, Wang M, Weber T, Darlington T (2005) Well-to-Wheels analysis of advanced fuel/vehicle systems: a North American study of energy use, greenhouse gas emissions, and criteria pollutant emissions. Argonne National Laboratory, Argonne, IL

Brown R, Orwig E, Nemeth J, Subietta Rocha C (2007) Economic potential for ethanol expansion in Illinois. Illinois Institute for Rural Affairs at Western Illinois University, Macomb, IL

Council of the European Union (2008) Presidency suggestions for a common scheme of sustainability criteria for biofuels. 9 September 2008, Brussels. Council of the European Union website, http://register.consilium.europa.eu/pdf/en/08/st12/st12157-re01ad01.en08.pdf (viewed 15 October 2008)

CRS (2007) Ethanol and other biofuels–potential for US–Brazil energy cooperation. Congressional Research Service, Washington, DC

Dehue B, Hamelinck C, Reece G, de Lint S, Archer R, Garcia E (2008) Sustainability reporting within the RTFO: framework report, ECOFYS. Commissioned by UK Department for Transport (January 2008)

Delucchi M (2006) Lifecycle analyses of biofuels. http://www.its.ucdavis.edu/publications/2006/UCD-ITS-RR-06-08.pdf

Denevan WM, Woods WI (2004) Discovery and awareness of anthropogenic Amazonian Dark Earths. University of Wisconsin–Madison, Southern Illinois University, Edwardsville, IL

Dohleman FG, Heaton EA, Long SP (2010) Perennial grasses as second-generation sustainable feedstocks without conflict with food production. In: Khanna M, Scheffran J, Zilberman D (eds) Handbook of bioenergy economics and policy. Springer, New York, pp 27–37

Dooley FJ, Cox M, Cox L (2008) Distillers grain handbook: a guide for Indiana producers to Using DDGS for animal feed. Department of Agricultural Economics, Purdue University, http://incorn.org/images/stories/IndianaDDGSHandbook.pdf. Accessed 15 May 2009

EAA (2007) Frequently asked questions about electrical vehicles. Electric Auto Association, http://www.pluginamerica.com/faq.shtml

Environmental Defense (2007) Potential impacts of biofuels expansion on natural resources: a case study of the Ogallala Aquifer region. http://www.environmentaldefense.org/documents/7011_Potential Impacts of Biofuels Expansion.pdf

Ezzati M, Kammen DM (2001) Quantifying the effects of exposure to indoor air pollution from biomass combustion on acute respiratory infections in developing countries. Environ Health Perspect 109:5481–5488

Faaij A, van Wijk A, van Doorn J, Curvers A, Waldheim L, Olsson E, Daey-Ouwens C (1997) Characteristics and availability of biomass waste and residues in the Netherlands for gasification. Biomass Bioenergy 12(4): 225–240

FAO (2003a) World agriculture: towards 2015/2030. Food and Agriculture Organization of the United Nations, Earthscan, London

FAO (2003b) Compendium of agricultural–environmental indicators 1989–91 to 2000. FAO Statistics Analysis Service, Statistics Division, Rome

FAO (2008) The state of food and agriculture 2008. Biofuels: prospects, risks and opportunities. FAO, Rome

FAO/PISCES (2009) Small-scale bioenergy initiatives: brief description and preliminary lessons on livelihood impacts from case studies in Asia, Latin America and Africa. Food and Agriculture Organization/Practical Action Consulting, Policy Innovation Systems for Clean Energy Security, January, http://www.fao.org/bioenergy/home/en

Fargione J, Hill JK, Tilman D, Polasky S, Hawthorne P (2008) Land clearing and the biofuel carbon debt. Science 319:1235–1238

Farrell AE, Plevin RJ, Turner BT, Jones AD, O'Hare M, Kammen D (2006) Ethanol can contribute to energy and environmental goals. Science 311:506–508

Fowles M (2007) Black carbon sequestration as an alternative to bioenergy. Biomass Bioenergy 31:426–432

Fraiture C de, Giordano M, Yongsong L (2008) Biofuels and implications for agricultural water use: blue impacts of green energy. Water Policy 10 [Suppl 1]:67–81

Fritsche UR, Wiegmann K (2008) Ökobilanzierung der Umweltauswirkungen von Bioenergie-Konversionspfaden. Expertise for the WBGU Report "World in Transition: Future Bioenergy and Sustainable Land Use". http://www.wbgu.de/wbgu_jg2008_ex04.pdf

Fritsche UR, Hünecke K, Hermann A, Schulze F, Wiegmann K (2006) Sustainability standards for bioenergy. WWF Germany, Berlin, November

Gallagher E (2008) Gallagher Review of the indirect effects of biofuels production. Renewable Fuels Agency, London. http://www.renewablefuelsagency.gov.uk/_db/_documents/Report_of_the_Gallagher_review.pdf

GAO (2007) "Biofuels: DOE lacks a strategic approach to coordinate increasing production with infrastructure development and vehicle needs", US Government Accountability Office, GAO-07-713, Washington, DC

Goolsby DA, Battaglin WA, Aulenbach BT, Hooper RP (2000) Nitrogen flux and sources in the Mississippi River basin. Sci Total Environ 248:75–86

Hill J, Nelson E, Tilman D, Polasky S, Tiffany D (2006) Environmental, economic, and energetic costs and benefits of biodiesel and ethanol biofuels. Proc Natl Acad Sci USA 103:11206–11210

Hill J, Polasky S, Nelson E, Tilman D, Huod H, Ludwig L, Neumann J, Zheng H, Bonta D (2009) Climate change and health costs of air emissions from biofuels and gasoline. Proc Natl Acad Sci USA 106:2077–2082

ICRISAT (2007) Pro-poor biofuels outlook for Asia and Africa: ICRISAT's perspective. Working paper, 13 March, International Crops Research Institute for the Semi-Arid Tropics, http://www.icrisat.org

IPCC (2007) Climate Change 2007: impacts, adaptation and vulnerability. Contribution of Working Group II to the Fourth Assessment Report of the IPCC. Intergovernmental Panel on Climate Change, Cambridge University Press, Cambridge

IWMI (ed) (2007) Water for food. Water for life. A comprehensive assessment of water management in agriculture. International Water Management Institute, Earthscan, London

Jacobson MZ (2008) Review of solutions to global warming, air pollution, and energy security. Energy Environ Sci 2:148–173

Kang S, Onal H, Ouyang Y, Scheffran J, Tursun D (2010) Optimizing the biofuels infrastructure: transportation networks and biorefinery locations in Illinois. In: Khanna M, Scheffran J, Zilberman D (Eds) Handbook of bioenergy economics and policy. Springer, New York, pp 151–173

Keeney D, Muller M (2006) Water use by ethanol plants: potential challenges, institute for agriculture and trade policy. October 2006, at http:// www.waterobservatory.org

Keeney D, Nanninga C (2008) Biofuel and global biodiversity. Institute for Agriculture and Trade Policy, Minneapolis, MN

Khanna M, Dhungana B, Clifton-Brown J (2008) Costs of producing miscanthus and switchgrass for bioenergy in Illinois. Biomass Bioenergy 32(6):482–493

Khanna M, Önal H, Chen X, Huang H (2010) Meeting biofuels targets: implications for land use, greenhouse gas emissions, and nitrogen use in Illinois. In: Khanna M, Scheffran J, Zilberman D (eds) Handbook of bioenergy economics and policy. Springer, New York, pp 289–208

Kim S, Dale BE (2004) Cumulative energy and global warming impact from the production of biomass for biobased products. J Ind Ecol 7:147–162

Kim Y, Mosier NS, Hendrickson R, Ezeji T, Blaschek H, Dien B, Cotta M, Dale B, Ladisch MR (2008) Composition of corn dry-grind ethanol by-products: DDGS, wet cake, and thin stillage. Bioresour Technol 99:5165–5176

Knappe F, Böß A, Fehrenbach H, Giegrich J, Vogt R, Dehoust G, Fritsche U, Schüler D, Wiegmann K (2007) Stoffstrommanagement von Biomasseabfällen mit dem Ziel der Optimierung der Verwertung organischer Abfälle. Im Auftrag des Umweltbundesamtes. UBA Texte 04/07. Institute for Energy and Environmental Research (IFEU), Heidelberg

Ladanei S, Vinterbäck J (2009) Global potential of sustainable biomass for energy. Department of Energy and Technology, Swedish University of Agricultural Sciences, Uppsala

Ladisch M, Dale B (eds) (2008) Cellulose conversion in dry grind plants. Bioresour Technol 99:5155–5260

Laird DA (2008) The charcoal vision: a win-win-win scenario for simultaneously producing bioenergy, permanently sequestering carbon, while improving soil and water quality. Agron J 100:178–181

Lehmann J (2007) A handful of carbon. Nature 447:143–144

Mackinnon L (2010) EU Commission rejects binding sustainability criteria for biomass, BIOMASS INTEL 2/26/10, http://www.biomassintel.com/eu-commission-rejects-binding-sustainability-criteria-biomass, accessed 14 March 2010

Marris E (2006) Black is the new green. Nature 442:624–626

Mathews JA (2007) Viewpoint biofuels: what a biopact between North and South could achieve. Energy Policy 35:3550–3570

Mueller S, Plevin R (2008) Global warming intensity of corn ethanol. BioCycle 49:50–53

National Commission on Energy Policy (2004) Ending the energy stalemate: a bipartisan strategy to meet America's energy challenges. National Commission on Energy Policy, Washington, DC

NRC (2007) Water implications of biofuels production in the United States. National Research Council, National Academies, http://www.nationalacademies.org/morenews/20071010.html

NRC (2009) Liquid transportation fuels from coal and biomass: technological status, costs, and environmental impacts. National Research Council, National Academy Press, Washington DC

NREL (2007) A national laboratory market and technology assessment of the 30x30 scenario. National Renewable Energy Laboratory Technical Report /TP-510-40942, January

OECD (2008) Economic assessment of biofuel support policies. Organisation for Economic Co-operation and Development Directorate for Trade and Agriculture, Paris

Perlack RD, Wright LL, Turhollow AF, Graham RL (2005) Biomass as feedstock for bioenergy and bioproducts industry: the technical feasibility of a billion-ton annual supply. Department of Energy / Department of Agriculture, Washington, DC

Pimentel D (2003) Ethanol fuels: energy balance, economics and environmental impacts are negative. Nat Resour Res 12:127–134

Post WM, Kwon KC (2000) Soil carbon sequestration and land-use change: processes and potential. Glob Change Biol 6:317–328

RFA (2010) Industry Statistics. Renewable Fuel Association, http://www.ethanolrfa.org/industry/locations

Rosillo-Calle F, Walter A (2006) Global market for bioethanol: historical trends and future prospects. Energy Sustain Dev 10:18–30

Runge CF, Senauer B (2007) How biofuels could starve the poor. Foreign Affairs (May/June):41–53

Scheffran J (2009) Biofuel Conflicts and human security: toward a sustainable bioenergy life cycle and infrastructure. Swords Ploughshares XVII(Summer):4–10

Scheffran J (2010a) The global demand for biofuels: technologies, markets and policies. In: Vertes A, Blaschek HP, Yukawa H, Qureshi N (eds) Biomass to biofuels: strategies for global industries. Wiley, New York, pp 27–54

Scheffran J (2010b) Bioenergy between sustainability and development: land use, food security and lifecycle analysis. In: Amann E, Baer W, Coes D (eds) Energy, biofuels and development: comparing Brazil and the United States. Routledge, Oxford (in press)

Scheffran J, Bendor T (2009) Bioenergy and land use: a spatial-agent dynamic model of energy crop production in Illinois. Int J Environ Pollut 39:4–27

Scheffran J, Battaglini A, Weber M (2004) Energie aus Biomasse und Bioabfällen—Brennstoff der Zukunft? In: Johnke B, Scheffran J, Soyez K (eds) Abfall, Energie und Klima. Schmidt, Berlin, pp 160–185

Searchinger T, Heimlich R, Houghton RA, Dong F, Elobeid A, Fabiosa J, Tokgoz S, Hayes D, Yu TH (2008) Use of US croplands for biofuels increases greenhouse gases through emissions from land use change. Science 319:1238–1240

Sheehan J, Aden A, Paustian K, Killian K, Bremer J, Walsh M, Nelson R (2004) Energy and environmental aspects of using corn stover for fuel ethanol. J Ind Ecol 7:117–146

Singh V, Johnston D, Naidu K, Rausch KD, Belyea RL, Tumbleson ME (2004) Effect of modified dry grind corn processes on fermentation characteristics and DDGS composition. In: Proceedings of the Corn Utilization and Technology Conference, 7–9 June 2004, Indianapolis, IN

Smeets E, Junginger M, Faaij A, Walter A, Dolzan P, Turkenburg W (2008) The sustainability of Brazilian ethanol—an assessment of the possibilities of certified production. Biomass Bioenergy 32:781–813

Smith T, Miller K, Lindenberg J (2009) Sustainable biofuel standards and certification. Swords Ploughshares XVII(Summer):26–31

Solomon BD, Barnes JR, Halvorsen KE (2007) Grain and cellulosic ethanol: history, economies, and energy policy. Biomass Bioenergy 31(6):416–425

Sylvester-Bradley R (2008) Critique of Searchinger (2008) & related papers assessing indirect effects of biofuels on land-use change. A study commissioned by AEA Technology as part of the Gallagher Biofuels Review, Version 3.2, 12-6-2008

Tilman D, Socolow R, Foley JA, Hill J, Larson J, Lynd L, Pacala S, Reilly J, Searchinger T, Somerville C, Williams R (2009) Beneficial biofuels—the food, energy, and environment trilemma. Science 17:270–271

Tyner WE (2009) The integration of energy and agricultural markets. Presented at the 27th International Association of Agricultural Economists Conference, Beijing, China, August 16–22, http://ageconsearch.umn.edu/bitstream/53214/2/Tyner 20IAAE 20paper 202009-3.pdf

UN (2007) Sustainable bioenergy: a framework for decision makers. United Nations, New York

UN (2008) High level task force on the global food crisis: elements of a comprehensive framework for action. United Nations, New York, Draft, 3 June 2008

US EPA (2007) Renewable fuel standard implementation. http://www.epa.gov/OTAQ/renewablefuels/index.htm Accessed 15 May 2009

Varghese S (2007) Biofuels and global water challenges. Institute for Agriculture and Trade Policy, Minneapolis, MN

Vidal J (2006) Cost of water shortage: civil unrest, mass migration and economic collapse. Guardian, August 2006. http://www.iwmi cgiar.org/press/coverage/pdf/guardianUnlimited.pdf

von Braun J, Pachauri RK (2006) The promises and challenges of biofuels for the poor in developing countries. Annual Report 2005–2006. International Food Policy Research Institute, Washington DC

Wallace R, Ibsen K, McAloon A, Yee W (2005) Feasibility study for co-locating and integrating ethanol production plants from corn starch and lignocellulogic feedstocks, revised January edn. NREL/TP-510-37092 USDA/USDOE/NREL

Wang M (2004) Fuel-cycle analysis of conventional and alternative fuel vehicles. In: Cleveland CJ (ed) Encyclopedia of energy, vol 2. Elsevier, New York

WBGU (2009) Future bioenergy and sustainable land use. German Advisory Council on Global Change, London: Earthscan. http://www.wbgu.de/wbgu_jg2008_engl.html

WDPA (2008) World database on protected areas. UNEP-WCMC website. http://www.wdpa.org

Wooley R, Ruth M, Sheehan J, Ibsen K, Majdeski H, Galvez A (1999) Lignocellulosic biomass to ethanol—process design and economics utilizing co-current dilute acid prehydrolysis and enzymatic hyrolysis—current and futuristic scenarios. Report No. TP-580-26157, National Renewable Energy Laboratory, Golden, CO

Worldwatch (2007) Biofuels for transportation, global potential and implications for sustainable agriculture and energy in the 21st century. Worldwatch Institute, Washington DC

Wright M, Brown RC (2007) Establishing the optimal sizes of different kinds of biorefineries. Biofuels Bioprod Biorefining 1(3):191–200

Xinhua (2007) China to produce liquid bio-fuel with non-food crops. http://news.xinhuanet.com/english/2007-09/04/content_6662806.htm

Index

A

Acetone-butanol-ethanol (ABE) fermentation process, 400, 401
Acid detergent fibre, 44, 45
Acid detergent lignin (ADL), 44–46
Agrobacterium, 94, 100, 101
Agro forestry, 172
Air pollution, 432
Alcaligenes, 77
Algae, 237–240
Alkali index, 43
Alkali metal
  calcium, 165
  magnesium, 165
  potassium, 165
  silicon oxide, 165
  sodium, 165
Alkaline metals, 30, 31
Amplified fragment length polymorphisms (AFLPs), 99
Arabidopsis, 68, 74, 77
Ash, 28–32, 43, 165, 176

B

Bacillus thuringiensis, 73
Bagasse, 413, 416, 417, 420
Baling, 348, 351
Bark, 164, 168, 169, 173
Barriers to biofuel production, 412
Biochar, 423
Biochemical conversion
  BioEnergy Science Center, 191
  fermentation, 176–177, 181
  recalcitrance, 191–193
  verenium, 178
  Zeachem, Inc, 178–179
Biodegradable plastics, 77
Biodiesel, 413, 415, 416, 421, 422, 424, 428, 434
Biodiesel and renewable diesel, 396–400
Biodiversity, 193, 195, 410, 417, 424, 426, 427, 430–431, 435, 437
Bioenergy infrastructure, 409–443
Bioenergy lifecycle, 409–443
BioEnergy Science Center, Oak Ridge National Laboratory, 191
Biofuel, 9, 10, 16–18
Biofuel crops, 263–281
  Arundo, 268, 278
  giant reed, 264, 268, 270, 275, 278
  harvesting, 272, 276, 278
  miscanthus, 268, 274, 275
  Miscanthus x giganteus, 264, 268, 270
  perennial rhizomatous grasses, 263, 272
  processing, 276, 278, 279
  Sorghum bicolor, 274
  storage, 265, 267, 272, 273, 275, 276, 278, 279, 281
  switchgrass, 264, 268, 270, 273–275
  transport, 263, 267, 272, 273, 275–279, 281
Biofuel lifecycle, bioenergy, 419, 420, 427, 432
Biofuels Emissions And Cost Connection (BEACCON) model, 419
Biofuels life-cycle analysis (LCA)
  corn ethanol pathway, 387
  GHGenius model, Canada, 391
  GREET model, Argonne National Laboratory, 389–390
  issues and uncertainties, 402–406

Biofuels life-cycle analysis (LCA) (cont.)
  LEM, University of California at Davis, 390–391
  life-cycle energy and GHG emission, 392–401
  Ludwig-Bölkow-Systemtechnik E3 database, 386, 387
  petroleum fuels, 388–389
  production pathways, 386, 387
  WTW analyses, 389
Biofuel supply chain, 267, 268, 276
Biogas, 414, 416, 417, 420, 436
Biogenic residues, 416–419
Biomass, 345–358, 412–423, 425, 427, 432, 434–437, 440, 441
  cellulosic, 414, 417
  lignocellulosic, 413, 418
Biomass conversion
  bio-oil, 12
  comparison, 15–16
  composition, 10
  diesel, 10
  to ethanol, 9
  higher temperature processes, 11–12
  liquid fuels, 9
  lower temperature processes, 13–15
  petroleum-based products, 4
Biomass Crop Assistance Program, 276
Biomass Technical Advisory Committee (BTAC), 319, 320
Bio-oil liquefaction, 11, 12
"Bio-pact" for a North–South Trade, 434
Biopower, 158–173
  co-fire, 159
Biorefinery, 410–415, 419
Biosynthesis, 241–247
Biotechnology, 177, 184, 185, 189, 196
Blending of ethanol, 415
Blending wall, 415
Brazil, 413, 414, 420, 424
Breeding programs, 69–71
British thermal unit (BTU), 4
Brown-midrib, 40, 41
Butanol
  Butamax Advanced Biofuels, LLC, 180
  ButylFuel, LLC, 180
  Gevo Inc., 180

C
Caffeic acid O-methyltransferase (COMT), 40, 41
Caffeoyl-CoA O-methyl transferase (CCoAOMT), 40, 41

Calorific value, 164, 193
Camelina, 231, 239
Canola, 212, 213, 215, 217, 219, 220, 226, 227, 231, 232, 237, 239, 246
Capacity of ethanol plants, 413
Cap and trade, 420
Carbon dioxide emission, 3–21
Carbon intensity, 420–423
Carbon neutral, 419
Carbon sequestration, 195, 422
Carbon stock, 420–423, 435–437
Carbon storage, 194, 421, 423, 431
Cartagena Protocol on Biosafety, 431
Castor, 232, 233, 235–237, 239, 246
Cellulose, 28–30, 32, 38–39, 41, 43–47, 160, 164, 173, 175–178, 180, 181, 186, 192, 193, 322–329, 331, 334
Cellulosic biomass, 4, 9–11, 14, 21, 414, 417
Cellulosic ethanol, 90, 96, 101, 385, 386, 392–396, 403, 405, 406, 412–414, 419, 420, 425
Cell wall, 96, 97, 101
Ceres expression & confinement system, 310–311
Certification of bioenergy, 410, 423, 433–443
Charcoal, 422, 423
Chia, 220, 234–235
Child labor, 433
China, 424, 428, 429
Chlorine, 30, 31, 44, 165
Chloroplast transformation
  homologous recombination, 303
  maternally inherited, 303
  paternal inheritance, 303
Cinnamyl alcohol dehydrogenase (CAD), 40, 41
Climate change, 410, 416, 417, 421–423, 426, 427, 429, 434, 443
Coal, 346, 355–359
Co-firing, 158–161, 182, 197, 420
Cogeneration, 157, 175, 416
Colonization pressure, 266, 272
Combined heat and power (CHP), 348, 353, 355, 357–359
Combustion, 420, 421
  co-combustion, 28
Commercial validation of the NIT calibration, 119–120
Computational general equilibrium (CGE) model, 402–403
Convention on BiologicaL Diversity (CBD), 430, 431, 435
Coppice, 168–171, 184, 190, 191

Index 451

Co-products, by-products, 418
Corn, 88, 89, 91, 94, 96–98, 100, 101, 111–123, 210–212, 214, 215, 219, 220, 232, 238, 240, 345–359
   butanol, 387, 400–401
   ethanol, 386–388, 392–395, 404–406
   ethanol industry, 111–123
   grain, 59–60
   grain fermentation potential, 114, 122–123
   stover, 412, 413, 416, 418, 420, 430, 431
Costs, 346–359
   cost of ethanol production, 413, 414, 420
Cottonseed, 212, 214, 219, 220, 222, 226–228
4-Coumarate: CoA ligase (4CL), 40, 41
Crambe, 231–232
Crop improvement methods, 68–69
Cropland, 423, 426, 430
Cytoplasmic male sterility (CMS), 78–79

D

Deforestation, 424, 426
Density, 164, 165, 170–172, 183, 184, 186, 190, 191, 193, 195
Desertification, 427
Development cooperation, 426, 435
Direct firing, 161–163
   boilers, 162
Discrimination, 433, 440
Distillation, 176, 179
Distillers dried grains and solubles (DDGS), 210, 211, 413
Distillers grains and soluables (DGS), 387, 388, 392, 393, 395, 400, 401, 404
Distribution, biofuel, 414
DNA sequences, 64–65
Domestication, 184–186
Drought and salt tolerance, 187, 189
   phospholipase Dα, 187
Drought tolerance, 170, 189
Dry, 413, 417, 418, 425, 438
Dry matter composition
   cellulose, 180–181
   hemicellulose, 180–181
   lignin, 180–181

E

Early detection and rapid response (EDRR), 279
Ecological footprint, 410
Economic sustainability, 195
Economies of scale, 412–415
Ecosystems, 421, 423, 427, 430, 431, 441

Efficiency, 412–415, 418–420, 425, 428, 429, 435, 442, 443
   energy efficiency, 419–420
EIA, stakeholder, 437
   environmental impact assessment, 442
Electric power, 357
Emissions
   carbon dioxide (CO_2), 161
   nitrogen oxides, 161
   sulfur dioxide, 161
Energy and Resource Groups Biofuel Analysis Meta-Model (EBAMM), 419
Energy balance, 166, 195, 413, 418–420, 427
Energy cane, 137, 139–141
Energy conversion efficiencies, 425, 436
   energy efficiency, 419–420, 425, 439
Energy density, 414, 423
Energy efficiency, 419–420, 439
Energy Independence and Security Act (EISA), 265, 267
Energy output, 166–167
Energy security, 434, 443
Environmental impact, 410, 412, 413, 419, 423, 430–431, 443
Environmental impact assessments (EIA), 431, 442
Enzymatic conversion, 416
Enzymatic hydrolysis, 414
Eradication
   biological, 280
   chemical, 279, 280
   cultural, 280
   mechanical, 279, 280
Ethanol, 87–91, 96, 97, 99, 101, 111–123, 210, 211, 215, 216, 323–325, 331–334, 339, 345–359
   E85, 415, 421–422, 425, 432
   production, 409, 413–415, 419, 420, 428, 429, 434
Eucalyptus
   E. amplifolia, 171
   E. globulus, 171
   E. grandis, 171, 189
   E. grandisx and *E. urophylla,* 190
   E. gunni, 193
   E. urophylla (E. urograndis), 171, 181
European Commission, European Union, 436
Expressed sequence tag (EST), 91, 99
Extrinsic moisture, 165

F

Fatty acid, 219–223, 225, 226, 230, 232, 233, 235, 237–240, 242, 243, 245–247

Feedstock, 345–347, 351, 353
Feedstock crops
  biomass production, 58, 59
  corn grain, 59–60
  sugarcane, 59–60
  sweet sorghum, 60
  switchgrass and miscanthus characteristics, 58, 59
  wood, 60
Fermentation, 175–177, 180, 181, 191
Fertilizer, 412, 417, 423–425, 428, 430, 431, 437, 438
Fischer-Tropsch (FT), 414
  diesel, 386
Flax, 229–231
Flex fuel vehicles (FFVs), 415
Food and Agricultural Organization (FAO), 424, 426, 427, 429, 432
Food crisis, 426, 429
Food price, 410, 424, 425
Food security, 410, 423–428, 435, 441, 442
Forest land, 417, 426
Forest management, 194, 195
Forest residues, 156, 157, 159, 182, 183
Forest Stewardship Council, 434
Fossil fuels, 8, 10, 19
Fourier transform infrared spectroscopy (FTIR), 42
Fragment length polymorphisms (RFLPs), 99
Freezing tolerance
  drought tolerance, 189
  Freeze Tolerant Eucalyptus, 190, 196, 197
  salt tolerance, 189
1-Fructosyltransferase (1-SST), 95, 101
Fuel-cycle analyses. *See* Well-to-wheels (WTW) analyses

G

Gasification, 27, 413, 414, 416, 422
  biomass, 11, 12, 14, 15
  fluidized bed gasifiers, 163
  gasifiers, 163
  integrated gas turbine, 163
GE biofuels and biomass crops
  biocontainment, 295
  commercialization, 294
  regulatory requirements, 295
  switchgrass, 294
Geneflow, 196, 197
Genetically engineered (GE) crops, 285, 286, 289
  ecological impact issues, 286
  risk assessments, 286, 289

Genetically modified organisms (GMOs), 430, 431
Genetically modified trees, 196
Genetic diversity, 431
Genetic use restriction technologies (GURTs)
  controversies, 304
  GeneSafe, 304–309, 312
  impact on seed viability, 305
  potential for socio economic impacts, 305
Genetic use restriction technology (GURT), 79
Genomic estimated breeding value (GEBV), 70, 71
Genomics, 66–67, 72, 80
German Advisory Council on Global Change (WBGU), 410, 417, 421–427, 429–431, 434, 435, 443
GHGenius model, 391
Global Bioenergy Partnership (GBEP), 434
Global warming potential (GWP), 421
Glucose, 322–325, 331–333
Glucuronoarabinoxylans (GAX), 29, 30
Glufosinate-resistant grasses
  bar gene, 287, 288
  creeping bentgrass, 287–294
  inter-specific hybridization, 288
  transgenic Zoysia grass, 288
Grass energy crops, 125–147
Grasses, 125, 126, 128–132, 134, 137, 140, 143, 146–147
Grassland, 421–423, 426, 431, 437
Green biomass fractionation, 334
Green biorefinery
  biomass fractionation and energy aspect, 334
  fibre-rich PC, 328–329
  grassland, 339
  green crop fractionation processes, 337–338
  mass and energy flows, 334–336
  multi-product system, 327
  nutrient-rich GJ, 328–329
  plant biomass, 338–339
  Trifolium pratense, 339
  wet-fractionation process, 328
Green crop fractionation processes, 337–338
Greenhouse gas (GHG) emissions, 7–10, 14, 18, 19, 385, 388, 390–399, 401, 405, 406, 410, 412, 416, 418, 420–423, 427, 436, 442
Greenhouse gases, 419
Green juice (GJ), 328, 329, 334–338

Index   453

GREET model, 389–390, 392, 394–396
Grinding, 349–351, 353, 354
Gulf of Mexico, 429

H
Hardwoods, 155, 158, 164, 167, 168, 171, 178, 180–182, 184, 189, 192
Health, 432–433, 439, 440
Heating value
  HHV, 30
  LHV, 30
Hemicellulose, 29, 32, 38, 39, 43–46, 322–325, 331
  arabinoxylan, 29, 30
  glucuronoarabinoxylans, 29
Herbaceous crops, 125, 146
Herbicide tolerance, 186
Heterosis, 67–68
Higher heating value, 30
High fermentable corn, 111–123
High fermentable corn hybrids, 111–123
High heating value, 166
Hybridization, 274
Hybrid total gene containment system, FLORICAULA/LEAFY, 310
Hybrid vigor. See Heterosis
Hydrogen, 414
Hydroxycinnamicacids, 43
  ferulic acid, 30
  p-coumaric acid, 30

I
Indirect land use, 424
Indirect land-use changes (iLUC), 421, 422, 427
Inductively coupled plasma mass spectrometry (ICP MS), 43
Inductively coupled plasma optical emission spectrometry (ICP OES), 43
Inhibitory compounds
  furan derivatives, 181
  phenolic compounds, 181
  weak acids, 181
Insect resistance, Bacillus thuringiensis (Bt) endotoxin genes, 186
Integrated assessment, 410
Integrated biomass analysis and logistics (IBSAL), 348–350
Integrated biorefineries
  BTAC, 319–320
  conversions steps, 321–322
  food–feed–fuel conflict, 321
  green biorefinery (see Green biorefinery)

LCF biorefinery, 322–324
  platform chemicals, 330–338
  research and development, 320
  two platform concept, 329–330
  whole crop biorefinery, 324–327
Integratedc pest management (IPM), 442
International Labour Organisation (ILO) standards, 432, 433, 440
International Sustainability and Carbon Certification (ISCC), 436, 437, 441, 442
International Water Management Institute (IWMI), 427–429
Intrinsic moisture, 165
Invasive species, 431
  Baker traits, 266
  biology, 263–281
  defines, 265
  ecology, 263–281
  Japanese knotweed, 266
  Johnsongrass, 264, 275
  Miscanthus sacchariflorus, 269
  Miscanthus sinensis, 269
Investments, 368–370, 375–382
Iowa, 428
Irrigation, 426–429, 437
Isomaltulose, 93, 95

J
Jatropha, 237–239, 428
Jojoba, 233–234

K
1-Kestose, 95
Klason lignin, 46

L
Labour conditions, working, 440
Lactic-acid fermentation, 337–338
Lag phases, 267, 275
Landfills, 416, 417, 439
Land use, 410, 415–417, 422–427, 430, 431, 436, 442
  change, 421, 422, 427
  conflicts, 425
Land use change (LUC) issue, 402–403
LCA issues and uncertainties
  co-product, 403–405
  direct and indirect LUC, 402–403
  environmental sustainability issues, 405–406
Learning curve, 412
Lesquerella, 232–233, 246
Lifecycle analysis (LCA), 410, 419

Lifecycle emission model (LEM), 390–391
Lifecycle energy and GHG emission
  biodiesel and renewable diesel, soybeans, 396–400
  corn and cellulosic ethanol, 392–395
  corn butanol, 400–401
  sugarcane ethanol, 395–396
Lignin, 29–32, 39–47, 164, 171, 175–178, 180, 188, 191–193, 322–325, 329–331, 334
Lignin reduction, 192
Lignocellulosic biomass, 176, 178
Lignocellulosic feedstock (LCF) biorefinery, 322–324, 338, 339
Linear programming model, 337
Liquefied petroleum gas (LPG), 16
Liquid biofuels
  C2 biofuels, 174–175
  cellulosic ethanol, 173–175
  Flambeau River Biorefinery, 174
  range fuels, 174, 175
Location of biorefineries, 411, 415
Logistics, 348
Low heating valve (LHV), 30, 400
Ludwig-Bölkow-Systemtechnik E3 database, 386, 387
Lysine fermentation, 338

M

Maize, 39–41
Male sterility
  ablation of tapetal cells, 301
  anther, 301
  anther-and tapetum-specific genes, 301
  barnase, 301, 302
  cytoplasmic male sterility, 301–304
  rice rts gene, 301
  rts gene, 301, 302
  tapetum, 301, 302
Marker-assisted selection (MAS), 98, 99, 101
Methanol, 179
Mexico, 424, 429
Microbial biomass-breakdown regime, 332
Midwest Consortium for Biobased Products, 418
Millennium Development Goals, 429
Minnesota, 428
Miscanthus, 31–35, 39, 41, 42, 44–47, 126–132, 413, 416, 419, 420, 422, 425
  advanced breeding, 127, 130–131
  cold temperature adaptation, 129
  conventional breeding, 128–131
  drought and water use efficiency, 129
  genetic improvement, 126–128
  growth, 127
  M. sacchariflorus, 32, 45
  M. sinensis, 32, 42, 45
  M. x giganteus, 32–37, 44–47, 77
  pest and disease resistance, 129–130
  phylogeny, 127
  trait targets, 128–130
  yield and flowering time, 128, 131
Mississippi River, 428
Moisture content, 163, 165–166, 170, 180, 191

N

National Renewable Energy Laboratory, 413
National Willows Collection, 36
Natural gas, 345, 346, 355–359
Near infrared spectroscopy (NIRS), 42, 45, 116
Nebraska, 428
Neutral detergent fibre, 44, 45
New crops, 229
Niche modeling
  BIOCLIM, 271
  bioclimatic envelope, 271
  climate-match modeling, 271
  CLIMEX, 271, 272
  GARP, 271
  Maxent, 271
NIT calibration, 116–120, 123
Nitrogen, 28, 30, 31, 33, 38, 43, 421, 428, 430, 431, 437, 438
Nutrient and water availability, 187

O

OECD, 420, 421, 424, 425
Oilseeds, 217, 218, 239, 241, 242, 245–247
Öko-Institut, 422
Organization of petroleum exporting countries (OPEC), 7, 20

P

Palm, 212–217, 219, 220, 222, 228, 229, 239
Palm oil, 424, 430, 434, 435
Panicum virgatum, 77
Peanuts, 226, 228, 229, 242
Pectin, 29, 30, 39
Pelletization, 347, 349–356, 359
Pennycress, 232
Perennial crops, 421, 425, 431
Perennial grasses, 412, 416, 418–420, 425
Perilla, 220, 234
Petroleum
  carbon dioxide accumulation, 7
  economic consequence, 18–19

GHG emissions, 19
Hubbert's peak, 6
LPG, 16
OPEC, 7
transportation, 4
US production, 4–6
Phosphate, 92, 95
Phosphorus, 430
Phosphotransferase (PFP), 95
Photosynthesis, 92, 93, 95
  C4, 88, 92, 100
Physical containment
  greenhouses, 301
  growth rooms, 301
  human management, 301
  underground facilities, 301
Pine
  loblolly, 155, 167, 172, 183, 184, 187–188, 190, 194
  Pinus echinata, 172
  Pinus elliottii, 172
  Pinus glabra, 172
  Pinus palustris, 172
  Pinus serotina, 172
  Pinus taeda, 155, 167, 172, 183
  sweetgum, 167, 184
Plant cell wall structure, 75–76
Plant designing
  feedstock crops, 58–61
  genomics, 66–67
  heterosis, 67–68
  molecular markers, 64–66
  molecular plant breeding, 68–71
  trait improvement, 61–64
  transgenic traits (*see* transgenic traits)
Plant genetic engineering
  biofuels-specific traits, 300
  cellulosic feedstock crops, 300
  impacts of gene flow, 300
  produced, 300
Plant improvement
  biomass composition, 62–63
  biomass yield, 61
  breeders, 61–62
  corn and wheat production, 63–64
  molecular plant breeding, 68–71
Plant protection product (PPP), 442
Platform chemicals
  biobased product flowchart, 330–332
  biotechnology, 332–333
  green biomass fractionation and energy aspects, 334

green crop fractionation processes, 337–338
  mass and energy flows, 334–336
Plenish, 223, 228
Pollen-mediated gene flow
  CP4 *EPSPS* positive seedlings, 291
  CP4 *EPSPS* gene, 291
  F1 hybrids, 290
Polyhydroxyalkanoates, 77
Poplar, 32, 36–37, 41, 159, 165–169, 177, 179, 181, 183, 194
Populus
  ArborGen, LLC, 185, 188
  aspen, 168
  cotton wood, 195
  GreenWood Resources, Inc., 169, 185
  International Paper, 185
  MeadWestvaco, 185
  P. alba, 168
  P. deltoide, 168, 184, 188
  P. grandidentata, 168
  P. maximowizcii, 168
  P. Nigra, 168
  poplar, 168
  P. tomentosa, 187
  P. tremula, 168, 187
  P. tremuloides, 168
  P. tricho-carpa, 185
Potential, waste, 417
Press cake (PC), 328, 329, 334–337
Pretreatment
  alkali, 177, 178
  ammonia fiber explosion, 178
  biological, 177
  chemical, 178
  dilute acid, 176–178
  electrical, 177
  organosolv, 178
  physical, 177
  steam explosion, 177–178
Process heat, 345–347, 355–357
Process optimization, 419
Producer gas, 11
Promoters, 94
Propagule pressure
  perennial rhizomatous grasses, 272
  rhizome fragments, 272, 273
  seeds, 272, 273
  stem nodes, 272, 273
Protected areas, 427, 431, 442
Pulp and paper industry, 156, 159, 189
Purpose grown trees, 155–198

Pyrolysis, 11–14, 26, 27, 43, 413, 423
Pyrophosphate fructose, 95

Q
Qualisoy, 223, 224, 228
Quantitative trait loci (QTL), 68–71, 98, 99
Quick Germ-Quick-Fiber process, 418

R
Rainforest, 430
Ralsonia, 77
Recycling, 417, 439
Reduced Emissions from Deforestation and Degradation (REDD), 434
Reed canary grass, 32, 38
Regenerative agriculture, 365–383
Regulated Emissions and Energy use in Transportation (GREET) model, 419
Regulation, 428, 429, 431, 432, 435, 438–442
Regulatory, 434, 435, 443
    regulation of bioenergy, 429
Release of GE plants
    deregulation, 311
    USDA APHIS, 311
    zero tolerance expectation, 312
Removal of transgenic cassettes
    CRE/loxP, 309
    FLP/FRT, 309, 310
    loxP-FRT fusion sequences, 309
Renewable Energy Directive, 436
Renewable Fuel Standard (RFS), 173, 174, 190, 191
Residues, 416–419, 423, 426, 430, 431, 435
    biogenic residues, 416–419, 431, 443
Resin, 164
Restriction, 99
Risk assessments, 263–281
Rotation cycles, 169, 188
Round Table on Responsible Soy, 434
Roundtable on Sustainable Biofuels, 434
Roundtable on Sustainable Palm Oil, 434
Round Up Ready bentgrass
    EPSPS (5-enolpyruvylshikimate-3-phosphate synthase) gene, 289
    genetically engineered line (ASR368), 288
    RoundUp resistant weeds, 290
    Willamette Valley, 288

S
Saccharification, 176, 192
Saccharum
    S. barberi, 91
    S. officinarum, 91, 95, 98
    S. robustum, 91
    S. sinense, 91
    S. spontaneum, 91, 92, 95, 98
Safety, 432–433, 439, 440
Safflower, 212, 219–222, 228, 243
Salix
    S. caprea, 170
    S. daphnoides, 170
    S. dasyclados, 170
    S. eriocephala, 170
    S. purpurea, 170
    S. schwerinii, 170
    S. triandra, 170
    S. viminalis, 170
    willow, 170, 184
Seed scatter
    gene flow, 293
    natural dispersal vectors, 293
Short rotation woody crops (SRWC), 165, 167–170, 183, 184, 186, 195
Silica, 165
Silvicultural management, 169, 170
Simple sequence repeats (SSRs), 99
Single nucleotide polymorphisms (SNPs), 99
Sink-source, 91–93, 95
Size reduction, 177, 183
Social Accountability in Sustainable Agriculture (SASA), 432
Social criteria, 432–433
    social standards, 432
Softwoods, 164, 172–173, 181
Soil conservation, 417
Soil fertility, 423
Soil microbial diversity, 430
Soil organic matter, 422, 424, 425, 430, 431, 437
Soil quality, 423, 430
Somatic embryogenesis, 188
Sorghum, 91, 92, 98–100, 102
    genetic improvement, 144–146
    growth, 142–144
    phylogeny, 142–144
Source-sink, 91–93, 95
Soy, 421, 422, 424, 430
Soybean, 210–215, 217, 219–226, 228–229, 232, 239, 242–247
Stakeholders, 433, 434, 437, 443
Starch, 87, 88, 97
Stover, 345–359
Stress tolerance, deoxyhypusine synthase, 187
Subsidies, 413, 425, 432, 434, 443
Sucrose, 87, 88, 90–95, 97, 98, 100–102
Sucrose isomerase (SI), 93, 95

Sucrose::sucrose, 95
Sugar and syngas platform, 330
Sugar beet, 99–102
Sugarcane, 59–60, 87–102, 413, 416, 417,
    420–422, 424, 427, 430
  genetic improvement, 139–141
  genetic improvement strategies, 141–142
  growth, 136–139
  phylogeny, 136–139
Sugarcane ethanol, 386, 387, 395–398, 403
Sugarcane sucrose, 91, 95, 97, 102
Sulphur, 28, 30
Sunflower, 212–215, 217, 219, 220, 222, 227,
    229, 239, 243
SuperCetane™ process, 397
Supply chain, 411–412, 415
Supply systems, 126, 146–147
Sustainability, 213, 216–223, 410, 412, 417,
    419, 420, 426, 430, 431, 434–437
Sustainability standards, 433–443
Sustainable Agriculture Practice Standard, 434
Sustainable Biofuel Meta-Standard, 435
Sustainable production, 158, 173, 193–197
Sweetgum, 167, 184
Sweet sorghum, 60, 100, 102, 428
Switchgrass, 32, 35–38, 46, 159, 174, 178, 181,
    387, 393–395, 406, 413, 420, 422, 425,
    428, 430
  advanced breeding, 136
  biomass yield, 135
  conventional breeding, 134–136
  genetic improvement, 133–134
  growth, 132–133
  pest and disease resistance, 135–136
  phylogeny, 132–133
  stand establishment, 134–135
  trait targets, 134–136
Switzerland, 435
Syngas, 11, 12
Syringyl lignin to guaiacyl lignin (S/G)
    ratio, 192

T
Thermal Conversion
  Clostridium ljungdahlii, 175
  partial oxidation, 175
  pyrolytic gasification, 175
  syngas, 175, 176
  two-stage thermal conversion, 175
Torrefaction
  briquettes, 166
  char, 166
Trade, biofuel, 412, 434

Transesterification process, 396–397
Trans Fatty, 222
Transgene, 94, 97
Transgene insertion, 186
Transgenic herbicide tolerance, 72–73
Transgenic traits
  co-products, 76–77
  endogenous traits, 74–76
  first generation, 72–73
  genetic confinement and prevention,
    77–79
  non-endogenous traits, 76
  US yield graph, 72
Transportation, 409–412, 414–417,
    419, 420
Transportation mode, 411, 415
  barge, 415
  pipelines, 415
  rail, 415
  truck, 415
Transportation network, 411, 415
Transport costs, 164, 183

U
UK Renewable Fuels Agency, 425
UK Renewable Transport Fuel Obligation
    (RTFO), 435
UN Commission on Sustainable Development
    (CSD), 429
United Nations Convention to Combat
    Desertification (UNCCD), 427
United Nations Framework Convention on
    Climate Change (UNFCCC), 435
US Energy Independence and Security Act,
    410
US Midwest, 413, 420
US Sustainable Biodiesel Alliance, 434

V
Verification of compliance, 442
Vistive, 223, 228

W
Waste, 412, 413, 416–420, 423, 426, 432,
    435, 439, 443
Waste management, 417, 418, 439
Wastewater, 428
Water crisis, 427–430
Water insecurity, 427
Water quality, 423, 425, 427–431
Water use, 412, 425, 427–429
Water use efficiency, 170

Weed Risk Assessment (WRA), 268, 271, 273, 277
Well-to-wheels (WTW) analyses, 389, 392
Wet distillers grains with solubles (WDGS), 418
Wet-fractionation, green biomass, 328, 334
Wet milling, 326–327
Wet mills, 413
Whole crop biorefinery, 322, 324–327, 329, 337
Wildlife, 430–431
Willow, 32, 36–37
Willow for Wales, 36, 37
Wood
    forest residues, 156, 157, 159, 182, 183
    pellet, 157, 162, 182
    residuals, 156, 157
Woody biomass, 155–198
    wood residuals, 156, 157
Working conditions, 433, 435, 439, 440
World Conservation Union (IUCN), 431
World energy resources
    biomass conversion (*see* Biomass conversion)
    commercial applications, 18–19
    energy crises, 7
    GHG emissions, 7–8
    oil and global climate change, 7
    petroleum (*see* petroleum)
    transportation, 8–10
World Water Council, 429

X

X-ray fluorescence (XRF), 44
Xylans, 76

Y

Yield
    β–1,4-endoglucanase, 187
    biomass, 165, 166, 185, 186, 189, 194
    energy, 166, 167, 193
    gene, GA2-oxidase, 187
    glutamate synthase, 187
    glutamine synthetase, 186
    mean annual increment, 172–173